Debbie Schenberger 2/03

AF711082

RF MEMS
and Their Applications

RF MEMS and Their Applications

Vijay K. Varadan
K.J. Vinoy
K.A. Jose
Pennsylvania State University, USA

JOHN WILEY & SONS, LTD

Copyright © 2003 John Wiley & Sons Ltd, The Atrium, Southern Gate, Chichester, West Sussex PO19 8SQ, England

Telephone (+44) 1243 779777

Email (for orders and customer service enquiries): cs-books@wiley.co.uk
Visit our Home Page on www.wileyeurope.com or www.wiley.com

All Rights Reserved. No part of this publication may be reproduced, stored in a retrieval system or transmitted in any form or by any means, electronic, mechanical, photocopying, recording, scanning or otherwise, except under the terms of the Copyright, Designs and Patents Act 1988 or under the terms of a licence issued by the Copyright Licensing Agency Ltd, 90 Tottenham Court Road, London W1T 4LP, UK, without the permission in writing of the Publisher. Requests to the Publisher should be addressed to the Permissions Department, John Wiley & Sons Ltd, The Atrium, Southern Gate, Chichester, West Sussex PO19 8SQ, England, or emailed to permreq@wiley.co.uk, or faxed to (+44) 1243 770571.

This publication is designed to provide accurate and authoritative information in regard to the subject matter covered. It is sold on the understanding that the Publisher is not engaged in rendering professional services. If professional advice or other expert assistance is required, the services of a competent professional should be sought.

Other Wiley Editorial Offices

John Wiley & Sons Inc., 111 River Street, Hoboken, NJ 07030, USA

Jossey-Bass, 989 Market Street, San Francisco, CA 94103-1741, USA

Wiley-VCH Verlag GmbH, Boschstr. 12, D-69469 Weinheim, Germany

John Wiley & Sons Australia Ltd, 33 Park Road, Milton, Queensland 4064, Australia

John Wiley & Sons (Asia) Pte Ltd, 2 Clementi Loop #02-01, Jin Xing Distripark, Singapore 129809

John Wiley & Sons Canada Ltd, 22 Worcester Road, Etobicoke, Ontario, Canada M9W 1L1

Library of Congress Cataloging-in-Publication Data

Varadan, V.K., 1943–
RF MEMS and their applications / Vijay K. Varadan, K.J. Vinoy, and K.A. Jose.
Includes bibliographical references and index.
ISBN 0-470-84308-X (alk. paper)
1. Radio circuits–Equipment and supplies. 2. Microelectromechanical systems. 3. Microwave circuits. I. Vinoy, K.J. (Kalarickaparambil Joseph), 1969– II. Jose K. Abraham. III. Title.
TK6560.V33 2002
621.384′13–dc21 2002071393

British Library Cataloguing in Publication Data

A catalogue record for this book is available from the British Library

ISBN 0-470-84308-X

Typeset in 10/12pt Times by Laserwords Private Limited, Chennai, India
Printed and bound in Great Britain by Biddles Ltd, Guildford and King's Lynn
This book is printed on acid-free paper responsibly manufactured from sustainable forestry in which at least two trees are planted for each one used for paper production.

Contents

Preface

The market for wireless personal communication devices has expanded so dramatically in the past two decades that the focus of research in the microwave and millimeter wave areas has shifted towards consumer applications, from the more traditional defense-related products. Accordingly, the production volume has increased manifold, while the power-handling capacity required for these systems has considerably reduced. These developments paved the way for increased application of microelectromechanical systems (MEMS) in many current and future radio frequency (RF), microwave and millimeter wave systems. Such devices are termed RF MEMS, although it encompasses *all* miniaturized devices; whether they are operated micromechanically, or fabricated by micromachining, or both. Fortunately enough, the processing techniques for MEMS systems have improved significantly over the years, and we are increasingly leaning towards their applications in microwave and millimeter wave systems, and even in optical systems. Apart from having the advantages of bulk production, and being miniaturized, these can often lead to more efficient systems compared with conventional ones.

The need for micromachining and MEMS based systems for RF and microwave applications arise from the inherent limitations of the existing devices. Motivations for incorporating MEMS based fabrication technologies in microwave and millimeter wave systems can broadly be classified into three. First, as the frequency increases, the *size* of the microwave components becomes smaller. Thus, for millimeter wave systems it is imperative that dimensions of most of the components are in the sub-millimeter range. This calls for *high-precision fabrication* technologies for which micromachining offers a viable route. In addition, this approach provides *system integration* capabilities. At lower frequencies (wavelength of the order of 1 to 2 cm) efforts have been made towards implementing micromachining techniques to concomitantly *reduce the effective dielectric constant* of the microstrip antenna substrate. Micromachining of these substrates not only does improve the radiation efficiency of the antenna, but also increases the bandwidth. Many MEMS based microwave components are aimed at *reducing insertion loss* and *increasing bandwidth.* This third aspect is valid for surface micromachined devices such as RF switches, tunable capacitors and micro inductors. Conventional RF switching systems such as PIN diodes tend to be inefficient at higher frequencies. MEMS based RF switches with very low actuation voltages have been reported recently. At microwave frequencies, micromachined lumped components can replace distributed components with flexibility in integration as well as improvement in bandwidth. Similarly, micromachined or MEMS based phase shifters can replace existing configurations, which tend to have higher insertion losses at high gigahertz frequencies. It may also be noted that micro fabrication technologies can help realize high Q micromechanical filters for frequencies up to and beyond 10 MHz, and micromachined surface acoustic wave (SAW) filters filling the gap up to 2 GHz. Fabrication processes for all these devices and their relative advantages are taken up extensively in this book. In addition, a brief description of packaging approaches that may be extended for these devices is also included for the sake of comprehensiveness in coverage.

We have endeavoured to present these topics so as to guide graduate students interested to do research in microfabrication techniques and their applications. It is therefore envisaged that parts of this books would form the curricula of electrical and mechanical engineering, applied physics or materials science departments. In addition, this would also serve as a reference book for practicing researchers in these areas for further widening the scope of their research.

Materials for this book have been taken from an advanced level course offered at Pennsylvania State University recently on RF MEMS, and many short courses presented across the world. Valuable comments from the participants of these courses have helped in evolving the contents of this book and are greatly appreciated. In particular we also wish to thank many of our colleagues and students, Taeksoo Ji, Yanan Sha, Roopa Tellakula, Hargsoon Yoon, and Bei Zhu, at the Center for Electronic and Acoustic Materials and Devices for their contributions in preparing the manuscript for this book.

We would like to thank Professors Vasundara V. Varadan and Richard McNitt for their support and encouragement. Our thanks are also due to our families, in particular Nisy John, for bearing with our preoccupation in preparing this manuscript. We are also indebted to various researchers for their valuable contributions cited in this book.

We are also grateful to the publisher's staff for their support, encouragement and willingness to give prompt assistance during this book project.

Vijay K. Varadan
K.J. Vinoy
K.A. Jose

1

Microelectromechanical systems (MEMS) and radio frequency MEMS

1.1 INTRODUCTION

During the past decade, several new fabrication techniques have evolved which helped popularize microelectromechanical systems (MEMS), and numerous novel devices have been reported in diverse areas of engineering and science. One such area is microwave and millimeter wave systems. MEMS technology for microwave applications should solve many intriguing problems of high-frequency technology for wireless communications. The recent and dramatic developments of personal communication devices forced the market to acquire miniaturized efficient devices, which is possible only by the development of radio frequency (RF) MEMS.

The term RF MEMS refers to the design and fabrication of MEMS for RF integrated circuits. It should not be interpreted as the traditional MEMS devices operating at RF frequencies. MEMS devices in RF MEMS are used for actuation or adjustment of a separate RF device or component, such as variable capacitors, switches, and filters. Traditional MEMS can be divided into two classes: MEMS actuators and MEMS sensors. The first one is a kind of moving mechanism activated by an electrical signal like Micromotor. Micro sensors are currently available for a large number of applications. Historically, owing to their ease of fabrication, these were the earliest microsystems. Another reason for the actuators not becoming popular is that the amount of energy generated by such tiny systems does not cause much impact in the associated systems. However, it can be seen later, for microwave and millimeter wave systems, these forces are sufficient to change the properties of overall systems. Passive devices include bulk micromachined transmission lines, filters and couplers. Active MEMS devices include switches, tuners and variable capacitors. The electromotive force used to move the structures on the wafer surface is typically electrostatic attraction, although magnetic, thermal or even gas-based microactuator structures have been developed.

Following the classical review paper by Brown (1998), the RF MEMS development to date can be classified into the following categories based on whether one takes an RF or MEMS view point: (1) **RF extrinsic** in which the MEMS structure is located outside the RF circuit and actuates or controls other devices in the RF circuit. In this class, one would consider the example of a tunable microstrip transmission line and associated phased shifters and arrays. Microstrip lines are extensively used to interconnect high-speed circuits and components because they can be fabricated by easy automated techniques. (2) **RF intrinsic** in which the MEMS structure is located inside the RF circuit and has both the actuation and RF-circuit function. In this class, one could consider traditional cantilever and diaphragm type MEMS which can be used as electrostatic microswitch and comb-type capacitors (Brown, 1998). With the invention of electroactive polymers (EAPs), multifunctional smart polymers and microstereo lithography, these types of RF MEMS can be easily conceived with polymer-based systems. They are also flexible, stable and long lasting. Moreover, they can be integrated with the organic thin film transistor. (3) **RF reactive** in which the MEMS structure is located inside, where it has an RF function that is coupled to the attenuation. In this class, capacitively coupled tunable filters and resonators provide the necessary RF function in the circuit. Microwave and millimeter wave planar filters on thin dielectric membrane show low loss, and are suitable for low-cost, compact, high-performance mm-wave one-chip integrated circuits.

One of the earliest reported applications of silicon-based RF MEMS technology for microwave applications is in the area of surface micromachined actuators for the realization of microwave switches. These possess very high linearity, low dc standby power and low insertion loss (Larson, 1999). These switches are based on electrostatic attraction counterbalanced by suitable mechanical forces on the beam to pull the switch into the right position. This switch can be designed to present nearly 50 Ω impedance across a broad range of frequencies when closed, and nearly an open circuit when there is no connection. This property makes this an attractive choice for microwave applications. Several new switch architectures have also been reported, including the air-bridge structure (Goldsmith, Eshelman and Dennston, 1998). This structure utilizes very high capacitance variation to achieve the switching action. This scheme, however, suffers from relatively high switching voltage requirements.

MEMs technology is also used for RF applications in the area of variable capacitors, as a replacement for varactor diodes as tuners (Wu *et al.*, 1998). Here, either a lateral or a parallel plate capacitance variation can be obtained with suitable fabrication approaches. The capacitance variation in the parallel plate version is over 3 : 1 making them attractive for wide-band tuning of monolithic voltage-controlled oscillators (VCOs). However their range is often limited by the low-frequency mechanical resonance of the structure.

1.2 MEMS

The term MEMS refers to a collection of microsensors and actuators which can sense its environment and have the ability to react to changes in that environment with the use of a microcircuit control. They include, in addition to the conventional microelectronics packaging, integrating antenna structures for command signals into micro electromechanical structures for desired sensing and actuating functions. The system also may need micropower supply, micro relay and microsignal processing units. Microcomponents make the system faster, more reliable, cheaper and capable of incorporating more complex functions.

In the beginning of the 1990s, MEMS emerged with the aid of the development of integrated circuit (IC) fabrication processes, where sensors, actuators and control functions are co-fabricated in silicon. Since then, remarkable research progresses have been achieved in MEMS under strong capital promotions from both government and industry. In addition to the commercialization of some less-integrated MEMS devices, such as microaccelerometers, inkjet printer heads, micro mirrors for projection, etc., the concepts and feasibility of more complex MEMS devices have been proposed and demonstrated for the applications in such varied fields as microfluidics, aerospace, biomedicine, chemical analysis, wireless communications, data storage, display, optics, etc. (Fujita, 1996, 1998). Some branches of MEMS, such as micro-opto-electromechanical systems (MOEMS), micro total analysis systems (μTAS), etc., have attracted a great deal of research interest since their potential application market. As of the end of the 1990s, most MEMS devices with various sensing or actuating mechanisms were fabricated using silicon bulk micromachining, surface micromachining and LIGA[1] processes (Bustillo, Howe and Muller, 1998; Guckel, 1998; Kovacs, Maluf and Petersen, 1998). Three dimensional microfabrication processes incorporating more materials were presented for MEMS recently when some specific application requirements (e.g. biomedical devices) and microactuators with higher output power were called for in MEMS (Fujita, 1996; Guckel, 1998; Ikuta and Hirowatari, 1993; Takagi and Nakajima, 1993; Taylor *et al.*, 1994; Thornell and Johansson, 1998; Varadan and Varadan, 1996; Xia and Whitesides, 1998).

Micromachining has become the fundamental technology for the fabrication of microelectromechanical devices and, in particular, miniaturized sensors and actuators. Silicon micromachining is the most mature of the micromachining technologies and it allows for the fabrication of MEMS that have dimensions in the submillimeter range. It refers to fashioning microscopic mechanical parts out of silicon substrate or on a silicon substrate, making the structures three dimensional and bringing new principles to the designers. Employing materials such as crystalline silicon, polycrystalline silicon and silicon nitride, etc., a variety of mechanical microstructures including beams, diaphragms, grooves, orifices, springs, gears, suspensions and a great diversity of other complex mechanical structures has been conceived (Bryzek, Peterson and McCulley, 1994; Fan, Tai and Muller, 1987; Middelhoek and Audet, 1989; Peterson, 1982; Varadan, Jiang and Varadan, 2001).

Sometimes many microdevices can also be fabricated using semiconductor processing technologies or stereolithography on the polymeric multifunctional structures. Stereolithography is a poor man's LIGA for fabricating high aspect ratio MEMS devices in UV-curable semi-conducting polymers. With proper doping, a semiconducting polymer structure can be synthesized and using stereo lithography it is now possible to make three-dimensional microstructures of high aspect ratio. Ikuta and Hirowatari (1993) demonstrated that a three-dimensional microstructure of polymers and metal is feasible using a process named IH Process (integrated hardened polymer stereolithography). Using a UV light source, XYZ-stage, shutter, lens and microcomputer, they have shown that microdevices such as springs, venous valves and electrostatic microactuators can be fabricated. In case of difficulty on the polymeric materials, some of these devices can be micromachined in silicon and the system architecture can be obtained by photoforming or hybrid processing (Ikuta and Hirowatari, 1993; Takagi and Nakajima, 1993; Tani and Esashi, 1995; Varadan, 1995; Varadan and Varadan, 1995, 1996). The photoforming or photo fabrication is an optical method such as the stereolithography, photo mask layering process and IH process which involves solidification of photochemical resin by light exposure. Takagi and Nakajima (1993) proposed new concepts of 'combined architecture'

[1] LIGA is a German acronym, for *Lithographie, Galvanoformung, Abformung* (lithography, galvanoforming, moulding)

and 'glue mechanism' using the photoforming process to fabricate complicated structures by combining components, each of them made by its best fabrication process. Batch processing of such hybrid silicon and polymer devices thus seems feasible.

The combined architecture may also result in sheets of smart skin with integrated sensors and actuators at the μm to mm scale. For some applications (say airfoil surface), the smart skin substrate has to be flexible to conform to the airfoil shape and at the same time it has to be compatible with the IC processing for sensor and smart electronics integration. It has been proposed by Carraway (1991) that polyimide is an excellent material for use as the skin because of its flexibility and IC processing compatibility. The control loop between the sensors and actuators employs the multifunctional materials which provide electrical functionality at selected locations using conductive polymers and electrodes that are connected to on-site antennas communicating with a central antenna. A related and difficult problem, and one which has been largely unaddressed, is the method for telemetry of the data. In some applications, stresses and strains to which the structure is subjected may pose a problem for conventional cabling. In others, environmental effects may affect system performance. Advances in ultra flat antenna technology coupled with MEMS sensors/actuators seems to be an efficient solution. The integration of micromachining and microelectronics on one chip results in so-called smart sensors. In smart sensors, small sensor signals are amplified, conditioned and transformed into a standard output format. They may include microcontroller, digital signal processor, application-specific integrated circuit (ASIC), self-test, self-calibration and bus interface circuits, simplifying their use and making them more accurate and reliable.

The basic MEMS utilize a diaphragm-based, a microbridge-based or a cantilever-based structure. Special processing steps commonly known as micromachining are needed to fabricate these membranes, cantilever beams, resonant structures, etc., which will be discussed later. For a given application, it may be necessary to have integrated MEMS employing one or more of the basic structures. These three structures provide some feasible designs for microsensors and actuators that eventually perform the desired task in most smart structures. However, the main issues with respect to implementing these structures are the choice of materials that are to be used in fabricating these devices and the micromachining technology that may be utilized. To address the first issue, we note that in all of the three structures proposed the sensing and actuation occur as a result of exciting a piezoelectric layer by the application of an electric field. This excitation brings about sensing and actuation in the form of expansion in the diaphragm, or in the free-standing beam in the microbridge structure, or in the cantilever beam. In the former two cases the expansion translates into upward curvature in the diaphragm or in the free-standing beam, hence resulting in a net vertical displacement from the unexcited equilibrium configuration. In the cantilever case, however, and upon the application of electric field, the actuation occurs by a vertical upward movement of the cantilever tip. Evidently, in all three designs the material system structure of the active part (diaphragm, free-standing beam, or cantilever beam) in the microactuator must comprise at least one piezoelectric layer as well as conducting electrodes for the application of electric field across this layer. Piezoelectric force is used for actuation for many of the applications mentioned above. Micromachining is employed to fabricate the membranes, cantilever beams and resonant structures.

Microsensors and actuators are fabricated using the well-known micromachining techniques in the microelectronics industry. Three-dimensional microactuators in polymer structures can be achieved using stereolithography on UV-curable backbone-type polymers (Ikuta and Hirowatari, 1993; Takagi and Nakajima, 1993; Tani and Esashi, 1995; Varadan, 1995; Varadan and Varadan, 1995, 1996). In the integrated MEMS device, we may use photoforming processing in achieving the combined sensor and actuator architecture as outlined by Takagi and Nakajima (1993). For large actuation, one could use a flex tensional transducer consisting of a piezoelectric diaphragm bridged into a cavity (Chin, Varadan and Varadan, 1994).

Silicon micromachining has been a key factor for the vast progress of MEMS in the past decade. This refers to the fashioning of microscopic mechanical parts out of silicon substrates and more recently other materials. It is used to fabricate such features as clamped beams, membranes, cantilevers, grooves, orifices, springs, gears, suspensions, etc. These can be assembled to create a variety of sensors. Bulk micromachining is the commonly used method but it is being replaced by surface micromachining which offers the attractive possibility of integrating the machined device with microelectronics which can be patterned and assembled on the same wafer. Thus power supply circuitry, signal processing using ASICs can be incorporated. It is the efficiency of creating several such complete packages using existing technology that makes this an attractive approach.

1.3 MICROFABRICATIONS FOR MEMS

Silicon micromachining has been a key factor for the vast progress of MEMS. Silicon micromachining refers to fashioning microscopic mechanical parts out of a silicon substrate or on a silicon substrate. Silicon micromachining comprises of two technologies: bulk micromachining, in which structures are etched into silicon substrate, and surface micromachining, in which the micromechanical layers are formed from layers and films deposited on the surface.

Bulk micromachining and surface micromachining are the two major micromachining processes of silicon; silicon wafer bonding is usually necessary for silicon microfabrication. LIGA and three-dimensional (3D) microfabrications have been used for high-aspect ratio and 3D microstructures fabrication for MEMS.

1.3.1 Bulk micromachining of silicon

Bulk micromachining technique was developed in 1960s and allows the selective removal of significant amounts of silicon from a substrate to form membranes on one side of a wafer, a variety of trenches, holes, or other structures (Figure 1.1). The bulk micromachining technique can be divided into wet etching and dry etching of silicon according to the phase of etchants. Liquid etchants, almost exclusively relying on aqueous chemicals, are referred to as wet etching, while vapor and plasma etchants are referred to as dry etching.

Bulk micromachining is the most mature of the two silicon micromachining technologies. It emerged in the early 1960s and has been used since then in the fabrication of different microstructures. It is utilized in the manufacturing of the majority of commercial devices – almost all pressure sensors and silicon valves and 90% of silicon accelerometers. The term bulk micromachining comes from the fact that this type of micromachining is

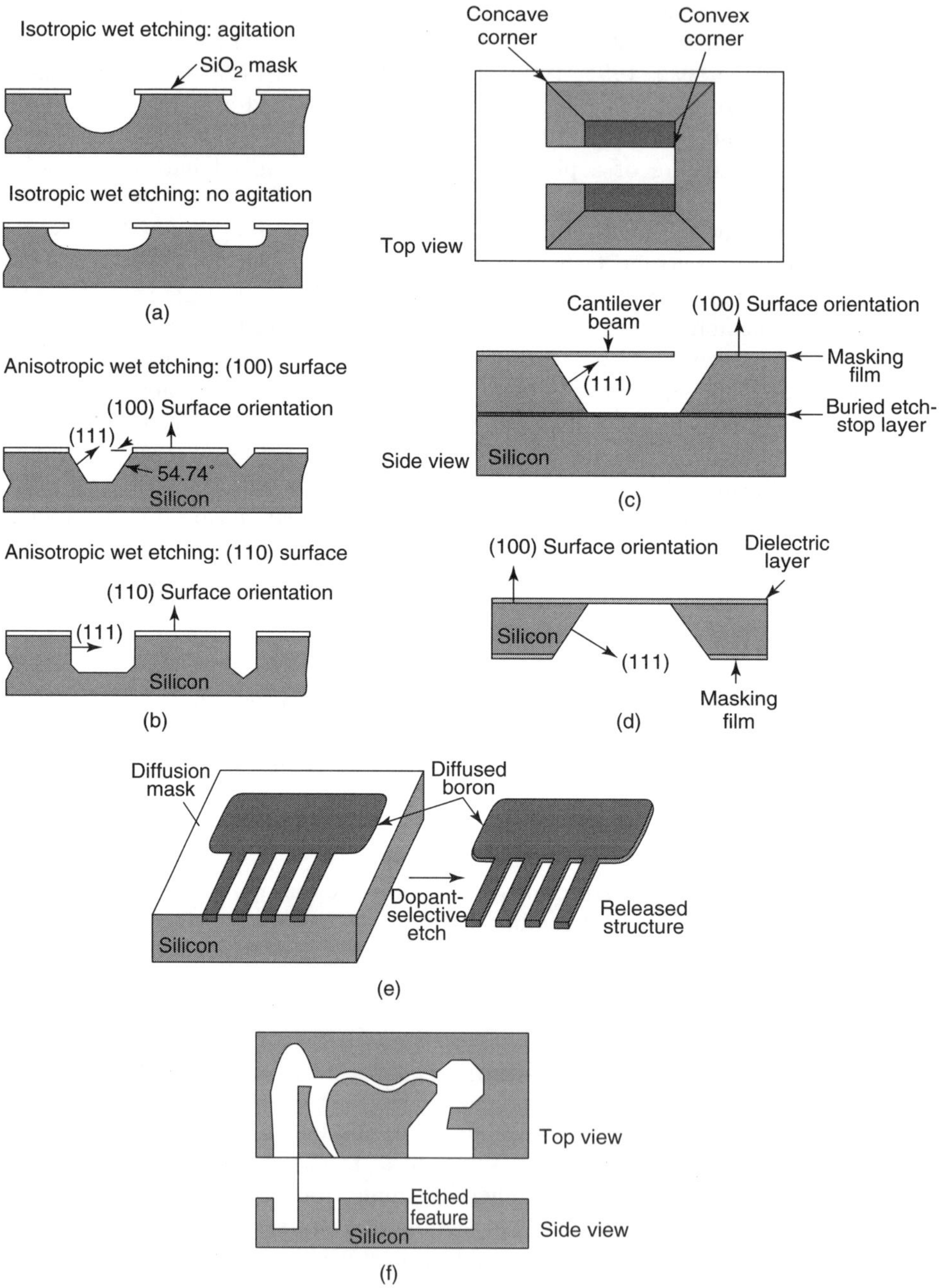

Figure 1.1 Bulk silicon micromachining: (a) isotropic etching; (b) anisotropic etching; (c) anisotropic etching with buried etch-stop layer; (d) dielectric membrane released by back-side bulk etching; (e) dopant dependent wet etching. (f) anisotropic dry etching. Reproduced from C.L. Goldsmith, S. Eshelman and D. Dennston, 1998, 'Performance of low loss RF MEMS capacitive switches', *IEEE Microwave and Guided Wave Letters* **8**: 269–271, by permission of IEEE, © 1998 IEEE

used to realize micromechanical structures within the bulk of a single-crystal silicon wafer by selectively removing ('etching') wafer material. The microstructures fabricated using bulk micromachining may cover the thickness range from submicron to full wafer thickness (200 to 500 μm) and the lateral size range from submicron to the lateral dimensions of a full wafer.

For etching such thick silicon substrate, anisotropic wet etchants such as solutions of potassium hydroxide (KOH), ethylenediamine pyrocatechol (EDP), tetramethylammonium hydroxide (TMAH) and hydrazine-water are used. These etchants have different etch rates in different crystal orientations of the silicon (Aeidel, 1987; Peterson, 1982). Wet etching in most case is done from the back side of the wafer while the plasma-etching is being applied to the front side. In recent years, a vertical-walled bulk micromachining technique known as SCREAM (single-crystal reactive etching and metallization), which is a combination of anisotropic and isotropic plasma etching, is used (Shaw, Zhang and MacDonald, 1994). The etch process can be made selective by the use of dopants (heavily doped regions etch slowly), or may even be halted electrochemically (e.g. etching stops upon encountering a region of different polarity in a biased p–n junction). A region at which wet etching tends to slow down or diminish is called an 'etch-stop'. There are several ways in which an etch-stop region can be created; doping-selective etching (DSE) and bias-dependent DSE (Petersen, 1982; Aeidel, 1982; Shaw, Shang and Macdonald 1994).

Wet etching occurs by dipping substrate into an etching bath or spraying it with etchants which may be acid or alkaline. Wet etching can either be isotropic etching or anisotropic etching depending on the structure of the materials or the etchants used. If the material is amorphous or polycrystalline, wet etching is always isotropic etching (Figure 1.1a). During isotropic etching (etchants used are acid solution), resist is always undercut, meaning the deep etching is not practical for MEMS. Single-crystal silicon can be anisotropically etched. The etching features are determined by the etching speed, which is dependent on the crystal's orientation. The etching slows down significantly at the (111) planes of silicon, relative to other planes. With the chosen wafers with different crystal orientation, different buck machined features can be achieved (Figures 1.1b and 1.1c). Most common etchants used for anisotropic etching of silicon include alkali hydroxide etchants (KOH, NaOH, etc.), ammonium-based solutions {NH_4OH, TMAH [$(CH_3)_4NOH$], etc.} and EDP (ethylene diamine pyrocatechol, and water). By combining anisotropic etching with boron implantation (P+ etch-stop), and electrochemical etch-stop technique, varied silicon microstructures can be bulk machined (Figure 1.1).

Dry etching occurs through chemical or physical interaction between the ions in the gas and the atoms of the substrate. Nonplasma, isotropic dry etching can be possible using xenon difluoride or a mixture of interhalogen gases and provides very high selectivity for aluminum, silicon dioxide, silicon nitride, photoresist, etc. The most common dry etching of bulk silicon are plasma etching and reactive ion etching (RIE) etching, where the external energy in the form of RF powder drives chemical reactions in low-pressure reaction chambers. A wide variety of chlorofluorocarbon gases, sulfur hexafluoride, bromine compounds and oxygen are commonly used as reactants. The anisotropic dry etching processes are widely used in MEMS because of the geometry flexibility and less chemical contamination than in wet etching sometimes. Arbitrarily oriented features can be etched deep into silicon using anisotropic dry etching (Figure 1.1f). Very deep silicon microstructures can be obtained by the deep RIE (DRIE) dry etching (Bryzek, Peterson and McCulley, 1994).

With bulk-micromachined silicon microstructures, the wafer-bonding technique is necessary for the assembled MEMS devices. Surface micromachining, however, can be used to build the monolithic MEMS devices.

1.3.2 Surface micromachining of silicon

Surface micromachining does not shape the bulk silicon but instead builds structures on the surface of the silicon by depositing thin films of 'sacrificial layers' and 'structural layers' and by removing eventually the sacrificial layers to release the mechanical structures (Figure 1.2). The dimensions of these surface micromachined structures can be several orders of magnitude smaller than bulk-micromachined structures. The prime advantage of surface-micromachined structures is their easy integration with IC components, since the wafer is also the working for IC elements. It should be noted that as miniaturization in immensely increased by surface micromachining, the small mass structure involved may be insufficient for a number of mechanical sensing and actuation applications.

Surface micromachining requires a compatible set of structural materials, sacrificial materials and chemical etchants. The structural materials must possess the physical and

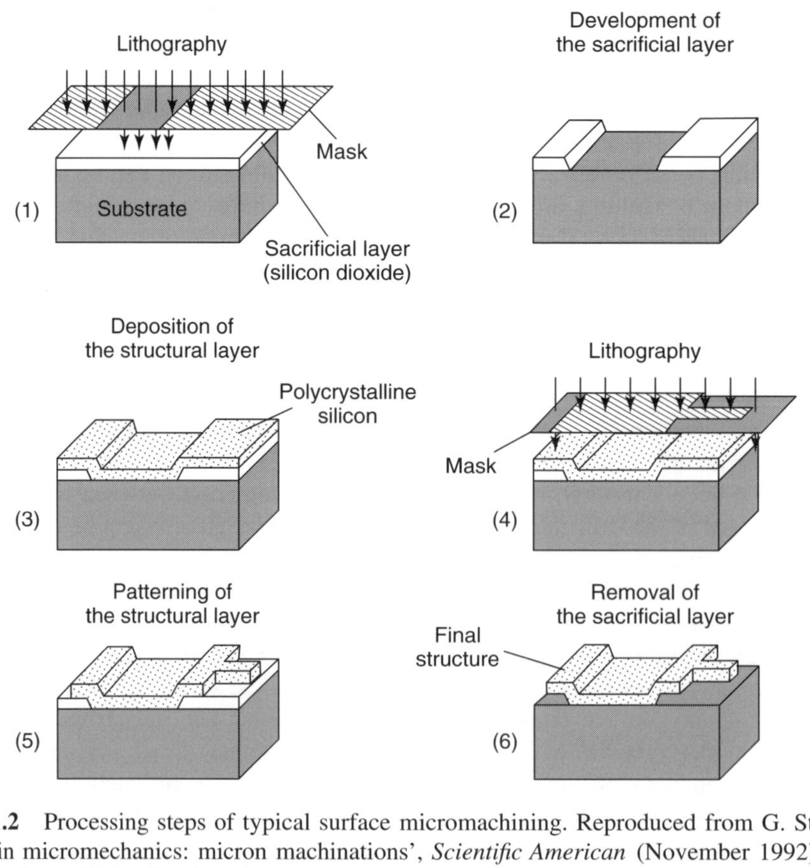

Figure 1.2 Processing steps of typical surface micromachining. Reproduced from G. Stix, 1992, 'Trends in micromechanics: micron machinations', *Scientific American* (November 1992): 72–80, by permission of Scientific American

chemical properties that are suitable for the desired application. In addition, they must have satisfactory mechanical properties; e.g. high yield and fracture stresses, minimal creep and fatigue and good wear resistance. The sacrificial materials must have good mechanical properties to avoid device failure during fabrication. These properties include good adhesion and low residual stresses in order to eliminate device failure by delamination and/or cracking. The etchants to remove the sacrificial materials must have excellent etch selectivity and they must be able to etch off the sacrificial materials without affecting the structural ones. In addition the etchants must have proper viscosity and surface tension characteristics. The common IC compatible materials used in surface micromachining are: (1) polysilicon/Silicon dioxide; low-pressure chemical vapor deposition (LPCVD) deposited polysilicon as the structural material and LPCVD deposited oxide as the sacrificial material. The oxide is readily dissolved in HF solution without the polysilicon being affected. Together with this material system, silicon nitride is often used for electrical insulation. (2) Polyimide/aluminum; in this case polyimide is the structural material and aluminum is the sacrificial material. Acid-based etchants are used to dissolve the aluminum sacrificial layer. (3) Silicon nitride/polysilicon; silicon nitride is used as the structural material, whereas polysilicon is the sacrificial material. For this material system, silicon anisotropic etchants such as KOH and EDP are used to dissolve polysilicon. (4) Tungsten/silicon dioxide; CVD deposited tungsten is used as the structural material with oxide as the sacrificial material. HF solution is used to remove the sacrificial oxide. Other IC-compatible materials such as silicon carbide, diamond-like carbon, zinc oxide, gold, etc. are also used.

Surface micromachining could also be performed using dry etching methods. Plasma etching of the silicon substrate with SF_6/O_2-based and CF_4/H_2-based gas mixtures is advantageous since high selectivities for photoresist, silicon dioxide and aluminum masks can be achieved. However, when using plasma etching, a large undercut of the mask is observed. This is due to the isotropic fluorine atom etching of silicon which is known to be high compared with the vertical etch induced by ion bombardment. In contrast, reactive ion etching of poly-Si using a chlorine/fluorine gas combination produces virtually no undercut and almost vertical etch profiles when using photoresist as a masking material. Thus, rectangular silicon patterns which are up to 30 mm deep can be formed using chlorine/fluorine plasmas out of polysilicon films and the silicon wafer surface.

Silicon microstructures fabricated by surface micromachining are usually planar structures (or are two dimensional). Other techniques involving the use of thin-film structural materials released by the removal of an underlying sacrificial layer have helped to extend conventional surface micromachining into the third dimension. By connecting polysilicon plates to the substrate and to each other with hinges, 3D micromechanical structures can be assembled after release. Another approach to 3D structures used the conformal deposition of polysilicon and sacrificial oxide films to fill deep trenches previously etched in the silicon substrate.

1.3.3 Wafer bonding for MEMS

Silicon micromachining has limitations in forming complex 3D microstructures in a monolithic format; multichip structures are then proposed for advanced MEMS, where wafer-to-wafer bonding is critical in the formation (Stix, 1992).

The wafer bonding for MEMS can be categorized into three major types: anodic bonding, intermediate-layer assisted bonding and direct bonding.

1.3.3.1 Anodic bonding

Anodic bonding is also called field-assisted thermal bonding, electrostatic bonding, etc. Anodic bonding is usually established between a sodium glass and silicon for MEMS. For the anodic bonding, a cathode and an anode are attached to the glass (or silicon with glass thin coating) and silicon wafer, respectively; voltages applied range from 200 V to 1000 V. At the same time, the anode is put on a heater providing the bonding temperature around 180 to ~500 °C (Figure 1.3). During the bonding, oxygen ions from the glass migrate into the silicon, resulting in the formation of a silicon dioxide layer between silicon wafer and glass wafer and form a strong and hermetic chemical bond.

The advantage of anodic bonding for MEMS is that the low temperature used can ensure the metalization layer (aluminum) could withstand this temperature without degradation.

1.3.3.2 Intermediate-layer assisted bonding

This type of bonding for MEMS requires an intermediate layer, which can be metal, polymer, solders, glasses, etc., to fulfill the bonding between wafers (Stix, 1992). One of the earliest wafer bonding – eutectic bonding – utilized gold as the intermediate layer for Si–Si bonding for pressure sensors (Ko, Suminto and Yeh, 1985). The Au–Si eutectic bonding takes place at 363 °C, well below the critical temperature of the metallized aluminium layer. But the stress generated during bonding was found to be significant and introduced sensor drift (Ko, Suminto and Yeh, 1985).

Polymers as an intermediate layer for bonding prevail at very low temperature, reasonable high strength, no metal ion presence and low stress because of the elastic property of polymers, etc. Usually, UV photoresists such as polyimide, AZ-4000, SU-8, polymethylmethacrylate (PMMA), and other UV-curable cross-linked polymers (Madou, 1997). The disadvantage is that the bonded device with polymer may not hold the hermetic sealing performance owing to the relatively high permittivity of polymers.

Glasses with low melting temperature as the intermediate layer for the bonding is also demonstrated, where a layer of glass frit is usually deposited on the silicon wafer. The flatness of the deposited frit layer is critical to obtaining uniform, strong, low-stress bonding. The screen printing of glass frit was used for pressure sensor bonding and exhibits good performance (Ko, Suminto and Yeh, 1985).

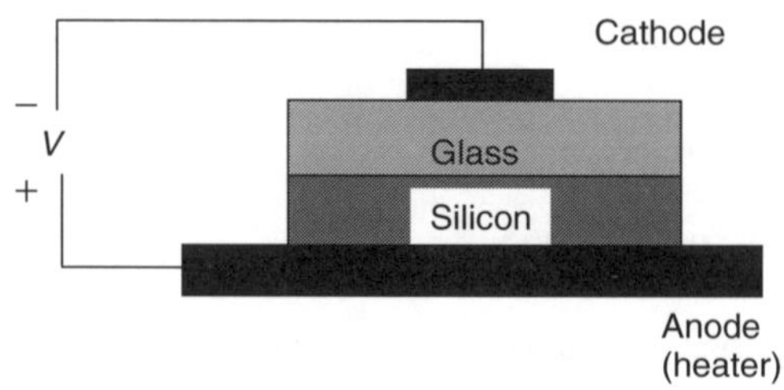

Figure 1.3 Anodic bonding

Other materials are also being developed as the intermediate layer for bonding with low temperature, high strength and low stress (Stix, 1992).

1.3.3.3 Direct bonding

Direct bonding is also called silicon fusion bonding, which is used for silicon–silicon bonding. Direct bonding is based on a chemical reaction between OH groups present at the surface of native silicon or grown oxides covering the wafers (Madou, 1997). The direct bonding usually follows three steps: surface preparation, contacting and thermal annealing.

The surface preparation step involves cleaning the surfaces of the two wafers to form a hydrate surface. The wafer surface should be mirror smooth, the roughness should be no greater than 10 Å, and the bow of a 4-inch wafer should be less than 5 micron to achieve the necessary flatness (Stix, 1992). Following this preparation, the wafers are aligned and contacted in a cleanroom environment by gently pressing the two wafers at the surface central point. The surface attraction of the two hydrated surfaces then brings the intimate contact over the entire wafer surfaces. The final step in direct bonding is to anneal the bonding from room temperature to 1200 °C. This annealing process increases the bond strength by more than one order of magnitude at temperatures as high as 800 to ∼1200 °C. But high-temperature annealing is not allowed for the metallized wafers. The direct bonding prevails in the high-strength bonding, and the devices' dimensions could be scaled down if direct bonding approaches are taken other than anodic bonding.

Some low-temperature direct bonding processes are to be further developed.

1.3.4 LIGA process

MEMS generally require complex microstructures that are thick and three-dimensional (Larson, 1999). Therefore, many microfabrication technologies have been developed to achieve high-aspect-ratio (height-to-width) and 3D devices. The LIGA process is one of those microfabrications.

LIGA is a German acronym for *Lithographie, Galvanoformung, Abformung* (lithography, galvanoforming, moulding). It was developed by the research Center Karlsruhe in the early 1980s in Germany using X-ray lithography for mask exposure, galvanoforming to form the metallic parts and moulding to produce microparts with plastic, metal, ceramics, or their combinations (Fujita, 1996). A schematic diagram of the LIGA process is shown in Figure 1.4. With the LIGA process, microstructures' height can be up to hundreds of microns to millimeter scale, while the lateral resolution is kept at the submicron scale because of the advanced X-ray lithography. Various materials can be incorporated into the LIGA process, allowing electric, magnetic, piezoelectric, optic and insulating properties in sensors and actuators with a high-aspect ratio, which are not possible to make with the silicon-based processes. Besides, by combining the sacrificial layer technique and LIGA process, advanced MEMS with moveable microstructures can be built (Figure 1.5). However, the high production cost of LIGA process due to the fact that it is not easy to access X-ray sources limits the application of LIGA. Another disadvantage of the LIGA process relies on that fact that structures fabricated using LIGA are not truly three-dimensional, because the third dimension is always in a straight feature. As we know, complex thick 3D structures are necessary for some advanced MEMS, which means other 3D microfabrication processes need to be developed for MEMS.

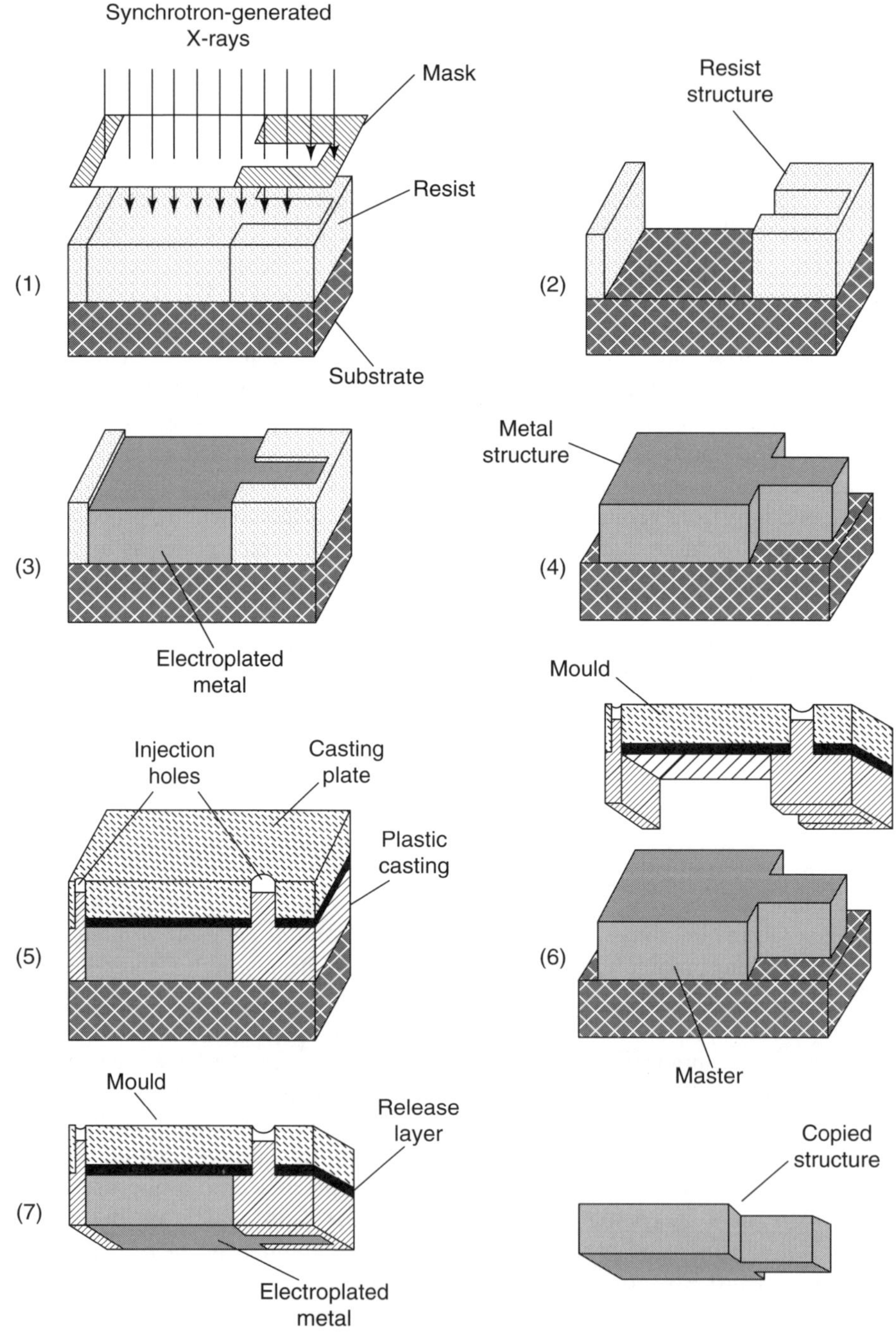

Figure 1.4 The LIGA process. Reproduced from G. Stix, 1992, 'Trends in micromechanics: micron machinations', *Scientific American* (November 1992): 72–80, by permission of Scientific American

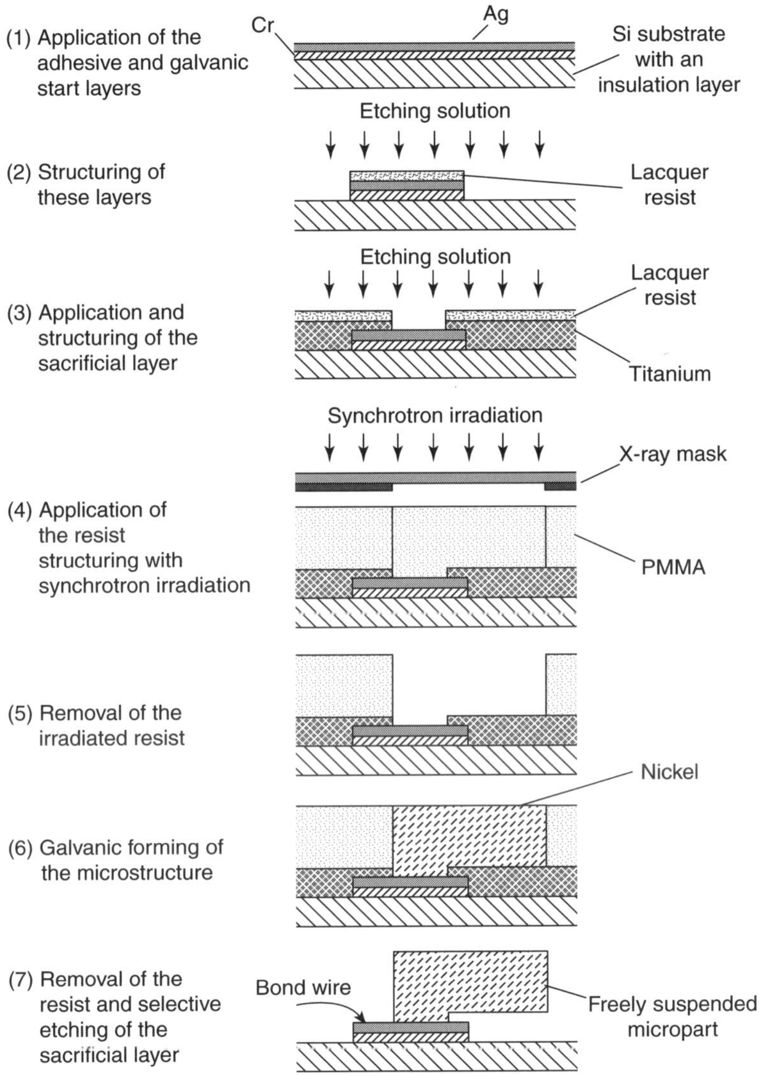

Figure 1.5 Combination of the LIGA process and the sacrificial layer process. Reproduced from J. Mohr *et al.*, 1991, 'Herstellung von beweglichen mikrostrukturen mit dem LIGA-verfahren', *KfK-Nachrichten, Jahrgang 23, Forschungszentrum Karlsruhe* **2–3**: 110–117, by permission of Forschungszentrum Karlsruhe

1.3.5 Micromachining of polymeric MEMS devices

In the micromachining concept for polymeric MEMS devices, two types of polymers are employed: one is the structural polymer and the other one is a sacrificial polymer. The structural polymer is usually a UV-curable polymer with urethane acrylate, epoxy acrylate and acryloxysilane as main ingredients. Its low viscosity allows easy processing through automatic equipment or manual methods without the addition of solvents or heat to reduce

Table 1.1 General properties of polymer

Physical properties:	
clarity	Transparent
flexibility	Good
adhesion (#600 Cellotape)	Excellent
weather resistance	Excellent
flammability (ASTM D635)	Self-extinguishing
Chemical Properties:	
fungus resistance (ASTM-G21)	Excellent
resistance to solvents	Excellent
resistance to chemicals	Excellent
resistance to water	Excellent
Thermal properties:	
continuous operating range (°C)	65–125
decomposition temperature (°C)	242
Mechanical properties:	
tensile Strength (psi; ASTM D 683)	3454
percentage elongation (ASTM D 683)	5.2
Dielectric properties:	
dielectric permittivity (200–1000 MHz)	1.9–2
loss tangent (200–1000 MHz)	0.023–0.05

the viscosity. It also complies with all volatile organic compound (VOC) regulations. It has excellent flexibility and resistance to fungus, solvents, water and chemicals. Other physical, chemical, mechanical and thermal properties are given in Table 1.1 (Varadan, Jiang and Varadan, 2001). This structural polymer may be used as a backbone structure for building the multifunctional polymer described below.

For 3D MEMS devices, the polymers need to have conductive and possibly piezoelectric or ferroelectric properties. For these polymers to be used for polymeric MEMS, they have to meet the following requirements: (1) interactions (chemical or physical) between functional polymer and nanoceramics; (2) strong interfacial adhesion between functional polymer and conducting polymer layers; (3) suitable elastic moduli to support the deformation initiated by MEMS devices; (4) excellent overall dimensional stability (allowing local mobility); (5) processes conducive to the attachment of nanoceramics and/or conductive phases and formation of a uniform coating layer; (6) long-term environmental stability. In addition, the multifunctionality of these polymers provides a large-scale strain under electric field and thus can be used as actuators for MEMS-based devices such as micropumps. In general, these polymers are biocompatible and thus useful for many medical devices. Other applications may include implanted medical delivery systems, chemical and biological instruments, fluid delivery in engines, pump coolants and refrigerants for local cooling of electronic components. The sacrificial polymer is an acrylic resin containing 50% silica and is modified by adding Crystal Violet (Varadan, Jiang and Varadan, 2001). This composition is UV curable and can be dissolved with $2\,\mathrm{mol\,l^{-1}}$ caustic soda at 80 °C. In principle this process is similar to the surface micromachining used for silicon devices. However, the process yields 3D structures.

1.3.6 Three-dimensional microfabrications

To fabricate 3D structures for MEMS, many novel 3D microfabrication techniques have been developed. Among them, microscale freeform fabrications are very impressive to achieve 3D MEMS devices.

Most of the freeform fabrications build 3D microstructures in an additive layer-by-layer fashion (Figure 1.6). The members of the freeform microfabrications family include microstereolithography (Ikuta and Hirowatari, 1993), electrochemical fabrication (EFAB; Cohen *et al.*, 1999), microphotoforming (Takagi and Nakajima, 1993), spatial forming (Taylor *et al.*, 1994), microtransfer moulding (Xia and Whitesides, 1998), localized electrochemical deposition (Madden and Hunter, 1996), etc. Complex 3D microstructures have been built using these techniques from the materials of polymer, ceramic, metal, etc. In principle, other smart materials can be incorporated into these microfabrications.

Another approach to build 3D MEMS devices is to combine the current micromachining processes, such as silicon micromachining, LIGA, precision mechanical machining, etc., or combining the new 3D microfabrication processes with the silicon micromachining and LIGA processes (Bertsch, Lorenz and Renaud, 1998; Takagi and Nakajima, 1994). The AMANDA process is one of these combined microfabrication processes, which combines the LIGA process (or precision machining) with silicon micromachining. AMANDA is the German acronym for *Abformung, Oberflachenmikromechanik und Membranubertragung*. The English translation could be surface micromachining, moulding, and diaphragm transfer. The AMANDA process is specially powerful in polymer MEMS fabrication (Schomburg *et al.*, 1998).

It is known that 3D MEMS devices are usually with multiple layers fabricated from many different structural and functional materials. At the beginning of MEMS fabrications, silicon and polysilicon are the major structural materials, but more structural materials need to be incorporated into MEMS devices since largely divergent applications require varied properties from the microstructures. For example, for the bio-MEMS devices, structural materials should be biocompatible; many polymers are then selected and used to build structures (Ruprecht *et al.*, 1998). While refractory materials may be required for the structures used in high temperature, harsh environments, structural ceramics may be the first choice for this type of application (Ayerdi *et al.*, 1997; Mehregany *et al.*, 1998). Metallic structures have a good reputation in the macro world, as some of the properties of metals are still maintained in microscale. Therefore, 3D microfabrications capable of processing a broad spectrum of structural materials were expected for MEMS,

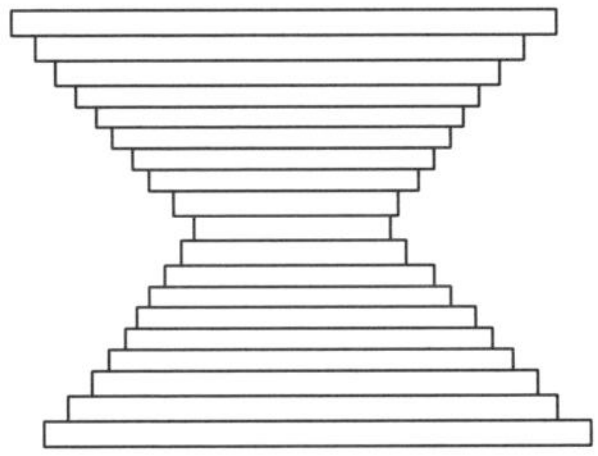

Figure 1.6 Freeform three-dimensional microfabrication (additive)

and much progress has been made (Cohen *et al.*, 1999; Ikuta and Hirowatari, 1993; Jiang, Sun and Zhang, 1999; Takagi and Nakajima, 1993; Taylor *et al.*, 1994; Zhang, Jiang and Sun, 1999).

On the other hand, most sensing and actuating structures used in MEMS are fabricated from thin films. For microsensors, it is advantageous using microelectronic compatible processes. But actuators are devices that modify their environment; they are fundamentally three-dimensional devices (Guckel, 1998). Thick and 3D structures are advantageous to provide higher output power, which has been expected for a while for MEMS since low power output MEMS have been less significant in actuation so far. So, incorporating more smart materials into MEMS devices acting as sensing and actuating structures has become the concern in MEMS development.

For 3D MEMS devices, the polymers need to have conductive and possibly piezo or ferroelectric properties. Such MEMS polymers involve the integration of conventional UV-curable polymers, optically transparent conductive polymers and nanopiezo or ferroelectric particles by chemical bonding as side-groups on the polymer backbone. The concept is to design a backbone with functional groups that will serve as anchor points for the metal oxides. The nanoparticles such as lead zirconate titanate (PZT), PLZT, etc. must have active surfaces or functional groups that can bond with the polymer chain. The nanoparticles provide the piezoelectric function in the polymer and the backbone provides the mechanical stability and flexibility if it is needed.

1.4 ELECTROMECHANICAL TRANSDUCERS

Mechanical filters and their micromechanical counterparts rely on the transformation of electrical energy to a mechanical form and vice versa, through a frequency-dependent transducer, for their operation. In this section basic principles of some of these common electromechanical transducers are discussed briefly. The energy conversion schemes presented here include piezoelectric, electrostrictive, magnetostrictive, electrostatic, electromagnetic, electrodynamic and electrothermal transducers.

Although some of these schemes are not amenable for micromechanical systems, an understanding of their operation would give some insights into their implementations in future RF-MEMS filters discussed later, in Chapter 5.

One important step in the design of these mechanical systems is to obtain their electrical equivalent circuit from their analytical model. This often involves first obtaining mechanical equivalent circuits using springs and masses, and then using the electromechanical analogies to reach the electrical equivalent circuits. Such conversions need not always be exact but would serve as an easily understood tool in their design. The use of these electrical equivalent circuits would also facilitate the use of vast resources available for modern optimization programs for electrical circuits in filter design.

A list of useful electromechanical analogies is given in Table 1.2 (Johnson, 1983). These are known as mobility analogies. These analogies become useful when one needs to replace mechanical components with electrical components which behave similarly, forming the equivalent circuit. As a simple example, the development of an electrical equivalent circuit of a mechanical transmission line component is discussed here (Johnson, 1983). The variables in such a system are force and velocity. The input and output variables of a section of a lossless transmission line can be conveniently related by

Table 1.2 Electromechanical mobility analogies [42]

	Mechanical parameter	Electrical parameter
Variable	Velocity, angular velocity	Voltage
	Force, torque	Current
Lumped network elements	Damping	Conductance
	Compliance	Inductance
	Mass, mass moment of inertia	Capacitance
Transmission lines	Compliance per unit length	Inductance per unit length
	Mass per unit length	Capacitance per unit length
	Characteristic mobility	Characteristic impedance
Immitances	Mobility	Impedance
	Impedance	Admittance
	Clamped point	Short circuit
	Free point	Open circuit
Source immitances	Force	Current
	Velocity	Voltage

ABCD matrix form as:

$$\begin{bmatrix} \dot{x}_1 \\ F_1 \end{bmatrix} = \begin{bmatrix} \cos \beta x & jZ_0 \sin \beta x \\ \dfrac{j}{Z_0} \sin \beta x & \cos \beta x \end{bmatrix} \begin{bmatrix} \dot{x}_2 \\ F_2 \end{bmatrix} \tag{1.1}$$

where

$$Z_0 = \frac{1}{A\sqrt{\rho E}} \sqrt{\frac{C_l}{M_l}} \tag{1.2}$$

$$\beta = \frac{\omega}{v_p} \tag{1.3}$$

$$v_p = \sqrt{\frac{E}{\rho}} = \frac{1}{\sqrt{C_l M_l}} \tag{1.4}$$

In these equations $j = \sqrt{-1}$, $\dot{x}_1$ and $\dot{x}_2$ are velocities, F_1 and F_2 forces at two ends of a transmission line, Z_0, β and v_p are the characteristics impedance, propagation constant and phase velocity of the transmission line, A is the cross-sectional area of the mechanical transmission line, E its Young's modulus, and ρ the density. Quantities C_l and M_l are compliance and mass per unit length of the line, respectively. Now, looking at the electromechanical analogies in Johnson (1983), the expressions for an equivalent electrical circuit can be obtained in the same form as Equation (1.1):

$$\begin{bmatrix} V_1 \\ I_1 \end{bmatrix} = \begin{bmatrix} \cos \beta x & jZ_0 \sin \beta x \\ \dfrac{j}{Z_0} \sin \beta x & \cos \beta x \end{bmatrix} \begin{bmatrix} V_2 \\ I_2 \end{bmatrix} \tag{1.5}$$

Table 1.3 Direct analogy of electrical and mechanical domains

Mechanical quantity	Electrical quantity
Force	Voltage
Velocity	Current
Displacement	Charge
Momentum	Magnetic flux linkage
Mass	Inductance
Compliance	Capacitance
Viscous damping	Resistance

Source: Tilmans, 1996.

In Equation (1.5) V and I are the voltage and current on the transmission line (with subscripts representing its ports). The other quantities in the matrix are also represented by equivalent electrical parameters as:

$$Z_0 = \sqrt{\frac{\mu}{\varepsilon}} = \sqrt{\frac{L_l}{C_l}} \tag{1.6}$$

$$v_p = \frac{1}{\sqrt{\mu\varepsilon}} = \frac{1}{\sqrt{L_l C_l}} \tag{1.7}$$

In Equations (1.6) and (1.7) L_l and C_l represent the inductance and capacitance per unit length of the line, and ε and μ are the permittivity and permeability of the transmission medium.

Apart from the above mobility analogy a direct analogy is also followed at times to obtain the equivalence between electrical and mechanical circuits. These result from the similarity of integrodifferential equations governing electrical and mechanical components (Tilmans, 1996). A brief list of these analogies are presented in Table 1.3.

An understanding of the underlying operational principle is essential in obtaining the equivalent circuit of these transducers. A brief description of the operational principles of some of these common transduction mechanisms used in electromechanical systems is provided below.

1.4.1 Piezoelectric transducers

When subjected to mechanical stress, certain anisotropic crystalline materials generate charge. This phenomenon, discovered in 1880 by Jaques and Pierre Curie, is known as piezoelectricity. This effect is widely used in ultrasonic transducers. Lead zirconate titanates (PZTs) are the most common ceramic materials used as piezoelectric transducers. These crystals contain several randomly oriented domains if no electric potential is applied during the fabrication process of the material. This results in little changes in the dipole moment of such a material when a mechanical stress is applied. However, if the material is subjected to an electric field during the cooling down process of its fabrication, these domains will be aligned in the direction of the field. When external stress is applied to such a material, the crystal lattices get distorted, causing changes in the domains

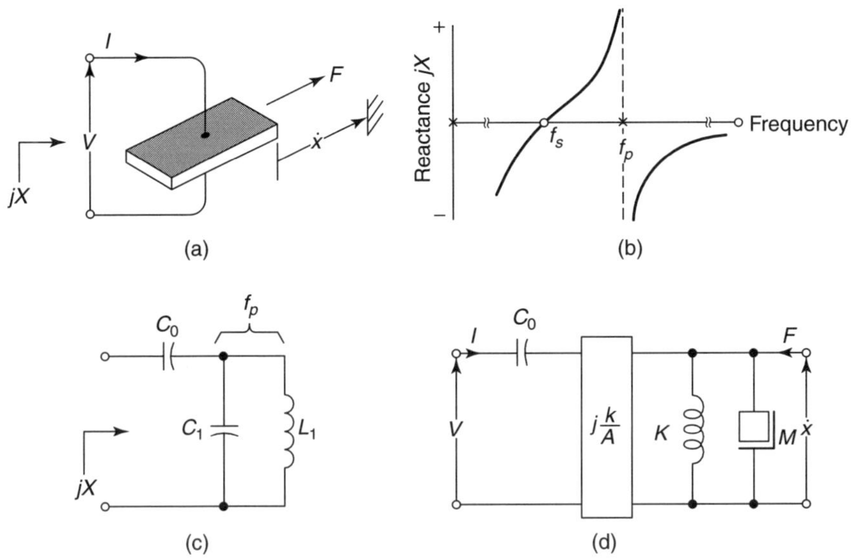

Figure 1.7 Development of equivalent circuit of a piezoelectric transducer. Reproduced from R.A. Johnson, 1983, *Mechanical Filters in Electronics*, Wiley Interscience, New York, by permission of Wiley, © 1983 Wiley

and a variation in the charge distribution within the material. The converse effect of producing strain is caused when these domains change shape by the application of an electric field.

The development of the equivalent circuit for a piezoelectric bar is illustrated in Figure 1.7 (Johnson, 1983). The bar vibrates in the direction (with force F and velocity $\dot{x}$) shown in the figure, by the application of an applied voltage (V). The reactance (jX) curve in Figure 1.7(b) can be obtained by ignoring higher order modes of vibration, and the losses. One circuit configuration that results in similar reactance characteristics is shown Figure 1.7(c). The electromechanical equivalent circuit can be constructed from this, incorporates a gyrator with a resistance A and an inverter of reactance $j\kappa$ in addition to the corresponding spring constant K and mass M. The gyrator represents the nonreciprocal nature of the piezoelectric transducer. The inverter is required here since the gyrator converts the parallel resonant circuit to a series circuit (Johnson, 1983). The series combination of inverter and gyrator functions as a transformer with an imaginary turns ratio $j\kappa/A$.

In general the piezoelectric transduction phenomenon is quadratic in nature, but may be assumed to be linear for small deformations. The electromechanical coupling can then be written as

$$Q = d_1 F \tag{1.8}$$

$$x = d_2 V \tag{1.9}$$

In these equations, d_1 and d_2 represent the piezoelectric charge modulus d in units 1 and 2, respectively. However, when both voltage and force are present, the following piezoelectric coupling equations are used:

$$Q = d_1 F + C_0 V \tag{1.10}$$

$$x = d_2 V + C_m F \tag{1.11}$$

where C_0 is the free capacitance and C_m the short-circuit compliance of the transducer. The electromechanical coupling coefficient is another important nondimensional quantity representing the performance of piezoelectric transducers. This is the ratio of mechanical work available to the electrical energy stored in the transducer (Hom *et al.*, 1994). The coupling coefficient depends on the type of material, mode of stress and the polarization of electric field. For a linear piezoelectric material, this is

$$\eta = \frac{d}{S\varepsilon} \tag{1.12}$$

where d is a constant for piezoelectric material, S is the elastic compliance and ε is the permittivity of the material.

PZT thin films have been developed using standard thin-film deposition techniques such as sputtering, and physical or chemical vapor deposition. Their use in sensors and actuators is inherently limited by the quality and repeatability of thin films obtained by these techniques. Compared with bulk material processing techniques thin-film performance is severely hampered by the surface properties where the film is deposited (Muralt, 2000). Nonferroelectric AlN thin films are also explored, for sensor applications where voltage output is required. However, PZT thin films are still preferred as actuators. Compared with other electromechanical conversion schemes these require low voltage input but have generally low electromechanical conversion efficiency.

1.4.2 Electrostrictive transducers

Electrostriction is the phenomenon of mechanical deformation of materials due to an applied electric field. This is a fundamental phenomenon present to varying degrees in all materials, and occurs as a result of the presence of polarizable atoms and molecules. An applied electric field can distort the charge distribution within the material, resulting in modifications to bond length, bond angle or electron distribution functions, which in turn affects the macroscopic dimensions of the material.

The electric field E and the electric displacement D in a material are related by

$$\mathbf{D} = \varepsilon_0 \mathbf{E} + \mathbf{P} \tag{1.13}$$

where ε_0 is the free space permittivity ($= 8.85 \times 10^{-12}\,\mathrm{F\,m^{-1}}$) and P is the polarization of the material.

Using conservation of energy, the first law of thermodynamics for a electrically deformable material is (Hom *et al.*, 1994):

$$\mathrm{d}U = \mathsf{T}_{\mathsf{ij}}\,\mathrm{d}\mathsf{S}_{\mathsf{ij}} + E_k\,\mathrm{d}D_k + T\,\mathrm{d}S \tag{1.14}$$

In Equation (1.14), U is the internal energy for unit volume of the material, **T** is the stress tensor, **S** is the infinitesimal strain tensor, T is the temperature and S is its entropy per unit volume.

The elastic Gibbs function of a material is defined as

$$G = U - \mathsf{T}_{ij}\mathsf{S}_{ij} - \tfrac{1}{2}\varepsilon_0 E_k E_k - TS \tag{1.15}$$

Taking the derivative of Equation (1.15) and making use of Equation (1.13) we get:

$$\mathrm{d}G = \mathrm{d}U - \mathsf{T}_{ij}\,\mathrm{d}\mathsf{S}_{ij} - \mathsf{S}_{ij}\,\mathrm{d}\mathsf{T}_{ij} - E_k(\mathrm{d}D_k - \mathrm{d}P_k) - T\,\mathrm{d}S - S\,\mathrm{d}T \tag{1.16}$$

Substituting for $\mathrm{d}U$ from Equation (1.14), this simplifies to:

$$\mathrm{d}G = -\mathsf{S}_{ij}\,\mathrm{d}\mathsf{T}_{ij} + \tfrac{1}{2}E_k P_k - S\,\mathrm{d}T \tag{1.17}$$

The derivative of the Gibbs function G can be obtained using the chain rule as:

$$\mathrm{d}G = \frac{\partial G}{\partial \mathsf{T}_{ij}}\mathsf{T}_{ij} + \frac{\partial G}{\partial P_k}P_k + \frac{\partial G}{\partial T}T \tag{1.18}$$

Comparing terms in Equation (1.17) and (1.18),

$$S_{ij} = -\frac{\partial G}{\partial \mathsf{T}_{ij}} \tag{1.19}$$

$$E_k = \frac{\partial G}{\partial P_k} \tag{1.20}$$

$$S = -\frac{\partial G}{\partial T} \tag{1.21}$$

Assuming isotropic dielectric behavior, the Gibbs energy function for an elastic material is given by (Hom *et al.*, 1994):

$$\begin{aligned} G = &-\frac{1}{2}s^P_{ijkl}\mathsf{T}_{ij}\mathsf{T}_{kl} - Q_{mnpq}\mathsf{T}_{mn}P_pP_q \\ &+ \frac{1}{2k}\left\{|\mathbf{P}|\ln\left[\left(1+\frac{|\mathbf{P}|}{P_s}\right)\left(1-\frac{|\mathbf{P}|}{P_s}\right)^{-1}\right]\right. \\ &\left.+ P_s\ln\left[1-\left(\frac{|\mathbf{P}|}{P_s}\right)^2\right]\right\} \end{aligned} \tag{1.22}$$

The first term on the right-hand side describes the elastic behaviour of the material, s^P being its elastic compliance at constant polarization. The electromechanical coupling is denoted in the second term with the electrostrictive coefficients forming the matrix **Q**. The last term is the dielectric behaviour of the material. P_s is the spontaneous polarization, and k is a material constant related to its dielectric constant. Since the material is assumed to be isotropic, the magnitude of polarization is given as:

$$|\mathbf{P}| = \sqrt{P_k P_k} \tag{1.23}$$

Temperature-dependent material coefficients used in Equation (1.22) such as s^P, $\mathbf{Q}$, P_s and k are obtained from electrical and mechanical measurements.

Substituting Equation (1.22) into Equation (1.19) we get the constitutive equations for electrostrictive materials as:

$$\mathsf{S}_{ij} = s^P_{ijkl}\mathsf{T}_{kl} + Q_{ijmn}P_m P_n \tag{1.24}$$

This shows the total strain in a material is the sum of elastic strain and polarization induced strain. The second term on the right-hand side of Equation (1.24) represents the electrostrictive effect. Thus this contribution is proportional to the square of the polarization in the material. This constitutive relation is valid even at large field intensities. Terms in the matrix $\mathbf{Q}$ are the electrostriction coefficients and are obtained from measurements.

The phenomenon of electrostriction is very similar to piezoelectricity. One of the fundamental difference between the two is the closeness of transition temperature of the material to the operating temperatures. This accounts for the improved strain and hysteresis properties for electrostrictive materials. However, a larger number of coefficients are required to model electromechanical coupling for electrostriction. The polarization in piezoelectric materials is spontaneous, while that in electrostrictive materials is field-induced. The properties of electrostrictive materials are more temperature-dependent, and the operating temperature range for these materials is narrower than for piezoelectrics (Chen and Gururaja, 1997).

Material compositions based on lead magnesium niobate [$Pb(Mg_{0.33}, Nb_{0.67})O_3$ (PMN)] are commonly used as electrostrictive transducers. Their properties have been studied extensively (Pilgrim, 2000). Practical thin-film transducers using this approach are yet to be realized. However, polymeric thin-film materials with compliant graphite electrodes are shown to have excellent electrostrictive properties (Pelrine, Kornbluh and Joseph, 1998). These materials are capable of efficient and fast response with high strains, good actuation pressures (up to 1.9 MPa), and high specific energy densities. In this case, the electrostriction phenomenon is not due to the molecular dipole realignment (Heydt *et al.*, 1998). In these silicone film actuators, the strain results from external forces caused by electrostatic attraction of their graphite compliant electrodes. Although their mechanism is electrostatics based, these actuators are shown to produce much larger effective actuation pressure than conventional air-gap electrostatics with similar electric field.

1.4.3 Magnetostrictive transducers

Certain ferromagnetic materials show deformation when subjected to a magnetic field. This phenomenon, commonly known as magnetostriction, is reversible and is also called the Joule and Villari effect. In their demagnetized form, domains in a ferromagnetic material are randomly oriented. However, when a magnetic field is applied these domains gets oriented along the direction of the field. This orientation results in microscopic forces between these domains resulting in the deformation of the material. By reciprocity, mechanical deformation can cause orientation of domains, resulting in induction at the macroscopic level (Rossi, 1988). The elongation is quadratically related to the induced magnetic field and hence is strongly nonlinear.

Apart from the ferroelectric bar, the magnetostrictive transducer consists of a coil and a magnet (Johnson, 1983). When a current I flows through the coil, the bar is deflected

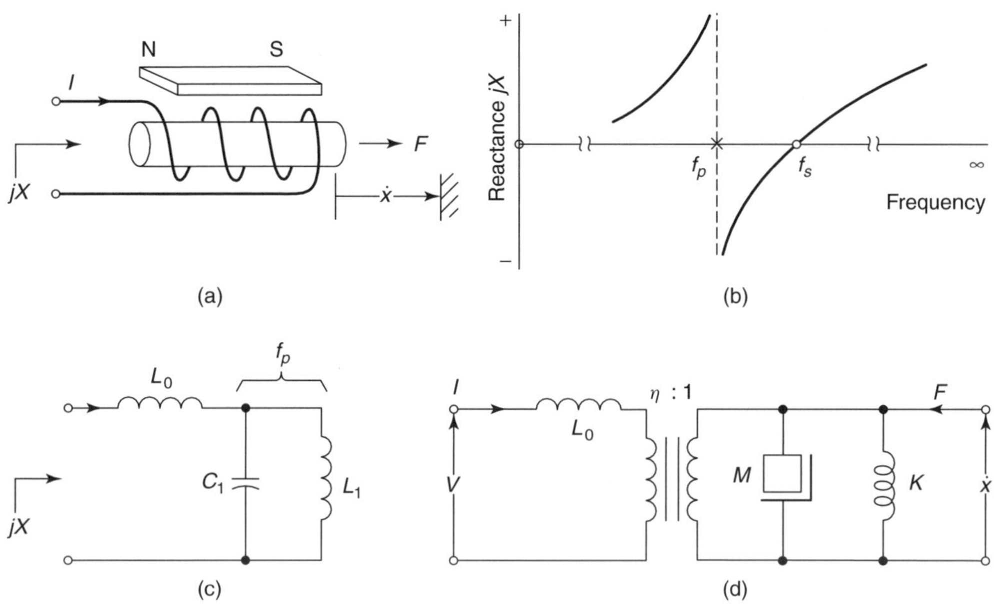

Figure 1.8 Equivalent circuit for a magnetostrictive transducer. Reproduced from R.A. Johnson, 1983, *Mechanical Filters in Electronics*, Wiley Interscience, New York, by permission of Wiley, © 1983 Wiley

in the direction shown with force F and velocity $\dot{x}$. The development of the equivalent circuit of such a transducer is shown schematically in Figure 1.8. The reactance (jX) diagram shown in Figure 1.8(b) is measured with no load. The pole and zero frequencies in this curve correspond to parallel and series resonances of the system. It is not very hard to obtain the component values of an LC circuit shown in Figure 1.8(c) which result in the same pole and zero frequencies as with the system in Figure 1.8(a). Therefore Figure 1.8(c) is an idealized electrical equivalent circuit for the transducer shown in Figure 1.8(a). This is an idealized model as it does not take into consideration the losses in the system.

It is now possible to translate this electrical equivalent circuit to the electromechanical circuit shown in Figure 1.8(d). This has electrical and mechanical components (mass M and spring K) connected with an electromechanical transformer. The turns ratio of this transformer is decided by the amount of coupling, known as the electromechanical coupling coefficient. This is defined as the ratio of the energy stored in the mechanical circuit to the total input energy.

The electromechanical coupling for a magnetostrictive transducer shown in Figure 1.8(a) relates the force at one end of the rod (the other end being constrained) to the current i in the coil as (Rossi, 1988):

$$F = \frac{g_\Delta EN}{R_m} i \tag{1.25}$$

where F is the magnetostrictive force, g_Δ is the magnetostrictive strain modulus, E is the Young's modulus of the material, R_m is the total reluctance of the magnetic circuit, and N is the number of turns in the coil. The ratio on the right-hand side of Equation (1.25) is the electromechanical coupling. The same value for the coefficient relates the induced voltage V at the terminals of the coil with the rate of change in displacement at the free end of the bar:

$$V = \frac{g_\Delta E N}{R_m}\dot{x} \tag{1.26}$$

Ferrites, and metallic alloys such as Permalloy (45% Ni + 55% Fe), Alfer (13% Al + 87% Fe) and Alcofer (12% Al + 2% Co + 86% Fe) are some of the common materials used in magnetostrictive transducers. Some of these materials can also be deposited as thin films thus making it possible to fabricate microactuators and sensors with them. Amorphous thin films such as $TbFe_2$, $Tb_{0.3}Dy_{0.7}Fe_2$ and $DyFe_2$ have been reported in the literature (Body, Reyne and Meunier, 1997). The realization of such thin films is more process dependent than their bulk counterparts, as the preparation conditions affect the homogeneity and growth process of the film as well as its stoichiometry. RF magnetron sputtering of METGLAS® 2605-SC ribbon with a chemical composition of $Fe_{81}Si_{3.5}B_{13.5}C_2$ on a GaAs substrate has been used in a pressure sensor with figure of merit comparable with that of conventional piezoresistive strain gauges (Karl *et al.*, 2000). Microelectromechanical filters using this technology have not been reported so far in the literature, but seem promising.

1.4.4 Electrostatic actuators

Electrostatic actuation is the most common type of electromechanical energy conversion scheme in micromechanical systems. This is a typical example of an energy-storage transducer. Such transducers store energy when either mechanical or electrical work is done on them (Crandall *et al.*, 1968). Assuming that the device is lossless, this stored energy is conserved and later converted to the other form of energy. The structure of this type of transducer commonly consists of a capacitor arrangement, where one of the plates is movable by the application of a bias voltage. This produces displacement, a mechanical form of energy.

To derive an expression for the electromechanical coupling coefficient, let us first consider a parallel plate capacitor. In Figure 1.9, the bottom plate is fixed, and the top one is movable. The constitutive relations of this structure for voltage (V) and force (F) are given in terms of displacement (x) and charge (Q). These relations can be obtained either analytically from electrostatics, or experimentally when a complicated system with various losses has to be modeled. Assuming that there are no fringing fields, the capacitance of this configuration at rest is widely known to be:

$$C_0 = \frac{\varepsilon A}{d_0} \tag{1.27}$$

However, when a voltage is applied across this system, the top plate moves towards the other, resulting in a net gap

$$d = d_0 - x \tag{1.28}$$

The capacitance with the plates at this new position is

$$C = \frac{\varepsilon A}{d} = \frac{\varepsilon A}{d_0 - x} = \varepsilon A\left[d_0\left(1 - \frac{x}{d_0}\right)\right]^{-1} = C_0\left(1 - \frac{x}{d_0}\right)^{-1} \tag{1.29}$$

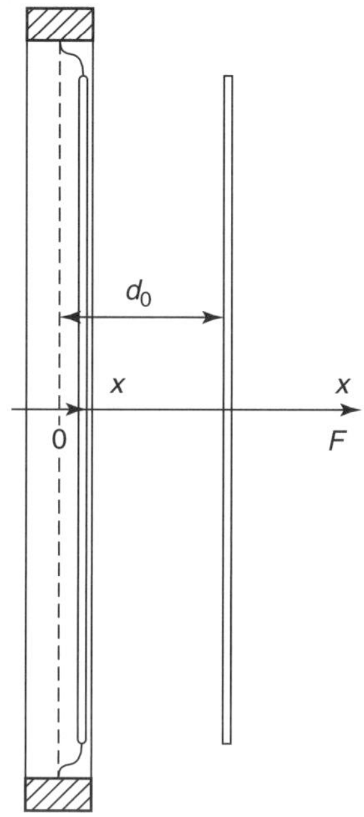

Figure 1.9 Schematic of an electrostatic transducer. Reproduced from M. Rossi, 1988, *Acoustics and Electroacoustics*, Artech House, Norwood, MA, by permission of Artech House, © 1988 Artech House

Since charge is conserved, the instantaneous voltage across these plates is given in terms of the charge (electrical quantity) and displacement (mechanical quantity) as:

$$V(t) = \frac{Q(t)}{C_0}\left[1 - \frac{x(t)}{d_0}\right] = \frac{Q(t)}{C_0} - \frac{Q(t)x(t)}{C_0 d_0} \tag{1.30}$$

Next we endeavor to derive the association between force with charge. The electrostatic force between the plates can be obtained from Coulomb's law. By the principle of conservation of energy, the mechanical work done in moving the plate should balance with an equal variation in electrical energy. Thus the net work done is

$$\mathrm{d}W = \mathrm{d}W_{\text{electrical}} + F_{\text{Coulomb}}\,\mathrm{d}x \equiv 0 \tag{1.31}$$

Therefore,

$$F_{\text{Coulomb}} = -\frac{\partial W_{\text{electrical}}}{\partial x} \tag{1.32}$$

where

$$W_{\text{electrical}} = \tfrac{1}{2}CV^2 \tag{1.33}$$

Substituting Equations (1.28)–(1.30) in Equation (1.33), and then back in Equation (1.32), the electrostatic force becomes:

$$F_{\mathrm{Coulomb}} = -\frac{1}{2}\frac{Q^2(t)}{C_0 d_0} \tag{1.34}$$

The nonlinearities in these electromechanical coupling equations – Equations (1.30) and (1.34) – are quite apparent. Such nonlinearities are significant in the realization of micro-switches. However, for applications in tunable capacitors, filters and resonators this may not be a desirable feature. However, for small variations about the rest position, these relationships can be assumed to be linear, as shown in the following simplification.

Equation (1.30) for the voltage across the plates can be expressed in terms of a static charge Q_0, and a dynamic component as:

$$V(t) = \frac{Q_0}{C_0} + \frac{Q_{\mathrm{d}}}{C_0} - \frac{Q_0}{C_0 d_0}x - \frac{Q_{\mathrm{d}}}{C_0 d_0}x \tag{1.35}$$

where

$$Q(t) = Q_0 + Q_{\mathrm{d}} \tag{1.36}$$

Considering only the dynamic component of voltage, and using the assumptions $Q_{\mathrm{d}} \ll Q_0$ and $x \ll d_0$, we get

$$V_{\mathrm{d}}(t) \approx \frac{Q_{\mathrm{d}}}{C_0} - \frac{V_0}{d_0}x \tag{1.37}$$

This electromechanical relation is obviously linear. A similar procedure would lead to the linearization of the other electromechanical coupling equation between the force and charge as:

$$(F_{\mathrm{Coulomb}})_{\mathrm{d}} = \frac{V_0}{d_0}Q_{\mathrm{d}} \tag{1.38}$$

It may, however, be reiterated that these linearized expressions are valid for a very small range of displacements around the rest position.

The electrostatic coupling equations in the sinusoidal state are written in the form (Rossi, 1988):

$$\tilde{V}_{\mathrm{ca}} = \frac{\tilde{I}}{j\omega C_0} - \frac{V_0}{j\omega d_0}\tilde{v} \tag{1.39}$$

$$\tilde{F}_{\mathrm{ca}} = \frac{V_0}{j\omega d_0}\tilde{I} \tag{1.40}$$

The coefficient on the right-hand side of Equation (1.40) is the electrostatic coupling coefficient. This being pure imaginary number the energy conversion is purely reactive. One of the equivalent circuits used to represent an electrostatic actuator is shown in Figure 1.10 (Rossi, 1988). The parameters appearing there are:

$$C'_{\mathrm{em}} = \frac{C_{\mathrm{m}}}{1 - C_0 C_{\mathrm{m}}\left(\frac{V_0}{d_0}\right)^2}\left(\frac{V_0 C_0}{d_0}\right)^2 \tag{1.41}$$

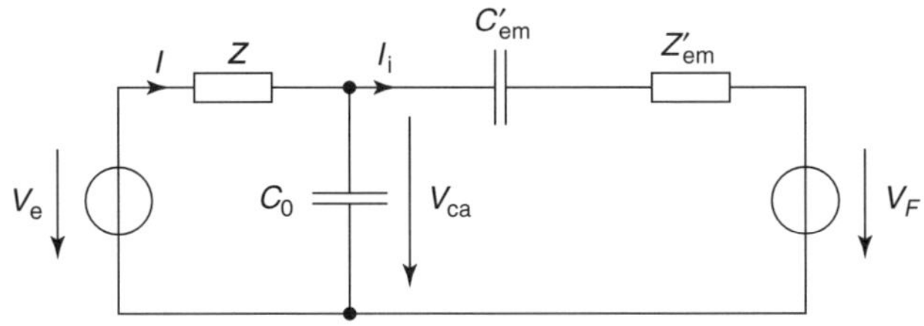

Figure 1.10 Equivalent circuit for electrostatic actuator. Reproduced from Rossi, 1988, *Acoustics and Electroacoustics*, Artech House, Norwood, MA, by permission of Artech House, © 1988 Artech House

$$Z'_{\text{em}} = Z'_{\text{m}} + \frac{1}{j\omega C''_{\text{m}}} \tag{1.42}$$

where

$$C'_{\text{m}} = \frac{C_{\text{m}}}{1 - C_0 C_{\text{m}} \left(\dfrac{V_0}{d_0}\right)^2} \tag{1.43}$$

and C_{m} and Z_{m} are the compliance of the moving plate and its mechanical impedance, respectively.

Fabrication of microsized devices with an electrostatic actuation scheme is relatively easy as it is independent of the properties of material systems. Therefore an electrostatic actuation scheme is the most preferred for microactuators. In addition to the parallel plate scheme described above, comb drives are also popular in these devices. Their operational mechanism will be discussed in Chapter 5.

1.4.5 Electromagnetic transducers

The magnetic counterpart of a moving plate capacitor is a moving coil inductor. This is yet another energy-storing transducer, the difference in this case being the forms of energy are magnetic and mechanical. A simplified sketch of such a transducer is shown in Figure 1.11 (Rossi, 1988). When a current i flows through the coil, the magnetic flux is φ. Neglecting nonidealities, such as electrical capacitance and resistance, and mechanical mass and friction, the constitutive relations for this device can be derived for the current (i) and force (F), in terms of displacement (x) and flux linkage (Crandall *et al.*, 1968). The conversion of energy takes place as a result of the interaction between these electrical and mechanical quantities in such a circuit.

In the transducer shown in Figure 1.11, the fixed armature has N turns of winding, and both this and the moving part are made of ferromagnetic materials.

Assuming infinite permeability for the ferromagnetic parts, the reluctance is confined only to the gap between them. Considering both the gaps, the total reluctance $\Re$ is approximately given by

$$\Re \approx \frac{2d(t)}{\mu_0 S} \tag{1.44}$$

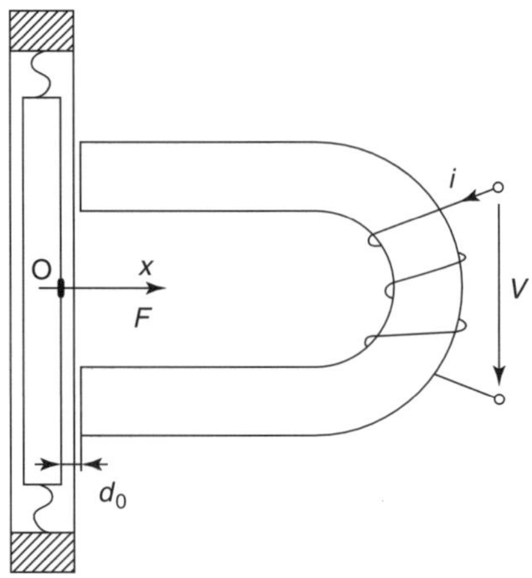

Figure 1.11 Schematic of an electromagnetic transducer. Reproduced from M. Rossi, 1988, *Acoustics and Electroacoustics*, Artech House, Norwood, MA, by permission of Artech House, © 1988 Artech House

where μ_0 is the permeability of air (medium in the gap) and S is its cross-sectional area. The reluctance at the rest position is

$$\Re_0 = \frac{2d_0}{\mu_0 S} \tag{1.45}$$

The position of the fixed element can, however, be expressed in terms of its rest position and the displacement as:

$$d(t) = d_0 - x(t) \tag{1.46}$$

Substituting Equations (1.45) and (1.46) into Equation (1.44), we get

$$\Re = \Re_0 \left[1 - \frac{x}{d_0}\right] \tag{1.47}$$

The inductance of the coil is expressed in terms of its reluctance as:

$$L = \frac{N^2}{\Re} = L_0 \left(1 - \frac{x}{d_0}\right)^{-1} \tag{1.48}$$

This may, however, be simplified for very small displacements using Taylor series expansions. Ignoring higher-order terms, the inductance of this coil becomes

$$L \approx L_0 \left[1 + \frac{x}{d_0}\right] \tag{1.49}$$

The voltage induced on the coil is

$$V = -\frac{\mathrm{d}(Li)}{\mathrm{d}t} \tag{1.50}$$

Substituting Equation (1.49) into Equation (1.50), the induced voltage is given as:

$$V \approx -L_0\frac{\mathrm{d}i}{\mathrm{d}t} - L_0 i\frac{v}{d_0} \tag{1.51}$$

where v is the velocity of the moving plate. This leads to a nonlinear relationship for the electromechanical coupling.

The stored magnetic energy is

$$W_{\mathrm{m}} = \frac{1}{2}Li^2 = \frac{\varphi^2}{2L} \tag{1.52}$$

Assuming the principle of conservation of energy, this balances with the mechanical energy spent on the displacement. At any instant of time dt, the magnetic force used up for generating the displacement is given by

$$F_{\text{magnetic}} = \frac{\partial W_{\mathrm{m}}}{\partial x} \tag{1.53}$$

As done with the electrostatic case, the nonlinearities of these expressions for electromechanical coupling can be linearized, by defining the components of the flux as:

$$\varphi = \varphi_0 + \varphi_{\mathrm{d}} \tag{1.54}$$

Assuming that the dynamic component $\varphi_{\mathrm{d}} \ll \varphi_0$, the relation between induced magnetic voltage and the dynamic component of current becomes [from Equation (1.51)]

$$V_{\mathrm{d}} = -L_0\frac{\mathrm{d}i_{\mathrm{d}}}{\mathrm{d}t} - L_0 I_0\frac{v}{d_0} = -L_0\frac{\mathrm{d}i_{\mathrm{d}}}{\mathrm{d}t} - \frac{\varphi_0}{d_0}v \tag{1.55}$$

The dynamic component of the magnetic force can be approximated as

$$(F_{\text{magnetic}})_{\mathrm{d}} = \frac{\varphi_0}{d_0}i_{\mathrm{d}} + \frac{\varphi_0^2}{L_0 d_0^2}x \tag{1.56}$$

Miniaturization of an electromagnetic actuator requires fabrication of magnetic thin films and current-carrying coils. Although few attempts have been made in this direction, the overall size of devices developed so far are not very small. Coupled with this is the difficulty in isolating magnetic field between adjacent devices, which makes fabrication of integrated microdevices challenging.

1.4.6 Electrodynamic transducers

This is one of the very common types of electromechanical actuation scheme. The primary component is a current-carrying moving coil such as the one commonly used in loud speakers. A schematic of such an actuator is shown in Figure 1.12. When a current flows through the coil it is deflected in the direction shown with a force F and velocity v. For simplicity in analysis, a small segment of the coil is shown along with the directions of the

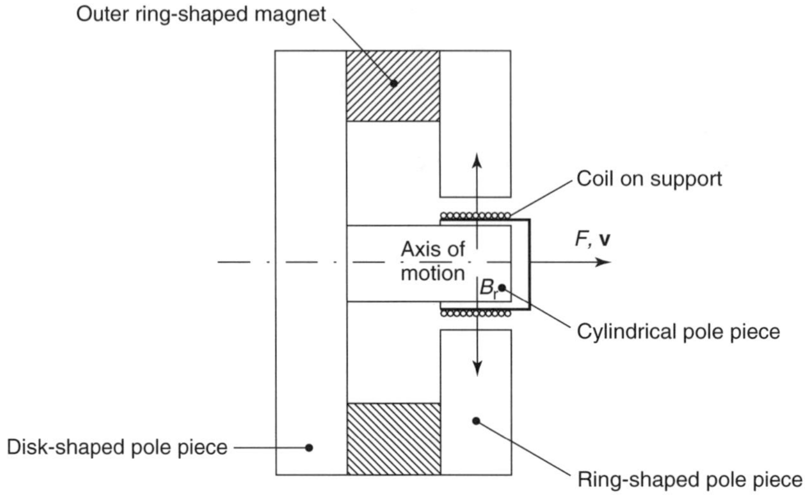

Figure 1.12 Schematic for an electrodynamic actuator. Reproduced from M. Rossi, 1988, *Acoustics and Electroacoustics*, Artech House, Norwood, MA, by permission of Artech House, © 1988 Artech House

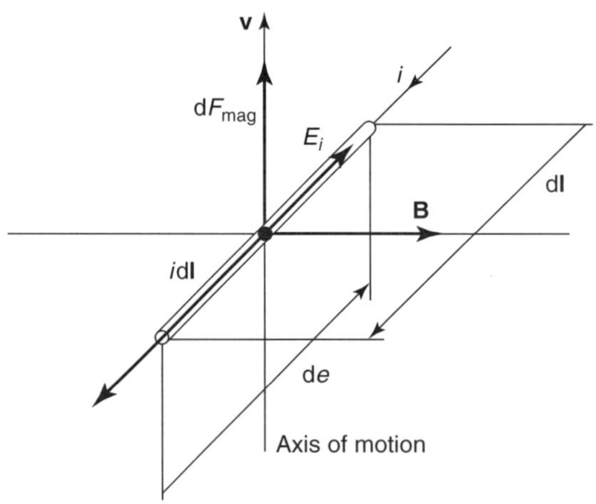

Figure 1.13 Field directions for a section of the coil. Reproduced from M. Rossi, 1988, *Acoustics and Electroacoustics*, Artech House, Norwood, MA, by permission of Artech House, © 1988 Artech House

field quantities in Figure 1.13. The element of length dl, carrying a current i, is further characterized by its velocity v and induction B. By Lenz's law for the electromotive force e,

$$\mathrm{d}e = (\mathbf{v} \times \mathbf{B}) \cdot \mathrm{d}\mathbf{l} \tag{1.57}$$

The magnetic force is given by Laplace's law:

$$\mathrm{d}F_{\mathrm{mag}} = i\,\mathrm{d}\mathbf{l} \times \mathbf{B} \tag{1.58}$$

Integration of Equation (1.58) along the length of the coil leads to the electromotive force across its terminals as:

$$e = \int_l \mathrm{d}e = -(Bl)v \tag{1.59}$$

According to Lenz's law, e is such that it opposes the current i. Similarly, the magnetic force is given by integrating Equation (1.58):

$$\mathbf{F}_{\mathrm{mag}} = \int_l \mathrm{d}\mathbf{F}_{\mathrm{mag}} = (Bl)i\hat{\mathbf{n}} \tag{1.60}$$

where $\hat{\mathbf{n}}$ is the positive unit vector in the direction of motion. The term (Bl) is called the electrodynamic coupling coefficient. In sinusoidal steady state these equations can be rewritten as (Rossi, 1988):

$$\tilde{\mathbf{V}} = (Bl)\tilde{\mathbf{v}} \tag{1.61}$$

$$\tilde{F}_{\mathrm{mag}} = (Bl)\tilde{I} \tag{1.62}$$

These equations can be represented in equivalent circuit form using an ideal transformer as shown in Figure 1.14(a). The external sources are represented with an ideal voltage source with an internal impedance Z on the electrical side, and ideal force source $\tilde{F}_e$ with an associated mechanical impedance Z_{m} on the mechanical side. The electrical resistance and self-inductance of the moving coil are incorporated into these impedances. From this equivalent circuit,

$$-\tilde{V}_{\mathrm{e}} + \tilde{Z}\tilde{I} + \tilde{V} = 0 \tag{1.63}$$

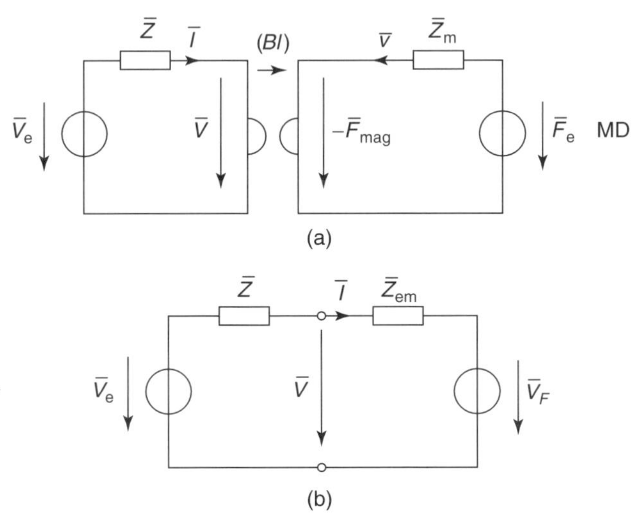

Figure 1.14 Simplified equivalent circuit of the transducer. Reproduced from Rossi, 1988, *Acoustics and Electroacoustics*, Artech House, Norwood, MA, by permission of Artech House, © 1988 Artech House

Substituting Equation (1.61) into Equation (1.63), we get the characteristic equation for the system as:

$$\tilde{V}_e = \tilde{Z}\tilde{I} + (Bl)\tilde{v} \tag{1.64}$$

Similarly, it is also possible to relate force and current in terms of another characteristic equation as (Rossi, 1988):

$$\tilde{F}_e = \frac{\tilde{v}}{\tilde{Y}_m} - (Bl)\tilde{I} \tag{1.65}$$

It is possible to simplify the equivalent circuit in terms of all electric quantities as shown in Figure 1.14(b). The need for a two-port coupling network is eliminated by combining Equations (1.64) and (1.65):

$$\begin{aligned}\tilde{V}_e &= \tilde{Z}\tilde{I} + (Bl)(\tilde{F}_e + (Bl)\tilde{I})\tilde{Y}_m \\ &= (\tilde{Z} + \tilde{Z}_{em})\tilde{I} + \tilde{V}_F\end{aligned} \tag{1.66}$$

where

$$\tilde{Z}_{em} = (Bl)^2\tilde{Y}_m \tag{1.67}$$

$$\tilde{V}_F = (Bl)\tilde{Y}_m\tilde{F}_e = (Bl)\tilde{v}_e \tag{1.68}$$

In Equation (1.68), v_e is the velocity of the open-circuited ($I = 0$) moving coil when an external force is applied to it. Z_{em} is called the motional impedance, representing the mechanical system in the electrical circuit.

As mentioned earlier, a very common form of electrodynamic transducer is found in loud speakers. However, because of the requirements of the coil and magnetic field, they are not so popular at the microscale. Electrodynamic micromotors have been successfully fabricated at reasonably smaller sizes (7 mm × 15 mm × 0.4 mm) (Frank, 1998). The resonant frequency of such a system is given as

$$f_0 = \frac{1}{2\pi}\left(\frac{BJ\eta_m}{\rho s}\right)^{1/2} \tag{1.69}$$

where η_m is the utilization factor of the rotor, ρ is the density of the material of the wire, J is the current density, and s the maximum displacement of the rotor.

As with the electrodynamic actuation scheme discussed previously, these devices also require fabrication of small sized magnets and current-carrying coils. In this case, however, the coil is also movable. This remains a fabrication challenge, as miniaturized devices are required for their eventual integration into RF circuits.

1.4.7 Electrothermal actuators

In electrothermal actuators, heat is applied to a bimorph beam the expansion of which is used to generate the mechanical moment required for the actuation. Large deflections and greater energy density are achievable with this scheme. But electrothermal transducers are too slow to be of any use at the frequencies of interest to us. However, for the sake

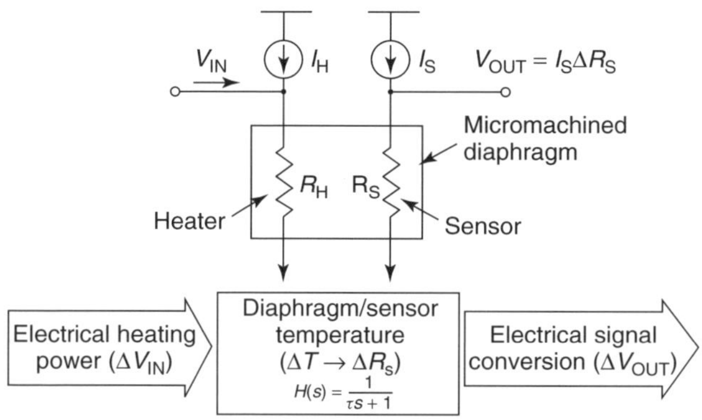

Figure 1.15 Principle of the electrothermal transducer. Reproduced from K.H. Lee, H.J. Byun, H.K. Lee, I.J. Cho, J.U. Bu and E. Yoon, 2000, 'An audio frequency filter application of micromachined thermally-isolated diaphragm structures', *13th Annual International Conference on Micro Electro Mechanical Systems, MEMS 2000*, IEEE, Washington, DC, 142–147, by permission of IEEE, © 2000 IEEE

of completion a brief discussion of their principles is given here. Several configurations are reported in the literature for this kind of actuator. In general, these consist of two transduction mechanisms, first an irreversible electrothermal process, and the second a reversible thermomechanical transduction. The configuration of an audio frequency filter with such an actuator is shown in Figure 1.15 (Lee *et al.*, 2000). This consists of a thermally isolated thin diaphragm fabricated by back-side etching, and a pair of metal resistors for actuator (heater) and sensor patterned by the liftoff process. In the figure, the heater element of resistance R_H is supplied with a current I_H. The heat generated is modulated with the input signal voltage V_{in}. The change in resistance of the sensor circuit is proportional to the change in temperature (ΔT), and results in a corresponding variation in the output voltage V_{out}.

The operation of this filter is fairly simple. Changes in ac input voltage at the heater terminals cause change in temperature on the diaphragm, which is detected by the sensing resistor and converted back to voltage variations. High-frequency components are filtered out, since the diaphragm does not respond fast enough to these variations. Hence this filter is useful for low audio frequencies only (Lee *et al.*, 2000).

Several other configurations of micromachined thermal actuators are also reported in literature (Huang and Lee, 1999; Riethmüller and Benecke, 1988). In Huang and Lee (1999), for example, the difference in electrical resistance of a wide and a narrow arm in a bimorph structure (see Figure 1.16; Huang and Lee, 1999) is used to generate the necessary deflection. This difference causes variation in the heat produced and hence thermal expansion of the two arms. Based on the dimensions of the structure shown in Figure 1.16(a), an equivalent model of the cantilever can be obtained in (b). The cross sectional view of the structure is shown in Figure 1.16(c) to indicate various layers of materials. It is possible to obtain a second-order differential equation to represent its model (Huang and Lee, 1999):

$$k_p \frac{d^2T}{dx_2} + J^2\rho = \frac{S}{h}\frac{T - T_s}{R_T} \qquad (1.70)$$

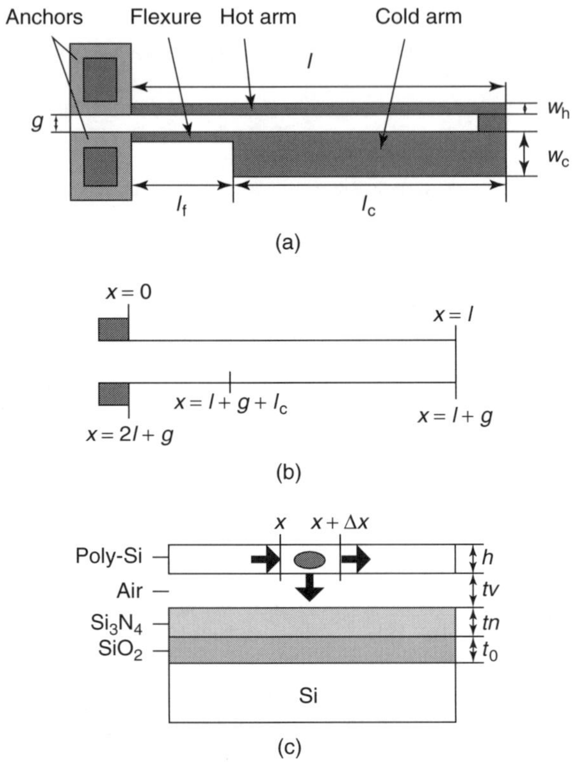

Figure 1.16 Bimorph electrothermal transducer. Reproduced from Q.A. Huang and N.K.S. Lee, 1999, 'Analysis and design of polysilicon thermal flexure actuator', *Journal of Micromechanics and Microengineering* **9**: 64–70, by permission of the Institute of Physics

where k_p is the thermal conductivity of polysilicon, T is the operating temperature, T_s is the substrate temperature, J is the current density, ρ is the resistivity of polysilicon, S is the shape factor, R_T is the thermal resistance between the polysilicon microbeam and the substrate, and h is the thickness of the beam.

The dissipation element in the transducer can be made of various materials including metallic thin films such as Cr/Au, Al and NiFe, or polysilicon, or, p or n doped areas in single-crystal silicon (Elwenspoek *et al.*, 1989). Piezoresistive effects in polysilicon can be used if the device is to be used as a detector also.

1.4.8 Comparison of electromechanical actuation schemes

A brief comparison of some of the electromechanical transducers discussed above is presented in Table 1.4. Owing to its simplicity, electrostatic actuation is the most preferred, especially in microdevices. The control signal here is voltage, which is easy to manipulate in electrical circuits. However, these devices require greater environmental protection as electrostatic fields are prone to attract dust, which could affect the performance of associated CMOS circuits. Electromagnetic and electrodynamic actuators are based on Lorentz force effects. The current-carrying coil is stationary in the former case, whereas

Table 1.4 Comparison of electromechanical transducers

Actuator	Fractional stroke (%)	Maximum energy density ($J\,cm^{-3}$)	Efficiency	Speed
Electrostatic	32	0.004	High	Fast
Electromagnetic	50	0.025	Low	Fast
Piezoelectric	0.2	0.035	High	Fast
Magnetostrictive	0.2	0.07	Low	Fast
Electrostrictive	4	0.032	High	Fast
Thermal	50	25.5	Low	Slow

Source: Wood, Burdess and Hariss, 1996.

in the latter it is moving. These are ideally suited when large currents are possible, even with lower voltages. However, they are prone to problems with power dissipation, but are tolerant to dust and humidity.

Actuators based on piezoelectricity, magnetostriction and electrostriction depend on changes in strain produced by an applied electric or magnetic field in some special materials used. The achievable strain is at a maximum for electrostrictive materials, but the force generated is at a maximum in magnetostrictive materials. Electrostrictive and piezoelectric materials deform with the application of an electric field, but whereas the relationship between the force produced and applied field is linear in piezoelectrics, it is quadratic in electrostrictive materials.

Most of these transduction schemes are nonlinear. That is, the transfer function between electrical (voltage or current) and mechanical (force or displacement) terms is not linear. Such nonlinearities distort the filtered signal and may cause loss of fidelity. One approach to overcome this difficulty is to restrict the signal to very small variations about a dc bias. It is fairly reasonable to assume that the response to these small signal variations is linear.

1.5 MICROSENSING FOR MEMS

Various microsensing and microactuation mechanisms have been developed for MEMS for diverse applications. Some of the commonly used sensing and actuating principles are introduced in this section. Many microsensors based on different sensing principles for MEMS have been developed (Fatikow and Rembold, 1997; Rai-Choudhury, 1997), including chemical sensors, gas sensors, optical sensors, biosensors, thermal sensors and mechanical sensors. Some of the major sensing mechanisms for mechanical microsensors are introduced in the following sections.

1.5.1 Piezoresistive sensing

Piezoresistive sensing utilizes resistors where the resistance is varied through external pressure, to measure such physical parameters as pressure, force and flow rate or to be used as accelerometer.

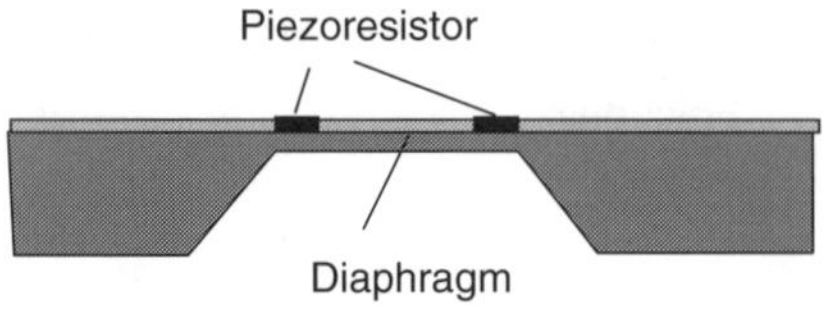

Figure 1.17 A piezoresistive sensing structure

A typical structure for piezoresistive microsensors is shown in Figure 1.17. The resistors are usually built on a silicon diaphragm. The deflection of the diaphragm leads to the dimension change of the resistors, resulting in the resistance changing as a result of the piezoresistive effect in silicon:

$$\frac{\Delta R}{R} = (1 + 2v)\frac{\Delta l}{l} + \frac{\Delta \rho}{\rho} \tag{1.71}$$

where ΔR is the change of the resistance, R is the original resistance, v is the Poisson ratio, Δl is the length change of the resistor, l is the original length of the resistor, and $\Delta \rho$ and ρ represent the resistivity change and resistivity of the resistor, respectively. It is easily found that the resistance of the resistors used for this type of piezoresistive microsensor is proportional to the external pressure when the resistivity change is ignored, since the dimension change is proportional to the applied pressure.

Another piezoresistive-type microsensor is schematically shown in Figure 1.18, where a semiconductor polymer foil is put on interdigital transducer (IDT) electrodes. If a voltage is applied to the electrodes and there is no pressure applied, the resistance is at the level of mega ohms. When a force is applied, the resistance decreases as a result of current that flows across the shunting polymer foil (Witte and Gu, 1992). Here, the sensing resistance is inversely proportional to the pressure applied.

The performance of the piezoresistive microsensors varies with temperature and the pressure. The sensitivity of the sensors decreases as temperature increases. Any residual

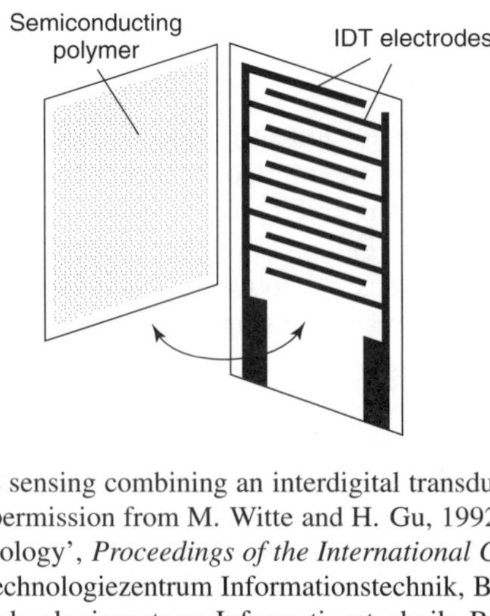

Figure 1.18 Piezoresistive sensing combining an interdigital transducer (IDT) and semiconductor polymer. Reproduced with permission from M. Witte and H. Gu, 1992, 'Force and position sensing resistors: an emerging technology', *Proceedings of the International Conference on New Actuators, Bremen, 1992*, VDI/VDE-Technologiezentrum Informationstechnik, Berlin, Germany, 168–170, by permission of VDI/VDE-Technologiezentrum Informationstechnik, Berlin

stress generated during the fabrication will also influence the sensitivity of the sensors. The nonlinear deflection of the diaphragm occurs when the high-pressure-induced deflection is over 10% of the diaphragm thickness.

1.5.2 Capacitive sensing

Capacitive sensing utilizes the diaphragm-deformation-induced capacitance change to convert the information of pressure, force, etc., into electrical signals such as changes of oscillation frequency, time, charge, and voltage. The structure of a typical capacitive microsensor is shown in Figure 1.19; an electrode on the flexible diaphragm and the other one on the substrate construct the sensing capacitor. The capacitive microsensors can be used for pressure, force, acceleration, flow rate, displacement, position and orientation measurement, etc.

For capacitive microsensors, the capacitance change is not linear with respect to the diaphragm deformation, and, also, the small capacitance (generally 1 to ~3 pF) requires the measurement circuit to be integrated on the chip. But the capacitive sensing was found to have potential for higher performance than piezoresistive sensing in applications requiring high sensitivity, low pressure range and high stability (Rai-Choudhury, 1997).

1.5.3 Piezoelectric sensing

Piezoelectric sensing is based on the piezoelectric effect of piezoelectric materials. The electrical charge change is generated when a force is applied across the face of a piezoelectric film. For a piezoelectric disc of a given thickness t, the voltage (V) generated across the electrode disc (Figure 1.20) when subjected to a stress (T) would be

$$V = gtT \tag{1.72}$$

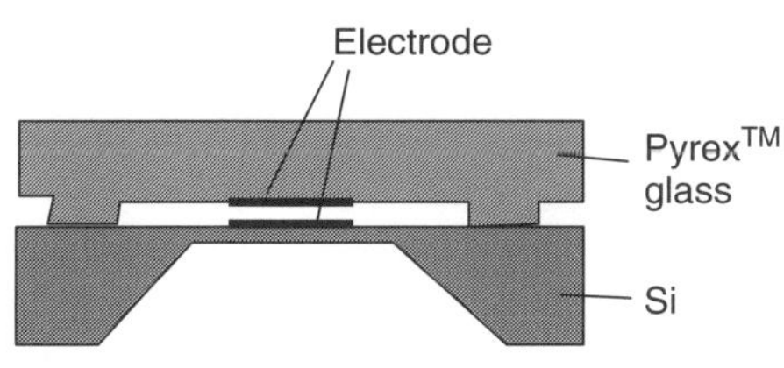

Figure 1.19 Capacitive sensing structure

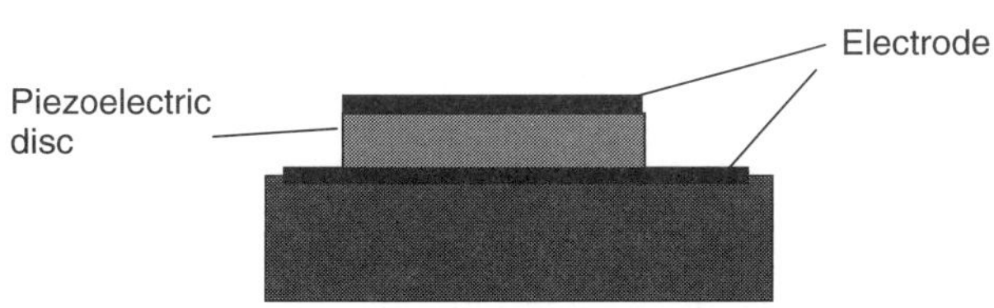

Figure 1.20 Piezoelectric sensing

where g is the piezoelectric voltage coefficient. Piezoelectric sensing is mostly used in sensors such as pressure sensors, force sensors, speedometers and accelerometers, hydrophones and microphones.

1.5.4 Resonant sensing

Resonant sensing is easily understood as the natural frequency of a string changing as a result of tensile force. In the developed resonant microsensor, strain caused by pressure on the diaphragm leads to the natural frequency of a resonator varying. By picking up the natural frequency variation of the resonator, the physical information that caused the strain will be sensed.

For an example, the natural resonant frequency of a flexure resonator with both ends fixed can be obtained from (Ikeda *et al.*, 1990)

$$f = \frac{4.73^2 h}{2\pi d^2}\left\{\frac{E}{12\rho}\left[1 + 0.2366\left(\frac{l}{h}\right)^2 \varepsilon\right]\right\}^{1/2} \tag{1.73}$$

where f is the natural frequency of the fundamental oscillating mode, l is the resonator length, h is the resonator thickness, E is the Young's modulus, ρ is the density of the diaphragm material and ε is the strain generated inside the resonator structure. Comparing resonant sensing with piezoresistive sensing, the resonator acts as a kind of strain gauge, the resonant strain gauge, which relates the strain to the resonant frequency. Therefore, the gauge factor of the above resonant strain gauge can be determined as:

$$k_{\text{gf}} = \frac{1}{2}\left[0.2366\left(\frac{l}{h}\right)^2\right]\left[1 + 0.2366\left(\frac{l}{h}\right)^2 \varepsilon\right]^{-1} \tag{1.74}$$

$$\frac{\Delta f}{f} = k_{\text{gf}}\varepsilon \tag{1.75}$$

If a strain is 100 ppm, for a 1.2-mm long, 20-micron wide and 5-micron thick resonator strain gauge, the gauge factor can be as high as 3000, whereas the piezoresistive strain gauge factor is only about 2. Since the gauge factor relates directly to the sensitivity of the sensor, the resonant sensing can be used to obtain highly sensitive microsensors. However, resonant sensing usually requires a more complex sensor structure than piezoresistive sensing does; the resonant strain gauges need to be encapsulated from the fluid (Harada *et al.*, 1999).

1.5.5 Surface acoustic wave sensors

Surface acoustic wave (SAW) based sensors form an important part of the sensor family and in recent years have seen diverse applications ranging from gas and vapor detection to strain measurement (Campbell, 1998). A new breed of SAW-based actuator modeled on MEMS-based microactuators have also been recently announced (Campbell, 1998). IDT and SAW devices were first used in radar and communication equipment as filters and delay lines and recently were found to be attractive sensors for various physical variables such as temperature, pressure, force, electric field, magnetic field, and chemical compounds. A SAW device usually is a piezoelectric wafer with IDT and reflectors on its surface. The IDT provides for the cornerstone of SAW technology. Its function is to convert the electrical energy into mechanical energy, and vice versa, for generating as well as detecting the SAW. The type of acoustic wave generated in a piezoelectric material depends mainly on the substrate material properties, the crystal cut and the structure of the electrodes utilized to transform the electrical energy into mechanical energy. The possibilities of various types of acoustic devices for sensor applications have been explored, focusing primarily on Rayleigh SAWs and shear horizontal surface acoustic waves (SH-SAWs), Love wave mode devices, the acoustic plate mode (APM) and flexural plate waves (FPW).

The Rayleigh wave has both a surface normal component and a surface parallel component, which is parallel to the direction of propagation. The Rayleigh wave has two particle-displacement components in the sagittal plane. Surface particles move in elliptical paths with a surface normal and a surface parallel component. The electromagnetic field associated with the acoustic wave travels in the same direction. The wave velocity is determined by the substrate material and the crystal cut. The energies of the SAW are confined to a zone close to the surface a few wavelengths thick (Campbell, 1998).

A selection of a different crystal cut can yield SH-SAWs instead of Rayleigh waves. The particle displacements of these waves would be transverse to the wave propagation direction and parallel to the plane of the surface. The frequency of operation is determined by the IDT finger spacing and the shear horizontal wave velocity for the particular substrate material. They have shown considerable promise in their application as sensors in liquid media and biosensors (Kondoh, Matsui and Shiokawa, 1993; Nakamura, Kazumi and Shimizu, 1977; Shiokawa and Moriizumi, 1987). In general the SH-SAW is sensitive to mass loading, viscosity, conductivity and the permittivity of adjacent liquid. The configuration of shear horizontal acoustic plate mode (SH-APM) devices is similar to the Rayleigh SAW devices, but the wafer is thinner, typically a few acoustic wavelengths. The IDTs generate shear horizontal waves that propagate in the bulk at angles to the surface. These waves reflect between the plate surfaces as they travel in the plate between the IDTs. The frequency of operation is determined by the thickness of the plate and the IDT finger spacing. SH-APM devices are used mainly in liquid sensing and offer the advantage of using the back surface of the plate as the sensing active area.

Lamb waves, also called acoustic plate waves, are elastic waves that propagate in plates of finite thickness and are used for health monitoring of structures and acoustic streaming.

An IDT consists of two metal comb-shaped electrodes placed on a piezoelectric substrate (Figure 1.21). An electric field created by the voltage applied to the electrodes induces dynamic strains in the piezoelectric substrate, which in turn launch elastic waves. These waves contain, among others, the Rayleigh waves, which run perpendicularly to the electrodes, with velocity v_R.

If a harmonic voltage, $v = v_o \exp(j\omega t)$, is applied to the electrodes, the stress induced by a finger pair travels along the surface of the crystal in both directions. To ensure constructive interference and in-phase stress, the distance between two neighboring fingers should be equal to half the elastic wavelength, λ_R.

$$d = \frac{\lambda_R}{2} \tag{1.76}$$

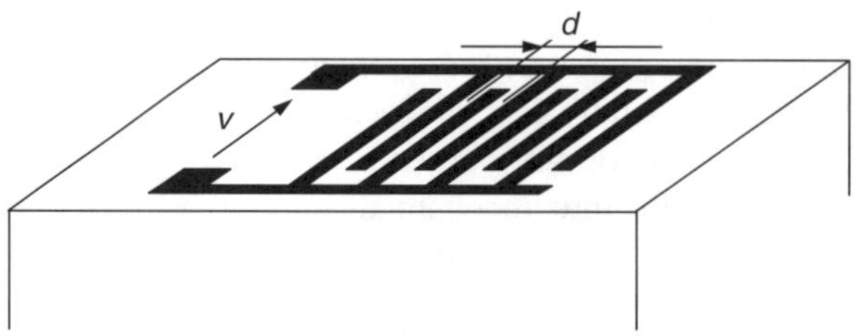

Figure 1.21 Finger spacings and their role in the determination of the acoustic wavelength. Reproduced from V.K. Varadan and V.V. Varadan, 1997, 'Microsensors, actuators, MEMS, and electronics for smart structures', in P. Rai-Choudhury (ed.), *Handbook of Microlithography, Micromachining, and Microfabrication, Volume 2: Micromachining and Microfabrication*, SPIE Optical Engineering Press, 617–688, by permission of SPIE Optical Engineering Press

The associated frequency is known as the synchronous frequency and is given by:

$$f_o = \frac{V_R}{\lambda_R} \tag{1.77}$$

At this frequency, the transducer efficiency in converting electrical energy to acoustical energy, or vice-versa, is maximized. The exact calculation of the piezoelectric field driven by the IDT is rather elaborate (Varadan and Varadan, 1997). For simplicity, the analysis of the IDT is carried out by means of numerical models. The frequency response of a single IDT can be simplified by the delta-function model (Varadan and Varadan, 1997).

The principle of SAW sensors is based on the fact that SAW traveling time between IDTs changes with the variation of physical variables.

One of these IDTs shown in Figure 1.22 acts as the device input and converts signal voltage variations into mechanical SAWs. The other IDT is employed as the output receiver to convert the mechanical SAW vibrations back into output voltages. These devices are reciprocal in nature; as a result, signal voltages can be applied to either IDT with the same end result.

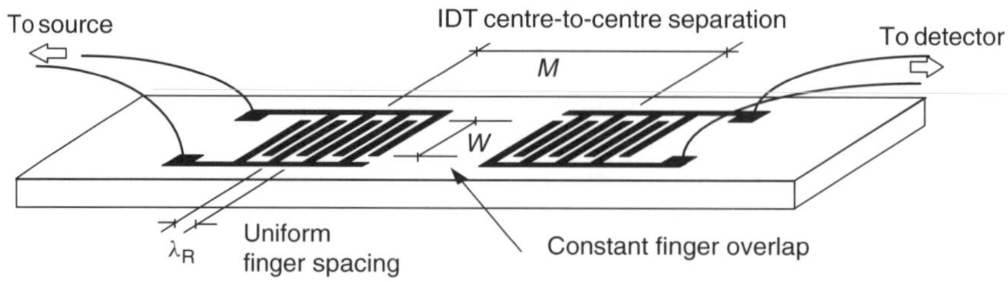

Figure 1.22 Schematic of a surface acoustic wave (SAW) device with interdigital transducers (IDTs) metallized onto the surface. Reproduced from V.K. Varadan and V.V. Varadan, 1997, 'Microsensors, actuators, MEMS, and electronics for smart structures', in P. Rai-Choudhury (ed.), *Handbook of Microlithography, Micromachining, and Microfabrication, Volume 2: Micromachining and Microfabrication*, SPIE Optical Engineering Press, 617–688, by permission of SPIE Optical Engineering Press

Acoustic sensors offer a rugged and relatively inexpensive platform for the development of wide-ranging sensing applications. A unique feature of acoustic sensors is their direct response to a number of physical and chemical parameters such as surface mass, stress, strain, liquid density, viscosity and dielectric and conductivity properties (Grate, Martin and White, 1993a). Furthermore, the anisotropic nature of piezoelectric crystals allows for various angles of cut, with each cut having unique properties. Applications such as a SAW-based accelerometer, utilizes a quartz crystal with an ST-cut, which has an effective zero temperature coefficient (Varadan and Varadan, 1996), with negligible frequency shift through changes in temperature. Again, depending on orientation of crystal cut, various SAW sensors with different acoustic modes may be constructed, with a mode ideally suited towards a particular application. Other attributes include very low internal loss, uniform material density and elastic constants and advantageous mechanical properties (Grate, Martin and White, 1993b). The principal means of detection of the physical property change involves the transduction mechanism of a SAW acoustic transducer, which involves transfer of signals from the physical (acoustic wave) to the electrical domain (Campbell, 1998). Small perturbations affecting the acoustic wave would manifest as large-scale changes when converted to the electromagnetic (EM) domain because of the difference in velocity between the two waves. Given that the velocity of propagation of the SAW on a piezoelectric substrate is 3488 m s^{-1} and the ac voltage applied to an IDT at a synchronous frequency of 1 GHz, the SAW wavelength is given by

$$\begin{aligned}\lambda_R &= \frac{v}{f}\\ &= \frac{3488\,\mathrm{m\,s^{-1}}}{1\times 10^9\,\mathrm{s^{-1}}}\\ &= 3.488\times 10^{-6}\,\mathrm{m}\\ &= 3.488\,\mu\mathrm{m}\end{aligned}$$

The EM wavelength in this case is

$$\lambda_C = \frac{c}{f}$$

where c is the velocity of light (equal to $3\times 10^8\,\mathrm{m\,s^{-1}}$). Thus,

$$\begin{aligned}\lambda_C &= \frac{3\times 10^8\,\mathrm{m\,s^{-1}}}{1\times 10^9\mathrm{s^{-1}}}\\ &= 0.3\,\mathrm{m}\end{aligned}$$

The ratio of the wavelengths is

$$\begin{aligned}\frac{\lambda_R}{\lambda_C} &= \frac{3.488\times 10^{-6}\,\mathrm{m}}{0.3\,\mathrm{m}}\\ &= 1.162\times 10^{-5}\end{aligned}$$

The sensing action of such transducers involves any influences that will alter the acoustic wave velocity and consequently the associated properties of the wave.

1.6 MATERIALS FOR MEMS

A broad spectrum of materials has been incorporated into MEMS. In addition to silicon materials, metal, metal alloys, ceramics and polymers are the four major material families used for MEMS (Larson, 1999).

1.6.1 Metal and metal alloys for MEMS

Thin-film metals have been used in IC chips for a long time; metal thick-film structures are required for some MEMS devices (Larson, 1999). Microelectroplating and photoforming are used to build thick-film metal structures (Romankiw, 1997; Taylor *et al.*, 1994). Nickel, copper and gold have been electroplated to form thick-film structures, and 3D stainless-steel microparts have been fabricated by photoforming (Taylor *et al.*, 1994). Most of the thick-film metals are used as structural materials of final devices, or as mould inserts for polymer on ceramic micromoulding.

Various metal alloys and the related processes have also been developed for MEMS. CoNiMn thin films were used as permanent magnet materials for magnetic actuation. NiFe permalloy thick films were electroplated onto a silicon substrate for magnetic microelectromechanical system devices, such as micromotors, microactuators, microsensors and integrated power converters, which could allow the production of new micropower magnetics on a chip with integrated circuits (Ahn and Allen, 1998). TiNi shape memory alloy films were sputtered on substrate for SMA sensing and actuating (Ohta *et al.*, 1998). TbFe and SmFe thinfilms were used for magnetostrictive actuation (Honda, Arai and Yamaguchi, 1997).

1.6.2 Polymers for MEMS

Polymers have been extensively used as both structural and functional materials for microdevices. As structural materials, the elasticity, optical properties and biocompatibility of polymers are utilized most in microdevices. Various polymer devices are made from thin polymer films, thick polymer films and 3D polymer microstructures. Some of the polymers and the related processes for polymer structure fabrication are listed in Table 1.5.

It should be pointed out here that the structural polymers listed in Table 1.5 can also be used to construct sensing and actuating components for MEMS, just silicon and polymer can be used to build microsensors and actuators, although they act as structural materials, too. Polymer strain gauges and capacitors can serve as sensing elements for piezoresistive and capacitive microsensors (Varadan and Varadan, 1995), and electrostatic polymer microactuators have been developed from polyimide bellow structures (Minami, Morishita and Esashi, 1999). Another important point is that wafer polymer microfabrication processes are being developed for polymer microdevices; the batch fabrication of polymer MEMS is not a concern any more.

Because the limited sensing and actuating mechanisms can be utilized to develop polymer MEMS just from structural polymers, a large number of species of functional polymers has been developed for MEMS during the past several years (Carraway, 1991). Some of the functional polymers used for MEMS are listed in Table 1.6.

Functional polymer–solid-powder composites with magnetic and magnetostrictive properties have been developed for microdevices. The polymer-bonded Terfenol D composites

Table 1.5 Polymers and fabrication processes for MEMS

Polymer name	Structure	Process	Property utilized	Reference
Polyimide	Thin film	Coating	Elasticity	Ruprecht *et al.*, 1998
Silicon rubber	Thick film	Moulding	Elasticity	Grosjean, Yang and Tai, 1999
Parylene C	Thin film	Coating	Vapor barrier	Grosjean, Yang and Tai, 1999
PMMA	Thick film	LIGA, hot embossing	Elasticity, optical	Becker and Heim, 1999; Guckel, 1998
Polycarbonate	Thick film	Hot embossing	Elasticity, optical transparency	Pan, Lin and Ni, 1999
PDMS	Thick film	Moulding	Elasticity, biomedical compatibility	Armani, Liu and Aluru, 1999
Epoxy resin	Thick film	Moulding	Encapsulation, resistance to large pH range	Armani, Liu and Aluru, 1999
Polyester	Thick film	Casting	Elasticity	Bohm, Olthuis and Bergveld, 1999
Polysulfone	Thick film	Moulding	Excellent mechanical and chemical resistance over broad temperatures (150 °F to 300 °F)	Varadan and Varadan, 1995
Acrylate, urethane, epoxy, etc.	Three-dimensional	Microstereo-lithography	Elasticity, optical transparency, UV-curable	Ikuta and Hirowatari, 1993; Shaw, Zhang and MacDonald, 1994; Takagi and Nakajima, 1993; Tani and Esashi, 1995; Varadan, 1995

Note: LIGA, *Lithographie, Galvanoformung, Abformung* (lithography, galvanoforming, moulding); PDMS, polydimethylsiloxare; PMMA, polymethylmethacrylate.

Table 1.6 Functional polymers for MEMS

	Functional property	Application
PVDF	Piezoelectricity	Sensor, actuator
Poly(pyrrole)	Conductivity	Sensor, actuator, electric connection
Fluorosilicone	Electrostrictivity	Actuator[a]
Silicone	Electrostrictivity	Actuator[a]
Polyurethane	Electrostrictivity	Actuator[a]

[a]Source: Pelrine *et al.*, 1997.
Note: PVDF, polyvinylidene fluoride.

showed excellent magnetostrictivity, which can be utilized for microactuation (Ruiz de Angulo, Abell and Harris, 1996). The polyimide-based ferrite magnetic composites were used as polymer magnets for magnetic microactuators (Lagorce and Allen, 1996).

In addition to being used as sensing and actuating materials, polymers have also been used for electronics materials. Polymer transistors have been developed. Therefore, integration of polymer sensors, actuators and electronics into polymer MEMS will be practical for some special applications.

1.6.3 Other materials for MEMS

Ceramics are another major species of material used for MEMS. In fact, SiO_2, Si_3N_4 thin films have ever been used for semiconductor devices and silicon MEMS devices. The thick ceramic film and 3D ceramic structures are also necessary for MEMS for special applications. For example, ceramic pressure microsensors have been developed for pressure measurement in the high-temperature environments (Jennifer and Allen, 1999), silicon carbide MEMS for harsh environments (Chin, Varadan and Varadan, 1994), etc. In addition to these structural ceramics, some functional ceramics such as ZnO, PZT, etc., have also been incorporated into MEMS. The fabrication processes developed for ceramic MEMS include screen printing, tape lamination, micromoulding, Sol-Gel and microstereolithography.

1.7 SCOPE OF THIS BOOK

The advancement of fabrication technologies is the key to the success of newer generation RF MEMS devices. There are two approaches in the fabrication of silicon-based MEMS devices: bulk micromachining and surface micromachining. Both these can be used in different contexts in the microwave and millimeter wave systems. Apart from the silicon-based technologies, polymer-based MEMS have recently been proposed. Materials systems and fabrication approaches of all these are presented in Chapter 2.

One of the earliest applications of MEMS technology for microwave applications was in the area of surface micromachined actuators for the realization of microwave switches with very high linearity, low dc standby power and low insertion loss. Several approaches for RF-MEMS switches are described in Chapter 3. Few innovative approaches to reduce this actuation voltage have been vigorously pursued in recent years and some of these are also introduced in that chapter.

Another use of MEMS technology for RF applications is in the area of microinductors and variable capacitors, as a replacement for varactor diodes for tuning. Several geometries are presented in Chapter 4 for micromachined high Q inductors. The important advantage of micromachined capacitors is their compatibility with the rest of the circuit, allowing fabrication of a fully integrated system.

High Q filters are widely used in most of the communication systems and in radars. For very low frequencies bulk mechanical filters are common. Their principles have been translated to smaller devices at higher frequencies. These microdevices, studied in Chapter 5, can be used for frequencies up to 10s of MHz, and can have Q values in the 1000s with proper packaging. Fabrication limitations restrict the extension of micromechanical filters for frequencies above $\sim$100 MHz. However, SAW filters and resonators can bridge this gap and provide high Q devices for frequencies up to 2 GHz. These planar devices can be accurately fabricated with modern microfabrication facilities. At microwave and millimeter wave frequencies planar filters on thin dielectric membranes show low loss and are suitable for low-cost, compact, high-performance monolithic microwave integrated circuits (MMICs). Some examples of these devices are also presented in Chapter 5.

Micromachined phase shifters are described in Chapter 6. These use either switching between sections of transmission line with different path lengths, with MEMS switches connecting between them, or a distributed array of capacitors over a transmission line. A polymer-based approach on similar lines is also introduced. Innovative micromachined phase shifters using nonlinear dielectric properties of ferroelectric materials such as barium strontium titanate are also addressed.

Micromachining for improving performance of microwave and millimeter wave transmission lines and components are presented in Chapter 7. Several examples of such devices for different frequency regimes are also reported.

In Chapter 8 the technology of micromachined antennas is covered. Many of the reported examples aim at improving the performance of microstrip antennas. Antennas with reconfigurable radiation characteristics are also fabricated by this approach. Micromachining is particularly useful at higher frequencies where horns and other antennas become so small that fabrication by conventional means is unreliable.

Integration and packaging are important for any MEMS component. Not many RF-MEMS devices have reached maturity, and so there are not many packaging concepts for such devices. However, several approaches followed for microelectronic devices may be adapted for these novel systems. Some such approaches are presented in Chapter 9.

REFERENCES

Aeidel, H., 1994, 'The mechanism of anisotropic silicon etching process for microelectromechanical structures', *Sensors and Actuators, A*, **40**, pp. 63–70.

Ahn, C.H., Allen, M.G., 1998, 'Micromachined planar inductors on silicon wafers for MEMS applications', *IEEE Transactions on Industrial Electronics* **45**(6): 866–876.

Armani, D., Liu, C., Aluru, N., 1999, 'Re-configurable fluid circuits by PDMS elastomer micromachining', in *Proceedings of IEEE MEMS '99*, IEEE, Washington, DC: 222–227.

Ayerdi, I., Castanō, E., Garciá-Alonso, A., Gracia, J., 1997, 'High-temperature ceramic pressure sensor', *Sensors and Actuators A: Physical* **60**: 72–75.

Becker, H., Heim, U., 1999, 'Silicon as tool material for polymer hot embossing', in *Proceedings of IEEE MEMS '99*, IEEE, Washington, DC: 228–231.

Bertsch, A., Lorenz, H., Renaud, P., 1998, 'Combining microstereolithography and thick resist UV lithography for 3D microfabrication', in *Proceedings of IEEE MEMS '98*, IEEE, Washington, DC: 18–23.

Body, C., Reyne, G., Meunier, G., 1997, 'Modeling of magnetostrictive thin films, application to a micromembrane', *Journal of Physique (France) Part III* **7**(1): 67–85.

Bohm, S., Olthuis, W., Bergveld, P., 1999, 'A plastic micropump constructed with conventional techniques and materials', *Sensors and Actuators A: Physical* **773**: 223–228.

Brown, E.R., 1998, RF MEMS switches for reconfigurable integrated circuits, *IEEE Transactions on Microwave Theory and Techniques* **46**: 1868–1880.

Bryzek, Peterson, K., McCulley, W., 1994, 'Micromachines on the march', *IEEE Spectrum* (May) 20–31.

Bustillo, J.M., Howe, R.T., Muller, R.S., 1998, 'Surface micromachining for microelectromechanical systems', *Proceedings of the IEEE* **86**(8): 1552–1574.

Campbell, C., 1998, *Surface Acoustic Wave Devices and their Signal Processing Applications*, Academic Press, London.

Carraway, D.L., 1991, 'The use of silicon microsensors in smart skins for aerodynamic research', in *International Congress on Instrumentation in Aerospace Simulation Facilities*, (December): IEEE, 413–422.

Chen, J., Gururaja, T.R., 1997, 'DC-biased electrostrictive materials and transducers for medical imaging', in *IEEE Ultrasonics Symposium 1997*, IEEE, Washington, DC: 1651–1658.

Chin, L.C., Varadan, V.V., Varadan, V.K., 1994, 'Hybrid finite element formulation for periodic piezoelectric arrays subjected to fluid loading', *International Journal for Numerical Methods in Engineering* **37**: 2987–3003.

Cohen, A., Zhang, G., Tseng, F., Frodis, U., Mansfeld, F., Will, P., 1999, 'EFAB: rapid, low-cost desktop micromachining of high aspect ratio true 3-D MEMS', in *Proceedings of IEEE MEMS '99*, IEEE, Washington, DC: 244–251.

Crandall, S.H., Karnopp, D.C., Kurtz, E.F. Jr., Pridmore-Brown, D.C., 1968, *Dynamics of Mechanical and Electromechanical Systems*, McGraw-Hill, New York.

Elwenspoek, M., Blom, F.R., Bouwstra, S., Lammerink, T.S.J., van de Pol, F.C.M., Tilmans, H.A.C., Popma, T.J.A., Fluitman, J.H.J., 1989, 'Transduction mechanisms and their applications in micromechanical devices', in *Proceedings of IEEE Micro Electro Mechanical Systems, 1989*, IEEE, Washington, DC: 126–132.

Fan, L.-S., Tai, Y.-C., Muller, R.S., 1987, 'Pin joints, gears, springs, cranks and other novel micromechanical structures', in *Tech Digest of 4th International Conference Solid State Sensors and Actuators A: Physical*, June, Tokyo: 849–852.

Fatikow, S., Rembold, U., 1997, *Microsystem technology and microrobots*, Springer, Berlin.

Frank, T., 1998, 'Two-axis electrodynamic micropositioning devices', *Journal of Micromechanics and Microengineering* **8**: 114–118.

Fujita, H., 1996, 'Future of actuators and microsystems', *Sensors and Actuators A: Physical* **56**: 105–111.

Fujita, H., 1998, 'Microactuators and micromachines', *Proceedings of the IEEE* **86**(8): 1721–1732.

Goldsmith, C.L., Eshelman, S., Dennston, D., 1998, 'Performance of low loss RF MEMS capacitive switches', *IEEE Microwave and Guided Wave Letters* **8**: 269–271.

Grate, J.W., Martin, S.J., White, R.M., 1993, 'Acoustic wave microsensors. Part I', *Analytical Chemistry* **65**: 940–948.

Grate, J.W., Martin, S.J., White, R.M., 1993, 'Acoustic wave microsensors. Part II', *Analytical Chemistry* **65**: 987–996.

Grosjean, C., Yang, X., Tai, Y.C., 1999, 'A practical thermopneumatic valve', in *Proceedings of IEEE MEMS '99*, IEEE, Washington, DC: 147–152.

Guckel, H., 1998, 'High-aspect-ratio micromachining via deep X-ray lithography', *Proceedings of the IEEE* **86**(8): 1586–1593.

Harada, K., Ikeda, K., Kuwayama, H., Murayama, H., 1999, 'Various applications of resonant pressure sensor chip based on 3-D micromachining', *Sensors and Actuators A: Physical* **73**: 261–266.

Heydt, R., Kornbluh, R., Pelrine, R., Mason, V., 1998, 'Design and performance of an electrostrictive-polymer-film acoustic actuator', *Journal of Sound and Vibration* **215**: 297–311.

Hom, C.L., Pilgrim, S.M., Shankar, N., Bridger, K., Massuda, M., Winzer, S.R., 1994, 'Calculation of quasi-static electromechanical coupling coefficients for electrostrictive ceramic materials', *IEEE Transactions on Ultrasonics, Ferroelectrics, and Frequency control* **41**: 542–551.

Honda, T., Arai, K.I., Yamaguchi, M., 1997, 'Basic properties of magnetostrictive actuators using Tb–Fe and Sm–Fe thin films', *IEICE Transactions on Electronics* **E80-C**(2): 232–238.

Huang, Q.A., Lee, N.K.S., 1999, 'Analysis and design of polysilicon thermal flexure actuator', *Journal of Micromechanics and Microengineering* **9**: 64–70.

Ikeda, K., Kuwayama, H., Kobayashi, T., Watanabe, T., Nishikawa, T., Yoshida, T., Harada, K., 1990, 'Silicon pressure sensor integrates resonant strain gauge on diaphragm', *Sensors and Actuators A: Physical* **21–23**: 146–150.

Ikuta, K., Hirowatari, K., 1993, 'Real three dimensional microfabrication using stereo lithography and metal molding', in *Proceedings of IEEE MEMS '93*, IEEE, Washington, DC: 42–47.

Jennifer, M., Allen, M.G., 1999, 'Wireless micromachined ceramic pressure sensors', *Proceedings of the IEEE Micro Electro Mechanical Systems (MEMS)* (17–21 Jan) 511–516.

Jiang, X.N., Sun, C., Zhang, X., 1999, 'Micro-stereolithography of 3D complex ceramic microstructures and PZT thick films on Si substrate', *ASME MEMS* **1**: 67–73.

Johnson, R.A., 1983, *Mechanical Filters in Electronics*, Wiley Interscience, New York.

Karl, W.J., Powell, A.L., Watts, R., Gibbs, M.R.J., Whitehouse, C.R., 2000, 'A micromachined magnetostrictive pressure sensor using magneto-optical interrogation', *Sensors and Actuators A: Physical* **81**: 137–141.

Kondoh, J., Matsui, Y., Shiokawa, S., 1993, 'New bio sensor using shear horizontal surface acoustic wave device', *Japanese Journal of Applied Physics* **32**: 2376–2379.

Kovacs, G.T.A., Maluf, N.I., Petersen, K.E., 1998, 'Bulk micromachining of silicon', *Proceedings of the IEEE* **86**(8): 1536–1551.

Ko, W.H., Suminto, J.T., Yeh, G.J., 1985, 'Bonding techniques for microsensors', *Studies in Electrical and Electronic Engineering* **20**: 41–61.

Lagorce, L.K., Allen, M.G., 1996, *Micromachined Polymer Magnets, ISAF '96*: Proceedings of IEEE Microelectromechanical Systems (MEMS), IEEE 85–90.

Larson, L.E., 1999, 'Microwave MEMS technology for next-generation wireless communications – invited paper', in *IEEE MTT-S International Symposium Digest*, IEEE, Washington, DC: 1073–1076.

Lee, K.H., Byun, H.J., Lee, H.K., Cho, I.J., Bu, J.U., Yoon, E., 2000, 'An audio frequency filter application of micromachined thermally-isolated diaphragm structures', in *13th Annual International Conference on Micro Electro Mechanical Systems MEMS 2000*, IEEE, Washington, DC: 142–147.

Madden, J.D., Hunter, I.W., 1996, 'Three-dimensional microfabrication by localized electrochemical deposition', *Journal of Microelectromechanical Systems* **5**(1): 24–32.

Madou, M., 1997, *Fundamentals of Microfabrication*, CRC Press, Boca Raton, FL.

Mehregany, M., Zorman, C.A., Rajan, N., Wu, C.H., 1998, 'Silicon carbide MEMS for harsh environments', *Proceedings of the IEEE* **86**(August): 1594–1610.

Middelhoek, S., Audet, S.A., 1989, *Silicon Sensors*, Academic Press, London.

Minami, K., Morishita, H., Esashi, M., 1999, 'A bellows-shape electrostatic microactuator', *Sensors and Actuators A: Physical* **72**3: 269–276.

Mohr, J., Bley, P., Burabaum, C., Menz, W., Wallarabe, U., 1991, Fabrication of microsensor and microactuator elements by the LIGA process. Transducers '91, Proceedings of the International Conference Solid State Sens and Actuators, IEEE Electron Dev Soc., pp. 607–609.

Muralt, P., 2000, 'PZT thin films for microsensors and actuators: where do we stand?', *IEEE Transactions on Ultrasonics, Ferroelectrics, and Frequency Control* **47**: 903–915.

Nakamura, N., Kazumi, M., Shimizu, H., 1977, 'SH-type and Rayleigh type surface waves on rotated Y-cut LiTaO3', in *Proceedings of IEEE Ultrasonics Symposium*. IEEE, Washington, DC: 819–822.

Ohta, A., Bhansali, S., Kishimoto, I., Umeda, A., 1998, 'Development of TiNi shape memory alloy film deposited by sputtering fro[illegible]eparate Ti and Ni targets', *Proceedings of SPIE* **3512**(September 21–22): 138–145.

Pan, L.W., Lin, L.W., Ni, J., 1999, 'Cylindrical plastic lens array fabricated by a micro intrusion process', in *Proceedings of IEEE MEMS '99*, IEEE, Washington, DC: 217–221.

Pelrine, R., Kornbluh, R., Joseph, J., Chiba, S., 1997, 'Electrostriction of polymer films for microactuators', in *Proceedings of IEEE MEMS '97*, IEEE, Washington, DC: 238–243.

Pelrine, R.E., Kornbluh, R.D., Joseph, J.P., 1998, 'Electrostriction of polymer dielectrics with compliant electrodes as a means of actuation', *Sensors and Actuators A: Physical* **64**: 77–85.

Petersen, K.E., 1982, 'Silicon as a mechanical material', *Proceedings of IEEE* **70**: 420–457.

Pilgrim, S.M., 2000, 'Electrostrictive ceramics for low-frequency active transducers', *IEEE Transactions on Ultrasonics, Ferroelectrics, and Frequency control* **47**: 861–876.

Rai-Choudhury, P., 1997, *Handbook of Microlithography, Micromachining and Microfabrication, Vol. 2: Micromachining and Microfabrication*, SPIE Optical Engineering Press.

Riethmüller, W., Benecke, W., 1988, 'Thermally excited silicon microactuators', *IEEE Transactions on Electron Devices* **35**: 758–763.

Romankiw, L.T., 1997, 'A path: from electroplating through lithographic masks in electronics to LIGA in MEMS', *Electrochimica Acta* **42**(20–22): 2985–3005.

Rossi, M., 1988, *Acoustics and Electroacoustics*, Artech House, Norwood, MA.

Ruiz de Angulo, L., Abell, J.S., Harris, I.R., 1996, 'Magnetostrictive properties of polymer bonded Terfenol-D', *Journal of Magnetism and Magnetic Materials* **157–158**(May): 508–509.

Ruprecht, R., Hanemann, T., Piotter, V., Husselt, J., 1998, 'Polymer materials for microsystem technologies', *Microsystem Technologies* **5**: 44–48.

Schomburg, W.K., Ahrens, R., Bacher, W., Goll, C., Meinzer, S., Quinte, A., 1998, 'AMANDA – low-cost production of microfluidic devices', *Sensors and Actuators A: Physical* **70**: 153–158.

Seidel, H., 1990, IEEE Solid State sensors and Actuator Workshop, IEEE, Piscataway, NJ, USA: pp. 86–91.

Shaw, K.W., Zhang, Z.L., MacDonald, N.C., 1994, 'SCREAM: a single mask, single-crystal silicon, reactive ion etching process for microelectromechanical structures', *Sensors and Actuators A: Physical* **40**: 63–70.

Shiokawa, S., Moriizumi, T., 1987, 'Design of SAW sensor in Liquid', *Proceedings of 8th Symposium on Ultrasonic Electronics*, Japan Society of Applied Physics, Tokyo.

Stix, G., 1992, 'Trends in micromechanics: micron machinations', *Scientific American*, (November): 72–80.

Takagi, T., Nakajima, N., 1993, 'Photoforming applied to fine machining', in *Proceedings of IEEE MEMS '93*, IEEE, Washington, DC: 173–178.

Takagi, T., Nakajima, N., 1994, 'Architecture combination by micro photoforming process', in *Proceedings of IEEE MEMS '94*, IEEE, Washington, DC: 211–216.

Tani, J., Esashi, M., (eds), 1995, *Proceedings of the International Symposium on Microsystems, Intelligent Materials and Robots* Tohoku University, Katahira, Sendai, Japan.

Taylor, C.S., Cherkas, P., Hampton, H., Frantzen, J.J., Shah, B.O., Tiffany, W.B., Nanis, L., Booker, P., Salahieh, A., Hansen, R., 1994, 'A spatial forming – a three dimensional printing process', in *Proceedings of IEEE MEMS '94*, IEEE, Washington, DC: 203–208.

Thornell, G., Johansson, S., 1998, 'Microprocessing at the fingertips', *Journal of Micromechanics and Microengineering* **8**: 251–262.

Tilmans, H.A.C., 1996, 'Equivalent circuit representation of electromechanical transducers: I. Lumped parameter systems', *Journal of Micromechanics and Microengineering* **6**: 157–176.

Varadan, V.K., (ed.), 1995, *Smart Electronics, SPIE Proceedings* **2448**, SPIE, Bellingham, WA.

Varadan, V.K. Varadan, V.V., 1995, '3D MEMS structures and their applications', in *Proceedings of the International Symposium on Microsystems, Intelligent Materials and Robots*, Tohoku University, Katahira, Sendai, Japan.

Varadan, V.K., Jiang, X., Varadan, V.V., 2001, *Microstereolithography and other fabrication techniques for 3D MEMS*, John Wiley, Chichester.

Varadan, V.K., Varadan, V.V., 1996, 'IDT, SAW and MEMS sensors for measuring deflection, acceleration and ice detection of aircraft', *SPIE* **3046**: 209–219.

Varadan, V.K., Varadan, V.V., 1996, 'Three dimensional polymeric and ceramic MEMS and their applications', in *Proceedings of SPIE* **2722**: 156–164.

Varadan, V.K., Varadan, V.V., 1997, 'Microsensors, actuators, MEMS, and electronics for smart structures', in Rai-Choudhury, P. (ed.), *Handbook of Microlithography, Micromachining, and Microfabrication, Volume 2: Micromachining and Microfabrication*, SPIE Optical Engineering Press: 617–688.

Witte, M., Gu, H., 1992, 'Force and position sensing resistors: an emerging technology', in *Proceedings of the International Conference on New Actuators, Bremen*, VDI/VDE-Technologiezentrum Informationstechnik, Berlin, Germany: 168–170.

Wood, D., Burdess, J.S., Harris, A.J., 1996, 'Actuators and their mechanisms in microengineering', in *Proceedings of the IEEE Colloquium on Actuator Technology: Current Practice and New Developments, 1996* **110**: 7/1–7/3.

Wu, H.D., Harsh, K.F., Irwin, R.S., Zhang, W., Michelson, A.R. and Lee, Y.C., 1998, 'MEMS designed for tunable capacitors', in *IEEE MTTS Symposium Digest* IEEE, Washington, DC: 129–130.

Xia, Y., Whitesides, G.M., 1998, 'Soft lithography', *Angew. Chem. International Ed.* **37**: 550–575.

Zhang, X., Jiang, X.N., Sun, C., 1999, 'Micro-stereolithography of polymeric and ceramic microstructures', *Sensors and Actuators A: Physical* **77**(2): 149–156.

2

MEMS materials and fabrication techniques

MEMS materials may be classified into five main types: metals, semiconductors, ceramics, polymers, and composites. This chapter first introduces the basic nature of each type of material and then discusses the different ways in which it can be prepared.

2.1 METALS

Metals are in general good thermal and electrical conductors. They are somewhat strong and ductile at room temperature, and they maintain good strength both at room and elevated temperatures. Table 2.1 provides atomic and crystal structure information on 12 selected metals and these illustrate the three principal lattice structures described here.

Metallization is a process whereby metal films are formed on the surface of a substrate. These metallic films are used for interconnections, ohmic contacts, etc. Metal films can be formed using various methods, the most important being physical vapor deposition (PVD). PVD is performed under vacuum using either an evaporation or sputtering technique.

2.1.1 Evaporation

Thin metallic films can be evaporated from a hot source onto a substrate as shown in Figure 2.1. An evaporation system consists of a vacuum chamber, pump, wafer holder, crucible and a shutter. A sample of the metal to be deposited is placed in an inert crucible, and the chamber is evacuated to a pressure of 10^{-6}–10^{-7} Torr. The crucible is then heated using a tungsten filament or an electron beam to flash-evaporate the metal from the crucible and condense onto the cold sample. The film thickness is determined by the length of time that the shutter is opened and can be measured using a QMB-based film thickness monitor. The evaporation rate is a function of the vapor pressure of the metal. Hence, metals that have a low melting point T_{mp} (e.g. 660 °C for aluminium) are easily evaporated, whereas refractory metals require much higher temperatures (e.g. 3422 °C for tungsten) and can cause damage to polymeric or plastic samples. In general, evaporated films are highly disordered and have large residual stresses; thus, only thin layers of the

Table 2.1 The atomic properties and crystal structures of selected metals

Atomic number, Z	Symbol	Atomic radius (Å)	Lattice structure	Interatomic distance (Å)
13	Al	1.43	FCC	2.86
22	Ti	1.47	HCP	2.90
24	Cr	1.25	BCC (α)	2.49
		1.36	HCP (β)	2.71
26	Fe	1.24	BCC (α)	2.48
		1.26	FCC (γ)	2.52
27	Co	1.25	HCP (α)	2.49
		1.26	FCC (β)	2.51
28	Ni	1.25	HCP (α)	2.49
		1.25	FCC (β)	2.49
29	Cu	1.28	FCC	2.55
30	Zn	1.33	HCP	2.66
47	Ag	1.44	FCC	2.97
78	Pt	1.38	FCC	2.77
79	Au	1.44	FCC	2.88
82	Pb	1.75	FCC	3.49

Note: BCC, base centered cubic; FCC, face centered cubic; HCP, hexagonal close packed.

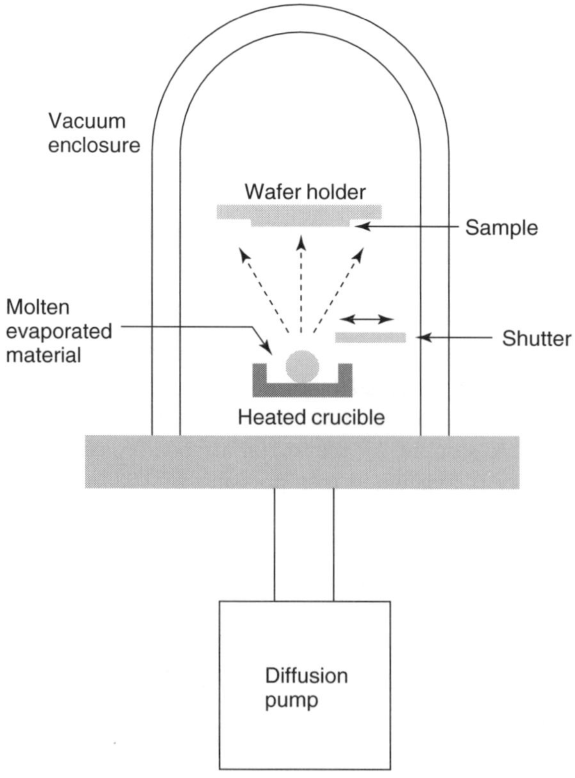

Figure 2.1 A thermal evaporation unit for depositing materials

metal can be evaporated. In addition, the deposition process is relatively slow, at a few nanometers per second.

2.1.2 Sputtering

Sputtering is a physical phenomenon involving the acceleration of ions via a potential gradient and the bombardment of a 'target' or cathode. Through momentum transfer, atoms near the surface of the target metal become volatile and are transported as a vapor to a substrate. A film grows at the surface of the substrate via deposition.

Figure 2.2 shows a typical sputtering system that comprises a vacuum chamber, a sputtering target of the desired film, a sample holder and a high-voltage dc or radio frequency (RF) power supply. After evacuating the chamber down to a pressure of 10^{-6} to 10^{-8} Torr, an inert gas such as helium is introduced into the chamber at a few mTorr of pressure. A plasma of the inert gas is then ignited. The energetic ions of the plasma bombard the surface of the target. The energy of the bombarding ions (~keV) is sufficient to make some of the target atoms escape from the surface. Some of these atoms land on the sample surface and form a thin film. Sputtered films tend to have better uniformity than evaporated ones, and the high-energy plasma overcomes the temperature limitations of evaporation. Most elements from the periodic table can be sputtered, as well as inorganic and organic compounds. Refractory materials can be sputtered with ease, whereas the evaporation of materials with very high boiling points is problematic. Also materials from more than one target can be sputtered at the same time. This process is referred to as co-sputtering.

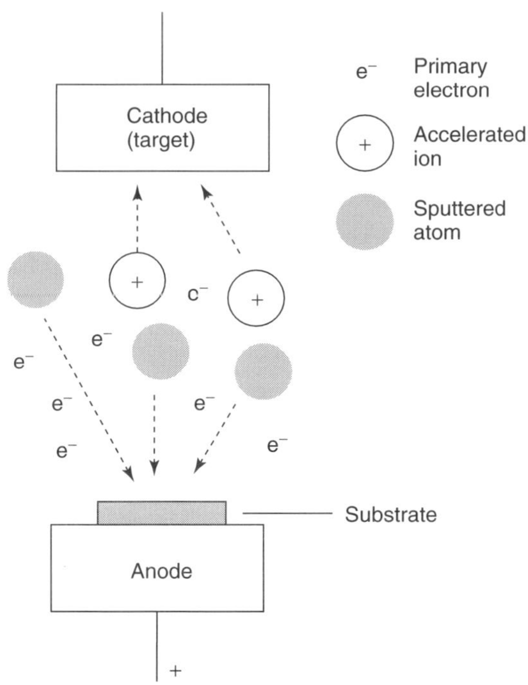

Figure 2.2 Schematic of sputtering unit for depositing materials

The structure of sputtered films is mainly amorphous, and its stress and mechanical properties are sensitive to specific sputtering conditions. Some atoms of the inert gas can be trapped in the film causing anomalies in its mechanical and structural characteristics. Therefore the exact properties of a thin film vary according to the precise conditions under which it was made.

2.2 SEMICONDUCTORS

2.2.1 Electrical and chemical properties

Semiconductors are commonly inorganic materials made from elements in the fourth column (group IV) of the periodic table. The most important among these elements is silicon, which can be modified in several ways to change its electrical, mechanical and optical properties. The use of silicon in solid state and microelectronics has shown a spectacular growth since the early 1970s, and this growth pattern is still continuing. Other semiconductor materials, from group IV elements in the periodic table, are germanium and carbon (diamond). Semiconductor materials can also be made from a combination of elements from either group III and group V or group II and group VI. Examples of these are gallium arsenide and zinc telluride materials. The name semiconductor is given to these materials because at certain regimes of temperatures they are able to exhibit good electrical conduction properties, and outside these temperature regimes they behave as insulators.

Semiconductor crystals can be made from single elements and from compounds. Semiconductors that are made from single elements are called elemental semiconductors. Elemental semiconductors are found in group IV of the periodic table [e.g. silicon (Si), and germanium (Ge)]. Compound semiconductors are made up of special combinations of group III and group V elements, or special combinations of group II and group VI elements as stated above. Table 2.2 lists a few of the elemental and compound semiconductors.

Table 2.2 Structure and lattice properties for some common elemental and compound semiconductors. The lattice constants and band-gaps are given at a temperature of 27 °C

Material	Lattice structure[a]	Lattice constant (Å)	Energy gap (eV)
Ge	Diamond structure	5.66	0.66
Si	Diamond structure	5.43	1.12
GaAs	Zinc-blende structure	5.64	1.44
GaSb	Zinc-blende structure	6.12	0.78
InSb	Zinc-blende structure	6.46	0.18
InAs	Zinc-blende structure	6.04	0.33
InP	Zinc-blende structure	5.86	1.25
PbSe	Zinc-blende structure	6.14	0.27
PbTe	Zinc-blende structure	6.34	0.30

[a]More precise classification of structures uses the alphanumeric system (i.e. A3 is diamond).

Among the elemental semiconductors, silicon is by far the most commonly used material. Silicon is the most important material for microelectronics and integrated circuit technology. Also silicon-based compounds and technologies are becoming the major cornerstones for the rapidly developing fields of MEMS and nanofabrication. For this reason we will be emphasizing silicon and using it to demonstrate the general properties of semiconductor materials. Table 2.3 lists a few of the mechanical, electrical and thermal properties of single crystalline silicon. GaAs is the most commonly used among the compound semiconductors, especially in fabricating optical and high-speed devices.

The crystal structure of many semiconductors, including silicon and GaAs, is based on the cubic crystalline system. Diamond itself could be thought of as a semiconductor with a wide band gap (~6 eV) and its structure is that of two interleaved face-centered cubic arrays with one array displaced a fraction of the interatomic distance from the other. In the GaAs type of compound, one of the two arrays is composed entirely of gallium atoms, whereas the other array is composed of arsenic atoms. This particular class of the diamond structure is called the zinc-blende structure. In the diamond lattice each atom has four nearest neighbors. In both elemental and compound semiconductors, there is an average of four valence electrons per atom. Each atom is thus held in the crystal by four covalent bonds with two electrons participating in each bond. In a perfect semiconductor crystal and at absolute zero temperature the number of electrons available would exactly fill the inner atomic shells and the covalent bonds. At temperatures above absolute zero some of these electrons gain enough thermal energy to break loose from these covalent bonds and become free electrons. Free electrons are responsible for electrical conduction across the semiconductor crystal. Some of the physical properties of selected semiconductor crystals are given in Table 2.3.

When impurities are intentionally added to a semiconductor the semiconductor is said to be 'doped'. Let us assume a hypothetical two-dimensional silicon crystal in which one silicon atom is replaced (or substituted) by an atom – in this example, a Group V element in the periodic table, namely phosphorus. Phosphorus has five valence electrons whereas silicon has only four. The phosphorus atom will share four of its electrons with

Table 2.3 Electrical, mechanical and thermal properties of crystalline silicon

Property	Value
Electrical:	
resistivity (Ω cm)	
(P-doped)	1–50
(Sb-doped)	0.005–10
(B-doped)	0.005–50
minority-carrier life-time	30–300
Mechanical:	
yield Strength ($N\,m^{-2}$)	7×10^9
Young's Modulus ($N\,m^{-2}$)	1.9×10^{11}
density ($g\,cm^{-3}$)	2.3
dislocations (cm^{-2})	<5000
Thermal:	
thermal conductivity ($W\,cm^{-1}\,°C^{-1}$)	1.57
thermal expansion ($°C^{-1}$)	2.33×10^{-6}

four neighboring silicon atoms in covalent bonds. The remaining fifth valence electron in phosphorus is loosely bound to the phosphorus nucleus. The ionization energy of an impurity atom of mass m in a semiconductor crystal can be estimated from a one-electron model. If this ionization energy is denoted by the symbol $E_{\rm d}$, then

$$E_{\rm d} = \left(\frac{\varepsilon_0}{\varepsilon_{\rm r}}\right)^2 \left(\frac{m^*}{m}\right) E_{\rm n} \tag{2.1}$$

where ε_0 is the permittivity of free space, $\varepsilon_{\rm r}$ is that of the semiconductor, m^* is the effective mass of the semiconductor material. $E_{\rm n}$ is the orbital energy of the donor electron. When the phosphorus atom in silicon is ionized, the released electron becomes a free electron which is available for conduction. The phosphorus atom is, hence, called a donor atom since it donates a free electron to the crystal. All atoms with five valence electrons (i.e. Group V elements) can behave in a similar manner to phosphorus in silicon, (i.e. donate a free electron to the semiconductor crystal). However, the amount of energy needed, $E_{\rm d}$, for this process to occur may differ from one type of donor atom to another. All Group V atoms will donate electrons if they substitute for host atoms in crystals of Group IV elemental semiconductors. Consequently, Group V elements, such as phosphorus or arsenic, are called donor atoms or simply donors and the doped semiconductor is now referred to as an extrinsic semiconductor. This may be contrasted to an intrinsic (undoped) semiconducting material.

Now imagine that we can introduce a large concentration of phosphorus atoms in an otherwise pure silicon crystal (e.g. a phosphorus atom concentration of $\sim 10^{15}\,\text{cm}^{-3}$). With a minimal energy supply each of these phosphorus atoms will donate an electron to the crystal amounting to a concentration of electrons in the conduction band of the order of $10^{15}\,\text{cm}^{-3}$ at room temperature. This concentration of electrons is to be contrasted to the concentration of conduction electrons in intrinsic silicon at room temperature, which is of the order of $10^{10}\,\text{cm}^{-3}$. Thus with this doping level a five order of magnitude increase in the free electron concentration has been achieved. Note that in a solid there are about 10^{22} to 10^{23} atoms cm^{-3} and a doping level of $10^{15}\,\text{cm}^{-3}$ is equivalent to merely replacing one silicon atom in every 10^7 to 10^8 atoms cm^{-3} by a phosphorus atom. Obviously, this level of doping introduces a very insignificant change in the overall crystal structure but its effect on the free electron concentration is clearly very significant. Note that conduction in this phosphorus-doped silicon will, therefore, be dominated by electrons. This type of extrinsic (Group IV) semiconductor, or more specifically silicon, is called an n-type semiconductor or n-type silicon. The term n-type indicates that the charge carriers are the negatively-charged electrons. The example discussed above was specific to silicon doped with phosphorus; however, the conclusions arrived at will apply generally to all elemental semiconductors doped with a higher group element. The values of the ionization energies $E_{\rm d}$ for several Group V donors in silicon are given in Table 2.4 together with those for some acceptors (see below).

Now consider the situation where a Group IV semiconductor is doped with atoms from an element in Group III of the periodic table (i.e. atoms that have only three valence electrons). To be more specific we will take as an example silicon doped with boron. The net effect of having a boron atom substituting for silicon is the creation of a free hole (an electron deficiency in a covalent bond). This hole is generated as follows: since boron has three valence electrons, three neighboring silicon atoms will be bonded covalently

Table 2.4 Common donor and acceptor atoms in silicon

Atom	Atomic number	Type	Ionization Energy in Si (eV)
Boron	5	Acceptor	0.045
Aluminium	13	Acceptor	0.057
Phosphorus	15	Donor	0.044
Gallium	31	Acceptor	0.065
Arsenic	33	Donor	0.049
Indium	49	Acceptor	0.16
Antimony	51	Donor	0.039

with boron. However, the fourth nearest neighbor silicon atom has one of its four valence electrons sitting in a dangling bond; that is, the whole system of the boron atom and the four neighboring silicon atoms is missing one electron. An electron from a neighboring Si–Si covalent bond may replace the missing electron thereby creating an electron deficiency (a hole) at the neighboring bond. The net effect is, hence, the generation of a free hole in the silicon crystal. Therefore this type of extrinsic semiconductor, silicon in this particular example, is called a p-type semiconductor or p-type silicon. It is p-type because electrical conduction is carried out by positively charged free holes. Common acceptor atoms to silicon are given in Table 2.4.

Diffusion and ion implantation are the two key processes used to introduce controlled amounts of dopants into semiconductors. These two processes are used to dope selectively the semiconductor substrate to produce either an n-type or a p-type region.

2.2.2 Growth and deposition

To demonstrate the methods of growing semiconductors we will consider crystal growth of silicon in details. We use silicon as an exemplar since it is the most utilized semiconductor in microelectronics and MEMS.

2.2.2.1 Silicon crystal growth from the melt

Basically, the technique used for silicon crystal growth from the melt is the Czochralski technique. The technique starts from a pure form of sand (SiO_2) called quartzite placed in a furnace with different carbon-releasing materials such as coal and coke. Several reactions take place inside the furnace and the net reaction that results in silicon is:

$$SiC + SiO_2 \longrightarrow Si + SiO\ (gas) + CO\ (gas) \tag{2.2}$$

The silicon so-produced is called metallurgical-grade silicon (MGS) which contains up to 2% impurities. Subsequently, the silicon is treated with hydrogen chloride to form trichlorosilane ($SiHCl_3$):

$$Si + 3HCl \longrightarrow SiHCl_3\ (gas) + H_2\ (gas) \tag{2.3}$$

$SiHCl_3$ is liquid at room temperature. Fractional distillation of the $SiHC1_3$ liquid removes impurities, and the purified liquid is reduced in a hydrogen atmosphere to yield electronic grade silicon (EGS) via the reaction:

$$SiHCl_3 + H_2 \longrightarrow Si + 3HCl \tag{2.4}$$

EGS is a polycrystalline material of remarkably high purity and is used as the raw material for preparing high-quality silicon wafers.

The Czochralski technique uses the apparatus shown in Figure 2.3 called the *puller*. The puller comprises three main parts:

- A furnace that consists of a fused-silica (SiO_2) crucible, a graphite susceptor, a rotation mechanism, a heating element and a power supply
- A crystal pulling mechanism, which is composed of a seed holder and a rotation mechanism
- An atmosphere control, which includes a gas source (usually an inert gas), a flow control and an exhaust system.

In crystal growing, the EGS is placed in the crucible and the furnace is heated above the melting temperature of silicon. An appropriately oriented seed crystal (e.g. [100]) is suspended over the crucible in a seed holder. The seed is lowered into the melt. Part of

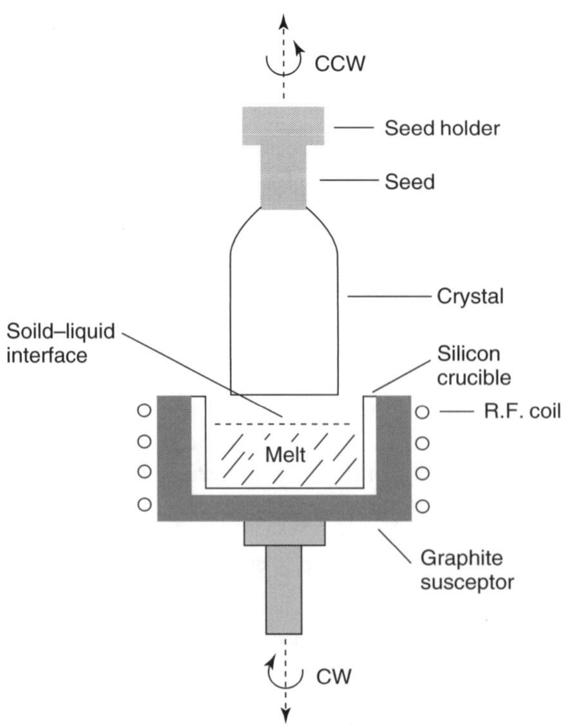

Figure 2.3 Czochralski crystal puller

it melts but the tip of the remaining seed crystal still touches the liquid surface. The seed is then gently withdrawn. Progressive freezing at the solid–liquid interface yields a large single crystal. A typical pull rate is a few millimeters per minute.

After a crystal is grown the seed is removed as well as the other end of the ingot, which is last to solidify. Next, the surface is ground so that the diameter of the material is defined. After that one or more flat regions are ground along the length of the ingot. These flat regions mark the specific crystal orientation of the ingot and the conductivity type of the material (Figure 2.4). Finally, the ingot is sliced by a diamond saw into wafers. Slicing determines four wafer parameters: surface orientation, thickness, taper (which is wafer thickness variations from one end to another) and bow (i.e. surface curvature of the wafer, measured from the center of the wafer to its edge). Typical specifications for silicon wafers are given in Table 2.5.

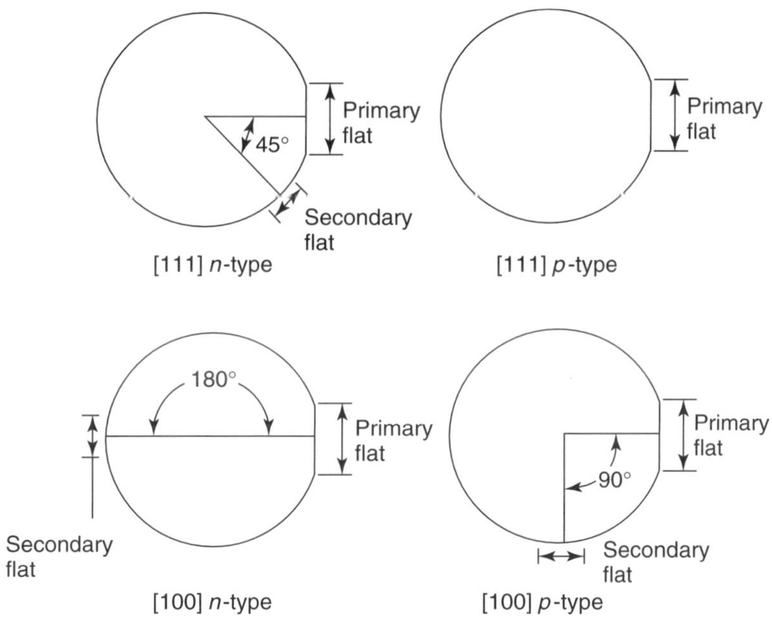

Figure 2.4 Crystal orientation and dopant type

Table 2.5 Specifications for silicon wafers

Parameter	Diameter (mm)		
	100	125	150
Thickness (mm)	0.50–0.55	0.60–0.65	0.65–0.7
Primary flat[a] length (mm)	30–35	40–45	55–60
Secondary flat length (mm)	16–20	25–30	35–40
Bow (mm)	60	70	60
Total thickness variation (μm)	50	65	50
Surface orientation	(1 0 0) or (1 1 1)	(1 0 0) or (1 1 1)	(1 0 0) or (1 1 1)

[a]Wafer flats are defined in Figure 3.20.

2.2.2.2 Epitaxial growth

The method for growing a silicon layer on a substrate wafer is known as an *epitaxial* process where the substrate wafer acts as a seed crystal. Epitaxial processes are different from crystal growth from the melt in that the epitaxial layer can be grown at a temperature very much below the melting point. Among various epitaxial processes, vapor phase epitaxy (VPE) is the usual process for silicon layer growth.

A schematic of the VPE apparatus is shown in Figure 2.5. The figure shows a horizontal susceptor made from graphite blocks. The susceptor mechanically supports the wafer and, being an induction-heated reactor, it also serves as the source of thermal energy for the reaction.

Several silicon sources are usually used: silicon tetrachloride ($SiCl_4$), dichlorosilane (SiH_2Cl_2), trichlorosilane ($SiHCl_3$) and silane (SiH_4). Typical reaction temperature for silicon tetrachloride is $\sim$1200 °C. The overall reaction in the case of silicon tetrachloride is reduction by hydrogen:

$$SiCl_4 \text{ (gas)} + 2H_2 \text{ (gas)} \longrightarrow Si \text{ (solid)} + 4HCl \text{ (gas)} \tag{2.5}$$

A competing reaction which would occur simultaneously is:

$$SiCl_4 \text{ (gas)} + Si \text{ (solid)} \longrightarrow 2SiCl_2 \text{ (gas)} \tag{2.6}$$

In Reaction (2.5) silicon is deposited on the wafer, whereas in Reaction (2.6) silicon is removed (etched). Therefore if the concentration of $SiCl_4$ is excessive, etching rather than growth of silicon will take place.

An alternative epitaxial process for silicon layer growth is molecular beam epitaxy (MBE) which is an epitaxial process involving the reaction of a thermal beam of silicon atoms with a silicon wafer surface under ultrahigh vacuum conditions ($\sim 10^{-10}$ Torr). MBE can achieve precise control in both chemical composition and impurity (if introduced intentionally) profiles. Single-crystal multilayer structures with dimensions of the order of atomic layers can be made using MBE.

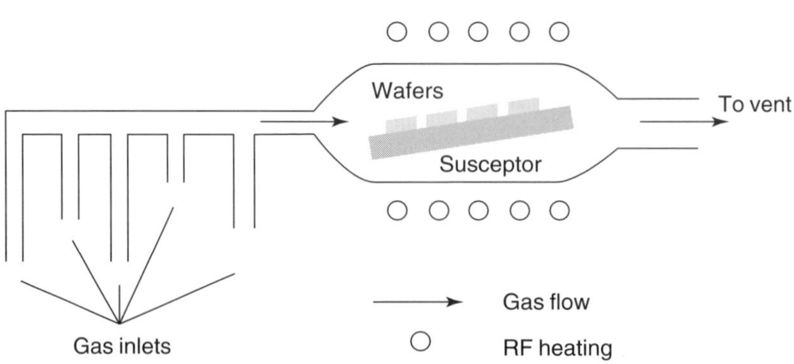

Figure 2.5 Silicon layer by vapor phase epitaxy

2.3 THIN FILMS FOR MEMS AND THEIR DEPOSITION TECHNIQUES

Many different kinds of thin films are used in the fabrication of MEMS. There are four important thin-film materials (or class of materials) in MEMS fabrication:

- thermal silicon oxide
- dielectric layers
- polycrystalline silicon (poly-Si)
- metal films (predominantly aluminium)
- ferroelectric thin films

The dielectric layers include deposited silicon dioxide (sometimes referred to as oxide), and silicon nitride. These dielectrics are used for insulation between conducting layers, for diffusion and ion implantation masks and for passivation to protect devices from impurities, moisture and scratches. Poly-Si is used as a gate electrode in metal–oxide–semiconductor (MOS) devices, as a conductive material for multilevel metallization and as a contact material for devices with shallow junctions. Metal films are used to form low-resistance ohmic connections both to heavily doped n^+/p^+ regions and poly-Si layers, and rectifying (nonohmic) contacts in metal–semiconductor barriers.

As shall become apparent in the following chapters, electronic materials are of major importance in MEMS devices. Therefore the methods used to grow thermal silicon dioxide and to deposit dielectric, poly-Si, metallic layers and ferroelectric thin films are reviewed in the following sections.

2.3.1 Oxide film formation by thermal oxidation

Thermal oxidation is the method by which a thin film of SiO_2 is grown on top of a silicon wafer. It is the key method of producing thin SiO_2 layers in modern integrated circuit (IC) technology. The basic thermal oxidation apparatus is shown in Figure 2.6. The apparatus comprises a resistance-heated furnace, a cylindrical fused-quartz tube containing the silicon wafers held vertically in slotted-quartz boat and a source of either pure, dry oxygen or pure water vapor. The loading end of the furnace tube protrudes into a vertical flow hood where a filtered flow of air is maintained. The hood reduces dust in the air surrounding the wafers and minimizes contamination during wafer loading.

Thermal oxidation of silicon in oxygen or water vapor can be described by the following two chemical reactions:

$$\text{Si (solid)} + \text{O}_2 \text{ (gas)} \xrightarrow{900 \text{ to } 1200\,^\circ\text{C}} \text{SiO}_2 \text{ (solid)} \tag{2.7}$$

and

$$\text{Si (solid)} + 2\text{H}_2\text{O (gas)} \xrightarrow{900 \text{ to } 1200\,^\circ\text{C}} \text{SiO}_2 \text{ (solid)} + 2\text{H}_2 \text{ (gas)} \tag{2.8}$$

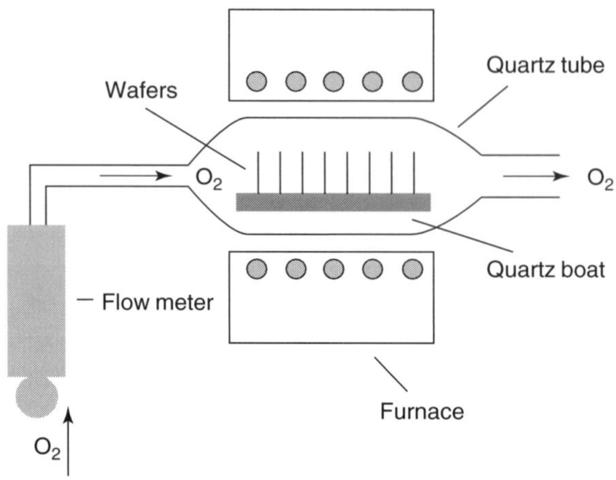

Figure 2.6 Furnace for thermal oxidation of silicon wafers

The silicon–silicon-dioxide interface transverses the silicon during the oxidation process. Using the densities and molecular weights of silicon and silicon dioxide, it can be shown that growing an oxide of thickness x consumes a layer of silicon that is $0.44x$ thick.

2.3.2 Deposition of silicon dioxide and silicon nitride

There are three deposition methods that are commonly used to form a thin film on a substrate. These methods are all based on chemical vapor deposition (CVD) and are:

- Atmospheric-pressure chemical vapor deposition (APCVD)
- Low-pressure chemical vapor deposition (LPCVD)
- Plasma-enhanced chemical vapor deposition (PECVD)

The latter method is an energy-enhanced CVD method. The appropriate method from among these three deposition methods is based on the substrate temperature, the deposition rate and film uniformity, the morphology, the electrical and mechanical properties and the chemical composition of the dielectric films.

A schematic diagram of a typical CVD system is shown in Figure 2.7 except that different gases are used at the gas inlet. Figures 2.7(a) and 2.7(b) show, respectively, a LPCVD and PECVD reactor. In Figure 2.7(a), the quartz tube is heated by a three-zone furnace, and gas is introduced at one end of the reactor (gas inlet) and pumped out at the opposite end (pump). The substrate wafers are held vertically in a slotted quartz boat. The type of LPCVD reactor shown in Figure 2.7(a) is a hot-wall LPCVD reactor where the quartz tube wall is hot because it is adjacent to the furnace, in contrast to cold-wall LPCVD, such as the horizontal epitaxial reactor, that uses radio frequency (RF) heating. Usually, reaction chamber LPCVD process parameters are in the ranges:

- pressure between 0.2 and 2.0 Torr
- gas flow between 1 to 10 $cm^3\ s^{-1}$
- temperatures between 300 and 900 °C

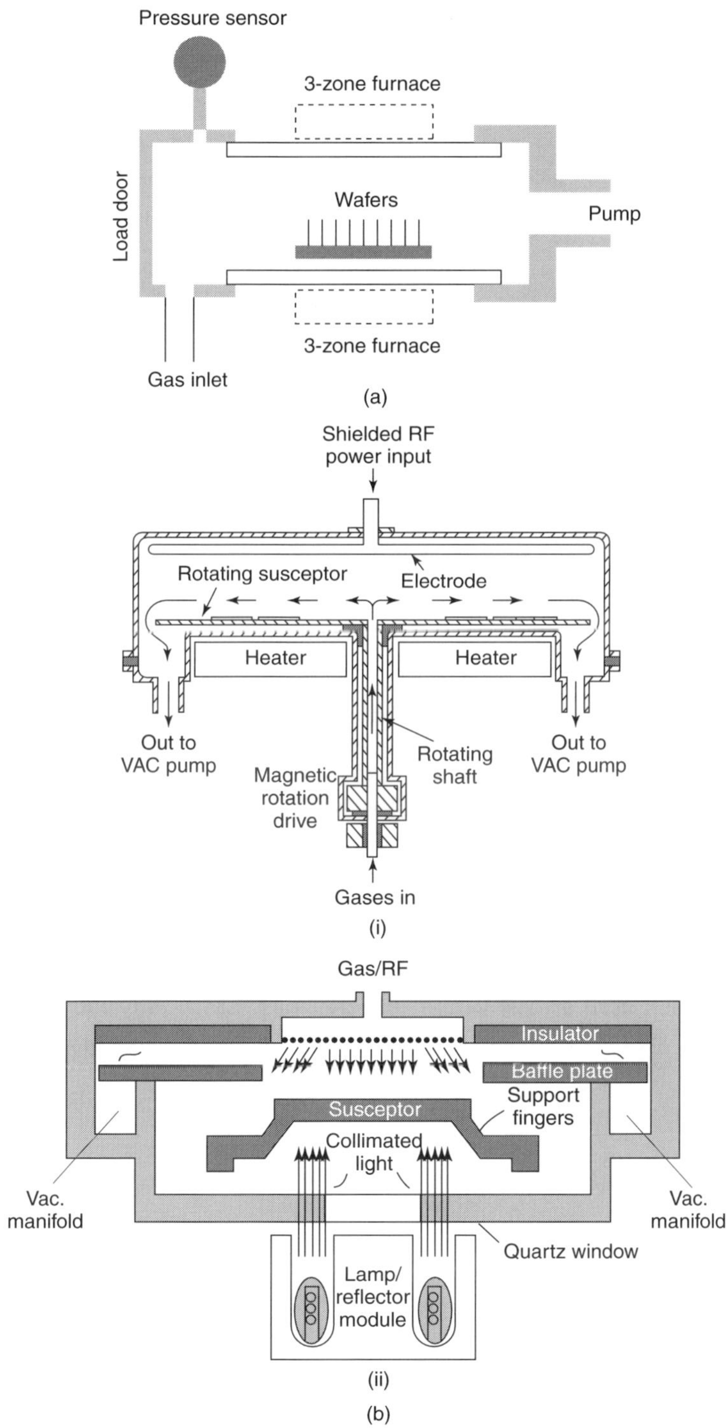

Figure 2.7 (a) Typical layout of low-pressure chemical vapor deposition reactor; (b) two plasma-enhanced chemical vapor deposition reactors

Figure 2.7(b) shows a parallel-plate, radial-flow PECVD reactor that comprises a vacuum-sealed cylindrical glass chamber. Two parallel aluminium plates are mounted in the chamber with an RF voltage applied to the upper plate while the lower plate is grounded. The RF voltage causes a plasma discharge between the plates (electrodes). Wafers are placed in the lower electrode, which is heated between 100 and 400 °C by resistance heaters. Process gas flows through the discharge from outlets located along the circumference of the lower electrode.

CVD is used extensively in depositing SiO_2, silicon nitride (Si_3N_4) and polysilicon. CVD SiO_2 does not replace thermally grown SiO_2, which has superior electrical and mechanical properties to CVD oxide. However, CVD oxides are used, instead, to complement thermal oxides and in many cases to form oxide layers that are much thicker in relatively very short times than thermal oxides. SiO_2 can be CVD deposited by several methods. It can be deposited from reacting silane and oxygen in an LPCVD reactor at 300 to 500 °C where

$$SiH_4 + O_2 \xrightarrow{500\,^{\circ}C} SiO_2 + 2H_2 \tag{2.9}$$

It can also be LPCVD deposited by decomposing tetraethylorthosilicate, $Si(OC_2H_5)_4$. The compound, abbreviated to TEOS, is vaporized from a liquid source. Alternatively, dichlorosilane can be used as follows:

$$SiCl_2H_2 + 2H_2O \xrightarrow{900\,^{\circ}C} SiO_2 + 2H_2 + 2HCl \tag{2.10}$$

Likewise, Si_3N_4 can be LPCVD deposited by an intermediate-temperature process or a low-temperature PECVD process. In the LPCVD process, which is the more common process, dichlorosilane and ammonia react according to the reaction

$$3SiCl_2H_2 + 4NH_3 \xrightarrow{\sim 800\,^{\circ}C} Si_3N_4 + 6HCl + 6H_2 \tag{2.11}$$

2.3.3 Polysilicon film deposition

Polysilicon is often used as a structural material in MEMS. Polysilicon is also used in MEMS for electrode formation and as a conductor or high-value resistor, depending on its doping level. A low-pressure reactor, such as the one shown in Figure 2.7(a), operated at a temperature of between 600 °C and 650 °C, is used to deposit polysilicon by pyrolyzing silane according to the following reaction:

$$SiH_4 \xrightarrow{600\,^{\circ}C} Si + 2H_2 \tag{2.12}$$

The most common low-pressure processes used for polysilicon deposition are the ones that operate at pressures between 0.2 and 1.0 Torr using 100% silane. Another process for polysilicon deposition involves a diluted mixture of 20% to 30% silane in nitrogen.

2.3.4 Ferroelectric thin films

New functional microsensors, microactuators and MEMS can be realized by combining ferroelectric thin films, having prominent sensing properties such as the pyroelectric effect, piezoelectric effect and electrooptic effect, with the microdevices and microstructures. There are so many ferroelectric materials including oxides and nonoxides and the selection depends on the application. Generally, ferroelectric oxides are superior to ferroelectric nonoxides for the MEMS applications considered in this book. One useful ferroelectric thin film studied for microwave components and RF MEMS is a low-loss barium strontium titanate (BST). We will concentrate on this material and preparation method in this section.

In general, BST is of interest for use in bypass capacitors and in dynamic random access memories and phase shifters for communication systems and adaptive antennas because of its high dielectric constant. The dielectric could be as high as 2500 at room temperature. For RF MEMS applications, the loss tangent of such materials should be very low. The loss tangent of BST can be reduced to 0.005 by adding a small percentage (1 to 4%) of Fe, Ni or Mn to the material mixture (Varadan and Teo, 2001a, 2001b; Varadan, selmi and Varadan, 1996).

(Ba-Sr)TiO_3 series, (Pb-Sr) TiO_3 and (Pb-Ca) TiO_3 materials and similar titanates for which the Curie temperature is in the vicinity of room temperature are well suited for MEMS phase shifter applications. The relative phase shift is obtained from the variation of the dielectric constant with dc biasing fields. For the frequency range from 400 MHz to 26 GHz, a different phase shift is obtained by a dc voltage controlled lumped BST capacitor in a coaxial line or strip line arrangement (Varadan *et al.*, 1992).

Ferroelectric thin films of BST usually have been fabricated by conventional methods, such as, RF sputtering (Won *et al.*, 1995), laser ablation (Li *et al.*, 1998), MOCVD (Levin, Leapman and Kaiser, 1998), hydrothermal methods (McCormick, Roeder and Slamovich, 2001), etc. Even though sputtering is an acceptable manufacturing process for the deposition of thin films, it has the potential for film degradation by neutral and negative-ion bombardment of the growing film. For BST, this resputtering can lead to off-stoichiometric films and degradation of electrical properties. In a recent study, Cukauskas, Kirchoefer and Pond (2000) have shown that inverted cylindrical magnetron (ICM) RF sputtering is superior for BST. This fabrication setup is discussed in the next section.

2.3.4.1 Inverted cylindrical magnetron radio frequency sputtering

Figure 2.8 illustrates the ICM sputter gun setup (Cukauskas, Kirchoefer and Pond, 2000). It consists of a water-cooled copper cathode which houses the hollow cylindrical BST target surrounded by a ring magnet concentric with the target. A stainless-steel thermal shield is mounted to shield the magnet from the thermal radiation coming from the heated table. The anode is recessed in the hollow-cathode space. It aids in collecting electrons and negative ions minimizing resputtering the growing film. Outside the deposition chamber, a copper ground wire is attached between the anode and the stainless-steel chamber. A dc bias voltage could be applied to the anode to alter the plasma characteristics in the cathode/anode space. The sputter gas enters the cathode region through the space surrounding the table.

Using the above set up, Cukauskas, Kirchoefer and Pond (2000) could deposit BST film at temperatures ranging from 550 to 8000 °C. The substrate temperature was maintained by two quartz lamps, a type-K thermocouple and a temperature controller. The films were deposited at 135 W to a film thickness of 7000 Å. The films were cooled to room temperature in 1 atm of oxygen before removing them from the deposition unit. This was

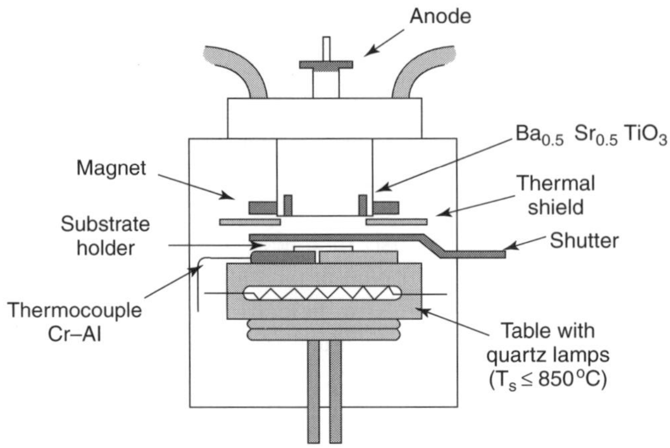

Figure 2.8 Schematic diagram of an inverted cylindrical magnetron sputter gun. Reproduced from E.J. Cukauskas, S.W. Kirchoefer and J.M. Pond, 2000, 'Low-loss $Ba_{0.5}$ $Sr_{0.5}$ TiO_3 thin films by inverted magnetron sputtering', *Journal of Applied Physics* **88**(5): 2830–2835, by permission of AVS publications, © 2000 AVS publications

then followed by annealing the films in 1 atm of flowing oxygen at a temperature of 7800 °C for 8 hours in a tube furnace.

2.3.4.2 Sol-gel processing technique

The sputtering techniques described above and other methods such as laser ablation, MOCVD and hydrothermal require much work, time, electricity and high-cost instruments, which leads to a high-cost product. The sol-gel method is one of the promising synthesis methods, which is being extensively used for the preparation of metal oxides in 'bulk', 'thin film' and 'single crystal'. The advantage of the sol-gel method is that the metal oxides can easily be doped accurately to change their stoichiometric composition because the precursors are mixing at the molecular level (Sedlar, Sayer and Weaver, 1995). Also, a large area of homogeneous film can be obtained at relatively low temperature of heat treatment. Nevertheless, the sol-gel is a technique for producing inorganic thin films without processing in vacuum. The sol-gel method was chosen because it not only offers a high purity, homogeneity of elements at the molecular level but also produces at low cost BST (Nazeri and Khan, 1995).

In the sol-gel method, the precursor solution of BSTs is prepared from barium 2-ethyl hexanoate, strontium 2-ethyl hexanoate and titanium tetraisopropoxide (TTIP). Methyl alcohol is used as a solvent along with acetyl acetonate. A known amount of barium precursor is dissolved in 30 ml of methyl alcohol. It is refluxed in a reflux condenser at a temperature of about 80 °C for 5 hr. Strontium 2-ethyl hexanoate is added to this solution and refluxed for 5 h to obtain a yellow colour solution. Acetylacetonate is added to the solution as a chelating agent, which prevents the precipitation. This solution is stirred and refluxed for another 3 h. Separately, a solution of titanium tetra isopropoxide (TTIP) is prepared in 20 ml methyl alcohol. TTIP solution is added to the barium strontium solution drop by drop, and, finally, refluxed for 4 h at 80 °C. Water is added to the BST solution drop by drop in order to initiate hydrolysis. This solution is refluxed for another 6 h with a vigorous stirring in a nitrogen atmosphere.

For thin-film deposition study, one could use a substrate such as Pt/Si. The substrate is immersed in the methanol and dried by nitrogen gas to remove the dust particles. The precursor solution is coated on the substrate by spin coating. The spin coating is done using a spinner rotated at a rate of 3100 rpm for 30 s. After coating on the substrate, films are kept on a hot plate for 15 min to dry and pyrolize the organics. This process will be repeated to produce multilayer films if needed. In such cases repeated heating after every spin coat is required in order to achieve successful burning of organics trapped in the films. It improves the crystallinity and leads to a dense sample after multiple coating. To obtain thicker films many depositions are required. The films are then annealed at 700 °C for 1 h in an air atmosphere.

2.4 MATERIALS FOR POLYMER MEMS

Recently, considerable effort has been focused on the use of polymers in microelectronic systems and MEMS. The particular features that make them attractive are: (1) mouldability, (2) conformability, (3) extreme ease in deposition in the form of thin and thick films, (4) semiconducting and even metallic behaviour in some selected polymers, (5) the wide choice of their molecular structure and the possibility of building charged particles and piezoelectric and pyroelectric effects in the side-chain.

Polymers are very large molecules (macromolecules) made up of a number of small molecules. The small molecules that connect with each other to build up the polymer are referred to as monomers, and the reaction by which they connect together is called polymerization.

The polymer processing may include photopolymerization, electrochemical polymerization, vacuum polymerization either stimulated by electron bombardment or initiated by ultraviolet irradiation, microwave-assisted polymerization, etc. These methods are also widely used for processing and curing thin and thick polymer films on silicon-based electronic components.

2.4.1 Classification of polymers

Polymers can be classified based on their structure, by the method of synthesis, physical properties and by end use as follows.

2.4.1.1 Linear, branched, and cross-linked polymers (based on structure)

A linear polymer is made up of identical units arranged in a linear sequence. This type of polymer has only two functional groups. Functional group is defined as the atom or group of atoms that defines the structure of a particular family of organic compounds and, at the same time, determines their properties. Some examples of functional groups are the double bond in alkenes, the triple bond in alkynes, the amino ($-NH_2$) group, the carboxyl ($-COOH$) group, the hydroxyl ($-OH$) group, etc. Functionality can be defined as the number of such functional groups per molecule of the compound. Branched polymers are those in which there are many side-chains of lined monomers attached to the main polymer chain at various points. These side-chains could be either short or long (Figure 2.9). When polymer molecules are linked with each other at points other than their ends to form a

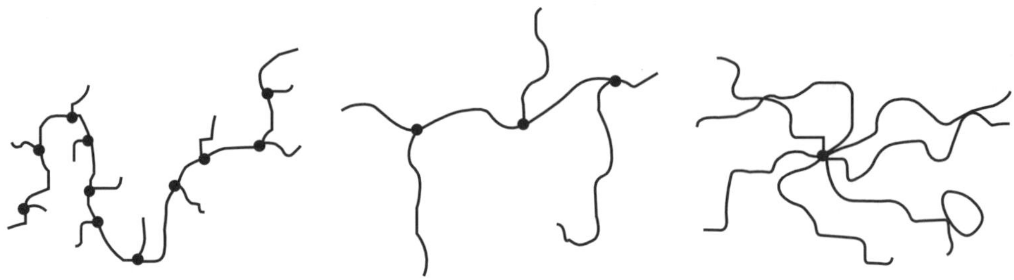

Figure 2.9 Various kinds of branching in polymers

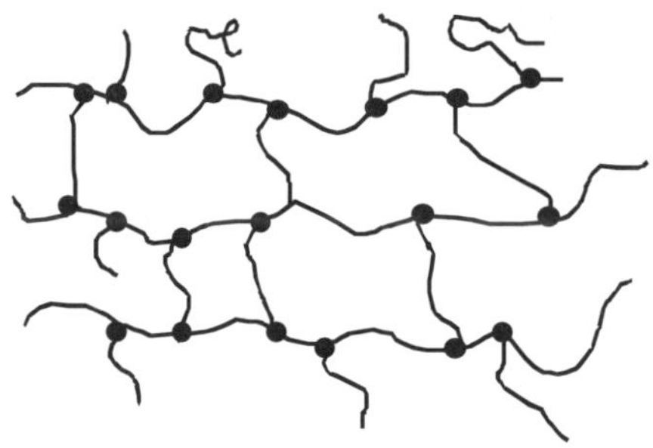

Figure 2.10 Cross-linking in polymers

network, the polymers are said to be cross-linked (Figure 2.10). Cross-linked polymers are insoluble in all solvents even at elevated temperatures.

2.4.1.2 Thermoplastic and thermosetting polymers (based on physical properties)

A polymer is said to be a thermoplastic if it softens (flows) when it is squeezed, or pulled, by a load, usually at a high temperature, and hardens on cooling. This process of reshaping and cooling can be repeated several times. Polyethylene [high-density polyethylene (HDPE), low-density polyethylene (LDPE), etc.], polyvinyl chloride (PVC), nylon, etc. are some of the examples of thermoplastic polymers.

Thermosets, on the other hand, are very hard, can flow and can be moulded when initially made, but then become set in their shape, usually through the action of heat and often pressure. This process of becoming an infusible and insoluble mass by the application of heat and pressure is called 'curing'. Reheating such a thermosetting polymer just results in the degradation of the polymer and will distort the article that has been made. Epoxy, phenol formaldehyde, etc. are some examples of thermosetting polymers.

2.4.1.3 Plastics, elastomers, fibers and liquid resins (based on end use)

Depending upon its final usage, a polymer can be classified as plastic, elastomer, fibre or liquid resin. When a polymer is formed into hard and tough articles by the application of heat and pressure then it is used as a plastic. When a polymer is vulcanized into rubbery materials, which show good strength and elongations, then it is used as an elastomer.

When the polymer is drawn into long filament-like material, whose length is at least 100 times its diameter, then it is used as a fibre. When the polymer is used in the liquid form, such as in sealants, adhesives, etc., they are said to be used as liquid resins.

2.4.1.4 Chain and step-growth polymerization (based on the method of synthesis)

There are basically two methods by which polymers can be synthesized, namely 'addition' or 'chain' polymerization and 'condensation' or 'step-growth' polymerization. When molecules just add on to form the polymer, the process is called 'addition' or 'chain' polymerization. The monomer in this case retains its structural identity when it is converted into the polymer (i.e. the chemical repeat unit in the polymer is the same as the monomer). When, however, molecules do not just add, but react with each other with the elimination of small molecules such as water, methane, etc., then the process is called step-growth polymerization. In this case, the chemical repeat unit is different when compared with that of the monomer.

2.4.1.5 Chain or addition polymerization

In this type of polymerization, a low molecular weight molecule (monomer) having a double bond is induced to break the double bond so that the resulting free valences will be able to bond to other similar molecules to form the polymer. Compounds containing a reactive double bond usually undergo chain polymerization. This polymerization takes place in three steps, namely, initiation, propagation and termination. This can be induced by a free radical, or ionic, or coordination mechanism. Depending on the mechanism, there are three types of chain polymerization, namely, free radical, ionic (cationic and anionic) and coordination polymerization.

Free radical polymerization

There are three steps in this type of polymerization:

1. initiation
2. propagation
3. termination

Initiation

In this type of polymerization, the initiation is brought about by the free radicals produced by the decomposition of initiators. Initiators function by breaking down to form free radicals. Each component has an unpaired (lone) electron with it and these species are called free radicals. This radical adds to a molecule of the monomer, and in doing so generates another free radical. This radical adds to another molecule of the monomer to generate a still larger radical, which in turn adds to another molecule of monomer, and so on.

$$\mathrm{PI} + h\nu \longrightarrow \mathrm{R_0}^{\bullet} \tag{2.13}$$

PI represents photoinitiator, $\mathrm{R_0}^{\bullet}$ is the reactive intermediate from the UV cleavage of PI.

The decomposition of the initiator to form these free radicals can be induced by heat, light energy or catalysts. Peroxides, many azo compounds, hydroperoxides, peracids, etc. are the most commonly used initiators. Initiators can also be decomposed by UV light. The rate of decomposition depends mainly on the intensity and wavelength of radiation and not so much on the temperature. Polymerization reactions initiated by UV light fall under the category of photoinitiated polymerization.

UV curing is based on photoinitiated polymerization that is mediated by photoinitiators. Photoinitiators are required to absorb light in the UV–visible spectral range, generally 250–550 nm, and convert this light energy into chemical energy in the form of reactive intermediates, such as free radicals and reactive cations, which subsequently initiate polymerization.

Propagation

In this step, the radical site on the first monomer unit reacts with the double bond of a fresh monomer molecule, which results in the linking up of the second monomer unit to the first and the transfer of the free radical onto the second monomer molecule. This process involving the attack on a fresh monomer molecule, which in turn keeps adding to the growing chain one after another, is called propagation. The chain keeps propagating until no monomer is present. This step can also end when the free radical site is killed by some impurities or by the termination process.

The propagation step can be represented schematically as

$$M_1^\bullet + M \longrightarrow M_2^\bullet \tag{2.14}$$

M represents the monomer molecule, $M_1^\bullet, \ldots, M_n^\bullet$ represents reactive molecules.

Termination

In this step, any further addition of the monomer units to the growing chain is stopped, and the growth of the polymer chain is inhibited. The decomposition of the initiator results in the formation of a large number of free radicals. Depending on factors such as temperature, time and monomer and initiator concentration, there exists a chance when the growing chains collide against each other. When this happens, two things can happen

- Termination by combination: here the chain terminates by the simple formation of a bond between two radicals.
- Termination by disproportionation: here, a proton is transferred and a double bond is formed.

These reactions can be represented schematically as follows:

$$M_x^\bullet + M_y^\bullet \longrightarrow M_{x+y} \text{ (combination)} \tag{2.15}$$

$$M_x^\bullet + M_y^\bullet \longrightarrow M_x + M_y \text{ (disproportionation)} \tag{2.16}$$

M_{x+y} is the stable polymer molecule with $x + y$ monomer units. M_x and M_y are also stable polymer molecules with x and y monomer units, respectively.

Table 2.6 Examples of monomers polymerized using free radical polymerization

Monomer	Formula
Ethylene	$CH_2{=}CH_2$
Butadiene	$CH_2{=}CH{-}CH{=}CH_2$
Styrene	$CH_2{=}CH{-}C_6H_5$
Vinyl chloride	$CH_2{=}CH{-}Cl$
Vinylidene chloride	$CH_2{-}CCl_2$
Acrylic acid	$CH_2{=}CH{-}COOH$
Methylmethacrylate	$CH_2{-}C\ (CH_3)\ COOCH_3$

Some of the common monomers that are polymerized using free radical polymerization are listed in Table 2.6.

Ionic polymerization

Ionic polymerization involves the attack on the π electron pair of the monomer. This attack is not by free radicals, but it is by either a positive or a negative ion.

Cationic polymerization

If the active site has a positive charge (i.e. a carbonium ion) then the process is called cationic polymerization. Monomers, which have R as an electron-donating group are the most suitable for cationic polymerization, for example, alkyl vinyl ethers, vinyl acetals, isobutylene, etc.

Initiation can be obtained by using protonic acids and Lewis acids. Lewis acids usually require a co-catalyst such as water or methyl alcohol. Here, a proton is introduced into the monomer. This proton pulls the π electron pair towards it and this is how the positive charge moves to the other end of the monomer resulting in the formation of a carbonium ion:

$$C + XH \rightleftharpoons \overset{\oplus\ominus}{HXC} \text{ (ion-pair formation)} \tag{2.17}$$

$$\overset{\oplus\ominus}{HXC} + M \longrightarrow \overset{\oplus\ominus}{HMXC} \text{ (initiation)} \tag{2.18}$$

$$\overset{\oplus\ominus}{HXC} + M \longrightarrow \overset{\oplus\ominus}{HMMXC} \text{ (propagation)} \tag{2.19}$$

$$HM_n\overset{\oplus\ominus}{MXC} + M \longrightarrow HM_nM + \overset{\oplus\ominus}{HXC} \text{ (termination)} \tag{2.20}$$

$$HM_n\overset{\oplus\ominus}{MXC} \longrightarrow HM_nM + \overset{\oplus\ominus}{MXC} \text{ (chain transfer to monomer)} \tag{2.21}$$

Here, C is the catalyst, XC is the co-catalyst and M is the monomer.

Propagation is when the carbonium ion attacks the π electron pair of the second monomer molecule. The positive charge is then transferred to the farther end of the second monomer, and thus a chain reaction is started.

Termination can occur by anion–cation recombination where an ester group forms. Termination can also occur by the splitting of the anion. It can also occur by the reaction with trace amounts of water.

Anionic polymerization

If the active site has a negative charge (i.e. a carbanion), then the process is called anionic polymerization. Monomers capable of undergoing anionic polymerization are isoprene, styrene, butadiene, etc.

Initiation takes place the same way as in cationic polymerization, except that here a carbanion is formed. The general initiators used here are the alkyl and aryl derivatives of alkali metals such as triphenyl methyl potassium and ethyl sodium. Propagation then proceeds with the transfer of the negative charge to the end of the monomer molecule. Termination is not always a spontaneous process, and unless some impurities are present or some strongly ionic substances are added, termination does not occur. So, if an inert solvent is used and if the impurities are avoided, the reaction proceeds till all the monomer is consumed. Once all the monomer is consumed, the carbanions at the chain end still remain active and it is considered as 'living' and polymers synthesized using this method are called 'living polymers'. This technique is useful for producing block polymers.

$$\mathrm{IA} \longrightarrow \overset{\oplus\ominus}{\mathrm{IA}} \text{ (ion-pair formation)} \tag{2.22}$$

$$\overset{\oplus\ominus}{\mathrm{AI}} + \mathrm{M} \longrightarrow \mathrm{A}\overset{\oplus\ominus}{\mathrm{MI}} \text{ (initiation)} \tag{2.23}$$

$$\mathrm{A}\overset{\oplus\ominus}{\mathrm{MI}} + \mathrm{M} \longrightarrow \mathrm{AM}\overset{\oplus\ominus}{\mathrm{MI}} \text{ (propagation)} \tag{2.24}$$

$$\mathrm{AM}_n\overset{\oplus\ominus}{\mathrm{MI}} + \mathrm{HA} \longrightarrow \mathrm{AM}_n\mathrm{MH} + \overset{\oplus\ominus}{\mathrm{AI}} \text{ (termination)} \tag{2.25}$$

where IA is the initiator and HA is a protonating agent.

2.4.1.6 Step-growth polymerization

Step polymerizations are carried out by the stepwise reaction between the functional groups of monomers. The reaction takes place in a stepwise manner. In such polymerizations, the size of the polymer chains increases at a relatively slow rate from monomer to dimer, trimer, tetramer, pentamer and so on.

$$\text{Monomer} + \text{Monomer} \longrightarrow \text{Dimer} \tag{2.26}$$

$$\text{Dimer} + \text{Monomer} \longrightarrow \text{Trimer} \tag{2.27}$$

$$\text{Dimer} + \text{Dimer} \longrightarrow \text{Tetramer} \tag{2.28}$$

$$\text{Trimer} + \text{Dimer} \longrightarrow \text{Pentamer} \tag{2.29}$$

$$\text{Trimer} + \text{Trimer} \longrightarrow \text{Hexamer} \tag{2.30}$$

Any two molecular species can react with each other throughout the course of the polymerization until eventually large polymer molecules consisting of large numbers of monomer molecules have been formed. These reactions take place when monomers containing more than two reactive functional groups react.

Typical condensation polymers include polyamides, polyesters, polyurethane, polycarbonate, polysulfide, phenol-formaldehyde, urea-formaldehyde, melamine formaldehyde, etc.

When a pair of bifunctional monomers (dicarboxylic acid/diamine or dialcohol/dihalide) undergoes polycondensation, it is called an AA–BB type polycondensation:

$$nA\text{–}A + nB\text{–}B \longrightarrow A\text{+}AB\text{+}_{2n-1}B + \text{byproduct} \tag{2.31}$$

When a single bifunctional monomer undergoes self-condensation, it is known as an A–B type polycondensation.

$$nA\text{–}B \longrightarrow B\text{+}AB\text{+}_{n-1}A + \text{byproduct} \tag{2.32}$$

If in the AA–BB type of polycondensation one of the monomers has a functionality of 3 or more, it forms a 3D network. Figure 2.11 shows the formation of a network in polymers with a functionality of 3 or higher. Table 2.7 shows some of the examples of functionality in compounds.

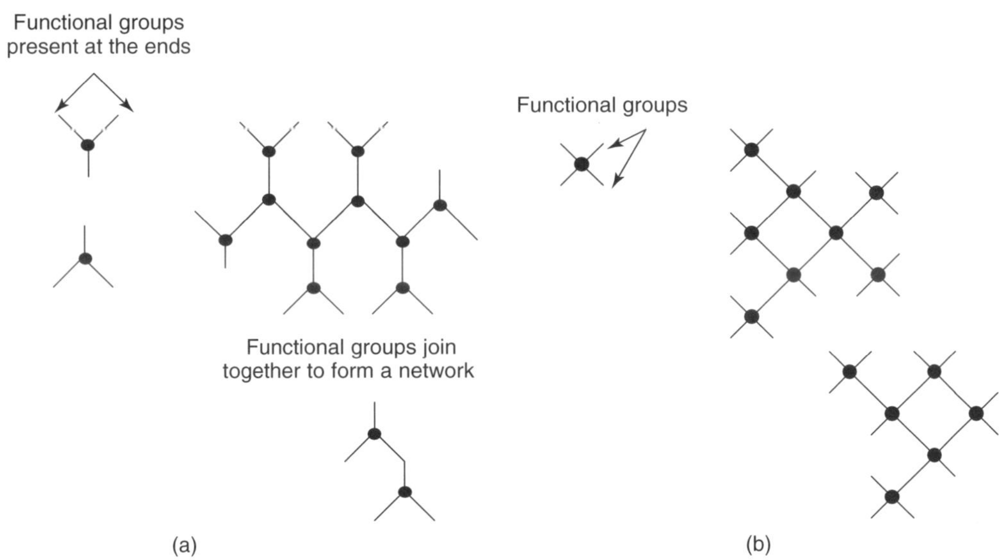

Figure 2.11 Schematic diagram showing the formation of networks in polymers with a functionality greater than 2: (a) trifunctionality; (b) tetrafunctionality

Table 2.7 Functionality of some compounds

Compound	Chemical formula	Functional group	Number of functional groups	Functionality
Ethyl alcohol	CH_3CH_2OH	$-OH$	1	Monofunctional
Hexamethylene diamine	$H_2NCH_2\ (CH_2)_4CH_2NH_2$	$-NH_2$	2	Bifunctional
Maleic acid	$HOOCCH_2CH(OH)COOH$	$-COOH$, $-OH$	3	Trifunctional
Gallic acid	$HOOCC_6H_2\ (OH)_3$	$-COOH$, $-OH$	4	Tetrafunctional

Table 2.8 Some of the polymers that can be prepared using step-growth polymerization

Polymer	Chemical formula
Nylon 6	$\left[NH-\overset{O}{\overset{\|\|}{C}}-(CH_2)_5 \right]_n$
Polycarbonate	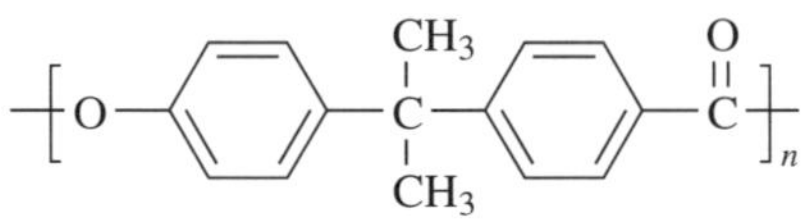
Polybutylene terephthalate	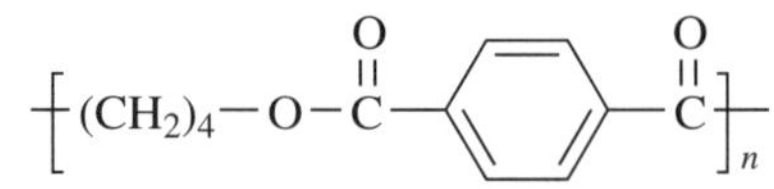

Some of the common monomers that are polymerized using step-growth polymerization are listed in Table 2.8.

2.4.2 UV radiation curing

Radiation curing refers to radiation as an energy source to induce the rapid conversion of specially formulated 100% reactive liquids into solids by the polymerizing and cross-linking of functional monomers and oligomers (usually liquid) into a cross-linked polymer network (usually solid; Fouassier, 1995).

Advantages of using radiation curing are as follows.

- Radiation curing has the advantage of high processing speed and hence high productivity.
- This is very convenient, economical and, since most of them are 'one-pack compositions', they can be dispensed automatically.
- There is very low heat generation. So, heat-sensitive substrates can be used.
- Lower energy and less space are needed than with conventional curing systems.
- Since the organic emission levels are very low, this is very eco-friendly.
- Capital costs (UV) are low.

The radiation energy could be from electron beams, X-rays, γ-rays, plasmas, microwaves and, of course, mostly used is UV light. UV radiation curing has also been extensively applied in MEMS, generating photoresist patterning for subsequent etching, building flexible polymer structures in both planar and 3D fashion (UV LIGA, microstereolithography, etc.). Therefore, it is necessary to take a look into the fundamentals of UV radiation curing before we introduce the novel 3D MEMS fabrication processes.

2.4.2.1 Relationship between wavelength and radiation energy

Planck developed his theory of black-body radiation on the basis of a postulate that radiation possessed particulate properties and that the particles, or photons, of radiation of specific frequency ν had associated with them a fixed energy ε given by the relation $\varepsilon = h\nu$, where h is called Planck's constant (6.626×10^{-34} J s; 9.534×10^{-14} Kcal s mol^{-1}), $\nu = c/\lambda$, where c is the speed of light (3×10^{17} nm s^{-1}), and λ is the wavelength in nanometers. Figure 2.12 shows the electromagnetic spectrum.

Typical average energies from homolytic cleavage of selected chemical bonds in organic molecules are shown in Table 2.9 (Kagan, 1993). The photons at wavelengths within UV range possess enough energy to break the bonds listed in table, and the bonds undergo rearrangements forming polymer networks (Haertling, 1989).

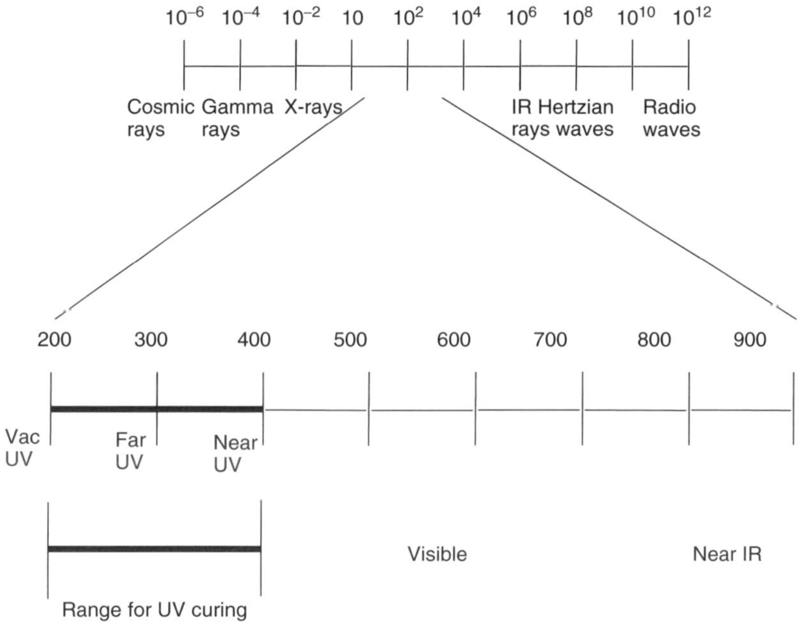

Figure 2.12 The electromagnetic spectrum; wavelength units are in nanometers. Reproduced from J. Kagan, 1993, *Organic Photochemistry: Principles and Applications*, Academic Press, London, by permission of Academic Press, Elsevier

Table 2.9 Energy and corresponding wavelength for homolytic fission of typical chemical bonds

Bond	Energy (kcal mol^{-1})	λ (nm)
C=C	160	179
C–C	85	336
C–H	95–100	286–301
C–O	80–100	286–357
C–Cl	60–86	332–477
C–Br	45–70	408–636
O–O	35	817
O–H	85–115	249–336

Source: Kagan, 1993.

2.4.2.2 Mechanisms of UV curing

UV curing is based on photoinitiated polymerization that is mediated by photoinitiators, which absorb UV light and convert the light energy into chemical energy in the form of reactive intermediates, such as free radicals and reactive cations, which subsequently initiate the polymerization. Typical photopolymer formulations contain a photoinitiator system, monomers and oligomers, a polymer or polymers to provide specific physical and/or processing properties and a variety of additives to modify the physical properties of the light-sensitive composition or the final properties of the cured photopolymer.

The photopolymerization reactions can then fall into two categories: radical photopolymerization and cationic photopolymerization. Both processes take place in three steps: photoinitiation, propagation and termination.

Generally, acrylates are associated with free radical polymerization whereas epoxies are typical of cationic curing. The reactive monomeric materials that are most commonly used are low molecular weight unsaturated acrylate or methacrylate monomers that can be made to cross-link with the use of a radical-generating photoinitiator. Cationic initiated cross-linking of monomeric materials with epoxy and/or vinyl ether functionality has increased in practicality with the development of new higher efficiency photoinitiators that generate cationic species (e.g. strong acids) upon UV exposure. Table 2.10 gives a comparison of cationic and free radical curing characteristics, showing each of them has certain advantages over the other. For instance, the speed of free radical curing is faster than that of cationic curing. However, through-cure of cationic systems is greater since free radicals have a limited lifetime. Moisture inhibition refers to the ability of a formulation to cure with atmospheric moisture presence, and post-irradiance cure refers to curing that takes place after the light source has been removed. For free radical curing in air, surface curing lags behind bulk curing, which is known as oxygen inhibition. The lag results from competition at the surface between oxygen molecules and free radicals for monomer sites. Also, an oxygen atmosphere increases chain terminations occurring at the surface.

Once photoinitiator (PI) absorbs light and is raised to an electronically excited state PI*, the lifetime of PI* is short, generally less than 10^{-6} s. During this time, PI* is partitioned among several processes including (1) decay back to PI (with emission of light and/or heat), (2) excited-state quenching by oxygen, monomer and other quenching

Table 2.10 A comparison of free radical curing with cationic curing

Property	Free radical curing	Cationic curing
Cure speed	Faster	Slower
Oxygen inhibition	Yes	No
Adhesion	Problematic	Excellent
Toxicity	Skin irritation	Acceptable
Moisture inhibition	No	Yes
Post-irradiation cure	No	Yes
Formulation latitude	Good	Limited
Through cure	Fair	Good
Viscosity	Higher	Lower
Cost	Moderate	Higher

Source: Haertling, 1989.

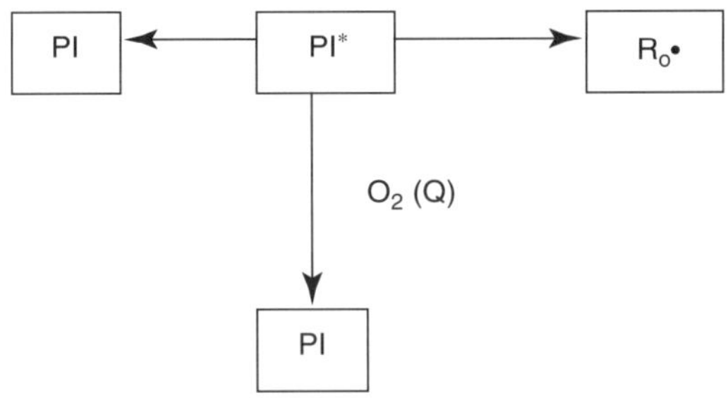

Figure 2.13 Conversion of excited photoinitiator (PI*)

agent (Q), (3) a chemical reaction yielding the initiator species, R_0 (shown in Figure 2.13; Pappas, 1992).

The rate of initiation (R_i) is expressed as the rate of formation of PI*, which corresponds to the number of photons absorbed by the PI per unit time.

$$R_i = I_{abs}Ff \tag{2.33}$$

where the term I_{abs} corresponds to the intensity of light absorbed by the PI. F is the fraction of PI* that yields initiator species, f is the fraction of initiator initiates polymerization. I_{abs} is related to the incident light intensity (I_o), the number of photons incident to the system per unit time and area, and the absorbance (A) of the PI:

$$I_{abs} = I_o(1 - 10^{-A}) \tag{2.34}$$

$$A = \varepsilon dc \tag{2.35}$$

where d is the path length of light (or film thickness), ε is the molar absorptivity of the PI and c is the PI concentration.

Generally, it is desirable that the rate of initiation R_i be high for efficient utilization of light energy and also be uniform throughout the system. For example, internal stresses may arise from nonuniform cross-linking resulting in adverse effects on adhesion to a substrate, as well as on mechanical properties, such as tensile strength.

From the above equations, one can see that the rate of initiation (R_i) increases proportionally with incident light intensity (I_o), and that as PI concentration increases the proportion of the incident light absorbed decreases exponentially per unit thickness from the initially exposed surface. The nonuniformity of the absorption increases with the absorbance (A). Therefore, the appropriate PI concentration, molar absorptivity of PI, and further the value of absorbance of the system are very important to optimize a monomer system for UV curing (Pappas, 1992).

2.4.2.3 Basic kinetics of photopolymerization

The rate of polymerization is an important parameter in characterizing the polymer curing; one can predict the curing profile by calculating the polymerization rate. The kinetics of photopolymerization presented below is helpful to understand how to calculate the rate of polymerization.

Radical photopolymerization is a chain reaction proceeding as:

$$PI + h\nu \xrightarrow{k_i} R^{\bullet} \text{ (photoinitiation)} \tag{2.36}$$

$$R^{\bullet} + M \xrightarrow{k_i} RM_1{}^{\bullet} \text{ (photoinitiation)} \tag{2.37}$$

$$RM_1{}^{\bullet} + M \xrightarrow{k_i} RM_2{}^{\bullet} \text{ (propagation)} \tag{2.38}$$

$$\vdots$$

$$RM_{n-1}{}^{\bullet} + M \xrightarrow{k_p} RM_n{}^{\bullet} \text{ (propagation)} \tag{2.39}$$

$$RM_n\bullet + RM_m\bullet \xrightarrow{k_t} RM_{m+n} \text{ (termination)} \tag{2.40}$$

where PI represents the photoinitiator, RM_{m+n} is the stable polymer molecule and k_i, k_p and k_t are the rate constants for initiation, propagation and termination, respectively.

The rate of photochemical initiation is expressed as

$$R_i = 2\Phi I_{abs} \tag{2.41}$$

where I_{abs} is the intensity of absorbed light in moles of light quanta per liter- second and Φ, referred to as the quantum yield for initiation, is the number of propagating chains initiated per light photon absorbed. The factor of 2 indicates that two radicals are produced per molecule undergoing photolysis. The factor of 2 should not be used for the photoinitiating systems where only one radical is generated per molecule. The maximum value of Φ is 1 for all photoinitiating polymerizations.

Monomers are consumed by the initiation reaction as well as by the propagation reactions. The rate of polymerization is obtained as

$$-\frac{d[M]}{dt} = R_i + R_p \tag{2.42}$$

where R_i and R_p are the rates of initiation and propagation, respectively. For a process producing high-molecular-weight polymers, the number of monomers reacting in the initiation step is far less than that in the propagation step. Thus Equation (2.42) can be simplified as

$$-\frac{d[M]}{dt} = R_p \tag{2.43}$$

Assume the rate constants for all the propagation steps are the same; the polymerization rate can be expressed by

$$R_p = k_p[M][M^{\bullet}] \tag{2.44}$$

where [M] is the monomers concentration and $[M^{\bullet}]$ is the total concentration of all chain radicals.

The polymerization rate cannot be directly obtained from Equation (2.44) since it is difficult to measure the radical concentrations quantitatively, which are very low ($\sim 10^{-8}$ M).

In order to eliminate [M•] from Equation (2.44), the steady-state assumption is made that the concentration of radicals increases initially, but reaches a constant, steady-state value within a very short time. This means that the rates of initiation R_i and termination R_t of radicals are equal, or

$$R_i = R_t = 2k_t[M^\bullet]^2 \tag{2.45}$$

The factor of 2 in the equation represents that the radicals are destroyed in pairs. Rearranging Equation (2.45), the concentration of radicals is

$$[M^\bullet] = \left(\frac{R_i}{2k_t}\right)^{1/2} \tag{2.46}$$

Substituting Equation (2.46) into equation (2.44), one gets

$$R_p = k_p[M]\left(\frac{R_i}{2k_t}\right)^{1/2} \tag{2.47}$$

Combination of Equation (2.41) and Equation (2.47) yields

$$R_p = k_p[M]\left(\frac{\Phi I_{abs}}{k_t}\right)^{1/2} \tag{2.48}$$

The absorbed light intensity can be expressed by

$$I_{abs} = I_0\left(1 - 10^{-\varepsilon c b}\right) \tag{2.49}$$

where I_0 is the incident light intensity, c is the species PI concentration that undergoes photoexcitation, ε is the molar absorptivity of the photoinitiator and b is the thickness of reaction system being irradiated. Thus the expression for R_p is

$$R_p = k_p[M]\left[\frac{\Phi I_0(1 - 10^{-\varepsilon c b})}{k_t}\right]^{1/2} \tag{2.50}$$

The process of cationic photopolymerization can be generalized as:

$$\text{PI} + h\nu \xrightarrow{k_i} \text{H}^+\text{X}^- \text{ (photoinitiation)} \tag{2.51}$$

$$\text{H}^+\text{X}^- + \text{M} \xrightarrow{k_i} \text{HM}_1{}^+\text{X}^- \text{ (photoinitiation)} \tag{2.52}$$

$$\text{HM}_1{}^+\text{X}^- + \text{M} \xrightarrow{k_p} \text{HM}_2{}^+\text{X}^- \text{ (propagation)} \tag{2.53}$$

$$\vdots$$

$$\text{HM}_{n-1}{}^+\text{X}^- + \text{M} \xrightarrow{k_p} \text{HM}_n{}^+\text{X}^- \text{ (propagation)} \tag{2.54}$$

$$\text{HM}_n{}^+\text{X}^- \xrightarrow{k_t} \text{HM}_n\text{X} \text{ (termination)} \tag{2.55}$$

The reaction rates for initiation, propagation and termination are expressed as, respectively:

$$R_\mathrm{i} = \Phi I_\mathrm{abs} \tag{2.56}$$

$$R_\mathrm{p} = k_\mathrm{p}[\mathrm{HM^+X^-}]\,[\mathrm{M}] \tag{2.57}$$

$$R_\mathrm{t} = k_\mathrm{t}[\mathrm{HM^+X^-}] \tag{2.58}$$

where $[\mathrm{HM} + \mathrm{X}^-]$ is the total concentration of total reactive centers. Supposing the steady-state assumption is also valid for cationic photopolymerization, one can get

$$[\mathrm{HM^+X^-}] = \frac{\Phi I_\mathrm{abs}}{k_\mathrm{t}} \tag{2.59}$$

Combination of Equations (2.57) and (2.59) yields

$$R_\mathrm{p} = \frac{k_\mathrm{p}\Phi I_\mathrm{abs}[\mathrm{M}]}{k_\mathrm{t}} \tag{2.60}$$

This is the rate of polymerization for cationic photopolymerization. R_p can also be expressed in terms of I_0, which is

$$R_\mathrm{p} = k_\mathrm{p}[\mathrm{M}]\frac{\Phi I_0(1 - 10^{-\varepsilon cb})}{k_\mathrm{t}} \tag{2.61}$$

2.4.3 SU-8 for polymer MEMS

SU-8 (first patented by IBM in 1989; Lee *et al.*, 1981) is a negative, epoxy-type, near-UV photoresist, which was specifically developed for applications requiring high aspect ratios in very thick layers (Despont *et al.*, 1998). Film with a thickness up to 2 mm and an aspect ratio larger than 20 has already been demonstrated with the standard contact lithography equipment.

Since SU-8 is lithographically patternable and quite stable after exposure to UV light, it has been widely used in the MEMS field for fabricating mechanical structures such as gears, coils, cantilevers and trenches (Dellman *et al.*, 1997; Despont *et al.*, 1998; Ding, Kuribayashi and Hashida, 1999). Recently it has been used to conceive polymer-based microinductors (Chomnawang and Lee, 2001), MOSFET-based hydrophones for underwater applications (Zhu and Varadan, 2002) and accelerometers (Zhu, Mehta and Varadan, 2002). In addition, owing to its relatively low-cost and low-temperature process, SU-8 is also used to form membranes with controllable thickness, which enables printed circuits to be lithographically fabricated on it (Liu, Steenson and Steer, 2001). SU-8 is also combined with microstereolithography technique to realize true 3D polymer structures (Bertsch, Lorenz and Renaud, 1998). Microstereolithography (MSL) is a process that makes it possible to build 3D complex in shape polymer structures (Varadan, Xiang and Varadan, 2001). MSL has been employed to fabricate polymer bridges for phase shifters, which are discussed in Chapter 6.

The SU-8 is unique since it offers several major advantages over commonly used photoresistors. Excluding the capability of being fabricated into ultra-thick structures, the

SU-8, as it is an epoxy-based resin, offers superior adhesion to most surfaces (Lorenz *et al.*, 1997). This property is very important when SU-8 is used as a structural or dielectric material.

2.4.3.1 Processing mechanism of SU-8

The SU-8 photoresist is prepared by dissolving an EPON resin SU-8 in an organic solvent GBL (gamma-butyloracton). The quantity of the solvent determines the viscosity and hence the range of the resist thickness. To induce the cross-linking of SU-8 under the exposure of UV light, a photoinitiator is added (10% of the EPON SU-8 weight) and mixed with the resin.

The EPON resin SU-8 is a multifunctional, highly branched polymeric epoxy resin, which consists of a bisphenol A novolac glycidyl ether. A typical molecular structure is shown in Figure 2.14 (www.microchem.com/su8.cfm/). On average a single molecule contains eight epoxy groups, from which comes the '8' in SU-8.

The photoinitiator, which consists of a triarylsulfonium salt, undergoes a photochemical transformation upon absorption of a photon and generates a photoacid. The reaction is described in Scheme 2.1 (www.microchem.com/su8.cfm/), where the photoacid is designated as H^+A^-.

The photoacid is produced only in the regions of the photoresist film that are directly exposed to light and acts as a curing agent in the subsequent cross-linking reaction that occurs during the postexposure bake (PEB). The temperature of the PEB is required to exceed the glass transition temperature (T_g) of the solid film, which is about 55 °C, since below T_g, the molecular motion is effectively frozen and only very little reaction can take place.

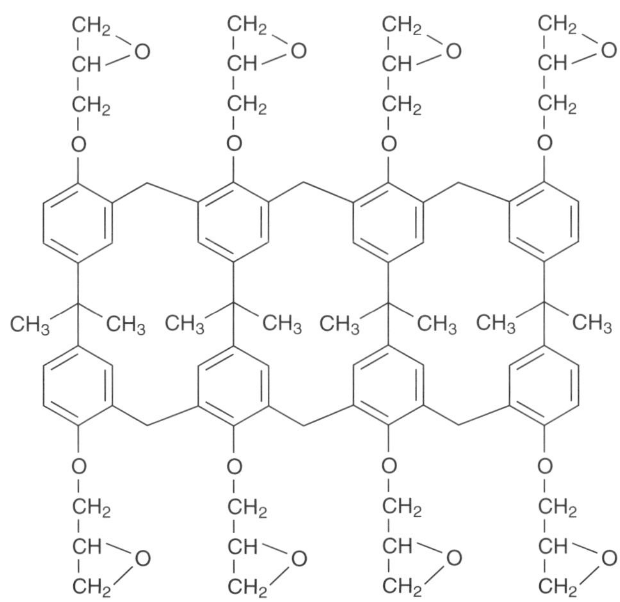

Figure 2.14 Typical molecular structure of SU-8 resin. Reproduced with permission from www.microchem.com/su8.cfm/

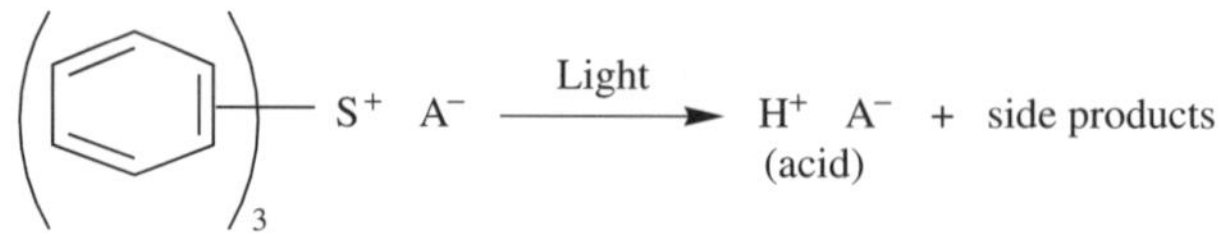

Scheme 2.1 Photochemical reaction of the photoinitiator. Source: www.microchem.com/su8.cfm/

During the PEB, the generated photoacid initiates the ring opening of the epoxy groups. This triggers the cross-linking reaction in a mechanism similar to that of cationic polymerization. Extensive cross-linking will yield a dense network that is insoluble in the organic developer, which is pure propylene glycol methyl ether acetate (PGMEA). However, the unexposed, uncross-linked resist dissolves in the developer, thus forming a negative image of the mask. The cross-linking reaction is illustrated in Scheme 2.2 (www.microchem.com/su8.cfm/).

2.4.3.2 Properties of SU-8

Several features of SU-8 make it attractive as a structural and dielectric material. First, it can be spun in thicknesses from 2 μm to over 1000 μm with a single coat on conventional spin-coating equipment. Second, it has exceptional optical clarity that allows maintaining the fine definition through the thickness and favours the pattern alignment in the following lithography steps. Third, highly cross-linked structure results in chemical resistance and high thermal characteristics with a processing temperature greater than 250 °C. Finally, it manifests excellent mechanical properties (Harriss *et al.*, 2000).

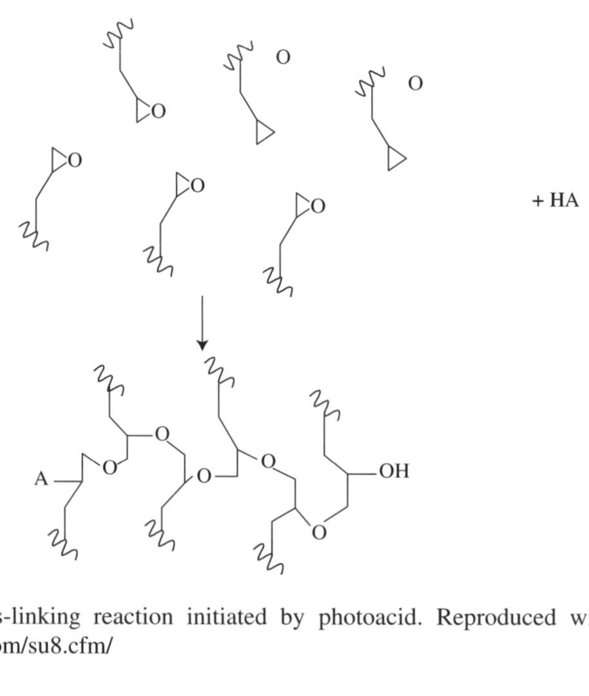

Scheme 2.2 Cross-linking reaction initiated by photoacid. Reproduced with permission from www.microchem.com/su8.cfm/

Table 2.11 Major properties of SU-8

Characteristic	Value	Remark	Reference
Modulus of elasticity, E (GPa)	4.95 ± 0.42	Hardbaked at 200 °C, beam deflexion test	Dellmann *et al.*, 1997
Poisson's coefficient	0.22	Postbaked at 95 °C	www.somisys.ch/
Glass temperature, T_g (°C)	≈55	Unexposed film (not cross-linked)	LaBianca and Delorme, 1995
Glass temperature, T_g (°C)	>200	Fully cross-linked film (exposed and post-hardbaked)	LaBianca and Delorme, 1995
Degradation temperature, T_d (°C)	≈380	Fully cross-linked film (exposed and post-hardbaked)	LaBianca and Delorme, 1995
Coefficient of thermal expansion (ppm °C)$^{-1}$	52.0 ±5.1	Postbaked at 95 °C, thermal cycling test on Si wafer	Despont *et al.*, 1998
Thermal conductivity ($Wm^{-1}K^{-1}$)	0.2	A general value for polymer, not for SU-8	Guerin *et al.*, 1997
Polymer shrinkage	0.075	Postbaked at 95 °C	Guerin *et al.*, 1997
Relative Permittivity, ε, at 10 MHz	3–4	Postbaked at 100 °C	Thrope, Steenson and Miles, 1998

Table 2.11 lists some mechanical, physical and electrical properties of the SU-8 photoresist.

2.4.3.3 Processing of SU-8

The SU-8 photoresist was commercially supplied by MicroChem Corp., Newton, Massachusetts. There are six standard grades of SU-8 photoresist, which are determined by the percent of solid SU-8 resin with respect to the solvent. The viscosity of each grade directly influences the thickness versus spin speed behaviour. Table 2.12 lists the layer thickness at a spin speed of 900 rpm for the different grade of SU-8 photoresist.

Another grade not listed in the table is the grade 2, which is the least viscous among the SU-8 photoresist series.

Table 2.12 Layer thickness and viscosity relationship at a spin speed of 900 rpm

SU-8	Solids (%)	Viscosity (cSt)	Thickness (μm)
5	51.8	265	12
10	59.1	989	30
25	63.3	2 646	58
50	69.1	14 953	150
100	72.9	52 407	320

Source: www.microchem.com/su8.cfm/

2.5 BULK MICROMACHINING FOR SILICON-BASED MEMS

The emergence of silicon micromachining has been an enabling factor for the rapid progress of the field of RF MEMS as discussed in Chapter 1. Both bulk micromachining and surface micromachining were briefly outlined before. For a detailed description with examples, one could refer to Gardner, Varadan and Awadelkarim 2001.

Bulk micromachining is the maturest of the two principal silicon micromachining technologies. It emerged in the early 1960s and has been used since then in the fabrication of many different microstructures. Bulk micromachining is utilized in the manufacture of the majority of commercial devices – almost all pressure sensors and silicon valves, and ~90% of silicon acceleration sensors. The term 'bulk micromachining' expresses the fact that this type of micromachining is used to realize micromechanical structures within the bulk of a single-crystal silicon wafer by selectively removing wafer material. The microstructures fabricated using bulk micromachining may cover the thickness range from submicrons to the thickness of the full wafer (200 to 500 μm), and the lateral size ranges from microns to the full diameter of a wafer (75 to 200 mm).

Etching is the key technological step for bulk micromachining. The etch process employed in bulk micromachining comprises one or several of the following techniques:

- wet isotropic etching
- wet anisotropic etching
- plasma isotropic etching
- reactive ion etching
- etch-stop techniques

Some of these etch processes have already been met as a standard technology employed in the microelectronics industry, such as reactive ion etching.

In addition to an etch process, bulk micromachining often utilizes wafer bonding and buried-oxide-layer technologies. However, the use of the latter in bulk micromachining is still in its infancy.

In the following sections we will describe the commonly-used bulk micromachining processes. Our discussion includes the important topics of etch-stops and wafer-to-wafer bonding.

2.5.1 Isotropic and orientation-dependent wet etching

Wet chemical etching is widely used in semiconductor processing. It is used for lapping and polishing to give an optically flat and damage-free surface and to remove contamination that results from wafer handling and storing. Most importantly, it is used in the fabrication of discrete devices and integrated circuits of relatively large dimensions to delineate patterns and to open windows in insulating materials. It is to be noted that most of the wet etching processes are isotropic, that is, unaffected by crystallographic orientation.

However, some wet etchants are orientation dependant, that is, they have the property of dissolving a given crystal plane of a semiconductor much faster than other planes (see Table 2.13). In diamond and zinc-blende lattices, the (1 1 1) plane is more closely

Table 2.13 Anisotropic etching characteristics of different wet etchants for single-crystalline silicon

Etchant	Temperature (°C)	Etch rate of Si($\mu m\,h^{-1}$)		
		(1 0 0)	(1 1 0)	(1 1 1)
$KOH{:}H_2O$	80	84	126	0.21
KOH	75	25–42	39–66	0.5
EDP	110	51	57	1.25
$N_2H_4H_2O$	118	176	99	11
NH_4OH	75	24	8	1

packed than the (1 0 0) plane and, hence, for any given etchant the etch rate is expected to be slower.

A commonly used orientation-dependent etch for silicon consists of a mixture of KOH in water and isopropyl alcohol. The etch rate is about 2.1 $\mu m\,min^{-1}$ for the (1 1 0) plane, 1.4 $\mu m\,min^{-1}$ for the (1 0 0) plane, and only 0.003 μm min for the (1 1 1) plane at 80 °C; therefore the ratio of the etch rates for the (1 0 0) and (1 1 0) planes to the (1 1 1) plane are very high, at 400 : 1 and 600 : 1, respectively.

2.5.1.1 Etch-stop techniques

Many different chemical etchants for silicon are known. The properties that make some of these etchants indispensable to micromachining of three-dimensional structures are selectivity and directionality. As etching processes in polar solvents are fundamentally charge-transport phenomena, it is not surprising that the etch rate may be dopant-type dependent, dopant-concentration dependent and bias dependent. Etch processes can be made selective by the use of dopants – heavily doped regions etch more slowly – or even halted electrochemically when observing the sudden rise in current through an etched n–p junction.

A region at which wet (or dry) etching tends to slow down (or halt) is called an 'etch-stop'. There are several ways in which an etch-stop region can be created. In the following subsections we discuss two such methods by which etch-stops are created. These methods are:

- doping-selective etching (DSE)
- bias-dependent DSE

2.5.1.2 Doping-selective etching

Silicon membranes are generally fabricated using the etch-stop phenomenon of a thin, heavily boron-doped layer, which can be epitaxially grown or formed by the diffusion or implantation of boron into a lightly doped substrate. This stopping effect is a general property of basic etching solutions such as KOH, NaOH, ethylenediamine pyrocatechol (EDP) and hydrazine (see Table 2.14). Owing to the heavy boron-doping the lattice constant of silicon decreases slightly. This leads to highly strained membranes that often show

Table 2.14 Dopant-dependent etch rates of selected silicon wet etchants for boron-doped silicon

Etchant[a]	Temperature (°C)	(1 0 0) etch-rate ($\mu m\ min^{-1}$)	
		B: $\ll 10^{19}\ cm^{-3}$	B: $\approx 10^{20}\ cm^{-3}$
EDP	115	0.75	0.015
KOH	85	1.4	0.07
NaOH	65	0.25–1.0	0.025–0.1

[a]With water as diluent.
Note: EDP, ethylenediamine pyrocatechol.

slip planes. They are, however, taut and fairly rugged even in a few micron thickness and about 1 cm diameter. The technique is not suited to stress-sensitive microstructures that could lead to the movement of the structures without an external load. In this case other etch-stop methods should be employed.

The main benefits of the high boron etch-stop are the independence of crystal orientation, the smooth surface finish and the possibilities it offers for fabricating released structures with arbitrary lateral geometry in a single etch step. However, the high levels of boron required are known to introduce considerable mechanical stress into the material, which may even cause buckling or even fracture in a diaphragm or other double-clamped structures. Moreover, the introduction of electrical components for sensing purposes into these microstructures, such as the implantation of piezoresistors, is inhibited by the excessive background doping. The latter consideration constitutes an important limitation to the applicability of the high boron dose etch-stop. Consequently, bias-dependent DSE, commonly referred to as an electrochemical etch-stop, is currently the most widely used etch-stop technique.

2.5.1.3 Conventional bias-dependent doping-selective etching or electrochemical etch-stop

In electrochemical etching of silicon, a voltage is applied to the silicon wafer (anode) by a counter-electrode (cathode) in the etching solution. The fundamental steps of the etching mechanism are:

1. injection of holes into the semiconductor to raise it to a higher oxidation state Si^+;
2. attachment of negatively-charged hydroxyl groups, OH^-, to the positively charged Si;
3. reaction of the hydrated silicon with the complexing agent in the solution;
4. dissolution of the reaction products into the etchant solution.

In bias-dependent etching, oxidation is promoted by a positive voltage applied to the silicon wafer, which causes an accumulation of holes at the silicon–solution interface. Under these conditions, oxidation at the surface proceeds rapidly while the oxide is readily dissolved by the solution. Holes as H^+ ions are transported to the cathode and released there as hydrogen gas bubbles. Excess hole–electron pairs can, in addition, be created at the silicon surface (e.g. by optical excitation), thereby increasing the etch rate.

The conventional electrochemical etch-stop technique is an attractive method for fabricating both microsensors and microactuators since it has the potential for allowing reproducible fabrication of moderately doped n-type silicon microstructures with good thickness control. However, a major limiting factor in the use of the conventional electrochemical etch-stop process is the effect of reverse-bias leakage current in the junction. Since the selectivity between n-type and p-type silicon in this process is achieved through the current-blocking action of the diode, any leakage in this diode will affect the selectivity. In particular, if the leakage current is very large it is possible for etching to terminate well before the junction is reached. In some situations, the etching process may fail completely because of this leakage. This effect is well known, and alternative biasing schemes employing three and sometimes four electrodes have been proposed to minimize this problem. To circumvent this weakness of the conventional electrochemical etch-stop technique, an alternative dopant-selective technique that uses pulsed anodizing voltages applied to silicon samples immersed in etching solutions has been developed (Wang *et al.*, 1992). This alternative technique is called selective etching by pulsed potential anodization and is described in the next section.

2.5.1.4 Selective etching of **n-type silicon by pulsed potential anodization**

The pulsed potential anodization technique selectively etches n-type silicon (Wang *et al.*, 1992). The difference in the dissolution time of anodic oxide formed on n-type and p-type silicon samples under identical conditions is used to create an etch selectivity. The mechanism responsible for this dissolution time difference is not fully understood at the present time. However, it is believed to be due to a difference in oxidation rates caused by the limited supply of holes in n-type samples (Wang *et al.*, 1992). This technique is applicable in a wide range of anodizing voltages, etchant compositions and temperatures. It differs from the conventional p-n junction etch-stop in that the performance of the etch-stop does not depend on the rectifying characteristics or quality of a diode. Using this technique, p-type microstructures of both low and moderate doping can be fabricated. Hence the pulsed potential anodization technique opens up the possibility for the creation of fragile microstructures in p-type silicon.

The main problems with the conventional electrochemical etch-stop and the pulsed potential anodization technique are related to the etch holders required for contacting the epitaxial layer (and the substrate for two, three or four electrodes) and for protecting the epitaxial side of the wafer from the etchant. Any leakage in these holders interferes with correct operation of the etch-stop. Moreover, mechanical stress introduced by the holder is known to reduce substantially production yield in many cases. Therefore, development of a reliable wafer holder for anisotropic etching with electrochemical etch-stop is not straightforward. The process of making contact with the wafer itself can also be critical and difficult to implement. Therefore single-step fabrication of released structures with either conventional electrochemical etch-stop or the pulsed potential anodization techniques may be troublesome. An alternative etch-stop technique which does not require any external electrodes (or connections to be made to the wafer) has been recently developed.

This new technique is what is referred to as the photovoltaic electrochemical etch-stop technique (PHET) (Peeters *et al.*, 1994).

2.5.1.5 Photovoltaic electrochemical etch-stop technique

The PHET approach is able to produce the majority of structures that can be produced by either the high-boron or the electrochemical etch-stop (Peeters *et al.*, 1994). PHET does not require the high impurity concentrations of the boron etch-stop and does not require external electrodes or an etch holder as in conventional electrochemical etch-stop or pulsed anodization techniques. Free-standing p-type structures with arbitrary lateral geometry can be formed in a single etch step. In principle, PHET is to be seen as a two-electrode electrochemical etch-stop where the potential and current required for anodic growth of a passivating oxide is not applied externally but is generated within the silicon itself. The potential essentially consists of two components, being the photovoltage across an illuminated p–n junction and the 'Nernst' potential of an n-Si/metal/etchant solution electrochemical cell.

2.5.2 Dry etching

As discussed above, bulk micromachining processes can yield single-crystal silicon (SCS) microstructures using crystal orientation dependent and dopant concentration dependent wet chemical etchants, such as EDP, KOH and hydrazine to undercut the SCS structures from a silicon wafer. However, the type, shape and size of the SCS structures that can be fabricated with the wet chemical etch techniques are severely limited. A dry-etch-based process sequence to produce suspended, SCS mechanical structures and actuators has been developed (Zhang and McDonald, 1992). The process is called SCREAM, for single-crystal reactive etching and metallization process. SCREAM uses reactive ion etching (RIE) processes to fabricate released SCS structures with lateral feature sizes down to 250 nm and with arbitrary structure orientations on a silicon wafer. SCREAM includes process options to make integrated, side-drive capacitor actuators. A compatible high step-coverage metallization process using metal sputter deposition and isotropic metal dry etch is used to form side-drive electrodes. The metallization process complements the silicon RIE processes used to form the movable SCS structures.

The SCREAM process can be used to fabricate complex circular, triangular structures in SCS. These structures can include integrated, high aspect ratio and conformable capacitor actuators. The capacitor actuators are used to generate electrostatic forces and so produce micromechanical motion.

2.5.3 Buried oxide process

The buried oxide process generates microstructures by means of exploiting the etching characteristics of a buried layer of silicon dioxide. After oxygen has been implanted into a silicon substrate using suitable ion implantation techniques, high-temperature annealing causes the oxygen ions to interact with the silicon to form a buried layer of silicon dioxide. The remaining thin layer of SCS can still support the growth of an epitaxial layer from a few microns to many tens of microns thick.

In micromachining, the buried silicon dioxide layer is used as an etch-stop. For example, the etch-rate of an etchant such as KOH slows down markedly as the etchant reaches the silicon dioxide layer. However, this process has the potential for generating patterned silicon dioxide buried layers by appropriately implanting oxygen.

2.5.4 Silicon fusion bonding

The construction of any complicated mechanical device requires not only the machining of individual components but also the assembly of the components to form a complete set. In micromachining, bonding techniques are used to assemble individually micromachined parts to form a complete structure. Wafer bonding, when used in conjunction with micromachining techniques, allows the fabrication of 3D structures that are thicker than a single wafer. Several processes have been developed for bonding silicon wafers. The most common bonding process is fusion bonding.

Several groups (Apel *et al.*, 1991; Lasky, 1986; Ohashi *et al.*, 1986) have demonstrated that the fusion of hydrophilic silicon wafers is possible for obtaining silicon-on-insulator (SOI) materials. Since then, wafer bonding techniques have found different applications in the field of microelectronics; several static random access memory (SRAM), CMOS and power devices have been fabricated on bonded SOI material. For micromechanical applications, fusion bonding rendered possible the fabrication of complex structures by combining two or more patterned wafers. This section describes the principles of wafer fusion bonding and presents fusion-bonding processes for MEMS device fabrication.

2.5.4.1 Wafer fusion

In its simplest form, the wafer fusion bonding process is the mating together of a pair of wafers at room temperature, followed by thermal annealing at temperatures of between 700 and 1100 °C. At room temperature, the wafers adhere via hydrogen bridge bonds of chemisorbed water molecules that subsequently react during the annealing process to form Si–O–Si bonds. Consequently, wafer pretreatment procedures that include hydrophilization steps (wet cleaning processes, plasma hydrophilization) support the process.

A major concern of all bonding processes is the presence of noncontacting areas, which are generally called voids. Voids are mainly caused by particles, organic residues, surface defects and inadequate mating. Therefore both the surfaces being fusion bonded have to be perfectly smooth and clean since the smallest of particles could cause large voids. Optimized processing includes wafer surface inspection, surface pretreatment (hydrophilization, cleaning) and mechanically controlled, aligned mating in a particle-free environment.

2.5.4.2 Annealing treatment

As discussed in the previous subsection, wafer bonding can involve a high-temperature annealing step that is to be performed after the room-temperature contacting of the surfaces. This annealing step is necessary to increase the strength of the bond. However, the high-temperature annealing step (usually at a temperature above 800 °C) may introduce problems, such as doping profile broadening, thermal stresses, defect generation and contamination. Annealing also prevents the use of bonding technology for compound semiconductor materials since their dissociation temperature is often low. In addition, postmetallization bonding also requires bonding temperatures that are less than ~450 °C since most of the common metals that are used in device fabrication melt below this temperature. Therefore, in order to make full use of the potential provided by wafer bonding for microstructures, low-temperature bonding methods have to be developed. Attempts to lower bonding temperatures and still achieve reasonable bond strength are currently underway.

Three annealing temperature ranges are of interest in wafer bonding:

- temperature <450 °C for postmetallization wafers;
- temperature <800 °C for wafers with diffusion dopant layers (e.g. p^+ etch-stop layers);
- temperature >1000 °C for wafer bonding before processing. According to the reaction mechanism, annealing at temperatures above 1000 °C for several hours should result in almost complete reaction of the interface. A 1000 °C anneal for about 2 h gives sufficiently high bond strength for all subsequent treatments (Harendt *et al.*, 1991); it is not possible to separate the two bonded silicon wafers without breaking the silicon.

2.5.4.3 Fusion of silicon-based materials

Fusion bonding of polysilicon, silicon dioxide or silicon nitride to silicon proceeds in a manner similar to silicon-to-silicon bonding. In the case of polysilicon bonding to silicon a polishing step for the two surfaces to be bonded is necessary. This polishing step produces two smooth defect-free surfaces. The bonding mechanism is most probably identical to silicon-to-silicon fusion bonding in that in both cases Si−OH groups are present at the surface. Thus pretreatment (hydrophilization) and annealing conditions are similar.

Because of the dissimilar mechanical characteristics of the different bonded materials, the yield of void-free wafers can be significantly reduced by wafer bow or defects caused by stress during thermal treatment. Bonding of wafers covered with a thin thermal oxide or a thin silicon nitride results in homogenous bonded wafers, while oxides with thicker oxide (or nitride films) were found to develop voids (Gardner, Varadan and Awadelkarim, 2001).

2.5.5 Anodic bonding

Silicon-to-silicon anodic bonding is a bonding technique used to seal silicon together by use of a thin sputter-deposited glass layer. The equipment used for anodic bonding is basically a heat chuck element with an electrode capable of supplying high voltage across the structure to be bonded. The system may automatically control the temperature and power supply during the bonding process.

After surface cleaning and polishing, one of the wafers (referred to here as the top wafer) is initially given a glass film a few microns thick. This glass film is sputtered onto the wafer surface. The top wafer is placed on top of a second silicon wafer, which is usually referred to as the support wafer; these two wafers are to be bonded. The support wafer rests on the aluminium chuck. The two wafers are usually sealed together by anodic bonding at temperatures less than 400 °C with an electrostatic dc voltage of 50 to 200 V. The negative electrode is connected to the top sputter-coated wafer. The voltage should be applied over a time long enough to allow the current to settle at the steady-state minimized level. Typically, the bonding process is terminated within 10 to 20 min. The bond process usually takes place in air at atmospheric pressure.

2.6 SILICON SURFACE MICROMACHINING

Since the beginning of the 1980s much interest has been directed towards micromechanical structures fabricated by a technique called 'surface micromachining'. The resulting $2\frac{1}{2}$-dimensional structures are located mainly on the surface of a silicon wafer and exist as a thin film – hence the half dimension. The dimensions of these surface micromachined structures can be an order of magnitude smaller than bulk-micromachined structures. The main advantage of surface micromachined structures is their easy integration with IC components, since the same wafer surface can also be processed for IC elements. However, as miniaturization is immensely increased by silicon surface micromachining, the small sizes or masses created are often insufficient for viable sensors and, particularly, actuators. The problem is most acute in capacitive mechanical microsensors (Section 6.4) and especially capacitively-driven microactuators because of the low coupling capacitances. Deep etching techniques, such as LIGA, have been developed in order to address this problem but are difficult to realize in silicon.

There are several common approaches to making MEMS devices using surface micromachining. The first of these approaches is sacrificial layer technology for the realization of mechanical microstructures. The second approach incorporates IC technology and wet anisotropic etching, and the third approach uses plasma etching to fabricate microstructures at the silicon wafer surface.

2.6.1 Sacrificial layer technology

Sacrificial layer technology uses, in most situations, polycrystalline rather than single-crystal silicon as the structural material for the fabrication of microstructures. Low-pressure chemical vapor deposition (LPCVD) of polysilicon is well known in standard IC technologies (see Chapter 4) and it has excellent mechanical properties which are similar to those of single-crystalline silicon. When polycrystalline silicon is used as the structural layer, sacrificial layer technology normally employs silicon dioxide as the sacrificial material, which is employed during the fabrication process to realize some microstructure but does not constitute any part of the final miniature device.

The key processing steps in sacrificial layer technology are:

1. deposition and patterning of a sacrificial silicon dioxide layer on the substrate;
2. deposition and definition of a polysilicon film;
3. removal of the sacrificial oxide by lateral etching in hydrofluoric acid (HF); that is, etching away of the oxide underneath the polysilicon structure.

Here we refer to polysilicon and silicon dioxide as the structural and sacrificial materials, respectively. The reason for doing this is that in almost all practical situations this is the preferred choice of material combination. However, several other material combinations are also being used in surface micromachining.

2.6.2 Material systems in sacrificial layer technology

An important consideration in the fabrication of an ideal mechanical microstructure is that it is without any residual mechanical stress, so that the films deposited have no significant residual strain. In particular, doubly supported free-standing structures will buckle in the presence of a relatively modest residual compressive strain in the structural material. By choosing the appropriate deposition conditions and by optimizing the annealing step an almost strain-free structural material layer can be obtained.

Surface micromachining requires a compatible set of structural materials, sacrificial materials and chemical etchants. The structural materials must possess the physical and chemical properties that are suitable for the desired application. In addition, the structural materials must have appropriate mechanical properties, such as high yield and fracture strengths, minimal creep and fatigue and good wear resistance. The sacrificial materials must also have good mechanical properties to avoid device failure during the fabrication process. These properties include good adhesion and a low residual stress in order to eliminate device failure by delamination and/or cracking. The etchants must have excellent etch selectivity and they must be able to etch off the sacrificial materials without affecting the structural materials. In addition, the etchants must also have appropriate viscosity and surface tension characteristics.

2.6.2.1 Polycrystalline silicon/silicon dioxide

This material system has already been mentioned in Chapter 1. The polysilicon/oxide material system is the most common one and it uses polysilicon deposited by LPCVD as the structural material and thermally grown (or LPCVD) oxide as the sacrificial material. The oxide is readily dissolved in HF solution without the polysilicon being affected. Silicon nitride is often used together with this material system for electrical insulation. The advantages of this material system include the following:

- Both polysilicon and silicon dioxide are used in IC processing and, therefore, their deposition technologies are readily available.
- Polysilicon has excellent mechanical properties and can be doped for various electrical applications. Doping not only modifies the electrical properties but can also modify the mechanical properties of polysilicon. For example, the maximum mechanically sound length of a free-standing beam is significantly larger for phosphorous-doped compared with undoped polysilicon. However, in most cases the maximum length attainable is limited by the tendency of the beam to stick to the substrate.
- The oxide can be thermally grown and deposited by CVD over a wide range of temperatures (from about 200 °C to 1200 °C), which is very useful for various processing requirements. However, the quality of oxide will vary with deposition temperature.
- The material system is compatible with IC processing. Both polysilicon and silicon dioxide are standard materials for IC devices. This commonality makes them highly desirable in sacrificial layer technology applications that demand integrated electronics.

2.6.2.2 *Polyimide/aluminum*

In this second material system, the polymer 'polyimide' is used for the structural material while aluminium is used for the sacrificial material. Acid-based aluminium etchants are used to dissolve the aluminium sacrificial layer. The three main advantages of this material system are:

- Polyimide has a small elastic modulus which is $\sim$50 times smaller than that of polycrystalline silicon.
- Polyimide can take large strains before fracture.
- Both polyimide and aluminium can be prepared at relatively low temperatures ($<400\,^{\circ}$C).

However, the main disadvantage of this material system lies with polyimide in that it has unfavorable viscoelastic characteristics (i.e. it tends to creep) and so devices may exhibit considerable parametric drift.

2.6.2.3 *Silicon nitride/polycrystalline silicon and tungsten/silicon dioxide*

In the third material system of silicon nitride/polysilicon, silicon nitride is used as the structural material and polysilicon as the sacrificial material. For this material system silicon anisotropic etchants such as KOH and EDP are used to dissolve the polysilicon.

In the fourth material system of tungsten/oxide, tungsten deposited by CVD is used as the structural material with the oxide as the sacrificial material. Here again, HF solution is used to remove the sacrificial oxide.

Finally, we give a worked example in which silicon nitride is employed as the structural material as before but, as a variant, and unusually, aluminium is used as the sacrificial layer instead of polysilicon.

2.6.3 Surface micromachining using plasma etching

Surface micromachining can also be realized using a dry etching rather than a wet etching process. Plasma etching of the silicon substrate, with SF_6/O_2-based and CF_4H_2-based gas mixtures is advantageous since high selectivities for photoresist, silicon dioxide and aluminium masks can be achieved. However, when using plasma etching a large undercut of the mask is generally produced. This is due to the isotropic fluorine atom etching of silicon that is known to be high compared with the vertical etch induced by ion bombardment. In contrast, reactive ion etching (RIE) of polysilicon using a chlorine/fluorine gas combination produces virtually no undercut and almost vertical etch profiles with photoresist used as the masking material. Thus, rectangular silicon patterns, which are up to 30 μm deep, can be formed using chlorine/fluorine plasmas out of polysilicon films and the silicon wafer surface. A deep etch process is essential for microactuators and so the deep RIE process is an attractive option.

2.6.4 Combined integrated-circuit technology and anisotropic wet etching

Anisotropic wet etching may be combined with an IC process to fabricate free-standing multilayer microstructures without additional masks. Its main merits are low cost and compatibility with standard IC processing. In the first phase, the multilayers are created using IC processing. Usually, the multilayer is composed of the standard insulating and passivating dielectric films, polysilicon layers and metal layers. The polysilicon and metal layers constitute the active layers and are usually sandwiched between the dielectric films that are necessary for electrical insulation and component passivation. By special design, windows are opened around the multilayer structures for removal of all dielectric layers thus exposing the silicon surface underneath.

In the second, so-called postprocessing, phase the wafers are immersed in anisotropic silicon etchants. Thus, the exposed silicon surface around the multilayer structure is removed and, by underetching, the microstructures finally become free-standing. Since the active layers are completely contained within the dielectric layers they are protected against the silicon etching process. An alternative approach is anisotropically to etch only the back side of the wafer (i.e. use a single-sided etching bath). This technique may be used to make certain structures but tends to be a more timing-consuming and therefore more costly process.

2.7 MICROSTEREOLITHOGRAPHY FOR POLYMER MEMS

MSL is also called microphotoforming and was first introduced to fabricate high aspect ratio and complex 3D microstructures in 1993 (Ikuta and Hirowatari, 1993). In contrast to conventional subtractive micromachining, MSL is an additive process and so enables the fabrication of high aspect ratio microstructures with novel smart materials. The MSL process is, in principle, compatible with silicon microtechnology and so post-CMOS batch fabrication is also feasible (Ikuta *et al.*, 1996; Zizzi *et al.*, 1996). For a detailed description of MSL and other fabrication techniques for polymeric MEMS, one could refer to Varadan, Xiang and Varadan (2001).

Different MSL systems have been developed in recent years to improve upon their precision and speed. Basically, scanning MSL (Ikuta *et al.*, 1996; Zizzi *et al.*, 1996) and projection MSL (Bertsch *et al.*, 1997; Monneret, Loubere and Corbel, 1999; Nakamoto and Yamaguchi, 1996) are the two major approaches that have been taken. Scanning MSL builds the solid microparts in a point-by-point and line-by-line fashion, while projection MSL builds one layer with each exposure thus speeding up the building process by a significant factor (Beluze, Bertsch and Renand, 1999). The details of the two approaches are presented in detail in Varadan, Xiang and Varadan (2001) and briefly described in Sections 2.7.1 and 2.7.2 below.

Bulk and surface micromachining of silicon devices were described before. These techniques are, however, not suitable for real 3D devices with high aspect ratios. In the traditional MEMS arena, 3D parts are fabricated by folding and assembling planar microfabricated silicon substrates. Even though many MEMS devices with integrated electronics have been achieved by using traditional micromachining techniques, some limitations have nevertheless to be underlined: (1) these techniques are very expensive and need specific installations as well as a cleanroom environment, (2) the materials that can be used up to now are restricted to silicon and metals and (3) the manufacture of 3D parts having curved surfaces or an important number of layers is not possible. For MSL techniques to be as successful as their silicon counterparts, one has to come up with a technique similar to silicon micromachining, which is described in Section 2.7.1. In addition, with the invention of organic thin-film transistors, now it seems possible to fabricate polymer-based MEMS devices with built-in-electronics (Varadan and Varadan, 2001).

2.7.1 Scanning method

Most MSL equipment for fabricating RF MEMS is based upon the *scanning* method (Figure 2.15). With the scanning method, a well-focused laser beam with beam spot size around 1 micron is directed onto the resin surface to initiate the polymerization process. A 3D microstructure is built up by the repeated scanning of either the light beam or the work piece layer by layer.

A diagram of the experimental setup is shown in Figure 2.15 and consists of a He-Cd laser of 442 nm wavelength, an optical shutter, a galvano scanner set, an X-Y-Z stage, an objective lens and a computer (Ikuta, Maruo and Kojima, 1998). The laser beam is focused inside the monomer volume by coordinating the beam scanning and Z stage movements, thus the 3D structures are formed inside the liquid. This process can make microparts with freely moving elements in a single step with no postprocess assembly necessary. This permits the fabrication of more complex MEMS devices than with silicon micromachining.

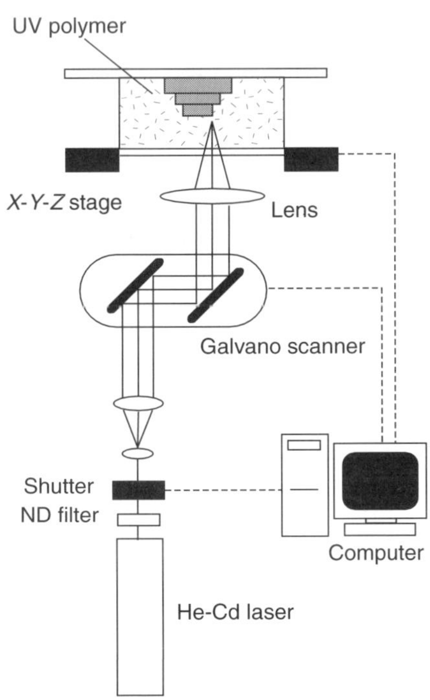

Figure 2.15 Experimental set up of scanning method

The resolution of this particular scanning process is excellent and is typically less than 1 μm. The fabrication speed can be increased by operating the galvano scanning mirror and *X*-*Y*-*Z* stages together.

2.7.2 Two-photon microstereolithography

As mentioned above, conventional MSL is limited in terms of the minimum thickness of the resin layers possible because of viscosity and surface-tension effects. In contrast, the two-photon MSL process does not have this problem because the resin does not need to be layered.

When a laser beam is focused on a point with a microscope objective lens as shown in Figure 2.16(a) (Maruo and Kawata, 1998) the density of photons decreases with the distance away from the focal plane, but the total number of photons in the beam at every cross-section remains the same [see Figure 2.16(b)]. Thereafter, the resin is solidified completely in the illuminated region even beyond focal point, leading to a poor resolution. This means that the linear response of the materials to the light intensity based on a single photon absorption does not have optical sectioning capability. However, if the material response is proportional to the square of the photon density, the integrated material response is enhanced greatly at the focal point [see Figure 2.16(c)] and therefore the two-photon absorption-based polymerization occurs only in a small volume within the focal depth. Normally, the beam power of the laser has to be extremely high (several kilowatts) in order to obtain two-photon absorption.

A two-photon MSL apparatus is shown in Figure 2.17 (Maruo and Kawata, 1998). The beam is generated by a mode-locked titanium sapphire laser and is directed by two galvanic scanning mirrors. The beam is then focused with an objective lens into the resin. A CCD camera is used to aid focusing and monitor the forming of the microstructure. A *Z* stage moves the resin container along the optical axis for multilayer fabrication. The objective lens used by Maruo had a numerical aperture of 0.85 (magnification of 40). The accuracy of the galvano scanner set (general scanning) and the *Z* stage (Sigma Optics) were 0.3 μm and 0.5 μm, respectively. The peak beam power in the resin was about 3 kW with a repetition of 76 MHz and pulse width of 130 fs at a wavelength of 770 nm.

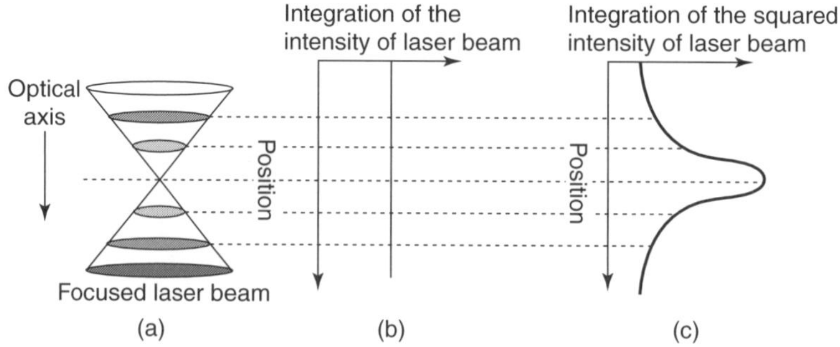

Figure 2.16 Two-photon absorption and one-photon absorption generated by a focused laser: (a) focused laser beam; (b) total one-proton absorption per transversal plane; and (c) total two-photon absorption per transversal plane

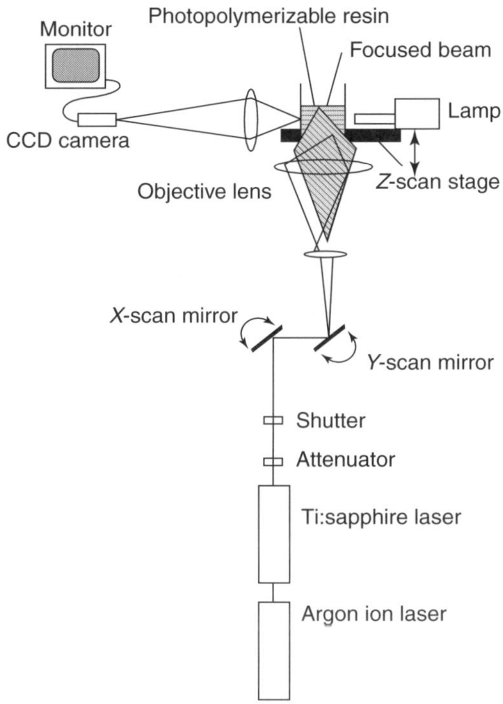

Figure 2.17 Optical setup for the two-photon microstereolithography system

2.7.3 Surface micromachining of polymer MEMS

Surface micromachining of polymer-based MEMS follows the same principle as silicon processing. However, the thin-film approach used in silicon is not followed here. The technique deals with solidifying both structural polymer and sacrificial polymer by the MSL process. Thus, one does not need a mask as in silicon processing. The structural polymer one could select for this technique can be either of the following family: (1) UV-curable electroactive polymer, (2) ionic conducting polymer, (3) UV001, from HVS Technologies (State College, PA 16803), which is a UV-curable polymer with urethene acrylate, epoxy acrylate and acryloxysilane or (4) UV-curable polymer with carbon nanotube chemically bonded. The sacrificial polymer is an acrylic resin containing 50% silica and is modified by adding crystal violet as given in Bertsch *et al.* (1997). This composition can be dissolved with $2\,\mathrm{mol\,l^{-1}}$ caustic soda at $80\,^{\circ}\mathrm{C}$. Some devices have been fabricated using this technique, as described in Ikuta, Maruo and Kojima (1998).

2.7.4 Projection method

As described in the preceding sections, the scanning MSL can be used for very fine, high-aspect-ratio 3D microstructure fabrication, but the fabrication speed is always a major concern – even with the galvano scanning method. Scanning MSL builds up the objects layer by layer, but each layer is itself built up line by line. Thus projection MSL has been proposed for the more rapid building of 3D microstructures; even though it is still building layer by layer, each layer is now written by just one UV exposure through a mask. The reintroduction of a photographic mask plate produces significant savings in time but does add the extra expense of preparing masks.

Basically, there are two types of projection MSL, one is the use of a real photographic mask to project the UV pattern for curing (Nakamoto and Yamaguchi, 1996), the other is to use a dynamic mask referred to here as the liquid crystal display (LCD) projection method (Bertsch *et al.*, 1997).

2.7.4.1 Mask projection microstereolithography

As in standard photolithography, an image is transferred to the liquid photopolymer by shining a UV beam through a patterned mask plate, as shown in Figure 2.18 (Taylor *et al.*, 1994). Then another fresh layer of liquid photopolymer is prepared on top of the patterned solid polymer. By repeating the above process, a multilayered 3D microstructure can be built by this mask projection MSL (Suzumori, Koga and Haneda, 1994).

Let us now consider the equations that govern the optics of mask projection MSL (Nakamoto and Yamaguchi, 1996). When a beam of uniform intensity I_0 passes through a square mask, with the center of the mask at $x = 0$ and $y = 0$, and the depth of the polymer along the z axis, and onto the surface of the liquid polymer [see Figure 2.19(a)] diffraction occurs and the intensity I_d may be expressed as follows:

$$I_\text{d}(x, y, z) = 0.25 I_0 (C_x^2 C_y^2 + C_x^2 S_y^2 + S_x^2 C_y^2 + S_x^2 S_y^2) \tag{2.62}$$

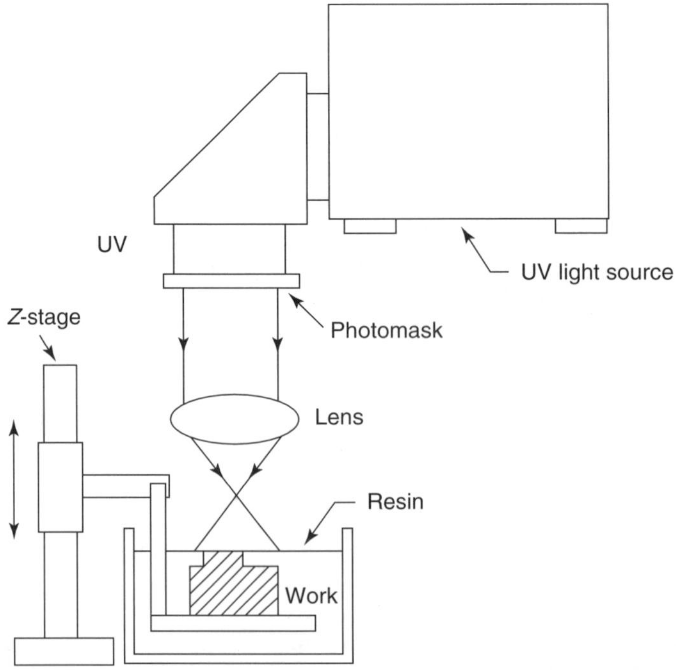

Figure 2.18 Apparatus for the mask projection method of microfabrication

where the various cosine and sine coefficients are defined by

$$C_x = \int_{p_1}^{p_2} \cos\left(\frac{\pi u^2}{2}\right) \mathrm{d}u \tag{2.63}$$

$$S_x = \int_{p_1}^{p_2} \sin\left(\frac{\pi u^2}{2}\right) \mathrm{d}u \tag{2.64}$$

$$C_y = \int_{q_1}^{q_2} \cos\left(\frac{\pi u^2}{2}\right) \mathrm{d}u \tag{2.65}$$

$$S_y = \int_{q_1}^{q_2} \sin\left(\frac{\pi u^2}{2}\right) \mathrm{d}u \tag{2.66}$$

and the limits on the integrals are defined by

$$p_1 = \frac{2}{\sqrt{m}}\left(\frac{x}{a} - 0.5\right) \tag{2.67}$$

$$p_2 = \frac{2}{\sqrt{m}}\left(\frac{x}{a} + 0.5\right) \tag{2.68}$$

$$q_1 = \frac{2}{\sqrt{m}}\left(\frac{y}{a} - 0.5\right) \tag{2.69}$$

$$q_2 = \frac{2}{\sqrt{m}}\left(\frac{y}{a} + 0.5\right) \tag{2.70}$$

$$m = \frac{2\lambda(h+z)}{a^2} \tag{2.71}$$

λ is the wavelength of the UV light, a is the length of each side of the square and h is the distance between the mask and the surface of the resin.

The light intensity inside the resin, $I(x, y, z)$, can be calculated according to the Beer–Lambert law

$$I(x, y, z) = I_{\mathrm{d}}(x, y, 0)\exp(-\alpha z) \tag{2.72}$$

where α is the absorption coefficient of the resin. When the polymer is irradiated for a time t, the irradiated energy $E(x, y, z)$ can be expressed as

$$E(x, y, z) = I(x, y, z) \cdot t$$

The polymer solidifies when the irradiated energy reaches its threshold value E_0. The values of α and E_0 are generally determined experimentally. Letting $E(x, y, z) = E_0$ allows the theoretical shape of the solidified polymer after projection to be determined.

Similar to scanning MSL, the fabrication precision is related to the exposure. In particular, the curing depth strongly depends upon the laser exposure, as shown in Figure 2.19(b) (Nakamoto and Yamaguchi, 1996). The lateral dimension is slightly influenced by the exposure and is determined mainly by the mask pattern in the case of a fixed distance between the mask and resin surface.

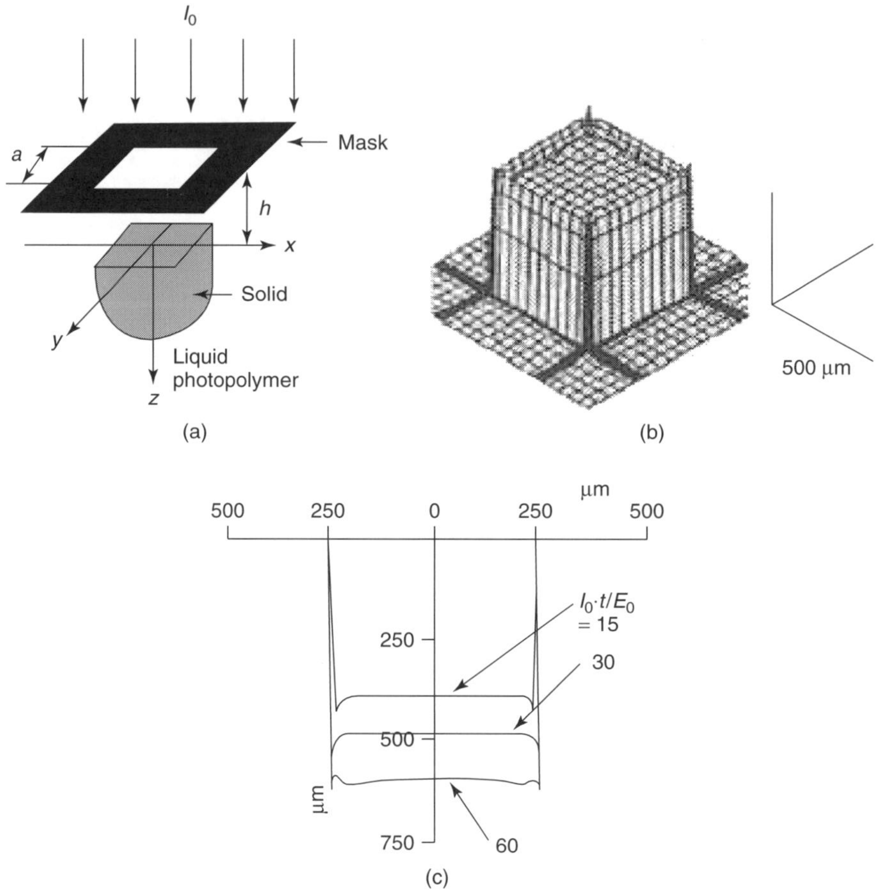

Figure 2.19 Model of mask projection microstereolithography and the simulation results: (a) theoretical model of the mask-based method; (b) simulated cross-section of a solidified polymer (a is 500 μm and h is 1000 μm); and (c) cross-section of the solidified polymer

Another important parameter in mask projection MSL is the distance between the mask and the resin surface. A large distance between the mask and resin surface results in a larger lateral dimension because of diffraction [Figure 2.19(c); Nakamoto and Yamaguchi, 1996]. Therefore, to minimize this effect and obtain the highest precision in mask projection MSL the mask should be located as close as possible to the resin surface.

2.7.4.2 Dynamic mask-projection microstereolithography

Dynamic mask projection MSL utilizes a dynamic mask generator rather than a 'static' photographic mask plate and so permits the rapid fabrication of complex 3D microobjects. A schematic view of a dynamic mask projection MSL is shown in Figure 2.20 (Bertsch *et al.*, 1997). In dynamic MSL, the mask pattern is produced by a computer-controlled LCD rather than by a chrome mask. Once the CAD (computer-aided design) file has been translated into a numerical control code that is sent to the LCD device via a computer, the

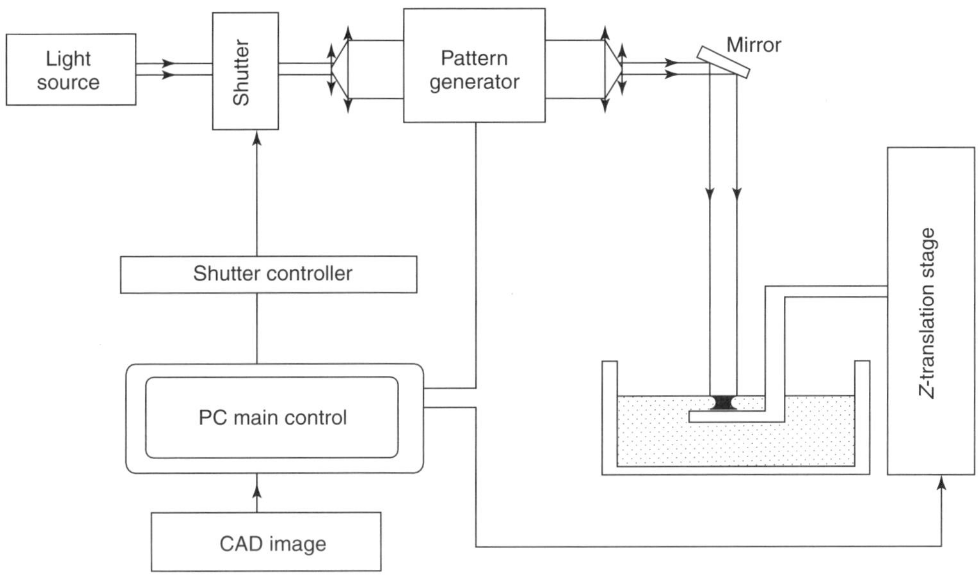

Figure 2.20 Principle of the integrated dynamic microstereolithography apparatus. Note: CAD, computer-aided design; PC, personal computer

LCD can then function as a dynamic mask to control the pattern of the layer. The light beam then passes through the LCD mask before being focused on the resin surface to allow the selective polymerization of the exposed areas corresponding to the transparent pixels of the LCD.

The rest of the apparatus used in dynamic mask projection MSL is the same as standard MSL; namely, layer preparation, beam on/off control, etc. It should be noted that the z-axis is the only moving element in the system and so it is simpler.

In dynamic mask MSL, the liquid–solid phototransformation can once again be described by the Beer–Lambert law. Because time t is now the most critical parameter in the process, the curing depth, d_c, is usually given in the form of

$$d_c = \frac{1}{\alpha C} \ln\left(\frac{t}{t_0}\right) \tag{2.73}$$

with

$$t_c = \frac{Q}{\alpha C F_0} \tag{2.74}$$

α is the absorption coefficient ($1\ \mathrm{mol}^{-1}\ \mathrm{cm}^{-1}$), C is the concentration of the photoinitiator ($\mathrm{mol\,l}^{-1}$), t is the exposure time (s), t_c is the exposure time (s) necessary to make the exposure reach the polymerization threshold energy (s), Q is the number of absorbed photons per unit volume ($\mathrm{photon\,m}^{-3}$), which is determined experimentally, and F_0 is the incident flux ($\mathrm{photon\,m}^{-3}\,\mathrm{cm}^{-1}$).

The liquid crystal matrix is inserted between four glass windows that are opaque to UV light; therefore, for this system, it is necessary to use a visible light source and a different set of chemical mixtures (Bertsch *et al.*, 1997). The lateral dimension is now determined by the LCD mask and so is limited by the resolution and contrast of current LCD displays.

2.7.5 Polymeric MEMS architecture with silicon, metal and ceramics

The fabrication of new MEMS devices requires the integration of various new functional materials such as polymers, ceramics, metals, and metal alloys. In this section, we described how the MSL process could be used to fabricate MEMS devices based upon these different materials.

2.7.5.1 Ceramic microstereolithography

Functional and structural ceramic materials possess useful properties such as high temperature or chemical resistance, high hardness, low thermal conductivity, ferroelectricity and piezoelectricity. Various novel approaches to ceramic microfabrication have been developed (for a review one could refer to Gardner, Varadan and Awadelkarim, 2001; Varadan, Xiang and Varadan, 2001). Unlike conventional silicon micromachining, MSL can be used to build the complex ceramic 3D microstructures in a rapid free-form fashion without the need for high pressures or high temperatures.

Ceramic MSL differs from polymeric MSL in the following aspects. First, the resin system for ceramic MSL is composed not only of the monomers and photoinitiators that are used in polymer MSL but also of ceramic powders, dispersants and diluents (Zhang, Jiang and Sun, 1999). Dispersants and diluents are used to obtain a homogeneous ceramic suspension with a relatively low viscosity. Upon UV polymerization, the ceramic particles are bonded together by the polymer and the ceramic body is formed. Generally, the viscosity of ceramic suspensions used for MSL is higher than the viscosity of most liquid polymers, leading to slow layer preparation. A precision blade has been designed for the layer preparation to solve this problem (Figure 2.21; Zhang, Jiang and Sun, 1999). Second, light transportation during MSL is more complicated in the solid–liquid two-phase medium. This is caused by light scattering off the solid ceramic particles and affecting both the curing depth and the line width.

The curing depth has been determined (for example, see Gardner, Varadan and Awadelkarim, 2001; Varadan, Xiang and Varadan, 2001) and is given by:

$$\phi_c = \frac{\phi}{Q\xi} \ln\left(\frac{E}{E_c}\right) \tag{2.75}$$

and

$$Q = \left(\frac{\Delta n}{n_0}\right)^2 \left(\frac{\lambda}{\phi}\right)^{-2} \tag{2.76}$$

where ϕ is the mean particle size of the ceramic powder, ξ is the volume fraction of the ceramic material in the suspension, n_0 is the refractive index of the monomer solution, Δn is the difference in refractive index of the ceramic solution and the monomer solution and λ is the wavelength of the UV light.

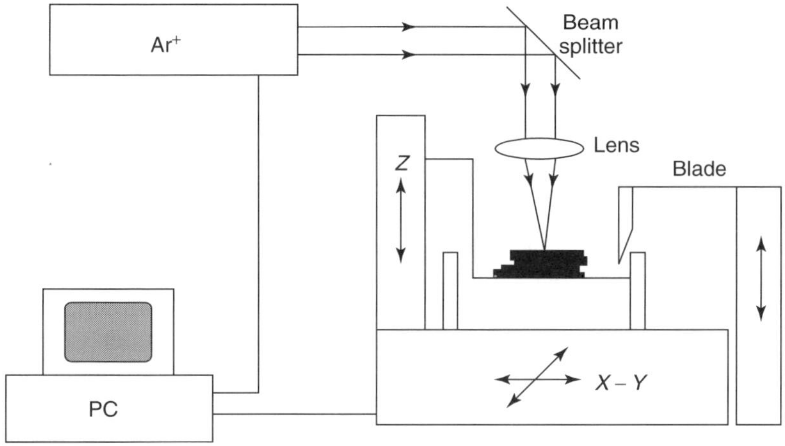

Figure 2.21 Microstereolithography apparatus for ceramics. Note: PC, personal computer

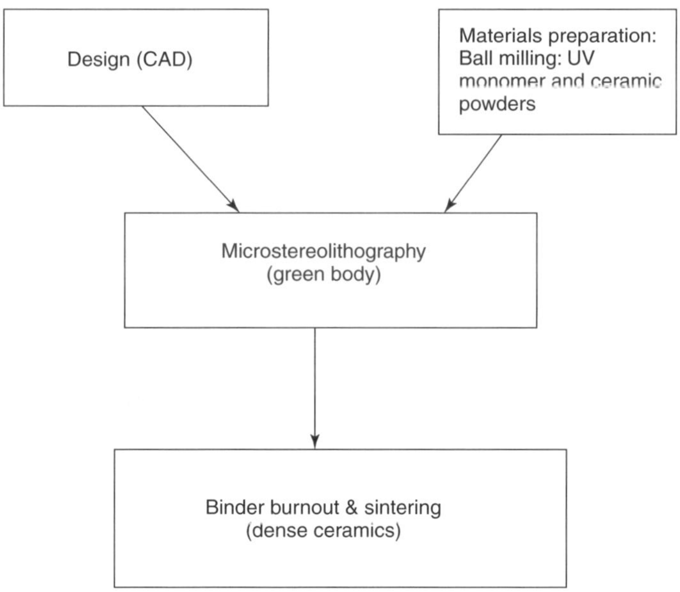

Figure 2.22 Processing steps for ceramic microstereolithography. Note: CAD, computer-aided design

The fabrication of ceramic microstructures using MSL typically follows the steps shown in Figure 2.22. First, the homogeneous ceramic suspension is prepared. Submicron ceramic powders are mixed with monomer, photoinitiator, dispersant, diluent, etc. by ball milling for several hours. The prepared ceramic suspension is then put into the vat and is ready for exposure defined by the CAD file. After which a (green) body ceramic micropart is obtained. Finally, the green body is put into a furnace first to burn off the polymer binders and is then sintered at higher temperatures to obtain the dense ceramic microparts. The temperatures of the binder burnout and sintering vary according to the choice of polymeric and ceramic materials. After sintering, the ceramic microstructures are ready for assembly and use.

2.7.5.2 Metallic microstructures

Metallic microstructures have been fabricated extensively for MEMS. 3D metallic microstructures have been built by spatial forming, electrochemical fabrication (EFAB), selective laser sintering and laser cladding processes (Cohen *et al.*, 1999; Kathuria, 1996; Taylor *et al.*, 1994). The spatial forming process is introduced in this section because of its relatively higher resolution and therefore its likelihood to succeed commercially.

Spatial forming combines several different technologies in order to generate solid metallic microstructures from fine powders (Taylor *et al.*, 1994). Similar to projection MSL, cross-section data from computer solid CAD models are used to define the patterns on a chrome mask. A custom-built offset printing press prints negative materials (space around the solid parts) on a ceramic substrate in multiple registered layers of pigmented organic ink averaging 0.5 μm thick, each layer being cured with UV light. After forming a number of layers of negative materials (e.g. 30), the positive ink, heavily loaded with metal powders (e.g. 50 vol%), is knifed onto the assembly filling the nonimage voids (Figure 2.23; Thornell and Johansson, 1998), followed by curing the filled material with UV light. By repeating the above steps until the desired thickness (e.g. 500 μm) is reached,

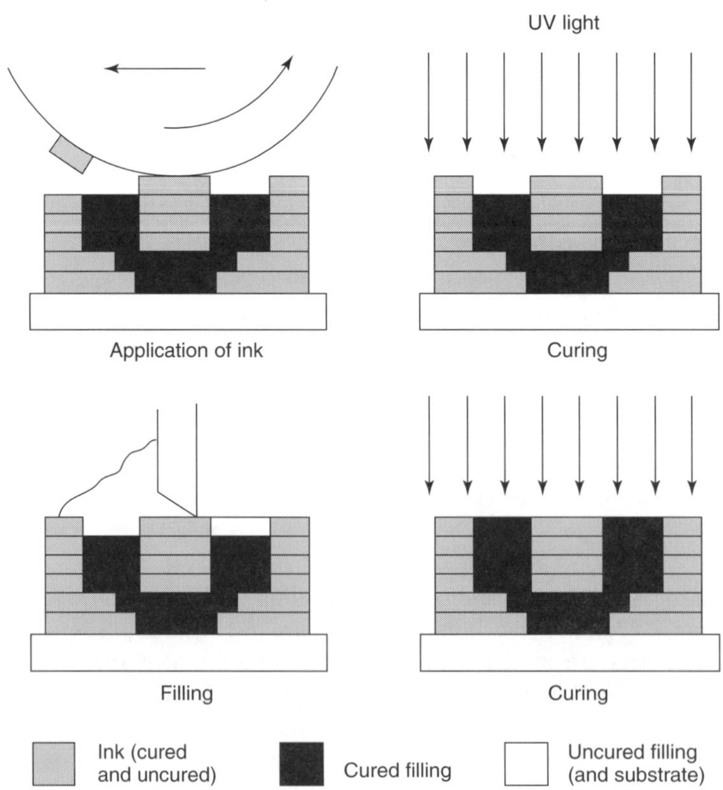

Figure 2.23 Schematic of the spatial-forming process

the green body metallic micropart is created. The green body part is then baked to remove the organic binders, and sintered in a controlled atmosphere furnace. The finished pure metallic microparts are thus finally obtained.

2.7.6 Microstereolithography integrated with thick-film lithography

Many micromechanical components have been fabricated using planar processes such as thin-film and bulk silicon micromachining. High aspect ratio micromachining is available through LIGA, deep RIE, and thick resist lithography with high resolutions. But these processes do not allow true 3D fabrications. However, MSL can be used to construct complex 3D microstructures, but with the constraints of a lower resolution and the problem associated with the manipulation and assembly of polymeric microstructures. An approach that seeks to combine MSL and thick resist lithography may provide a technique to build new 3D microstructures with more functionality (Bertsch, Lorenz and Renaud, 1998). The material used for such technique is EPON SU-8 whose properties are given in Section 2.4.3. The use of SU-8 for phase shifters is given in Chapter 6.

2.8 CONCLUSIONS

This chapter has introduced the topic of electronic materials and polymers and the fabrication techniques for conceiving MEMS devices. Bulk and surface micromachining of silicon are described with various etch-stop and bonding techniques. In addition, we have reviewed the emerging field of microstereolithography and its combination with other process technologies. MSL is attractive in that it can be used to make in batch process truly 3D microparts in a wide range of materials, polymers, metals and ceramics at a modest cost. Surface micromachining aspects of polymer-based MEMS were also discussed.

REFERENCES

Apel, U., Graf, H.G., Harendt, C., Hofflinger, B., Ifstrom, T., 1991, 'A 100-V lateral DMOS transistor with a 0.3 micron channel in a 1 micron silicon-film-on-insulator-on silicon', *IEEE Transactions on Electronic Devices* **38**: 1655–1659.

Beluze, L., Bertsch, A., Renaud, P., 1999, 'Microstereolithography: a new process to build complex three-dimensional objects', *Proceedings of SPIE* **3680**: 808–817.

Bertsch, A., *et al.*, 1997, 'Microstereolithography using a liquid crystal display as dynamic mask generator', *Microsystem Technology* **3**: 42–47.

Bertsch, A., Zissi, S., Jezequel, J.Y., Corbel, S., Andre, J.C., 1997, 'Microstereolithography using a liquid crystal display as dynamic mask generator', *Microsystem Technologies* **3**: 42–47.

Bertsch, A., Lorenz, H., Renaud, P., 1998, 'Combining microstereolithography and thick resist UV lithography for 3D microfabrication', in *IEEE Proceedings of MEMS '98*, IEEE, Washington, DC: 18–23.

Chomnawang, N., Lee, J.-B., 2001, 'On-chip 3D air core microinductor for high frequency applications using deformation of sacrificial polymer', *Proceedings of SPIE* **4334**: 54–62.

Cohen, A., Zhang, G., Tseng, F.-G., Frodis, U., Mansfeld, F., Will, P., 1999, 'EFAB:rapid low-cost desktop micromachining of high aspect ratio true three-dimensional MEMS', in *Proceedings of IEEE MEMS*, IEEE, Washington, DC: 244–251.

Cukauskas, E.J., Kirchoefer, S.W, Pond, J.M., 2000, 'Low-loss Ba0.5 Sr0.5 TiO_3 thin films by inverted magnetron sputtering', *Journal of Applied Physics* **88**(5): 2830–2835.

Dellmann, L., Roth, S., Beuret, C., Racine, G.A., Lorenz, H., Despont, M., Renaud, P., Vettiger, P., De Rooij, N.F., 1997, 'Fabrication process of high aspect ratio elastic structure for piezoelectric motorapplications', *Proceedings of the International Conference on Solid State Sensors and Actuators, Transducers '97*, Chicago, Volume 1, 641–644.

Despont, M., Lorenz, H., Fahrni, N., Brugger, J., Renaud, P., Vettiger, P., 1998, 'High aspect-ratio, ultra-thick, negative-tone near-UV photoresist for MEMS applications', *Sensors and Actuators A* **64**: 33–39.

Ding, X., Kuribayashi, K., Hashida, T., 1999, 'Development of a planar type micro electro magnetic distance sensor using high aspect ratio photoresist and its application', in *International Symposium on Micomechatronics and Human Science*, IEEE Nagoya, Japan: 227–233.

Fouassier, J.-P., 1995, *Photoinitiation, Photopolymerization and Photocuring: Fundamentals and Applications*, Hanser/Gardener, New York.

Gardner, J.W., Varadan, V.K, Awadelkarim, O., 2001, *Microsensors, MEMS and Smart Devices*, John Wiley, Chichester, UK.

Guerin, L., Bossel, M., Demierre, M., Calmes, S., Renaud, P., 1997, 'Simple and low cost fabrication of embedded microchannels by using a new thick film photoplastic', *Proceedings of the International Conference on Solid State Sensor and Actuators*, Part 2. IEEE, NJ: 1419–1422.

Haertling, C.L., 1989, 'Patterned ceramics for multilayer packaging using ultraviolet curabe pastes', M.Sc. thesis, Pennsylvania State University, university park, PA.

Harendt, C., Graf, H.G., Hollinger, B., Penteker, E., 1991, 'Silicon direct bonding for sensor applications – characterization of the bond quality', *Sensors and Actuators A* **25**: 87–92.

Harriss, J.E., Pearson, L.W., Wang, X., Barron, C.H, Pham, A.V., 2000, 'Membrane-supported Ka band resonator employing organic micromachined packaging', *IEEE MTT-S International Microwave Symposium* **2**: 1225–1228.

Hashimoto, T., Kikuchi, T., Watanabe, K., Ohashi, N., Saito, T., Yamaguchi, H., Wada, S., Natsuaki, N., Kondo, M., Kondo, S., Homma, Y., Owada, N., Ikeda, T., 1998, 0.2-μm bipolar-CMOS technology on bonded SOI with copper metallization for ultra high-speed processors, *Proceedings of the International Electron Devices Meeting*, San Francisco, CA, USA, IEEE, Piscataway, NJ: 209–212.

Ikuta, K., Hirowatari, K., 1993, 'Real three dimensional microfabrication using stereolithography and metal molding', in *Proceedings of IEEE MEMS*, IEEE, Washington, DC: 42–47.

Ikuta, K., Ogata, T., Tsubio, M., Kojima, S., 1996, 'Development of mass productive microstereolithography (mass-IH process)', in *Proceedings of IEEE MEMS*, IEEE, Washington, DC: 301–305.

Ikuta, K., Maruo, S., Kojima, S., 1998, 'New microstereolithography for freely moved three dimensional microstructure–super IH process with submicron resolution', in *Proceedings of IEEE MEMS*, IEEE, Washington, DC: 290–295.

Kagan, J., 1993, *Organic Photochemistry: Principles and Applications*, Academic Press, New York.

Kai, C.C., 1994, 'Three-dimensional rapid prototyping technologies and key development areas', *Computing & Control Engineering Journal* **5**(4): 200–206.

Kathuria, Y.P, 1996, 'Rapid prototyping: an innovating technique for microfabrication of metallic parts', in *Proceedings of the 1996 7th International Symposium on Micro Machine and Human Science, MHS '96*, October 2–4, Nagoya, Japan, IEEE, Piscataway, NJ: 59–65.

LaBianca, N., Delorme, J., 1995, 'High aspect ratio resist for thick film applications', *Proceedings of SPIE* **2438**: 846–852.

Lasky, J.B., 1986, 'Wafer bonding for silicon-on-insulator technologies', *Applied Physics Letters* **48**: 78–80.

Lee, K., LaBianca, N., Rishton, S., Zohlgharnain, S., 1981, 'Micromachining applications for a high resolution ultra-thick photoresist', *Journal of Vacuum Science and Technology B* **13**: 3012–3016.

Levin, I., Leapman, R., Kaiser, D.L., 2000, 'Microstructure and chemistry of non-stoichiometric (Ba,Sr)TiO_3 thin films deposited by metalorganic chemical vapor deposition', *Journal of Materials Research* **15**: 1433.

Li, H.C., Si, W., Wang, R.L., Xuan, Y., Liu, B.T., Xi, X.X., 1998, 'Oxide thin films for tunable microwave devices', *Materials Science and Engineering B* **56**: 218.

Liu, W.Y., Steenson, D.P, Steer, M.B., 2001, 'Membrane-supported CPW with mounted active devices', *IEEE Microwave and Wireless Components Letters* **11**: 167–169.

Lorenz, H., Despont, M., Fahrni, N., Labianca, N., Renaud, P., Vettiger, P., 1997, 'SU-8: a low cost negative resist for MEMS', *Journal of Micromechanics and Microengineering* **7**: 121–124.

Maruo, S., Kawata, S., 1998, 'Two-photon-absorbed near-infrared photopolymerization of three dimensional microfabrication', *Journal of Microelectromechanical Systems* **5**: 24–32.

McCormick, M.A., Roeder, R.K, Slamovich, E.B., 2001, 'Processing effects on the composition and dielectric properties of hydrothermally derived $Ba_xSr_{1-x}TiO_3$ thin films', *Journal of Materials Research* **16**: 1200.

Monneret, S., Loubere, V., Corbel, S., 1999, 'Microstereolithography using a dynamic mask generator and noncoherent visible light source', *Proceedings of SPIE* **3680**: 553–561.

Nakamoto, T., Yamaguchi, K., 1996, 'Consideration on the producing of high aspect ratio microparts using UV sensitive photopolymer', in *Proceedings of the 1996 7th International Symposium MicroMachine and Human Science, MHS '96*, October 2–4, Nagoya, Japan, IEEE, Piscataway, NJ: 53–58.

Nazeri, A., Khan, M., 1995, 'Strontium-barium-titanate thin films by sol-gel processing', *Journal of Material Science Letters* **14**: 1085.

Pappas, S.P., 1992, *Radiation Curing: Science and Technology*, Plenum, New York.

Peeters, E., Lapadatu, D., Puers, R., Sansen, W., 1994, 'PHET, an electrodeless photovoltaic electrochemical etch-stop technique', *Journal of Microelectromechanical Systems* **3**: 113–123.

Sedlar, M., Sayer, M., Weaver, L., 1995, 'Sol-gel processing and properties of cerium doped barium strontium titanate thin films', *Journal of Sol-gel Science and Technology* **5**: 201.

Suzumori, K., Koga, A., Haneda, R., 1994, 'Microfabrication of integrated FMAs using stereolithography', in *Proceedings of IEEE MEMS*, IEEE, Washington, DC: 136–141.

Taylor, C.S., Cherkas, P., Hampton, H., Frantzen, J.J., Shah, B.O., Tiffany, W.B., Nanis, L., Booker, P., Salahieh, A., Hansen, R., 1995, 'IEEE micro electro mechanical systems', in *Proceedings of MEMS '95*, IEEE, Washington, DC: 203–208.

Thornell, G., Johansson, S., 1998, Microprocessing at the fingertips', *Journal of Micromechanics and Microengineering* **8**: 251–262.

Thrope, J., Steenson, D., Miles, R., 1998, 'High frequency transmission line using micromachined polymer dielectric', *Electronics Letters* **34**: 1237–1238.

Varadan, V.K., Ghodgaonkar, D.K., Varadan, V.V., Kelly, J.F., Glikerdas, P., 1992, 'Ceramic phase shifters for electronically steerable antenna systems', *Microwave Journal* **35**: 116–127.

Varadan, V.K., Selmi, F., Varadan, V.V., 1996, 'Voltage tunable dielectric ceramics which exhibit low dielectric constants and applications thereof to antenna structure', US Patent 5,557,286.

Varadan, V.K., Xiang, J., Varadan, V.V., 2001, *Microstereolithography and Other Fabrication Techniques for 3D MEMS*, John Wiley, Chichester, UK.

Varadan, V.K, Varadan, V.V., 2001, 'Microstereolithography and fabrication of 3D MEMS and their applications', *Proceedings of SPIE* **4592**: 9–20.

Varadan, V.K., Teo, P.T., 2001a, 'Tunable dual-based ferroelectric antenna', US Patent 6,329,959.

Varadan, V.K., Teo, P.T., 2001b, 'Tunable electromagnetic coupled antenna', US Patent 6,333,719.

Wang, S.S., McNeil, V.M., Schmidt, M.A., 1992, 'An etch-stop utilizing selective etching of n-type silicon by pulsed potential anodisation', *Journal of Microelectromechanical Systems* **1**: 187–192.

Won, J.H., Paek, H., Huang, Y.S., Kim, K.K., Cho, Y.S., 1995, 'Phase formation and characteristics of rf sputtering of barium strontium titante thin films on various bottom layers', *Journal of Material Science: Materials in Electronics* **6**: 161–164.

Zhang, Z.L, McDonald, N.C., 1992, 'A RIE process for submicron, silicon electromechanical structures', *Journal of Micromechanics and Microengineering* **2**: 3138.

Zhang, X., Jiang, X.N, Sun, C., 1999, 'Microstereolithography of polymeric and ceramic microstructures', *Sensors and Actuators* **77**: 149–156.

Zhu, B., Mehta, A., Varadan, V.K., 2002, 'Integrated MOSFET based accelerometer', *Smart Materials and Structures*. In press.

Zhu, B., Varadan, V.K., 2002, 'Integrated MOSFET based hydrophone device for underwater applications', *Proceedings of SPIE* **4700**: In press.

3

RF MEMS switches and micro relays

3.1 INTRODUCTION

Perhaps the most extensively studied RF MEMS component is the RF switch. Switches and relays are simple but vital components of all automated systems. Switching provides for an interface between a system and devices with the capability for automatic redirection of signals, enhancing their flexibility and expandability. For example, in test and measurements, switching allows minimization of instruments and hence simplifies the test system. One voltmeter, for instance, can be used to measure voltages at different points on a complex circuit by switching the test points. A very fundamental definition for such a component is: *a switch is a device for making or breaking an electric circuit.* It is well known that voltages and currents in an electric circuit obey Kirchoff's voltage and current laws. In very simple terms, these laws are: (i) *the rise and drops in a voltage around any closed loop must sum to zero*; and (ii) *the total current flow into any one junction must sum to zero.* These laws are to be followed at all states of the system, and during interruptions. In other words, if there is an interruption of current in a circuit, it must do so in accordance with these laws. Although it may sound simple to interrupt a circuit, or break a conduction path, analysis of the switching process is anything but simple. This analysis becomes all the more complicated at higher frequencies. A switch in an RF signal path can introduce resistance and capacitance in signal-to-signal and signal-to-ground paths as well as cross-talk. Because of their mechanical operation, switches generally have a limited lifetime and are prone to failure. The finite time to toggle a switch is the limitation in many RF applications. This time ranges from milliseconds to a few hundred nanoseconds, depending on the type of switch.

Figure 3.1 presents some of the typical applications of RF switches (Chang, 1994). As shown in Figure 3.1(a), the switch can be used to share an antenna between a transmitter and a receiver. In the transmitting mode, the switch has to be in position 1 and in the receive mode it has to switch to position 2. In digital modulation in communication systems [Figure 3.1(b)] the switch serves as a gate to pass and stop the signal so that the desired waveform can be generated. Figure 3.1(c) presents the design of a wideband signal generator using few narrow band sources. The use of a switch in a wideband receiver for channel selection is explained in Figure 3.1(d). A switch can be used to control and select different measurement systems. For example to monitor a signal using a power meter as well as a spectrum analyzer, it is necessary to use a switch [Figure 3.1(e)]. Figure 3.1(f) shows yet another application of a switch, in radar systems.

This chapter presents various designs of RF switches using both mechanical and solid-state technology. The switch design starts with the selection of the actuation technique and the optimization of various parameters.

The RF switch is more than just a few series or shunt connections of diodes or monolithic microwave integrated circuits (MMICs). It is an integral part of any RF system. Proper selection of the right RF switch can make the difference between a marginal performance and meeting the intended design goals. Through careful selection of the key parameters, an RF switch can be optimized to achieve the desired values for specific applications. These specification parameters are described in Section 3.2. The basic of electronic switching is explained in Section 3.3. Several switching systems currently used in RF and microwave applications are introduced in Section 3.4. This section is provided to view the emergence of RF MEMS switches in the right perspective, with regard to

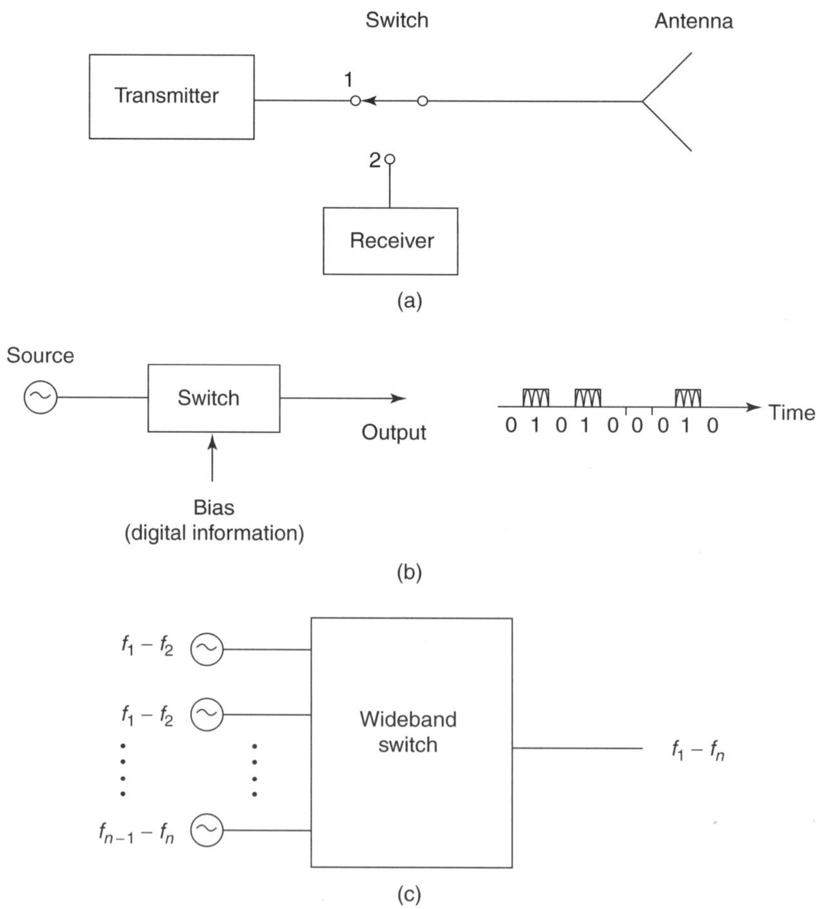

Figure 3.1 Examples of typical applications of switches. Note: f, frequency; BPF, bandpass filter. Reproduced from K. Chang, 1994, *Microwave Solid-state Circuits and Applications*, John Wiley, Chichester, UK, by permission of Wiley, © 1994 Wiley

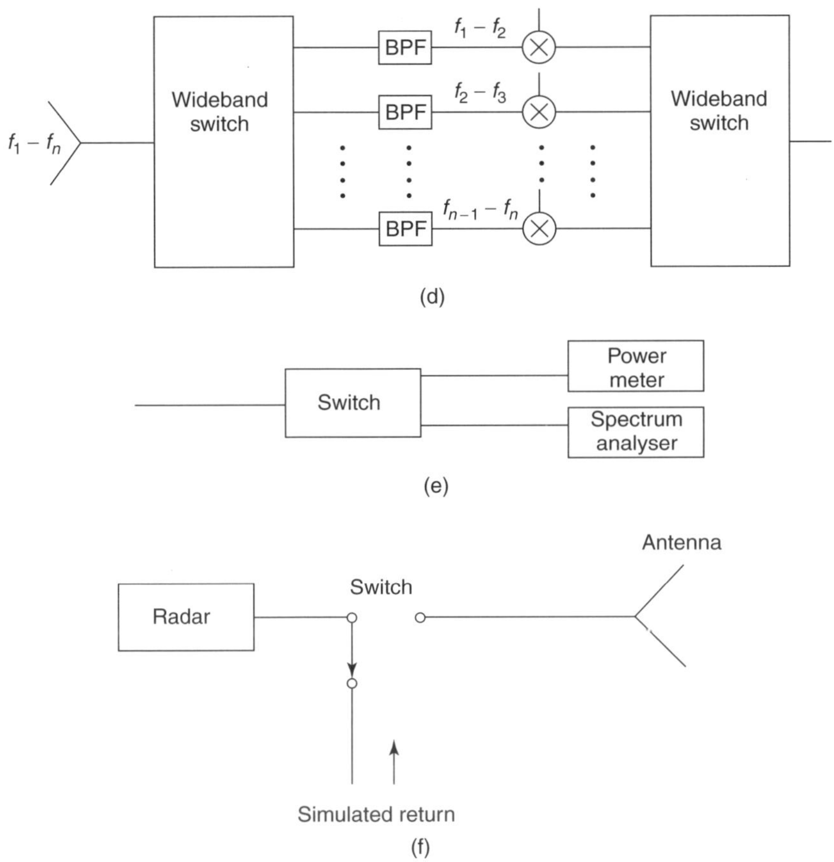

Figure 3.1 (*continued*)

the existing technology. This is followed by a description of actuation schemes useful for MEMS switches in Section 3.5. Section 3.6 reviews some of the technology behind the miniaturization of electromagnetic relays along with microactuation schemes, which is followed by the dynamics of switch operation in Section 3.7. Section 3.8 is devoted to some of the modeling aspects, where fundamental concepts to mechanical and microwave modeling of RF MEMS switches are introduced. The main concerns in MEMS switches for RF application are summarized in Section 3.9 and the chapter is concluded with Section 3.10. A detailed case study of any single device is not intended, but most of the necessary fundamental aspects of the design and modeling of RF MEMS switches are covered in these sections.

3.2 SWITCH PARAMETERS

Electrical energy is easily transported by means of conductors such as wires or bus bars, which can be controlled by relays or switches. In a simple electric circuit, the principal parts are a source of electrical energy, a load or an output device and a complete path for the flow of current. If any one of the above requirements is not fulfilled, current cannot flow in the circuit and the energy from the source cannot be delivered to the output device. Various parameters to be considered in the design of RF switches are (a) transition time; (b) switching rate; (c) switching transients; (d) RF power handling; (e) matching with circuit; (f) bandwidth; (g) insertion loss; (h) isolation; (i) series resistance; (j) actuation voltage; (k) lifetime; (l) resonant frequency; (m) interception and level of distortion; (n) phase and amplitude tracking. Apart from these, switches based on mechanical actuation schemes have a few additional parameters to be considered. Life-cycle and resonant frequency of the mechanical component are the most important of these.

Transition time

The transition time is a measure of speed with which the position of a switch can be toggled. This is defined as the time required for the output RF signal to rise from 10% to 90% of its value for off-to-on transition and 90% to 10% for on-to-off transition. In other words, it is the time taken for the output voltage to change to within 1 dB of the final state. In a simple mechanical switch, the transition time is the time required for the moving contact to leave one stationary contact and strike the opposite stationary contact.

Switching rate

The switching rate also represents the time for toggling from one state of the switch to another. However, in this case, the time is measured from 50% on the *control voltage* to 90% of the RF envelop when the switch is turned on. Similarly, when the switch is turned off, the time is measured till the RF signal voltage reaches 10% of the original. Hence, the switching rate is the time required for the switch to respond at the output due to the change in control voltage. Various delays such as driver delay and driver rise time are added to the mechanical switching time or the transition time. Therefore, a semiconductor-controlled switch is much faster than a mechanical one. The switching rate, also referred to as switching speed, is always larger than the transition time of a switch.

Switching transients

Switching transients are the exponentially decaying voltage spikes at the input, output or both of an RF signal path, due to a change in the control voltage. These switching transients are often called sidebands due to switching, and it shows important indications of the performance of a switching system. It is often required to monitor the output RF spectrum during the design of an RF system, and hence components of the RF chain, such as amplifiers and switches, must be tested with a known stimulus. Both electromechanical and electromagnetic transients exist during the switching process. While the electromechanical transient is due to mechanical motion (wherever present) of the switch element, the electromagnetic transient is due to energy exchange between electric fields and magnetic fields of the electric equipment in the network.

It may be noted that these transients arise from nonlinearities in the network. The switching transients in PIN diode switches are due to the stored charge in the intrinsic region being quickly discharged by the control voltage.[1] In balanced Schottky barrier designs, the charge stored by the diode is very small and the majority of the transients are caused by the mismatch within the drive circuits. However, the switching transient mechanism of the galium arsenide field effect transistor (GaAs FET) MMIC circuits results when the rapidly changing gate voltage is coupled to the switch output through the gate-to-channel capacitance of the FET, thus experiencing a greater feedthrough because of its faster switching speed.

RF power handling

RF power handling is a measure of how efficiently a switch passes the RF signal. This is commonly specified in terms of a 1 dB compression point, which is adopted from the amplifier characterization industry. It is commonly assumed that the output power level follows the input power with a linear ratio. But in many devices there is a maximum power above which this linearity does not hold. The 1 dB compression point is defined as the maximum input power level at which the output power differs by 1 dB with respect to linearity. The 1 dB compression points and the power handling of many devices such as PIN diodes and MMIC switches are functions of frequency.

Impedance matching

Impedance matching is a critical element in all high-frequency design. The switching device should be matched at both input and output sides, for both the on and the off state of the switch, to minimize its impact on the performance of the rest of the system. An improperly matched component results in unwanted reflections within the circuit, which can cause major damage to other systems. Although an ideal match is seldom achieved, care should be taken to minimize the reflections within acceptable limits.

Available bandwidth

Although most of the switching systems do not have a limit on the lowest frequency of operation, they do have an upper limit. For semiconductor devices this is due to the finite time in carrier mobility. The losses incurred from resistance and parasitic reactances are the main cause limiting the performance of electromechanical switches at higher frequencies.

Insertion loss

The insertion loss of an RF device is a measure of its efficiency for signal transmission. In the case of a switch, the insertion loss is specified only when its state is such that signal is transmitting or when the switch is in the on state. This is specified in terms of the transmission coefficient, S_{21}, in decibels, between the input and output terminals of the switched circuit. Usually specified in decibels, one of the design goals for most

[1] For the origin of the name PIN diode, see Section 3.4.2.

of the RF switches is to minimize the insertion loss. The insertion loss tends to degrade with increase in frequency for most of the solid-state switching systems. Compared with these, RF MEMS switches can be designed to operate with a small insertion loss at several gigahertz. Resistive losses at lower frequencies and skin-depth effects at higher frequencies are the major causes for losses.

Isolation

The isolation of a switching system is specified when there is no signal transmission. This is also measured as S_{21} between the input and output terminals of the switched circuit, under the no-transmission state or when the switch is in the off condition. A large value (in decibels) indicates very small coupling between input and output terminals. Thus the design goal is to maximize the isolation. In RF MEMS switches isolation may degrade as a result of proximity coupling between the moving part and the stationary transmission line as a result of leakage currents.

Series resistance

In many instances, the switch is connected in series with the transmission path. Any resistance offered by the switch during signal transmission (on state) would result in loss of signal level. One way of representing this, especially at lower frequencies, is to use the series resistance of the switch while it is conducting. At higher frequencies, this is often represented by the insertion loss.

Actuation voltage

All automated systems require a control signal for actuation. Depending on the scheme and its efficiency, these voltages vary significantly. Although this is not a big problem with semiconductor-based switching systems, one of the design objectives of state-of-the-art electromechanical switching systems is to bring down these voltages to levels compatible with the rest of the circuit.

Life-cycle

This is also not a significant issue with semiconductor-based switches, but in all schemes that involve moving parts, the lifetime may have to be considered. The breakdown of such moving components because of fatigue and environmental effects limits the lifespan of these systems.

Resonant frequency

The moving parts in mechanical switches have resonant frequencies that can be modeled in terms of their effective spring constants and resonating mass. At this frequency, the potential energy and the kinetic energy tend to resonate. This frequency limits the maximum rate at which the switch can be toggled, but this virtually has no bearing on the frequency of the actual RF signals the switch carries.

In electrical circuits, resonance occurs when the reactance of an inductor balances with the reactance of the capacitance for a given frequency. In a resonant circuit, when it is in series resonance, the current will be at a maximum, offering a minimum impedance and vice versa in parallel resonance.

Intercept points

The intercept point is the extrapolation of the distortion power to the power level of the drive signals, assuming the switch has no compression of the signals. It is usually assumed that the intercept points are related to the frequency of the minority carrier's lifetime in PIN diodes. The ratio of the stored charge to diode series resistance is the common driving factor in PIN diode distortion.

Phase and amplitude tracking and matching

Phase and amplitude matching specifications are important in multi-throw switches since, depending upon the design of the device, the individual throws can have different electrical lengths and losses. This will result in different phase and amplitude characteristics for each throw.

3.3 BASICS OF SWITCHING

Figure 3.2 presents general switch configuration in an electric circuit. The contact mechanism can be either mechanical switches such as relays, rotary switches, plunger/snap-acting switches or solid-state switches such as FETs, MMICs, MEMS, etc. Though the make-and-break arrangement to interrupt the current flow in a circuit using a mechanical switch is simple, it is not in microscopic view. In the case of an opening switch with a steady-state current flow, the flow of current from source to load tends to change when the switch contact begins to move. At microscopic distances, the current density become high enough such that portions of the metallic surface actually melt owing to resistive heating; and the liquid metal vapor plasma state continues the electrical conducting path as the contact physically parts. This can cause electrical breakdown or an arc when the contacts are fall apart, around few microns, which is critical in the case of MEMS microswitches,

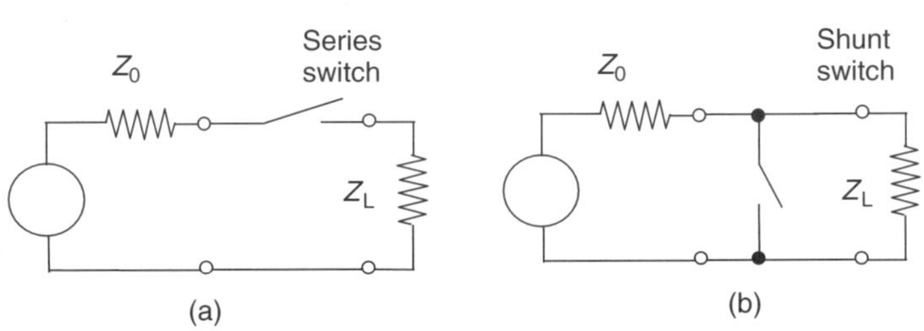

Figure 3.2 Typical switch configurations: (a) series switch, (b) shunt switch. Note: Z, impedance

which is discussed in Section 3.6. However, semiconductor switches do not need to arc to break a circuit, since they supply their own conducting medium. A semiconductor can conduct current only as long as mobile carriers, either electrons or holes, are provided from the supply devices. By turning off the switch, the semiconductor material will revert to an insulating state and block the flow of current by blocking the flow of mobile carriers – that is, the switch will turn off.

3.3.1 Mechanical switches

Examples of mechanical switches are toggle switches, push-button switches, relays, etc. In relays, the electromagnetic force arising from the flow of current through a coil causes a metallic contact physically to open or close. The most widely used switching elements are armatures and are available in various forms for a wide variety of specifications. Because of the physical movement of the contact points, the relays suffer from low lifetime, degradation of electrical contacts and mechanical wearout. All these cause a very high rate of failure in operation.

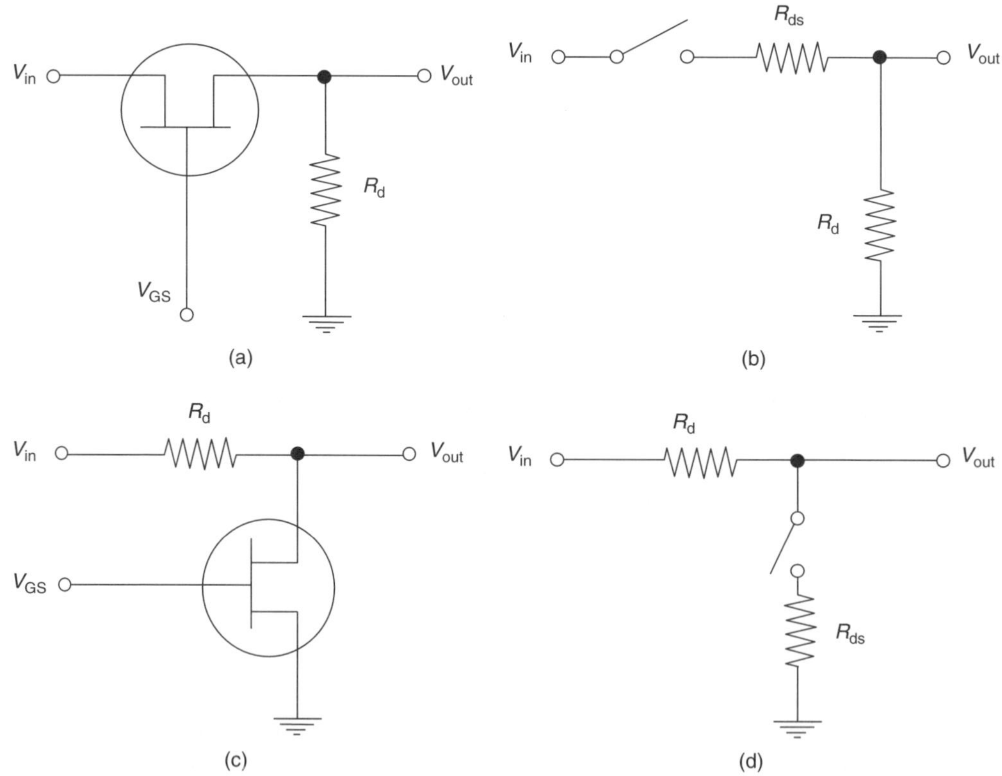

Figure 3.3 Typical configuration of FET (field effect transistor) switching circuits: (a) series switch; (b) series switch equivalent circuit; (c) shunt switch; (d) shunt switch equivalent circuit. Note: V_{GS}, gate voltage; V_{in} and V_{out}, input and output voltage, respectively; R_d, R_{ds}, bias resistance of the circuit and the drain-source resistance of the FET, respectively

3.3.2 Electronic switches

Varieties of switching components are used in electronic circuits; examples are diodes, transistors and FETs. Figure 3.3 presents the basic switching configuration of FET switches. The FET can be turned on and off by controlling the gate voltage V_{GS}. In a typical configuration, the resistance R_d is much greater than R_{ds}, and V_{in} is less than 100 mV. As shown in the equivalent circuit, when V_{GS} is zero, the switch is closed and the output is approximately equal to the input. When V_{GS} is more negative than $V_{GS(off)}$, the FET is open and V_{out} is zero. In shunt configuration, when V_{GS} is zero, the switch is closed. Because of the voltage divider, the output voltage is low. When V_{GS} is more negative than $V_{GS(off)}$, the FET is cut off and the switch in Figure 3.3(d) is open, which will give the same output as input voltage.

3.4 SWITCHES FOR RF AND MICROWAVE APPLICATIONS

The main uses of RF switches in the telecommunication industry are for signal routing, in impedance matching networks and for changing the gain of amplifiers. Telecommunication covers a broad range of frequencies, from below HF through VHF. Several new applications at frequencies in the microwave and millimeter wavebands are also well established. These include AM band (at the low end of MHz), commercial FM band (88–108 MHz), military handheld radio transceivers and cellular radios (900 MHz and 2.4 GHz) and Bluetooth™ (2.45 GHz). In addition to these, there are also other applications at frequencies ranging from Ku-band (12.4–18 GHz) to the upper side of W band (75–110 GHz), which require the use of high-quality RF switches. The wide frequency spectrum used for telecommunications demands different switch technologies for various frequency bands of application (Yao, 2000). Furthermore, increased use of personalized communication terminals requires downsizing of mobile systems and their accessories.

Along with varying frequency requirements, the power handling capability of the device may also differ for various applications. In the case of silicon FETs, for example, it can handle high power at low frequency, but the performance drops off dramatically as frequency increases (Ota *et al.*, 1995). In the case of GaAs metal–semiconductor field-effect transistors (MESFETs) (Ayali, 1982; Caverly, 1993; Gopinath and Rankin, 1985; Slobodnik *et al.*, 1989) and PIN diodes (Alekseev and Pavlidis, 1998; Kobayashi *et al.*, 1993; Putnam *et al.*, 1994) the high-frequency operation is fairly well with small signal amplitudes. In short, when the signal frequency is greater than a few gigahertz these solid-state switches have large insertion loss (typically 1–2 dB) and poor isolation (~ -20 to -25 dB). The need for an alternative switch is inevitable and MEMS exhibits promising characteristics as the new technology for integrated switching devices.

The selection of a switch largely depends on the signal level and speed of operation that the application demands. Both mechanical and solid-state switches have their own advantages and disadvantages. Owing to the integration compatibility and low manufacturing cost, most present switching techniques for RF and microwave applications are met by the solid-state semiconductor devices. This section is therefore intended to present different RF switching systems in perspective, before discussing the RF MEMS switches themselves.

3.4.1 Mechanical RF switches

Electromechanical switches have been widely used for high-power applications in TV, AM, FM, HF and other broadcast systems. Owing to the high power, these devices are made available with either waveguide or coaxial connector ports. Electromechanical RF and microwave switches with various specifications and frequency bands are available from several commercial vendors,[2] for power levels ranging from kilowatts to megawatts. For example Model RP-A82 from Microlab/FXR (www.microlab.fxr.com) is a waveguide mechanical switch, which controls 50 kW average power and 5 MW peak power from 2.9 GHz to 3.1 GHz. Since these switches should handle very high power, it should achieve very good voltage standing wave ratio (VSWR) and insertion loss. Model RP-A82 has been designed with an insertion loss of 0.07 dB and VSWR 1.1 : 1 with an isolation of 70 dB across 200 MHz bandwidth. Figure 3.4 shows a typical waveguide electromechanical switch from Advanced Switch Technology, Canada, which can be driven by a dc voltage. The low-frequency coaxial switch from Micro Communications (Manchester, NH) can handle 1500 kW peak power from dc to 800 MHz. Figure 3.5 presents some of the examples of coaxial switches operating from dc to 26 GHz, with switching time around 15 ms. The space programs Pathfinder, Geosat, Immersat III and Intelsat are a few examples of where Dow-Key (www.dowkey.com/products/) microwave switches are used. An extensive list of manufactures offering high-frequency switches is provided in the microwave and RF product data directory (www.mwrf.com/products/). The selection of the RF switch for a circuit depends mainly on the type of platforms and applications.

Mechanical switching is done through a make or break in the transmission line or electrical path by a control signal activating an electromagnetic relay. These mechanical switches can be designed to turn on and off in different ways. For example, in a latching design the switch remains in a preselected position when the actuating voltage is removed, holding the switch in that position until the next actuating voltage is applied. Accordingly,

Figure 3.4 Waveguide switches. (Reproduced courtesy of Advanced Switch Technology, Ontario, Canada www.astswitch.com)

[2] See www.microlab.fxr.com, www.mcibroadcast.com, www.elmika.belt.net, www.walmba.org, www.dowkey.com, www.astswitch.com and www.macom.com/products.

Figure 3.5 Various coaxial switches operating from dc to 26.5 GHz. The switching time is 15 ms. (Reproduced courtesy of DBP Microwave, Pasadena, CA. www.dbp4switches.com)

some switches are built *normally open* with all output ports disconnected from the input until the actuating voltage is applied. In other designs, the switch port opens with the actuation voltage and returns to the predetermined *closed position* upon the removal of the voltage.

A typical fail-safe electromechanical switch uses an electromagnet for actuation to move an arm, and a spring to pull it back to the initial position. Such a switch is always in the normal position until the application of current to the coil actuating it. The switch returns to the normal position as the power is removed from the coil. This device finds several applications where the switch should be in the normal position in the event of a power failure.

In many switching configurations, when the switch is off (or the electrical path open), there is no alternative path for the RF power to propagate. To prevent high RF power from reflecting back to the source in such instances, circulators or coupling devices must be incorporated in the design to direct the RF energy to a dummy load. Internal terminations, generally 50 Ω resistors, are added to the switch ports to absorb the RF power while the electrical path is interrupted (Losee, 1997).

Even though many of these electromechanical switches exhibit excellent RF characteristics such as low insertion losses and high isolation, typically up to several hundreds of megahertz, they have a very slow switching speed. In general, these switches are operating at a speed of 2 to 50 ms and are rated for several million switching operations. This is because the switching is performed by physically blocking or opening the transmission path in a device. The mechanical resonant frequency of these moving parts determines the maximum frequency of operation (switching speed) of these devices. For higher-speed operation semiconductor-based switching devices, discussed next, are preferred.

3.4.2 PIN diode RF switches

Electromechanical switches perform poorly when used as high-speed switches. For applications where the operational speed is more important than the power handling, solid-state switches are preferred. These electronic switches have speeds orders of magnitude faster than mechanical switches. Furthermore, they can be housed in packages measuring a fraction of the size of the equivalent mechanical switches operating at the same frequency. Electronic switches utilize semiconductors – either diodes or FETs – as the switching element. The most common electronic switch is the PIN diode, with diodes placed in series, shunt or combinations of them. Since the diode is in the RF path, the series designs exhibit low insertion loss over a wide frequency range, while shunt design provides high isolation. When a compromise of good isolation and low insertion loss over the broad frequency range is required, a series–shunt configuration of PIN diodes may be resorted to.

Semiconductor devices such as PIN diodes have a semiconductor junction acting as an electronic control element. This junction can be switched on or off by controlling its bias voltage. The series impedance of a PIN diode changes from low resistance under forward bias to a low-loss capacitor at reverse bias. Thus by switching between forward and reverse bias the PIN diode can be used as an RF switch to route the signal or to switch within the circuit to give the desired RF outputs. This simple switching mechanism is utilized in many RF circuits and applications. This makes a PIN diode an attractive component for RF switching and phase shifting applications. One of the important applications of such PIN diodes is the design of phased array antennas.

The name of PIN diodes comes from its unique doping profile, which consists of a highly doped intrinsic i-region sandwiched between doped p and n regions. When the diode is forward biased, electron–hole pairs are injected into the i-region. These charge-carrier pairs do not recombine immediately, causing the charge to be stored in the i-region and hence reducing the resistivity. The amount of charge stored in the i-region is the product of the forward bias current I_f, and the minority carrier life time τ:

$$Q = I_f \tau \tag{3.1}$$

The PIN diode is similar to p–n diodes but has a very small junction capacitance because of its wider depletion region. This effect is very useful for a diode to be used as a microwave switch because the lower the capacitance, the higher the impedance of the diode under reverse bias and the more the isolation when the device is in open-circuit (off) configuration. Figure 3.6(a) presents the voltage–current characteristics of a PIN diode. The equivalent circuit for a PIN diode when it is off or on are shown in Figure 3.6(b). When the diode is on the series resistance R_{on} has a very small value. When it is off the capacitance is very high. It may be noted that the resistance in the latter case is in series, which is contrast to the equivalent circuit of a FET switch where the capacitance and resistance are in parallel.

Depending on the structure and fabrication process, PIN diodes are divided into two categories: bulk and epitaxial diodes. In a bulk diode, a lightly doped n-type substrate is diffused with p-type and n-type material to form the bottom and top, which constitutes the PIN structure. A typical i-region width for a bulk diode is about 100–250 nm, with a

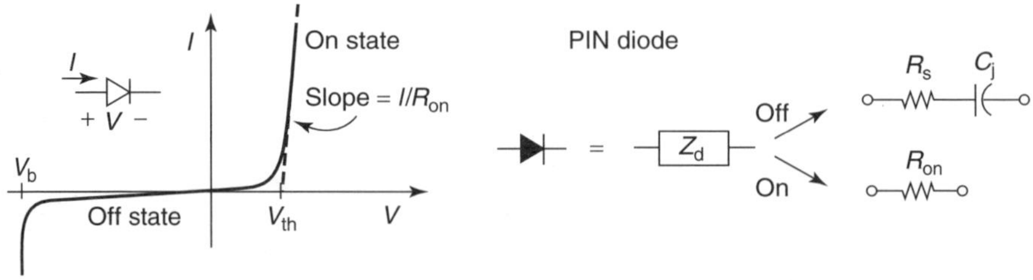

Figure 3.6 (a) Current–voltage characteristics of a PIN diode; (b) PIN diode on-state and off-state characteristics. Note: I, current; C_j, capacitance; R_{on} and R_s, series resistance when the diode is on and off, respectively; V, voltage; V_b, voltage; V_{th}, voltage; Z_d, impedance

carrier lifetime of 300–6000 ns. In an epitaxial diode, on the other hand, a thin epitaxial layer (i-region) is grown on a heavily doped n-type substrate. A thin layer of heavily doped p-type material is diffused onto its top layer. The small width (3–20 nm) and the imperfections on the crystal lattice of the i-region results in a shorter lifetime (5–300 ns). The comparatively low resistance of the i-region of an epitaxial diode makes it an attractive choice for low-power switching applications. The lowest operating frequency depends on the minority carrier lifetime τ (Chang, 1990):

$$f_c = \frac{1}{2\pi\tau} \tag{3.2}$$

In most of the microwave applications, the control voltages are supplied by drivers operated with TTL (transistor–transistor logic) circuits. Parameters such as actuating voltage and operational speed of the driver circuits determine the speed of operation of the diode. Although PIN diode switches can achieve a transition time of 5 ns, delays through the drivers are typically about 30 ns, thus making the switching rate much slower than the transition time.

PIN diodes can be configured as both series and shunt switches as shown in Figure 3.7. The switches can be designed with diodes series mounted or shunt mounted with respect to the transmission direction. Figure 3.8 shows the series mounted circuits on a microstrip line in single-pole single-throw (SPST) and single-pole double-throw (SPDT) configurations. In an SPST switch [Figure 3.8(a)], the signal passes through the line if the PIN diode is forward biased. In the SPDT case, however, the signal passes through the line from port 1 to port 2 if the diode is positively biased, and from port 1 to port 3 if the bias voltage is negative [Figure 3.8(b)]. In the case of a series switch [Figure 3.7(a)], forward biasing the diode (on) results in low impedance and the signal flows from input to output (RF on). When the diode is reverse biased (off), it is in a high impedance state causing the input signal to be reflected (RF off). In case of a shunt switch [Figure 3.7(b)], when the diode is forward biased (on), the diode is in a low impedance state tending to be a short circuit for the signal line (RF off). When the diode is reverse biased (off) state, the high shunt impedance loading in the signal line is negligible and the signal passes from input

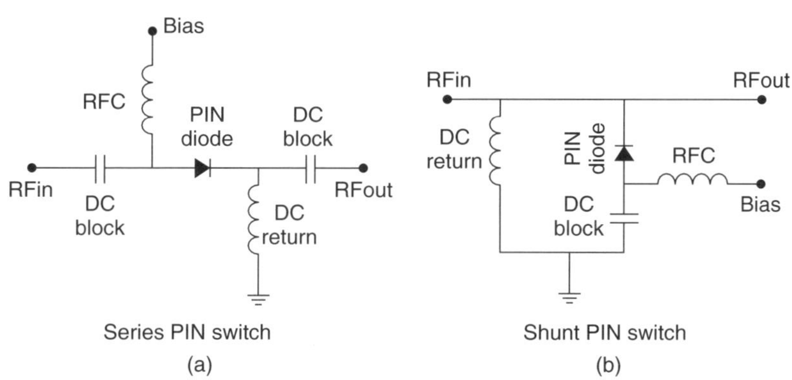

Figure 3.7 Configuration of (a) series and (b) shunt switch with PIN diode. Reproduced from A.M. Street, 2000, 'RF switch design', *IEEE Training Course 2000: How to Design RF circuits*, publication 2000/027, IEEE, Piscataway, NJ, USA: 4–4/7, by permission of IEEE, © 2000 IEEE

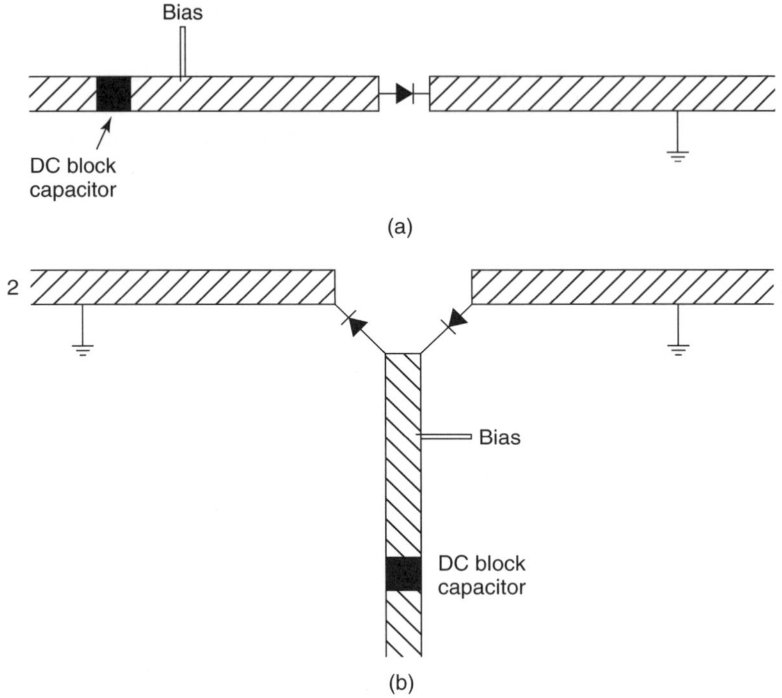

Figure 3.8 Series mounted PIN diode switches in microstrip circuits: (a) single-pole single-throw; (b) single-pole double-throw. Reproduced from K. Chang, 1994, *Microwave Solid-state Circuits and Applications*, John Wiley, Chichester, UK, by permission of Wiley, © 1994 Wiley

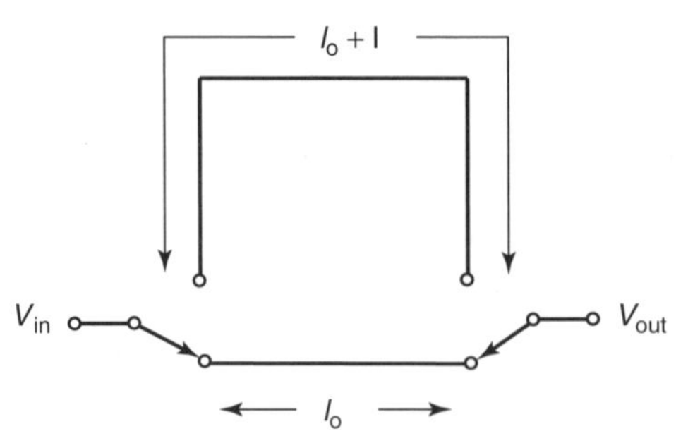

Figure 3.9 Schematic representation of use of PIN diode as a switched line phase shifter

to output (RF on). The series diode offers superior isolation since the diode inductive reactance increases the impedance in the off state. However, the diode inductance will also degrade the insertion loss.

We end the discussion on PIN with a listing of some of its applications (Bahl and Bhartia, 1988; Chang, 1990; Street, 2000). It is used to:

- switch different lines of a phase shifter in phased array antennas, as shown in Figure 3.9; the phase shift is obtained by switching the signal between the two different path lengths l_o and $(l_o + l)$ and corresponds to the additional path delay βl, where β is the propagation constant of the medium

- protect the receiver from the transmitter in radar applications by using the switch as an isolation device
- connect many narrow-band devices to create a wideband system
- control the gate signals in a digital modulator

3.4.3 Metal oxide semiconductor field effect transistors and monolithic microwave integrated circuits

PIN diode switches were the only choice for high-speed microwave applications until mid-1980s. However, with the emergence of GaAs FET devices, switches much faster than PIN diodes have been developed making use of the very high mobility of the carriers in GaAs. This has triggered a growing market for GaAs MMIC switches, generally based on MOSFETs as switching elements. In general, GaAs FETs are characterized by higher 'on' resistance and larger 'off' capacitance compared with PIN diodes. As a result, switches and phase shifters using GaAs FETs must account for their finite 'on' resistance (Bahl and Bhartia, 1988).

The FET switch is a three-terminal device with the gate voltage acting as the control signal. The low and high impedance states required for switching applications are obtained by making the gate voltage equal to zero and greater than the pinch-off voltage, respectively (Bahl and Bhartia, 1988; Browne, 1989). In either case the switch requires virtually no dc bias power. Hence the analysis of these devices can be simplified by considering only a passive mode of operation (Ayali, 1982). The switch on state resistance R_{on} depends the current path and the cross-sectional area (Ayali *et al.*, 1982). FET switches are regarded as important MMIC elements with applications such as in switched lines (Ayali *et al.*, 1982; Sokolov *et al.*, 1983), loaded lines (Andricos, Bahl and Griffin, 1985), high/low pass phase shifters (Schindler and Miller, 1988) and transmit/receive (Ayali *et al.*, 1983; Lau *et al.*, 1988) and SPDT switches (Schindler and Morris, 1987). Figures 3.10 and 3.11 show some of the packaged chip and MMIC SPDT switches. Model SW-229 [Figure 3.10(a)] draws only 1 mA current from ±5 V supply with 175 ns switching time.

Series and shunt configurations of FET switches are given in Figure 3.3. In both cases, the FET can be turned on or off by controlling the gate voltage V_{GS}. Silicon FETs can handle high power at low frequencies, but the performance drops off dramatically as frequency increases. However GaAs MOSFETs and PIN diodes have good high-frequency performance characteristics for small signal amplitudes.

In a hybrid microwave integrated circuit (MIC), the active and lumped components are connected to distributed circuit components on a planar transmission line, typically microstrip, by soldering or wire bonding techniques. MMICs, however, are becoming widespread because both active and passive components are fabricated simultaneously on a semi-insulating semiconductor substrate (e.g. GaAs). This approach eliminates the need for attaching discrete components and thus reducing losses due to wire bond interconnects. Major advantages of MMICs are low cost, small size, simplified packaging, better reliability, capability for high volume of manufacturing, and multifunctional performance

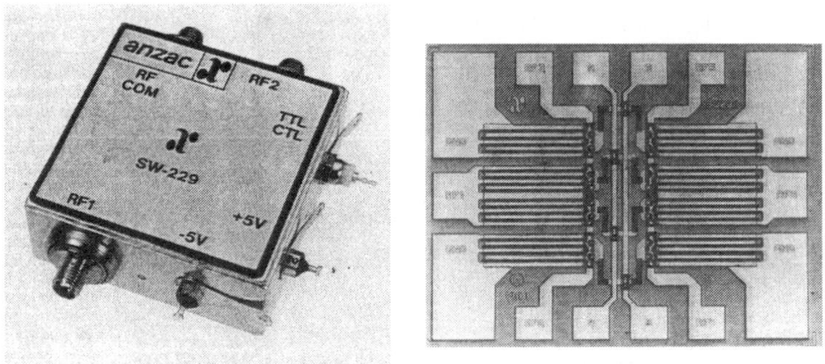

(a) (b)

Figure 3.10 (a) Packaged chip and (b) monolithic microwave integrated circuit switches. The typical packaged unit single-phase double-throw model SW-229 shown in part (a) draws only 1 mA current from ±5 Vdc, with 20 ns transition time and 175 ns switching time from dc to 2 GHz. Reproduced from J. Browne, (1989), 'Switches perform high-frequency signal routing', *Microwaves and RF* (July): 125–132, by permission of Penton Media, Adams-Russel Electronics, Anzac Div, Burlington, MA

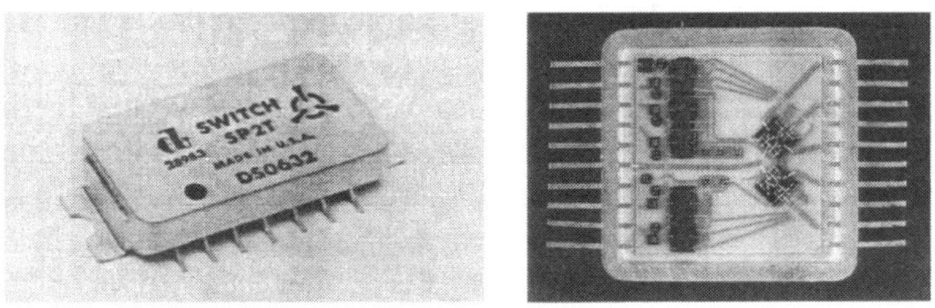

(a) (b)

Figure 3.11 (a) Model DSO632 Single-Phase double-throw switch with TTL driver from Daico Industries; (b) monolithic microwave integrated circuit SP5T switch from Tachonics Corporation. Reproduced from J. Browne, (1989), 'Switches perform high-frequency signal routing', *Microwaves and RF* (July): 125–132, by permission of Penton Media, Adams-Russel Electronics, Anzac Div, Burlington, MA

on a single chip. Compared with typical PIN diode switches, MMIC switches operate over a broad bandwidth. Their power consumption is much lower. MMICs have faster switching speed and lower transients compared with PIN diodes. However, MMICs have a higher insertion loss.

For signal frequencies above 1 GHz, these solid-state switches have large insertion loss at on state (typically 1–2 dB) and poor isolation at off state (−20 to −25 dB). These limitations of solid-state technology have led to increased research for MEMS-based RF switches. The fundamentals of these devices are discussed in the next section.

3.4.4 RF MEMS switches

Advances in silicon-based processing technology in the second half of the twentieth century resulted in a rapid improvement in the performance of computers. These have millions of transistors, the on or off position of which being controlled by a control signal. These can be thought of as a highly integrated array of switches. But for RF applications, these fast-acting solid-state switches continue to have disadvantages such as low power handling and high resistive losses. Electromechanical switches, in contrast, are high power devices, but useful only at lower RF frequencies, and operate at a much slower speed. A technology that has emerged making use of the advantages of both solid-state and electromechanical systems, and overcoming most of their disadvantages, is based on microelectromechanical systems (MEMS).

Most MEMS devices use silicon as the basic material and the technology is derived largely from advances in silicon processing. It has been widely established that silicon also possesses unique useful mechanical properties that facilitate fabrication of miniaturized high-precision mechanical devices and components. Advances in MEMS technology in the late twentieth century made possible the design and fabrication of micromechanical switches. MEMS switches have low resistive loss, negligible power consumption, good isolation and high power handling capability compared with semiconductor switches. The development of substrate-independent MEMS switches could introduce affordable microwave components and systems into the marketplace, with a dramatic improvement in both performance and cost.

From the fabrication point of view, the main drawback of conventional electromechanical switches is the impossibility of batch production. By adopting the highly repeatable nanofabrication techniques for MEMS, they can be easily reproduced and integrated with the existing silicon technology used for their control circuits. In addition to being smaller, lighter, faster and consuming less power, micromechanical switches and relays have a high off-state to on-state impedance ratio. The first reported work on MEMS switches is by Petersen (1978, 1979). This switch is fabricated on silicon, with an electrostatically movable SiO_2 membrane as the switching component.

The method of actuation is the critical parameter for RF MEMS switches because the greater the degree of mechanical complexity, the more issues of wear and tear will there be. Actuation mechanisms used in MEMS include electrostatic, electromagnetic, magnetic, piezoelectric and thermal. However, the electrostatic actuation mechanism seems to be the most common method for MEMS switches because of its low power consumption.

Compared with PIN diode switches, electrostatic MEMS switches have several advantages. These MEMS switches have very low power consumption (of the order of microjoules) for the switching process. They also possess a very large on–off capacitance ratio. However, one of the serious disadvantages of the MEMS switch is its low switching speed. Switches with speeds as fast as 1–10 μs have been fabricated using the current technology. This is much slower than current solid-state switches. Other key research issues include high actuation voltages and low RF power handling. A brief comparison of the properties of various switches are provided in Table 3.1. Figure 3.12 shows a comparison of the cost, power consumption and insertion loss of most commonly used RF switches with those of MEMS switches.

The MEMS switches can be characterized based on four topologies: (1) the co-planar cantilever series switch, (2) the shunt cantilever switch, (3) the membrane switch and (4) the mercury contact switch.

3.4.5 Integration and biasing issues for RF switches

The widespread application of MEMS can yield reflective changes in the architecture of modern communication systems. The technology of RF switches promises to provide

Table 3.1 Comparison of RF switches with MEMS switches

	Relay	Schottky	MOSFET	P-I-N diode	MMIC	MEMS switches
Transition time (ns)	N/A	5	N/A	300	5	N/A
Switching speed (S)	2×10^{-3}	N/A	3×10^{-9} to 10^{-8}	650×10^{-9}	25×10^{-9}	$\sim 10^{-6}$
Voltage (V)	100–200	N/A	5–50	3–5	N/A	3–30
Current (μA)	1–2	N/A	<10	10 000	N/A	<10
Contact resistance (Ω)	0.005–0.075	N/A	<1	1	N/A	3–5
Figure of merit (GHz)[a]	N/A	N/A	300–400	1500–2000	N/A	>3000
Life-cycles	10^7–10^8	$>10^9$	$>10^9$	N/A	N/A	$>10^{13}$
Loss at 1 GHz (dB)	0.25	0.85	0.5–1.0	0.5–1.0	1.1	0.1
Isolation at 1 GHz (dB)	70	62	20–40	40	60	>40
Size (mm^2)	N/A	N/A	1–5	0.1	N/A	<0.1
Total current (mA)	~60	180	N/A	30	0.15	N/A
Supply voltage	+12	+5	N/A	±5	+5	N/A
Transient signals (mV)	N/A	10	N/A	210	100	N/A
Bandwidth, MHz	dc to 1200	2–500	N/A	20–2000	5–4000	dc–40 000
Design flexibility?	N/A	N/A	N/A	Yes	N/A	N/A

N/A, not applicable.

Note: MEMS, microelectromechanical systems; MMIC, monolithic microwave integrated circuit; MOSFET, metal oxide semiconductor field effect transistor; PIN, see Section 3.4.2; RF, radio frequency.

[a]Figure of merit, f, is equal to $\dfrac{1}{2\pi R_{on} C_{off}}$.

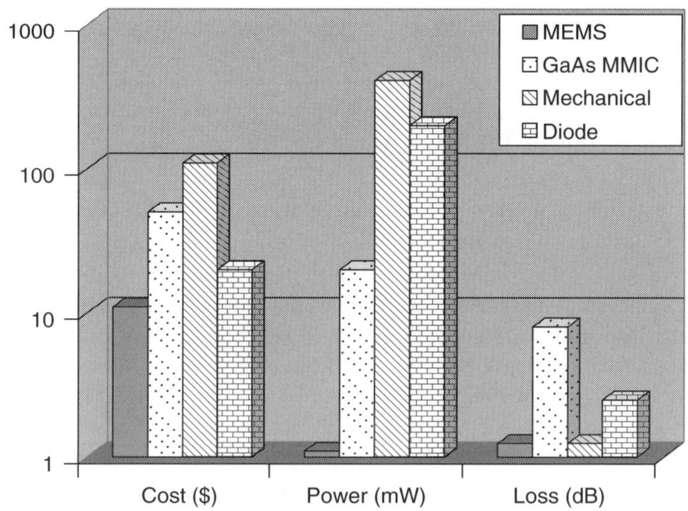

Figure 3.12 Comparison of cost, power consumption and loss in MEMS devices with that of popular RF switching devices. Note: MEMS, microelectromechanical systems; MMIC, monolithic microwave integrated circuit; RF, radio frequency

dramatically improved switching and filtering capabilities at the front end of next generation (3G) communication radio receivers and wireless technology. For a typical cellular phone at a PCS frequency, the switch is required to have an isolation of 60 dB and an insertion loss of 0.3 dB. In addition the intermodulation and distortion due to RF MEMS switches are extremely low, rendering their dynamic range more than adequate for most of high-tech receiver applications.

Most of the switching devices require bias voltage for turning the device on or off. The bias must be applied such that its path does not influence the RF signal. In practice the bias is generally applied through additional components (e.g. an RF choke) that blocks RF while acting as a dc short. In three-terminal devices such as FETs this may not be a significant issue, but many designs of electrostatic RF MEMS switches do require a well-designed layout for isolating the dc from RF transmission lines. Unwanted dc in an RF signal path can damage many RF devices and test systems. In these devices, dc can also be blocked in the RF path. Such dc blocks are commercially available for different connector configurations. These dc blocks can also be integrated into the circuit, by means of coupled lines.

Practical aspects such as integration of RF MEMS components with MMIC technologies and their packaging are discussed in Chapter 9.

3.5 ACTUATION MECHANISMS FOR MEMS DEVICES

Advances in micro and nano fabrication techniques influenced the outlook of MEMS. Recently, a microelectromechanical system has been defined as a miniature device or an array of combined electrical and mechanical components fabricated with IC batch-processing techniques (Bryzek, Petersen and McCulley, 1994). The key advantage of the MEMS device is its ability for bulk reproduction and batch fabrication. Most mature fabrication technologies such as bulk micromachining and most recent techniques such as LIGA (German acronym for *Lithographie, Galvanoformung, Abformung*, meaning lithography, galvanoforming, moulding) have their own merits and demerits while adapting for MEMS fabrication. It is still early for a time-tested categorization of RF MEMS devices because the development of MEMS devices for RF applications is yet to mature. For the RF design engineer, even though the MEMS devices are mechanically actuating with prescribed electric fields, it gives clear demarcation in the functionality inside the RF circuit. In an RF circuit, switching devices such as bipolar junction transistors (BJTs) or FETs can be replaced with MEMS switches or it is possible to use the MEMS device to re-route the RF signal between different transmission paths. However, in many RF circuits, the most important consideration should be its reactive elements and the induced inductive or capacitive elements because each have their own clear operational functions in an RF circuit. The actuating and control circuits should not load the circuit, and the dc voltage has to be isolated from the RF path.

The actuation of the switch can be electrostatic, magnetic or electromagnetic: each has its pros and cons. The advantage of electrostatic actuation is that there is no current consumption; its drawback is that it requires a higher actuation voltage, typically 5–100 V. The advantage of electromagnetic actuation is the lower voltage, but, with significantly higher current consumption. Electrostatic switches offer the most promise as configuration switches, where low power consumption is the key factor.

3.5.1 Electrostatic switching

3.5.1.1 Series contact switches

The importance of silicon in the early 1980s revolutionized integrated circuit technology. Later, silicon also proved to be a material for the development of precision, miniature mechanical systems. In the late 1970s Petersen (1978, 1979) developed a new class of micromechanical membrane switch on silicon. This electrostatically actuated microcantilever, was demonstrated to switch low-frequency electrical signal and has operating characteristics in between the conventional silicon transistors and electromagnetic relays. These extremely small devices, typically less than 100 μm long, are controlled and actuated by electrostatic fields. This device is composed of thin (0.35 μm) metal-coated insulating membrane attached to the silicon substrate at one end and is suspended over a shallow rectangular pit. The pit is made by etching the silicon out from under the deposited insulating film in a carefully controlled etching procedure.

The compatibility with the conventional silicon circuitry triggered further extensions of the above switch and since then MEMS switches have demonstrated improved performance at microwave frequencies using cantilever (Bozler *et al.*, 2000; Chang and Chang, 2000; Hyman *et al.*, 1999a, 1999b; Larson, Hackett and Melendes, 1991; McNie *et al.*, 2000; Pacheco, Nguyen and Katehi, 1998; Park *et al.*, 2000; Randall *et al.*, 1996; Shen and Feng, 1999; Tangonan *et al.*, 1999; Yao and Chang, 1995; Zavracky, Majumdar and McGruer, 1997; Zavracky *et al.*, 1999), capacitive (Brown, 1998; Goldsmith *et al.*, 1998; Majumdar *et al.*, 1997; Muldavin and Rebeiz, 2000a, 2000b; Pacheco, Katehi and Nguyen, 2000; Pacheco, Peroulis and Katehi, 2001; Park *et al.*, 2001; Poltz *et al.*, 2001; Rizk *et al.*, 2001; Santos *et al.*, 1997; Sovero *et al.*, 1999; Ulm *et al.*, 2000; Yao *et al.*, 1997) and membrane (Goldsmith *et al.*, 1995, 1996, 2001; Hiltmann *et al.*, 1997;

Muldavin and Rebeiz, 1999; Tan and Rebeiz, 2001; Yao *et al.*, 1999) topologies. These switches have shown that the moving metal contact possesses low parasites at microwave frequencies and could be able to achieve low on-state (resistive switching) or high off-capacitive (capacitive switching) impedances. These result in switches with very low loss, low voltage electrostatic actuation with no dc current and with the potential of ultra-linear small-signal operation. The advantage of this approach is that a bending beam switch can be designed to nearly 50 Ω impedance across a broad range of frequencies, yet is nearly an open circuit when there is no actuation. Several new switch architectures have been presented; the most promising one is an air bridge structure. Two common forms of these switches are resistive (metal–metal) contact to achieve an ohmic contact and capacitive (metal–insulator–metal) contact, which gives a capacitance ratio between on and off states. Each of these switches has its own advantages and disadvantages. Detailed analyses of each of these topologies are presented in preceding sections.

The cantilever structure consists of a thin strip of metal fixed at one end and is suspended over the metallic transmission line with a gap of a few micrometers. The cantilever can be connected in series with the transmission line as shown in Figure 3.13 or the metallic contact can be on the top of the line as in Figure 3.14. In between the transmission line and the fixed end of the cantilever there is a metallic electrode that serves as a pull-down electrode. In both cases, the actuation voltage will pull down the strip to close the gap in the transmission line and make a conducting path. In the bridge-type or doubly supported cantilever configuration, as shown in Figure 3.15, a thin membrane of metal is suspended over the free space in the middle.

The switch based on a suspended cantilever structure has two different mechanical elements which contribute its operation. The first armature is a thin beam attached at the anchor region at the left of the actuation electrode as shown in Figure 3.16, which determines the suspension of the remaining components. The second major element is the wide cantilever suspended over the actuation electrode towards the RF transmission line.

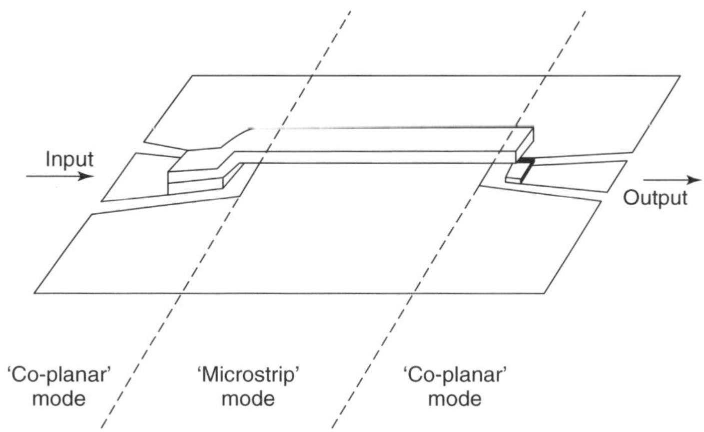

Figure 3.13 Cantilever switch in a transmission line. Reproduced from L.E. Larson, R.H. Hackett, M.A. Melendes and R.F. Lohr, 1991, 'Microactuators for GaAs based microwave integrated circuits', in *IEEE Transducers '91 Conference on Solid State Sensors and Actuators*, IEEE, Piscataway, NJ, USA: 743–746, by permission of IEEE, © 1991 IEEE

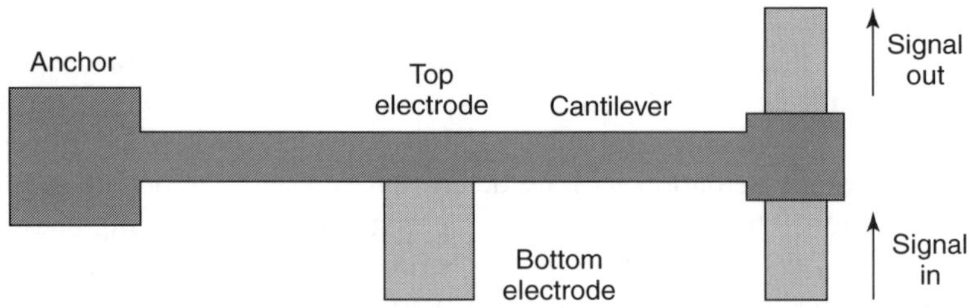

Figure 3.14 Cantilever switch with actuator electrode. Reproduced from H.J. de los Santos, Y.-H. Kao, A.L. Caigoy and D. Dirmars, 1997, 'Microwave and mechanical considerations in the design of MEM switches for aerospace applications', in *Proceedings of the IEEE Aerospace Conference, Aspen, CO*, Volume 3, IEEE, Piscataway, NJ, USA: 235–254, by permission of IEEE, © 1997 IEEE

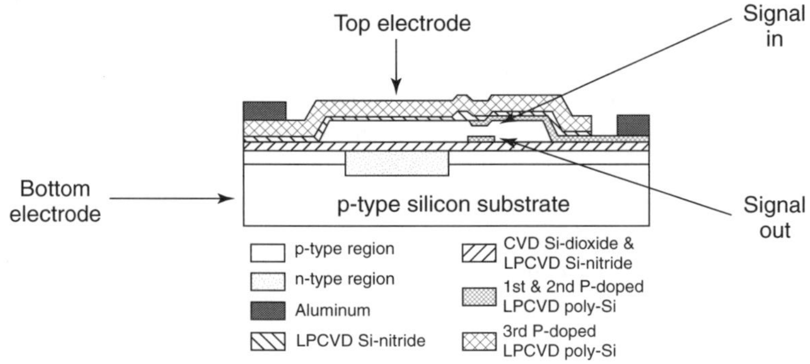

Figure 3.15 Doubly supported cantilever beam. Note: CVD, chemical vapor deposition; LPCVD, low-pressure chemical vapor deposition. Reproduced from H.J. de los Santos, Y.-H. Kao, A.L. Caigoy and D. Dirmars, 1997, 'Microwave and mechanical considerations in the design of MEM switches for aerospace applications', in *Proceedings of the IEEE Aerospace Conference, Aspen, CO*, Volume 3, IEEE, Piscataway, NJ, USA: 235–254, by permission of IEEE, © 1997 IEEE

The boundary conditions at the electrode and the amount of deflection of the suspended beam determine the contact force applied to the metallic contacts. The mechanical equations used to explain the wide-bending beams (Timoshenko and Krieger, 1987) can be used to describe the actuation mechanism of the MEMS switches. The actuation mechanism of the electrostatic MEMS switches can be explained from the equivalent parallel plate capacitor shown in Figure 3.16(b). The metallic parts of the switches are suspended over a bottom metal contact such that the two contacts form a capacitor. When bias voltage is applied between the contacts, charge distributes in such a way that electrostatic force develops between the metals. These charges force the top freely suspended contact to move down towards the bottom electrode, independent of the charge polarity. This creates an opposing tensile force on the cantilever as the structure starts bending. When the applied force reaches a threshold value, the tensile force is no longer balanced by the electrostatic force, so the cantilever abruptly falls to the bottom electrode and makes electrical contact [Figure 3.16a]. The cantilever releases back when the applied voltage is reduced below the threshold value but typically at a much lower voltage than

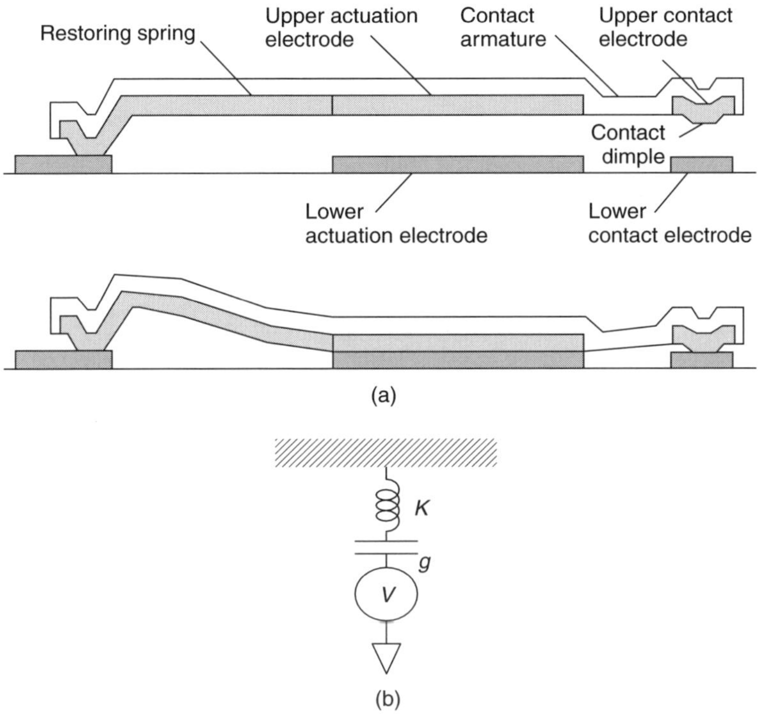

Figure 3.16 (a) Schematic view of an open and closed switch. The contact force is generated by bending of the armature at the free end due to the dimple height. Typical film thicknesses are 1–2 μm and gaps are 2 μm. (b) Equivalent mechanical model. Note: g, gap between electrodes; K, spring constant; V, voltage. Part(a) reproduced from D. Hyman and M. Mehregany, 1999, 'Contact physics of gold microcontacts for MEMS switches', *IEEE Transactions on Components and Packaging Technology* **22**(3): 357–364, by permission of IEEE, © 1999 IEEE

the pull-down voltage. This creates a hysteretic behavior (Brown, 1998; Zavracky and Morrison, 1984), which is inherent for all MEMS switches.

The deflection of the tip of the beam, δ, can be written as (Hyman *et al.*, 1999a)

$$\delta = 6\frac{F(1-\nu^2)L^3}{E}\frac{L^3}{Wt^3} \tag{3.3}$$

where W, L and t are the width, length and thickness of the beam, respectively. E and ν are the Young's modulus and Poisson's ratio, respectively. F is the electrostatic force.

The cantilever switch equivalent circuit in Figure 3.16(b) approximates the switch as a single rigid parallel plate capacitor suspended above a fixed ground plate by an ideal linear spring. The single degree of freedom is the gap between the top movable plate and bottom fixed plate. The advantage of this model is its ability to predict the pull-in voltage of the system as a function of applied voltage since the bottom plate is modeled as fixed and the top plate is held by a spring constant K. The spring constant is determined from the Young's modulus and Poisson ratio of the upper plate and the residual stress within the switch body. The applied voltage V creates an electrostatic force on the top plate given by

$$F = \frac{\varepsilon_0 A V^2}{2g^2} \tag{3.4}$$

where ε_0 is the permittivity of the free space, A is the effective area of the capacitor and g is the physical separation between the contacts. When the voltage exceeds the threshold voltage, the plates touch each other and this force is counteracted by the strong repulsive force which arises from the solid-state compression at the bottom layer of the cantilever. This force can be approximated by $f\theta(-g)$, where f is the force constant and θ is the unit step function. Using Hooke's law, the upward force acting due to the springs can be written as $K(g_0 - g)$, where g_0 is the relaxed gap. The size of the gap can be found by balancing these two forces for an applied voltage and can be written as (Brown, 1998)

$$\frac{\varepsilon_0 A V^2}{2g^2} - f\theta(-g) - K(g_0 - g) = 0 \tag{3.5}$$

The spring constant for a double clamped beam (Figure 3.15) can be written as

$$K = 16EW\left(\frac{t}{L}\right)^3 \tag{3.6}$$

where E is the Young's modulus, and W, L, t are the width, length and thickness of the beam, respectively. The force versus gap dimension is calculated for a dimension of $W = 100\,\mu\text{m}$, $t = 0.5\,\mu\text{m}$, and $E = 8 \times 10^{10}\,\text{N m}^{-2}$ (gold) so that $A = 4 \times 10^4\,\mu\text{m}^2$ and $K = 0.25\,\text{N m}^{-1}$. The relaxed gap dimension in this case is $4\,\mu\text{m}$ and it is assumed that there is a $0.2\,\mu\text{m}$ layer of Si_3N_4 is on the top of the bottom electrode. The electrostatic force is calculated for the above structure and is plotted as force against the gap dimension in microns in Figure 3.17. The dashed curve represents the spring force and the solid lines represent the sum of the electrostatic and compressional forces.

In the unactuated state, the cantilever switch exhibits high impedance due to the air gap between the bottom and top metal plates. Owing to the capacitance nature of the actuation in RF MEMS switches, it does not require continuous dc current for operation. The electrostatic energy required to pull the switch to the on or off state is just $0.5CV^2$. In general, independent of the type and operation of the switch, the switch state with control electrode drawn down is dominant both in capacitance and applied voltage. Hence the power dissipated is approximately $0.5CV^2 fs$, where fs is the switching rate. The switch explained above has a down capacitance of 13 pF and, if we assume a down-state bias voltage of 4 V and switching frequency of 10 kHz, the power dissipation is approximately $1\,\mu\text{W}$.

The low power dissipation and low bias current in RF MEMS switch operation clearly simplifies the RF isolation from the biasing circuits. The RF isolation from the dc biasing circuits can be done with resistors. In contrast, the traditional solid-state RF switches draw much larger dc currents and hence the isolation has to be done with inductors because, in those circuits, the resistors would create a voltage drop and power consumption. The chip resistors are much smaller and inexpensive compared with inductors and its fabrication can be done monolithically, whereas the RF MEMS switch is fabricated on silicon.

The above analysis gives only the physical insight and ignores practical difficulties such as the effect of stress in the material and stiction between the top and bottom electrodes.

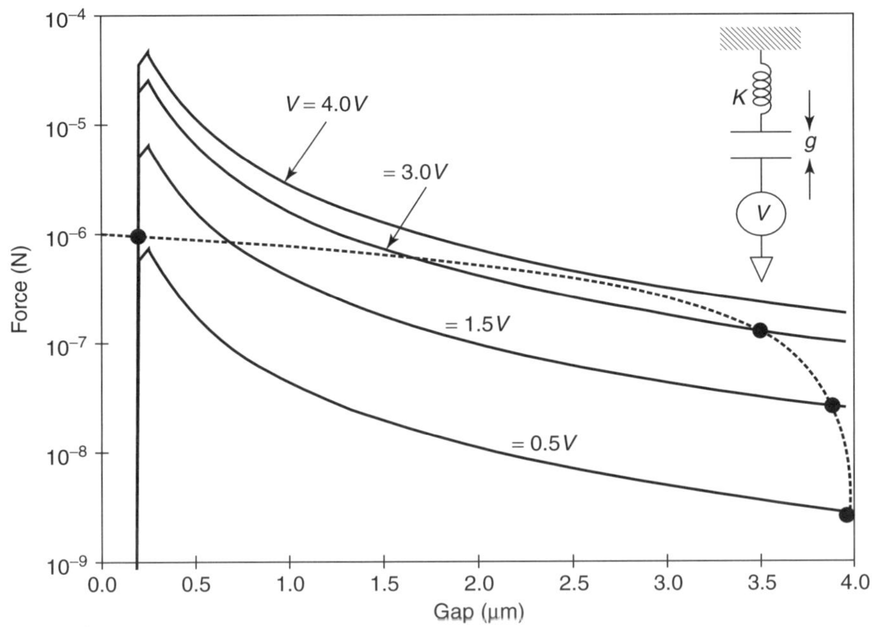

Figure 3.17 Plot of equilibrium force versus the gap dimension in a typical bridge switch. Reproduced from E.R. Brown, 1998, 'RF-MEMS switches for reconfigurable internal integrated circuits', *IEEE Transactions on Microwave Theory and Techniques* **41**: 1323–1328, by permission of IEEE, © 1998 IEEE. Note: dashed curve, spring force; solid lines, sum of electrostatic and compressional forces

For a metal air bridge, it can be easily see that the stress is generally tensile which is at the level of 10^7 Pa. This will affect the spring constant and hence an increase in threshold voltage. The stiction is described as a process of bonding the top and bottom electrodes together by a microscopic surface force due to the planar nature of the electrodes. The surface morphology has a strong influence on stiction and is a serious problem in particular in metal-to-metal switches.

One of the critical electromechanical concerns of the MEMS switch is the switching speed. Mechanical switches are inherently slower than electronic switches, with switching speeds in the microsecond to millisecond range, depending on the material and switch construction. The low mass of membrane switches makes it suitable for relatively fast mechanical switches compared with cantilever-style switches.

When the applied voltage reaches the threshold value, the cantilever abruptly falls to the bottom electrode and the switch is in the on state. It can be seen that if the magnitude of the voltage is reduced, the cantilever releases back (off state) at a much lower voltage than the threshold voltage, which creates the hysteresis. Figure 3.18 shows the typical measured change in gap due to the applied voltage for a cantilever switch (Schiele *et al.*, 1998). The lowest actuation voltage for this cantilever beam is 20 V with an actuation current of 50 nA, caused by leakage. Hence the power consumption in this case is 1.0 μW. Micromechanical switches fabricated in electroplated nickel using a four-level surface micromachining process (Zavracky, Majumdar and McGruer, 1997) have proved that the device is nonhysteritic. The devices can be configured with three terminals; a source,

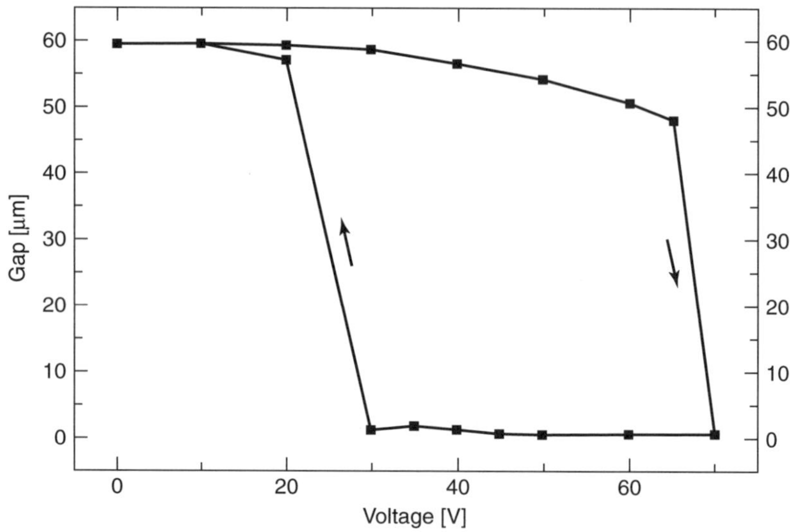

Figure 3.18 Measured change in gap due to the applied voltage when the voltage increases and decreases. Reproduced from I. Schiele, J. Huber, B. Hillerich and F. Kozlowski, 1998, 'Surface micromachined electrostatic micro relay', *Sensors and Actuators A* **66**: 345–354, with permission from Elsevier Science, © 1998 Elsevier Science

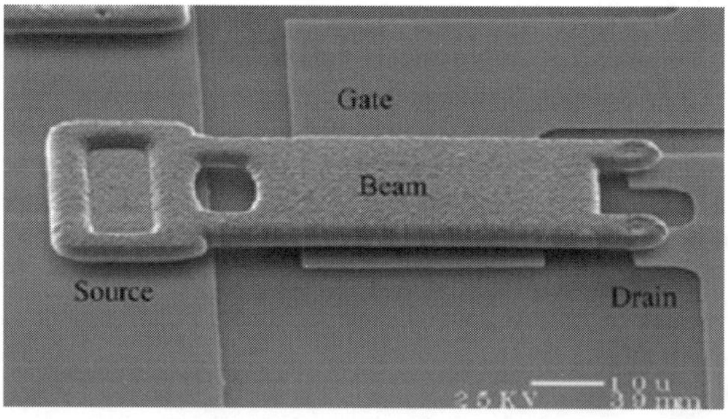

Figure 3.19 Scanning electron microscope photograph of a MEMS switch, showing the source, gate and drain concept. Reproduced from P.M. Zavracky, S. Majumdar and N. McGruer, 1997, 'Micromechanical switches fabricated using nickel surface micromachining', *Journal of Microelectromechanical systems* **6**(1): 3–9, by permission of IEEE, © 1997 IEEE

a drain and a gate 30 μm wide, 1 μm thick and 65 μm long, respectively. The base of the cantilever beam is attached to the source as shown in Figure 3.19. The cantilever extends over the gate and drain. A small contact protrusion is defined at the end of the beam, over the drain. When a voltage is applied to the gate electrode, the electrostatic force pulls down the beam until the switch closes. When the gate voltage is removed, the restoring force of the beam returns it to its original position. In the case of this device, if the contact touches the drain before the beam bends to the unstable point, the device will have a single well-defined threshold voltage. If, however, the device is designed such that the beam can move beyond its stable deflection point, the switch will snap closed. It will not open again until the voltage is lower than the threshold voltage for actuation; the device will exhibit hysteresis. This device is nonhysteretic because the threshold voltage is a function of beam geometric parameters and can be easily varied.

3.5.1.2 Shunt capacitive switches

The MEMS series contact switches perform a 'make and break' operation in an electrical path typically with metal–metal contacts and hence show good insertion loss and exhibit excellent linearity. Many applications demand the lowest attainable insertion loss and low operating voltages. The high insertion loss from a switch increases the required power from the source and negatively impacts the noise figure of the system. Metal membrane switches show good insertion loss, switching voltages compatible with the IC technology, fast switching speeds and excellent linearity (Goldsmith *et al.*, 1995).

The cross-sectional view of the switch is shown in Figure 3.20. The input RF signal traverses into a via interconnect to the top of the membrane. The output line, which is underneath the membrane, is connected using a thin metal strip.

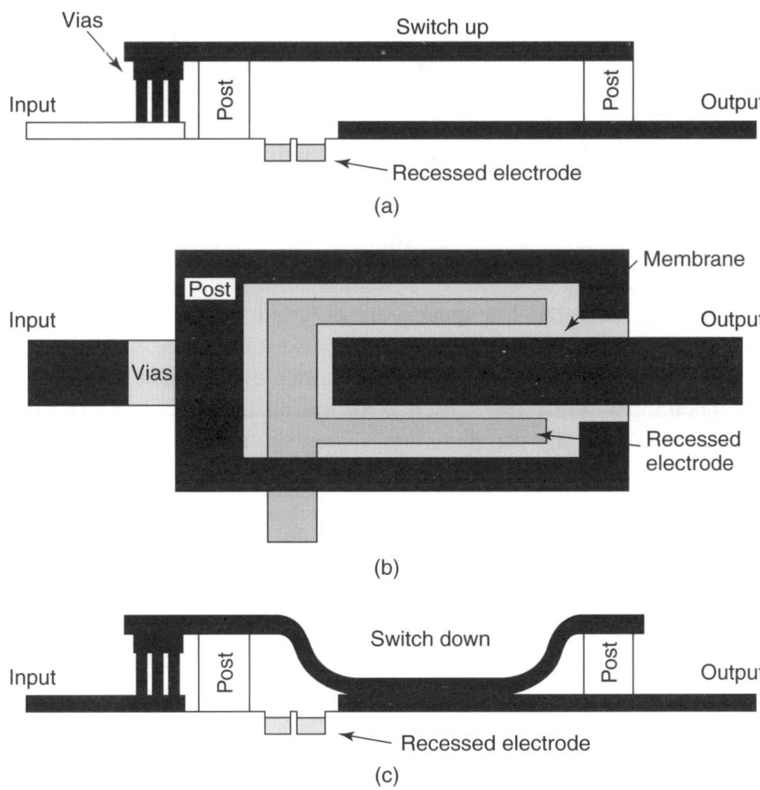

Figure 3.20 Simple single-phase single-throw membrane switch. Reproduced from C. Goldsmith, H.T. Lin, B. Powers, W.R. Wu and B. Norvell, 1995, 'Micromachined membrane switches for microwave applications', in *Proceedings of IEEE Microwave Theory and Technology Symposium, 1995*, IEEE, Piscataway, NJ, USA: 91–94, by permission of IEEE, © 1995 IEEE

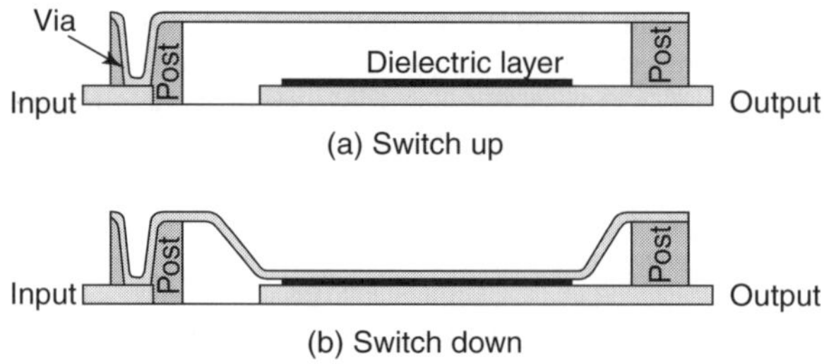

Figure 3.21 Cross-sectional view of the switch with dielectric layer. Reproduced from C. Goldsmith, J. Randall, S. Eshelman, T.H. Lin, D. Denniston, S. Chen and B. Norvell, 1996, 'Characteristics of micromachined switches at microwave frequencies', in *Proceedings of IEEE Microwave Theory and Technology Symposium, 1996*, IEEE, Piscataway, NJ, USA: 1141–1144, by permission of IEEE, © 1996 IEEE

Figure 3.20(b) shows the top view of the membrane switch. When an actuating voltage is applied to the control electrode, the electrostatic force causes the membrane to pull downwards. With enough voltage, the membrane deforms and comes in contact with the bottom transmission line closing the two poles, as shown in Figure 3.20(c). The isolation of the switch in the off state is determined by the parasitic capacitance between the top membrane and bottom output line. The isolation can be further improved by the addition of a thin dielectric layer, which will also reduce the stiction between the lines.

As shown in Figure 3.21, the improved design of the switch incorporates a dielectric layer of 1000 Å silicon nitride or STO (strontium titanate oxide). In unactuated state, the switch exhibits high impedance owing to the air gap between the bottom and top metal plates. The electrostatic force causes the upper membrane to deflect downwards and, when the potential exceeds the threshold voltage, the membrane deflects into the actuated position shown in Figure 3.21(b). In this state, the top membrane rests directly on the dielectric layer and RF is capacitively coupled to the bottom output line. This capacitive coupling causes the switch to exhibit low impedance between the upper and lower electrodes. The ratio of the off-to-on impedance of the switch is directly related to the ratio of capacitances while the switch is in the on and off state. The thin dielectric layer also serves to reduce the problems associated with stiction between the two metal layers, which is common in all direct metal contact switches.

Electromechanical characteristics A one-dimensional model, explained in Figure 3.16(b), can be used to approximate the electromechanical behavior of a shunt switch. Same as a cantilever switch, the membrane switch is modeled as a single rigid parallel-plate capacitor suspended above a fixed ground plane by an ideal spring. This single-degree-of-freedom model is able to predict correctly the pull-in of the membrane as a function of applied voltage. The motion of the switch can be described by the pressure balance equation (Goldsmith *et al.*, 1996)

$$P(g) = K_{\mathrm{s}}(g_0 - g) - \frac{\varepsilon_0 V^2}{2g^2} \tag{3.7}$$

where P is the total pressure of the mechanical body of the switch, g is the gap between the membrane and the bottom electrode, g_0 is the initial height of the membrane with no applied field, V is the applied electrostatic potential and ε_0 is the permittivity of free space. The spring constant K_s is determined by the Young's modulus and Poisson ratio of the membrane and the residual stress. When the electrostatic field is applied to the switch, the membrane starts to deflect downwards by decreasing the gap and increasing the electrostatic pressure on the membrane. At a critical gap height of $2/3g_0$, the mechanical system goes unstable, causing the membrane to suddenly snap down onto the bottom plate. The pull-down voltage of the system can be written as

$$V_p = \left(\frac{8K_s g_0^3}{27\varepsilon_0} \right)^{1/2} \tag{3.8}$$

It can be seen from the Equation (3.8) that the gap height depends on the applied voltage, and the variation of gap height as a function of applied voltage is shown in Figure 3.22. When the electrostatic field is removed from the switch, the tension in the metal membrane pulls it back into the unactuated state. The actuation voltage of different membrane sizes are plotted in Figure 3.23.

Goldsmith *et al.* (1998) provided a design with significant improvements in operating frequency and switching speed. The switch circuit is on the top of the silicon dioxide using a 4.0-μm thick aluminum co-planar waveguide transmission line which is built on a highly resistive silicon substrate (>10 kΩ cm) as shown in Figure 3.24. The transmission lines have a width of 120 μm and a gap of 80 μm. Since the thick aluminum metallization system is compatible with CMOS circuitry, it exhibits low losses of the order of 0.6 dB mm^{-1} at high frequencies. Good physical contact is achieved by a smooth surface finish between the membrane and the lower electrode and hence minimizes the air gap. The top of the lower electrode has a thin film of silicon nitride, which blocks the dc control signal from shorting out during the switching activation and capacitive coupling of the RF signals from upper membrane to lower electrode. The metallic switch membrane is thin aluminum, less than 0.5 μm in thickness, and has high conductivity for RF with good mechanical properties. A series of holes of 2 μm diameter are patterned in

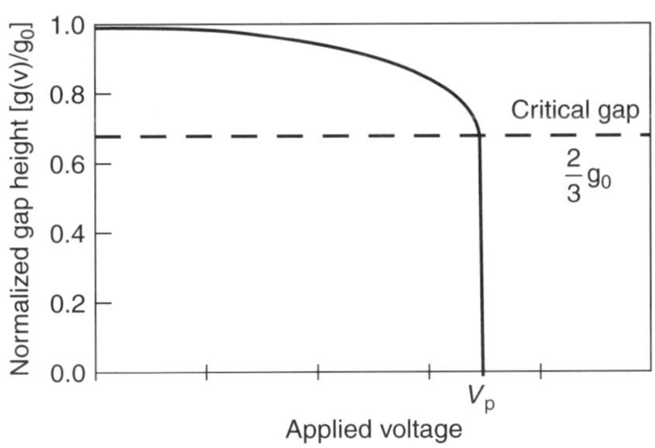

Figure 3.22 Plot of gap height against applied voltage. Reproduced from C. Goldsmith, J. Randall, S. Eshelman, T.H. Lin, D. Denniston, S. Chen and B. Norvell, 1996, 'Characteristics of Micromachined switches at microwave frequencies', in *Proceedings of IEEE Microwave Theory and Technology Symposium, 1996*, IEEE, Piscataway, NJ, USA: 1141–1144, by permission of IEEE, © 1996 IEEE

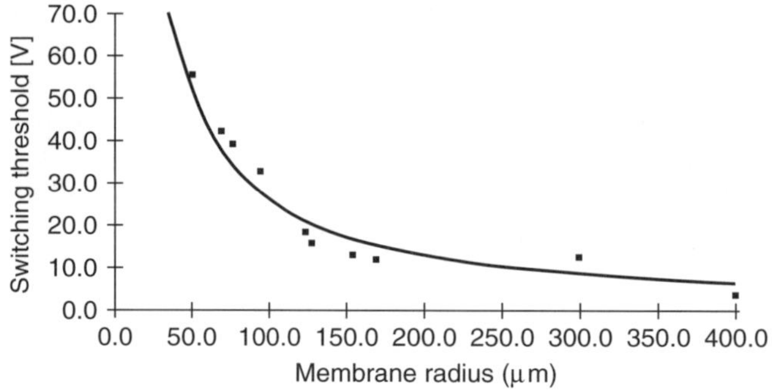

Figure 3.23 Measured actuation voltage for different membrane sizes. Reproduced from C. Goldsmith, J. Randall, S. Eshelman, T.H. Lin, D. Denniston, S. Chen and B. Norvell, 1996, 'Characteristics of Micromachined switches at microwave frequencies', in *Proceedings of IEEE Microwave Theory and Technology Symposium, 1996*, IEEE, Piscataway, NJ, USA: 1141–1144, by permission of IEEE, © 1996 IEEE

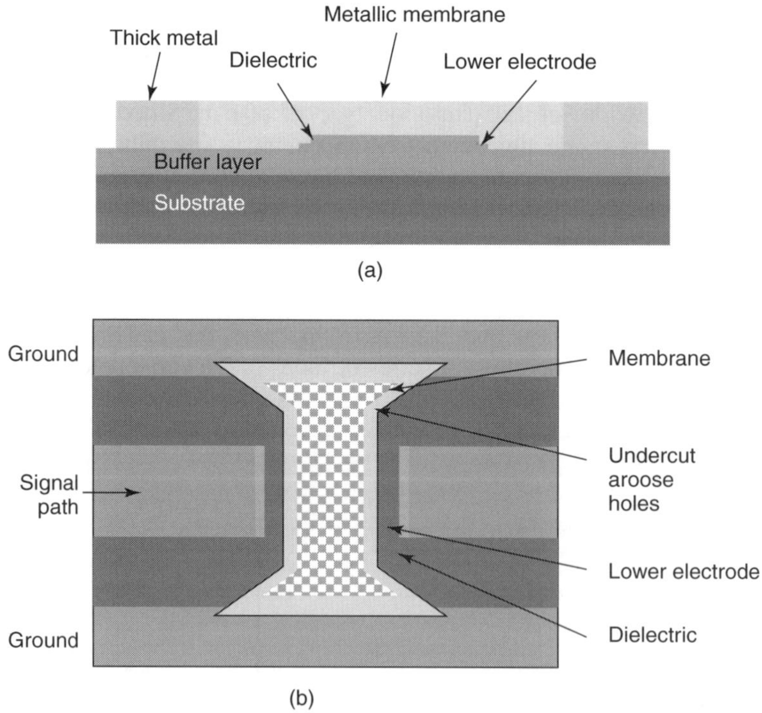

Figure 3.24 (a) Cross-sectional view and (b) top view of a MEMS capacitive switch. Reproduced from C.L. Goldsmith, Z. Yao, S. Eshelman and D. Denniston, 1998, 'Performance of low-loss MEMS capacitive switches', *IEEE MW and Guided wave Letters* **8**(8): 269–271, by permission of IEEE, © 1998 IEEE

throughout the membrane to decrease the squeeze film damping of air and to allow access for micromachining. By removing the sacrificial polymer, the membrane becomes free to move up and down onto the lower electrode in response to the applied electrostatic field.

The residual tensile stress of the membrane keeps it suspended above the RF path when the actuation voltage is zero. The RF signal in the co-planar transmission line will experience a capacitive reactance due to the grounded metal membrane. The capacitance for the present design is of the order of 20–50 fF. When an electric field is applied between the membrane and the lower electrode, the membrane starts to deform owing to the formation of positive and negative charges on the metallic surfaces. These charges exhibit an attractive force, which, when strong enough, causes the suspended metal membrane to snap down onto the lower electrode forming a low-impedance RF path to the ground. The typical capacitance in this state is 3–4 pF. This micromechanical variable capacitor serves as a high-performance RF switch by efficiently transmitting and blocking the RF signal by the membrane deformation.

The measured S parameters of an RF MEMS switch is shown in Figure 3.25 for a frequency range from 0.13 GHz to 40 GHz. The high return loss is caused by the proximity of the parasitic capacitance in the transmission line path to the suspended ground metal membrane. The RF measurements demonstrate that these devices provide efficient switching of high-frequency signals. The switching speed of the device is <5 μS.

3.5.1.3 *Effect of surface roughness of the dielectric layer*

When the membrane is snapped down, the capacitance can be easily calculated from a parallel plate approximation, neglecting the fringing capacitances,

$$C_{\mathrm{d}} = \frac{\varepsilon_0 \varepsilon_{\mathrm{r}} A}{g_0} \tag{3.9}$$

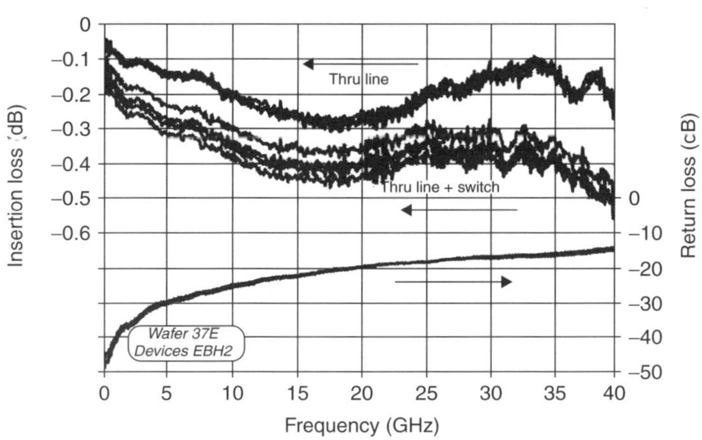

Figure 3.25 Measured insertion loss and return loss RF MEMS switch. Reproduced from C.L. Goldsmith, Z. Yao, S. Eshelman and D. Denniston, 1998, 'Performance of low-loss MEMS capacitive switches', *IEEE MW and Guided wave Letters* **8**(8): 269–271, by permission of IEEE, © 1998 IEEE

When the membrane is in the up state, the capacitance of the system can be approximated using the above equation by considering the dielectric layer between them. It can be written as

$$C_u = \varepsilon_0 w W \left(g_0 + \frac{t_d}{\varepsilon_r} \right)^{-1} \tag{3.10}$$

where w is the width of the membrane and W is the width of the centre conductor of the co-planar waveguide (CPW) line and g_0 is the gap between the membrane and the bottom transmission line. The term t_d/ε_r is added to take account of the finite thickness, t_d, of the dielectric layer between the membrane and the bottom electrode where ε_r is the relative permittivity. For a switch with a layer of dielectric constant 7.6 and a thickness of 1500 Å, bridge gap g_0 of 4 μm, neglecting t_d/ε_r, will result in an error of 3% in the estimation of the capacitance. The error will increase to 10% for a gap, g_0, of 1.5 mm [58].

The down-state/up-state capacitance ratio can be written as

$$\frac{C_d}{C_u} = \left(\frac{\varepsilon_0 \varepsilon_r A}{t_d} \right) \left[\varepsilon_0 A \left(g_0 + \frac{t_d}{\varepsilon_r} \right)^{-1} + C_f \right]^{-1} \tag{3.11}$$

For a dielectric thickness of 1000 Å, area 80 × 100 μm and $\varepsilon_r = 7.6$, the capacitance ratio is 60 : 1 for $g_0 = 1.5$ μm and 120 : 1 for $g_0 = 4$ μm [58].

It can be easily seen from Equation (3.11) that the thickness of the dielectric will control the capacitance ratio and hence a very thin layer can achieve a very high capacitance ratio. However, the deposition techniques of the dielectric layer make it impossible to obtain a thickness less than 1000 Å. Also, the dielectric layer should withstand actuation voltages in the range of 5–50 V without any dielectric breakdown. The down-state capacitance also depends on the smoothness of the dielectric layer or the metallic bridge layer. The calculated reduction in down-state capacitance assuming a perfectly flat dielectric layer and different surface roughness is shown in Figure 3.26. The contact area assumed is 50% of the total area.

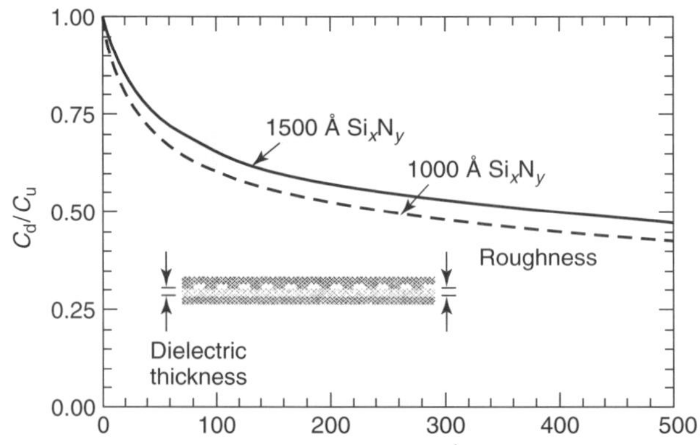

Figure 3.26 Computed down-state capacitance against the roughness in the overlying layer. Reproduced from J.B. Muldavin and G. Rebeiz, 2000a, 'High isolation CPW MEMS shunt switches, part I: Modeling', *IEEE Transactions on Microwave Theory and Techniques* **48**(6): 1045–1052, by permission of IEEE, © 2000 IEEE

From Figure 3.26, it is clear that a capacitance degradation of 65% occurs for a surface roughness of 100 Å. Therefore, for an optimal capacitance ratio, it is essential that the roughness of the MEMS bridge and of the dielectric layer should be kept to less than 40 Å.

3.5.2 Approaches for low-actuation-voltage switches

3.5.2.1 Cantilever switches

The high-voltage actuation of MEMS switches makes them far beyond the compatibility of standard IC technology because for applications in RF and microelectronics systems the voltages should be around 5 V. It is clear from Equation (3.8) that pull-in voltage can be reduced by three different methods: (a) increasing the area of actuation, (b) decreasing the gap between the switch and the bottom electrode, (c) by designing the structure with a low spring constant. Increasing the area is not a practical solution because the compactness is the prevailing issue and adoption of MEMS technology is to achieve the miniaturization. In the second case, the return loss associated with the RF signal restricts the size of the gap. The most flexible route is the third one, in which the design of the spring does not considerably impact the size, weight or RF performance. It is proven that by designing switch structures with extremely compliant folded suspension springs and a large electrostatic area the actuation voltage can be reduced. Two novel designs of micromechanical switches using serpentine and cantilever springs (Pacheco, Nguyen and Katehi, 1998) and hinged cantilever structures (Shen and Feng, 1999) have proved that the voltage can be reduced to 14 to 16 V.

Figure 3.27 presents the micrographs of the serpentine and cantilever spring designs. These switches are basically identical except for the configuration of the springs attached to the cantilever. The serpentine contains two meanders of 220 μm length and the cantilever beams are 250 μm long. Both springs are 4 μm wide and the switches are 4.2 μm above the finite ground co-planar waveguide (FGCPW) (Brauchler *et al.*, 1996). The

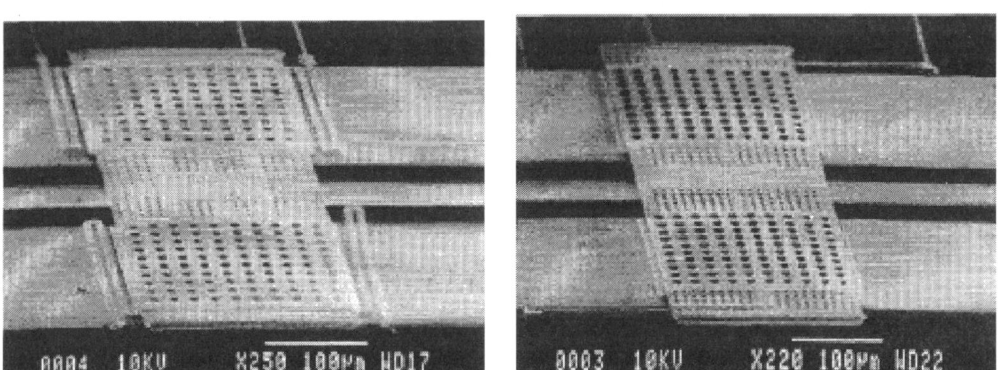

(a) (b)

Figure 3.27 Micromechanical switches with (a) serpentine and (b) cantilever springs. Reproduced from S. Pacheco, C.T.-C. Nguyen and L.P.B. Katehi, 1998, 'Micromechanical electrostatic K-band switch', in *Proceedings of IEEE MTT-S*, IEEE, Piscataway, NJ, USA: 1569–1572, by permission of IEEE, © 1998 IEEE

Table 3.2 Mechanical parameters of switches shown in Figure 3.26

	Serpentine	Cantilever
Mass (Kg)	1.48×10^{-9}	1.19×10^{-9}
Spring constant ($N\,m^{-1}$)	0.478	0.654
Damping coefficient ($N\,m^{-1}\,s^{-1}$)	6.76×10^{-7}	6.76×10^{-7}
Actuation voltage (V)	4.95	5.79
Resonant frequency (kHz)	17.97	20.95

Source: Pacheco, Nguyen and Katehi, 1998.

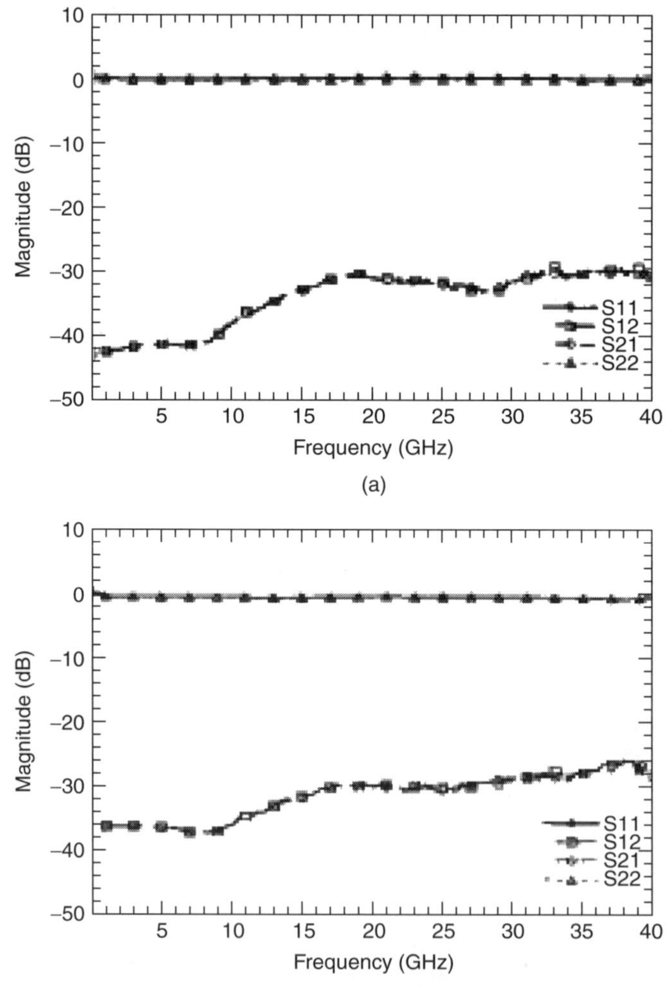

Figure 3.28 Measured *S* parameters of the (a) serpentine switch on; (b) the cantilever switch on. Reproduced from S. Pacheco, C.T.-C. Nguyen and L.P.B. Katehi, 1998, 'Micromechanical electrostatic K-band switch', in *Proceedings of IEEE MTT-S*, IEEE, Piscataway, NJ, USA: 1569–1572, by permission of IEEE, © 1998 IEEE

actuation pads are 220 μm × 220 μm in dimension. The mechanical parameters of serpentine and cantilever design are shown in Table 3.2.

The extremely low spring constant of these switches can cause the cantilever to move in high-pressure and acceleration conditions. Placing an electrode above the switch in an SPDT configuration circumvents this problem. The applied voltage between the upper electrode and the cantilever maintains the switch clamped to the top electrode preventing any unwanted movement of the beam. Removing the voltage between the top electrode and the cantilever frees the cantilever and the necessary pull-in voltage is applied between the ground plane and the switch. The measured *S* parameters of the switches are shown in Figure 3.28. The actuation voltage is between 14–16 V, which is higher than the design voltages presented in Table 3.2. This is mainly because of the compressive internal stress of the gold, which is not accounted for in the design.

Low-voltage actuation can also be achieved in 'hinged' RF MEMS switch design (Shen and Feng, 1999), which shows 0.5 dB insertion loss and an isolation better than 27 dB from 0.25 GHz to 40 GHz. The schematic cross-sectional view of the switch is shown in Figure 3.29. A CPW is used to guide the RF signals. The conductive pad inserted in between the top and bottom electrodes hangs across the signal line and the ground plane. The up and down movement of the cantilever is guided using brackets as shown in Figure 3.29. The actuation voltage applied between the top and bottom electrodes facilitates movement of the conducting pad up and down in a SPDT configuration. When the voltage is applied to the bottom electrode, the pad will pull down and touch the signal line and the ground plane, which eventually shorts the RF path, making the switch off state. When a voltage is applied to the top electrode, the conducting pad will be attracted towards the top plate. This removes the RF short and the signal flows through the output port, which corresponds to the switch on state. Hence the switching operation can be realized by applying two out-of-phase pulses one to the top and the other to the bottom actuation electrode. The minimum electrostatic force required for actuation is the sum of the weight of the cantilever structure and the air friction of the conducting pad. The

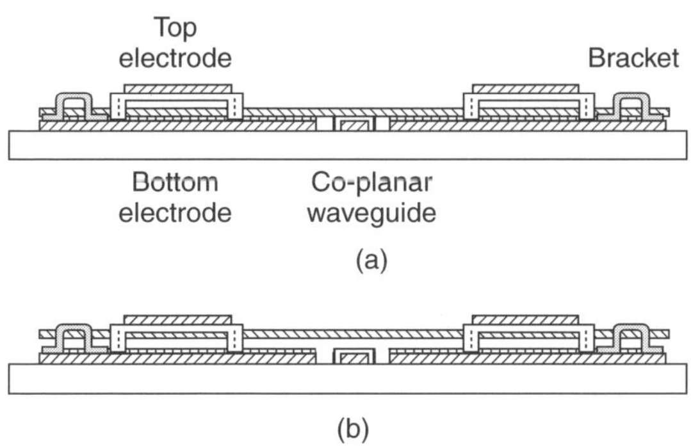

Figure 3.29 Schematic cross-section of switches with hinge configuration: (a) switch off and (b) switch on. Reproduced from S.-C. Chen and M. Feng, 1999, 'Low actuation voltage RF MEMS switches with signal frequencies from 0.25 to 40 GHz', in *Proceedings of IEEE International Electron Devices Meeting*, IEEE, Piscataway, NJ, USA: 689–692, by permission of IEEE, © 1999 IEEE

minimum voltage required for the switch operation can be calculated from

$$V_{\mathrm{min}} = \left[\frac{2(mg + F_{\mathrm{friction}})}{\varepsilon_0 A_{\mathrm{pad}}} d \right]^{1/2} \tag{3.12}$$

where d is the spacing between the conductive pad and the electrodes, mg is the weight of the conductive pad, F_{friction} is the frictional force, A_{pad} is the area of actuation pad and ε_0 is the permittivity of the air.

In an ideal case, for a conductive pad of size $100 \times 400\,\mu\mathrm{m}^2$ and d of $4\,\mu\mathrm{m}$, the minimum voltage will be less than 1 V, ignoring the air friction.

3.5.2.2 Shunt switches

The design of folded suspension is of crucial importance in realizing MEMS switches with low actuation voltages because the folded suspension has the ability to provide very low values of spring constant in a compact area. It also provides high cross-axis sensitivity between vertical and lateral dimensions. The spring constant in the z-direction, k_z, for folded suspensions can be written as (Pacheco, Katehi and Nguyen, 2000)

$$k_z = Ew\left(\frac{t}{L_{\mathrm{c}}}\right)^3 \left\langle 1 + \frac{L_{\mathrm{s}}}{L_{\mathrm{c}}} \left\{ \left(\frac{L_{\mathrm{s}}}{L_{\mathrm{c}}}\right)^2 + 12(1+\nu)\left[1 + \left(\frac{w}{t}\right)^2\right]^{-1} \right\} \right\rangle^{-1} \tag{3.13}$$

where E and ν are the Young's modulus and Poisson's ratio for the metal, respectively L_{c} and L_{s} are the length of the cantilever and the spring, respectively, w, width of the cantilever and t, thickness of the cantilever. The design values of the parameters used in Equation (3.13) are shown in Table 3.3. The total spring constant, K_z, is the sum of all four suspensions attached to the structure:

$$K_z = \frac{4k_z}{N} \tag{3.14}$$

Table 3.3 Physical dimensions of MEMS switch used in the design of Pacheco and co-workers

Dimension	Value
Length of spring, L_{s} (μm)	250
Length of cantilever, L_{c} (μm)	50
Thickness of the cantilever, t (μm)	2
Width of the cantilever, w (μm)	5
Number of meanders, N	4
Mass (kg)	$3,23 \times 10^{-9}$
Length in x direction, L_x (μm)	250
Length in y direction, L_y (μm)	250
Width of conductor, w_{cond} (μm)	60
Gap, g_0 (μm)	3
Total spring constant, K_z (N m^{-1})	0.521
Actuation voltage, V_{pi} (V)	1.94

Source: Pacheco, Katehi and Nguyen, 2000.

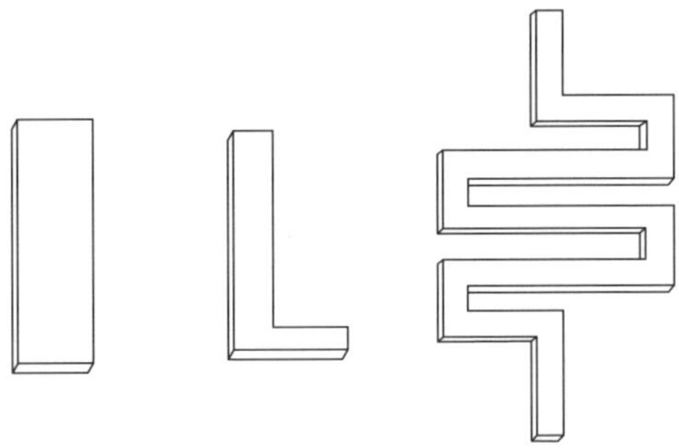

Figure 3.30 Schematic diagram of various hinges used in RF MEMS switch suspension. Reproduced from J.Y. Park, G.H. Kim, K.W. Chung and J.U. Bu, 2001, 'Monolithically integrated micromachined RF MEMS capacitive switches', *Sensors and Actuators A* **89**: 88–94, with permission from Elsevier Science, © 2001, Elsevier Science

Table 3.4 Measured actuation (pull-in) voltages, V_{pi}, for different number of meanders, N

N	V_{pi}	(V)
	design	measured
1	3.90	35
2	2.75	28
3	2.24	20
4	1.94	15
5	1.74	9

Source: Pacheco, Katehi and Nguyen, 2000.

where N is the number of meanders in the suspension. The spring constant decreases linearly with successive addition of meanders to the folded suspension. Figure 3.30 shows various geometries for the hinges used to obtain low spring constants.

Table 3.4 shows the predicted and measured pull-in voltages, V_{pi}, for each suspension with different meanders. The actuation voltage is dropped with increasing number of meanders. However, the measured values are considerably higher than the calculated values because of the fabrication problems (Pacheco, Katehi and Nguyen, 2000).

Series–shunt absorptive switches In general, a series MEMS switch is of a metal–metal contact type and the shunt switch is capacitive in nature. To achieve good isolation performance at low microwave frequencies, the shunt switch must also be a metal–metal contact type because the isolation of the switch is determined by the shunt configuration (Tan and Rebeiz, 2001). The series and shunt switches must be individually biased and connected with matched resistances. The configuration of a wide-band absorptive switch using two MEMS dc-connected series switches and an in-line shunt dc contact switch is shown in Figure 3.31. When the switch is actuated, the resistor is shorted and a low-loss path to the input signal is provided.

To ensure a good dc contact, a 'dimple' in the bridge (6000 Å deep) is designed for each contact point. When the bridge is pulled down, the dimple makes contact with the ground

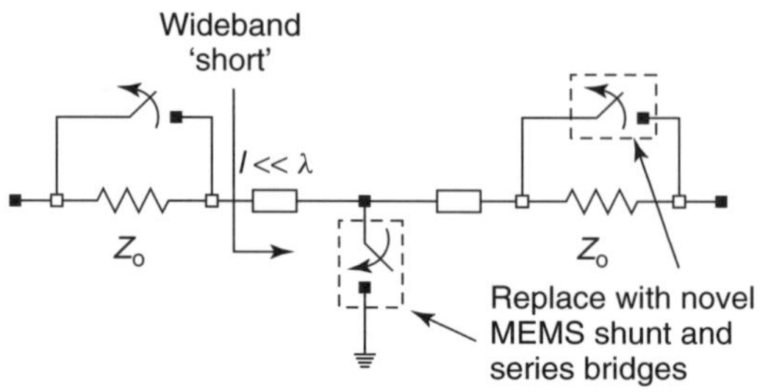

Figure 3.31 Configuration of series–shunt absorptive switch. Note: ℓ, length; Z_0, impedance; λ, wavelength. Reproduced from G.L. Tan and G.M. Rebeiz, 2001, 'DC-26 GHz MEMS series–shunt absorptive switches', in *Proceedings of IEEE Microwave Symposium*, Volume 1, IEEE, Piscataway, NJ, USA: 325–328, by permission of IEEE, © 2001 IEEE

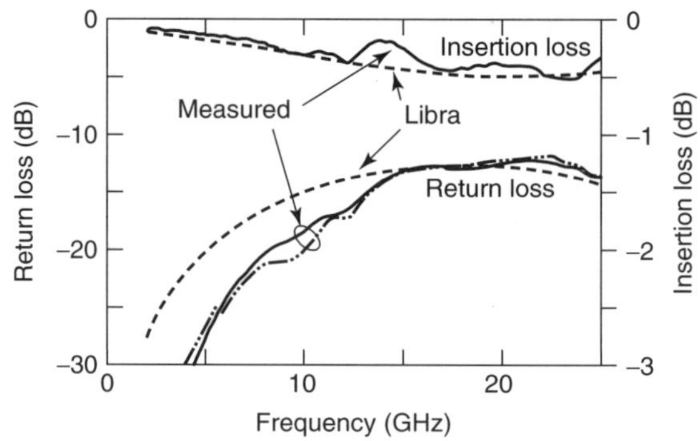

Figure 3.32 Simulated and measured performance of the absorptive switches. Note: details of Libra may be found in Agilent Technologies (http://eesof.tm.agilent.com/). Reproduced from G.L. Tan and G.M. Rebeiz, 2001, 'DC-26 GHz MEMS series–shunt absorptive switches', in *Proceedings of IEEE Microwave Symposium*, Volume 1, IEEE, Piscataway, NJ, USA: 325–328, by permission of IEEE, © 2001 IEEE

plane (shunt switch) or the transmission line (series switch), but there is only minimal contact at the pull-down electrodes, unless the applied voltage is very much higher than the actuation voltages. This helps to reduce the stiction problem. The measured results of the switch are shown in Figure 3.32 along with the simulated results using Libra (Agilent Technologies, (http://eesof.tm.agilent.com/)). The switch shows an insertion loss of 0.5 dB or less from dc to 26 GHz.

3.5.3 Mercury contact switches

An electrostatic micromechanical relay with a stationary mercury micro-drop at the point of contact is realized on polysilicon through the MCNC (Microelectronics Center of North Carolina) multi-user MEMS process (MUMPs) technique (Saffer, Simon and Kim, 1996; Simon, Saffer and Kim, 1997). A 2-μm thick cantilever is made on polysilicon with a width from 2 to 3 μm and 300 to 500 μm length. Figure 3.33 presents the schematic concept of the relay with a polysilicon beam, a fixed driving electrode and a fixed signal electrode. The process flow of fabrication of the cross-section marked AA′ in Figure 3.33

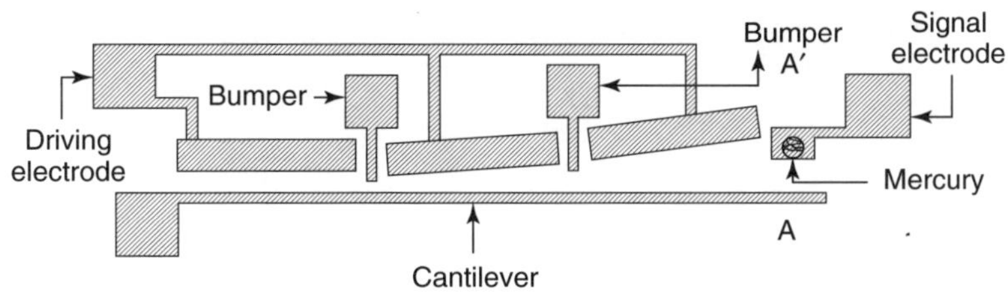

Figure 3.33 Schematic diagram of the mercury contact micro relay. Reproduced from S. Saffer, J. Simon and C.J. Kim, 1996, 'Mercury contact switching with gap-closing microcantilever', *Proceedings of SPIE*, 2882: 204–209, by permission of SPIE

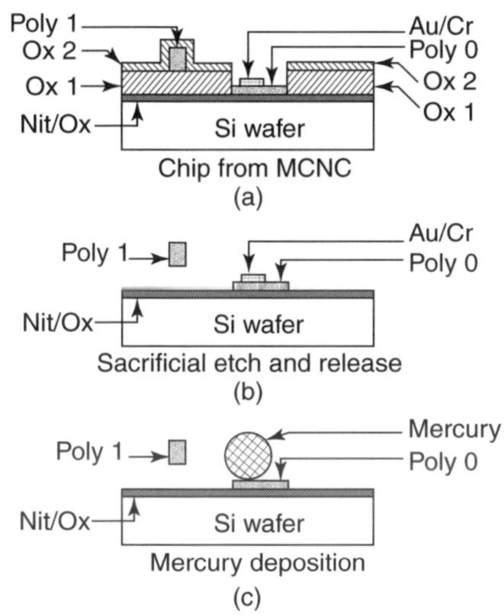

Figure 3.34 Process flow at cross-section AA′ of Figure 3.33. Note: MCNC, Microelectronics Center of North Carolina. Reproduced from S. Saffer, J. Simon and C.J. Kim, 1996, 'Mercury contact switching with gap-closing microcantilever', *Proceedings of SPIE*, 2882: 204–209, by permission of SPIE

is summarized in Figure 3.34. The beams are made from 2-μm thick Poly 1 and the signal electrode is made from 0.5-μm thick Poly 0. The 2-μm sacrificial oxide layer is etched in HF and the beams are freed by realizing methods such as CO_2 critical point drying, HF vapor release and *p*-dichlorobenzene sublimation. The device is then exposed to mercury vapor and the vapor is allowed selectively to condense on the prescribed sites of the device. Nucleation sites were made with materials that react only with mercury so that the mercury could condense only in those sites. This avoided forming mercury uniformly all over the device. The diameter of the mercury ball is 10 μm and is controlled by the time the nucleation site is exposed to the mercury vapor and the temperature of the mercury reservoir. Figure 3.35 presents the fabricated device with mercury ball.

The measured off resistance of the device was greater than 200 MΩ and the total resistance of the device varied from 1.9 kΩ to 3.2 kΩ, with a current-carrying capacity of 10 mA.

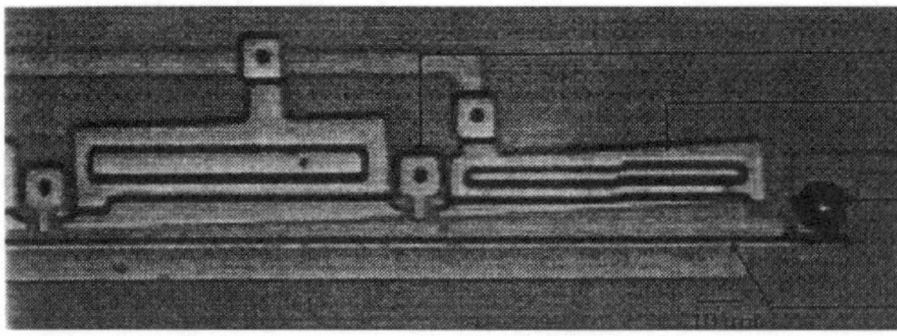

Figure 3.35 Fabricated device with mercury ball. Reproduced from S. Saffer, J. Simon and C.J. Kim, 1996, 'Mercury contact switching with gap-closing microcantilever', *Proceedings of SPIE*, 2882: 204–209, by permission of SPIE

3.5.4 Magnetic switching

Magnetic actuators can perform latching without any energy from outside by using permanent magnets or semihard ferromagnetic materials. The high magnetic energy density allows these actuators to provide long-range actuation and large force. Even though the magnetic force decreases with miniaturization a magnetic actuator can provide a larger force than an electrostatic actuator when the actuation gap is more than 1 μm (Busch-Vishnic, 1992). Therefore magnetic actuators are useful for submillimetre actuators. The high-performance permanent magnets allow these electromagnetic motors to be made with a diameter less than 1 mm (Ioth, 1993). Efforts are also being made to make planar electromagnetic actuators by employing plated magnetic material (Ahn and Allen, 1993; Ahn, Kim and Allen, 1993; Guckel *et al.*, 1993; Hosaka, Kuwano and Yanagisawa, 1993; Tilmans *et al.*, 1999). The major problem in reducing the size of the electromagnetic actuator is to reduce the size of the coils because of its fabrication difficulties. A detailed discussion of the micro relays using magnetic actuation is presented in Section 3.6.

3.5.5 Electromagnetic switching

There is a steadily growing demand for the traditional electromechanical relays (EMRs), including armature relays, reed relays and solid-state relays, which were introduced in 1980s because of their high reliability and long lifetimes, with negligible mechanical wear-out (Coutrot *et al.*, 2001; Sadler, Liakopoulos and Ahn, 2000). However, EMRs outperform solid-state relays in aspects like very low on resistance, very low off-state leakage, low output capacitance and high resistance to electromagnetic interference (EMI). However, not shared by the solid-state relays, the EMRs are rather large in size, have difficulties in integrating with electronics, have low resistance to shock and vibration, require large driving power and produce high acoustical noise. The micromachined EMRs, called *micro relays*, are very well suited to remove these shortcomings while preserving the merits of relays. The introduction of MEMS technology to mechanical relays has improved the size, cost and switching time along with the possibility of integrating them with the electronic components. Many telecommunication network systems require a matrix of switches with high switching speed. At the same time these network switches do not require to switch large electrical loads. Micromachining allows for easy integration of multiple relays on the same substrate, thus reducing the manufacturing cost. Compared with switching devices such as diodes and transistors, an integrated 8×8 matrix switch can reduce the control lines from 64 to 16.

The actuation mechanism of a conventional EMR is generally but rather limited to electromagnetic means because of the use of a coil and a ferromagnetic armature. Micromachining allows the implementation of other actuation principles such as electrostatic and electromagnetic actuators. The electromagnetic actuator is superior to the electrostatic type because the former can be controlled with low-cost and low-voltage control devices. However, electrostatic actuators need low current but voltages ranging from 5–100 V, which requires special design of control devices and high-voltage isolation circuits. The permanent magnets or semihard magnetic materials allow addition of a self-latching mechanism in electromagnetic actuators.

A fully integrated electromagnetically actuated micromechanical relay enclosed in ceramic and plastic packages in presented by Tilmans *et al.* (1999). A multilayer copper coil is used to actuate a movable NiFe armature, thereby closing the contacts.

The design approach includes micromachining fabrication of actuator, electrical contacts and housing, and packaging. The contacts and the armature are housed in a hermetically sealed cavity. The electromagnetic chip shown in Figure 3.36(a) uses a ferromagnetic substrate comprising a U-core electromagnet, which consists of a double-layer copper coil of 127 turns with $6 \times 8\,\mu m^2$, electroplated nickel–iron NiFe (50/50) poles of $1 \times 0.5\,mm^2$ and lower electrical gold contact of $0.5\,\mu m$. The armature upper chip shown in Figure 3.36(b) is fabricated on an oxidized silicon substrate. The chip accommodates a 20-mm thick electrodeposited NiFe (80/20) armature which has a keeper plate ($2 \times 1.8\,mm^2$) and two supporting beams ($1.6 \times 0.15\,mm^2$) which act as springs. The keeper and the beams are suspended $1\,\mu m$ above the silicon substrate. The upper gold contacts of $1.5\,\mu m$ thick are deposited on a keeper plate as shown in Figure 3.37(b). The size of the contacts is $0.20 \times 1.5\,mm^2$. The contacts and the armature are hermetically sealed inside the cavity as shown in Figure 3.37(a). The sealing ring consists of a layer of nickel covered with SnPb. The contact gap and the actuation (pole) gap differ by the total thickness of the contacts, which is approximately $2\,\mu m$. The size of the micro relay is approximately $5.3 \times 4.1\,mm^2$ and the overall thickness of the assembly is approximately 1 mm [Figure 3.37(b)].

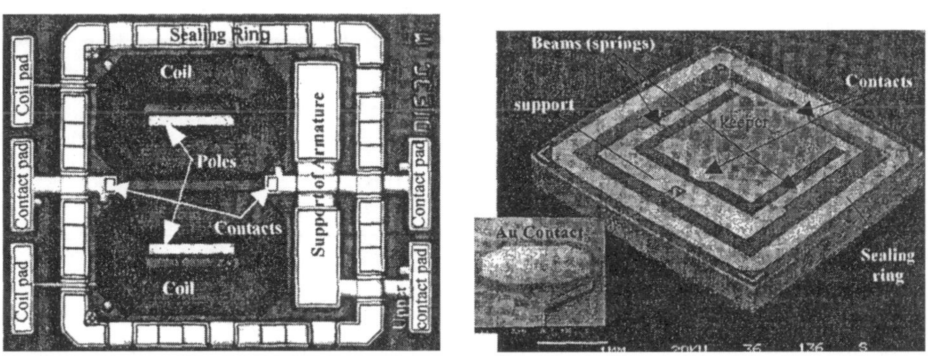

(a) (b)

Figure 3.36 (a) SEM photograph of micro relay (b) the photograph of the armature chip. Reproduced from H.A.C. Tilmans, E. Fullin, H. Ziad, M.D.J. Van de Peer, J. Kesters and E. Van Geffen, 1999, 'A fully packaged electromagnetic micro relay', in *Proceedings of IEEE International Conference on Microelectromechanical Systems, MEMS '99*, IEEE, Piscataway, NJ, USA: 25–30, by permission of IEEE, © 1999 IEEE

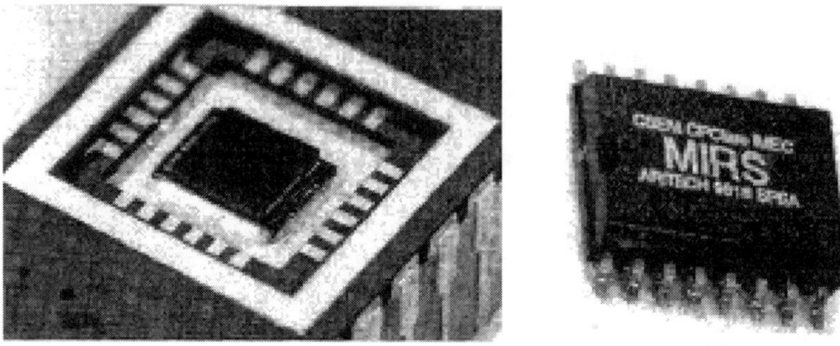

(a) (b)

Figure 3.37 (a) Micro relay mounted on a ceramic package. (b) SOIC package. Reproduced from H.A.C. Tilmans, E. Fullin, H. Ziad, M.D.J. Van de Peer, J. Kesters and E. Van Geffen, 1999, 'A fully packaged electromagnetic micro relay', in *Proceedings of IEEE International Conference on Microelectromechanical Systems, MEMS '99*, IEEE, Piscataway, NJ, USA: 25–30, by permission of IEEE, © 1999 IEEE

Table 3.5 Typical performance parameters of the micro relay

Parameter	Value
'On' resistance (Ω)	0.4
'Off' resistance at 50 V dc (TΩ)	>10
dc breakdown over contacts (kV, dc)	0.2
Open-gap capacitance at 10 kHz (pF)	4
Operating voltage (V)	1.9
Release voltage (V)	0.7
Minimum driving power (mW)	1.6
Minimum driving current (mA)	8.4
Coil resistance (kΩ)	0.22
Coil inductance at 1 kHz, open state (mH)	0.18
Insulation resistance of coil contacts (TΩ)	>10
ac breakdown of coil contacts (kV, ac)	0.4
Capacitance of coil contacts at 10 kHz (pF)	5
Mechanical life at 100 Hz	$>10^6$
Turn-on time at 10 Hz, 1 atm (ms)	
at 8 V	1
at 4 V	2
Turn-off time at 10 Hz, 1 atm, 4–8 V (ms)	0.2
Switching speed (Hz)	>500

Source: Wright and Tai, 1999.

An analytical model based on finite element calculations (Fullin *et al.*, 1998) shows that the magnetic force F_m acting on the keeper is in general limited by the magnetic saturation of the keeper and/or by the residual pole gap spacing. The calculated F_m is around 2 mN for a magnetomotive force of 0.8 AT for a permeability $\mu_r = 2000$, and a saturation induction of 1 T of the keeper material. The average keeper length is 1.6 mm with a residual gap of 1 μm. The contact force F_c, which is due to pull-in, is limited by the maximum magnetic force minus the spring force. Thus F_c is less than 1 mN.

The typical performance characteristics of the micro relay are shown in Table 3.5.

In conventional EMRs the electrical contacts are hermetically sealed inside the capsule, which is generally filled with nitrogen or vacuum to increase the breakdown voltage and to improve the life expectancy of the switching contacts. The samples of micro relay assembly have been mounted in ceramic 24-pin side-brazed packages, as shown in Figure 3.37(a).

3.5.6 Thermal switching

In addition to the advantages of micromechanical relays, they suffer disadvantages; for instance the coils used to construct the electromagnetic actuators should be planar and the fabrication of the coil makes the process more complex. One way to simplify the magnetic actuator is to eliminate the coil structure and have a coil-free magnetic actuator on a silicon substrate. The thermally controlled magnetic actuator (TCMA) is actuated by changing the local magnetization of the structure by spot heating using an infrared laser beam. Eliminating the coil structure not only simplifies the magnetic actuator but enables microfabrication.

The principle of operation of a TCMA is shown in Figure 3.38. It consists of permanent magnets, an armature made of a soft magnetic material and stators made of thermosensitive

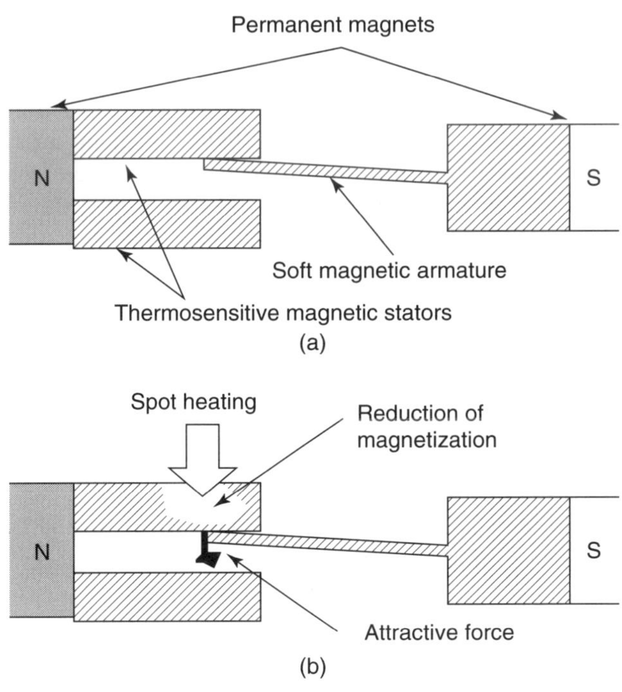

Figure 3.38 Principle of operation of thermally controlled magnetization micro relay. (a) without heat; (b) with heat. Note: N, north; S, south. Reproduced from E. Hashimoto, H. Tanaka, Y. Suzuki, Y. Uensishi and A. Watabe, 1994, 'Thermally controlled magnetic actuator (TCMA) using thermo sensitive magnetic materials', in *Proceedings of IEEE Microelectromechanical Systems Workshop, 1994*, IEEE, Piscataway, NJ, USA: 108–113, by permission of IEEE, © 1994 IEEE

magnetic materials with a low Curie point. When one of the stators is heated to reduce its magnetization, the force between the two stators changes because of the gradient of the magnetic field and the other stator attracts the armature beam. This is done by using a spot-heating infrared laser and micrometer-focusing lens system.

3.6 BISTABLE MICRO RELAYS AND MICROACTUATORS

Merging of several complementary technologies such as silicon micromachining, flip chip assemblies and PbSn solder attachments, have triggered highly miniaturized microwave multifunction devices, which are essential for today's cost-conscience industry. The heart of this low-cost, high-performance and high-profit microwave circuit technology is RF micromachining, which has led to successful fabrication of microwave circuits integrated with silicon packages. It has been proved that, to increase the packaging density and at the same time to reduce the size and cost, the most viable solution is the integration of analog and digital circuits on a single chip. When reducing the size of the overall system, high-density integration and packaging become critical. The miniaturization of electromagnetic relays in particular is very difficult because of the difficulties in fabricating small magnetic coils. However, the miniaturization could be achieved using electrostatic actuation technology. In 1993, it was proved that relays of $1\,mm^3$ in size are those for which electromagnetic actuation is the most practical and economic choice (Hosaka, Kuwano and Yanagisawa, 1994). Electrostatic actuated micro relays (Gretillat *et al.*, 1997; Hosaka, Kuwano and Yanagisawa, 1994), thermally controlled micro relays (Hashimoto, Uensishi and Watabe, 1995, Hashimoto *et al.*, 1994), MEMS relays for miniaturization of brushless dc motor controllers (Wright and Tai, 1999), magnetic microactuators (Fullin *et al.*, 1998; Tilmans *et al.*, 1999), polysilicon switches and micro relays (Drake *et al.*, 1995), mercury drop micro relays (Saffer, Simon and Kim, 1997; Simon, Saffer and Kim, 1997) are some of the trends in micro relay technology. The actuation mechanisms are presented in detail in the following sections.

3.6.1 Magnetic actuation in micro relays

The introduction of MEMS technology to mechanical relays reduces their size, cost and improves their switching time along with the advantage of integration with silicon technology. As shown in Figure 3.39, relays of size $1\,mm^3$, which is two to four orders of magnitude smaller than conventional relays, were found to be most practical and economical choice. The contact electrode plays a key role in the development of a micro relay. The contact resistances for three typical materials used in relays are studied for different forces and are presented in Figure 3.40. It can be seen that the resistance decreases as the contact force increases and converges to around $2\,\Omega$. The residual value is the sum of the resistance at the contact point and due to the external circuits. The resistance values using silver is found to vary owing to the formation of nonconductive compounds such as AgO on its surface.

The micro relay springs are made of micromachined flat symmetric springs as shown in Figure 3.41. Ion sputtering permalloy onto a silicon substrate is used to fabricate cross and spiral springs. The contact force and switching speed of a relay depend mainly on

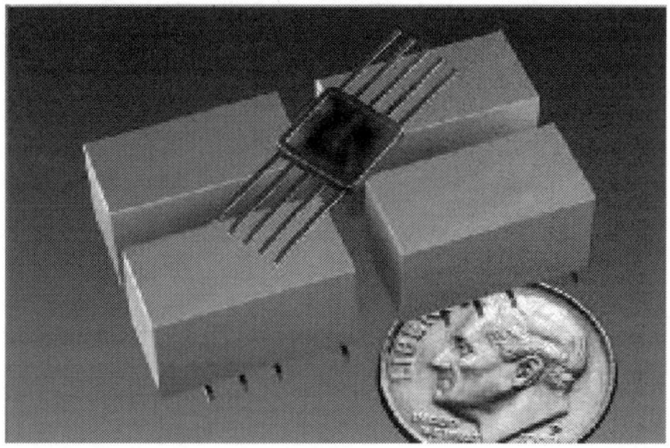

Figure 3.39 A single package containing four Cornos micro relay devices can replace four traditional relays. Photo reproduced with permission from www.memsrus.com

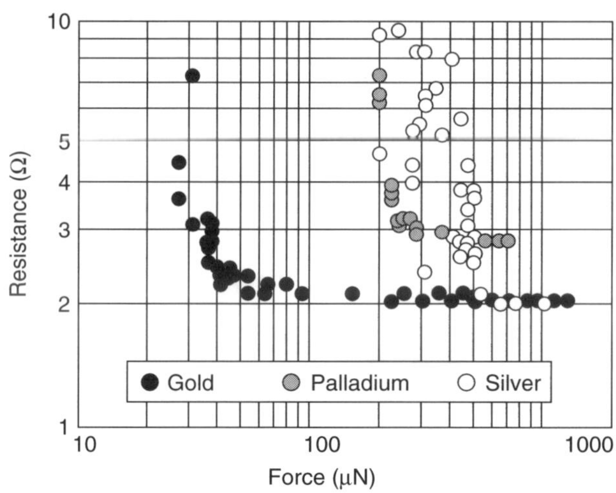

Figure 3.40 Variation of contact resistance with contact force for three common materials. Reproduced from H. Hosaka, H. Kuwano and K. Yanagisawa, 1993, 'Electromagnetic micro relays: concepts and fundamental characteristics', in *Proceedings of IEEE Symposium on Microelectromechanical Systems MEMS '93*, IEEE, Piscataway, NJ, USA: 12–15, by permission of IEEE, © 1993 IEEE

the stiffness and the resonance frequency of the spring. Measured and calculated static deflections of the spring when the forces are applied to the centre of the structure are shown in Figure 3.42. Linear plate theory is used for the spiral springs, and both linear and geometric nonlinear beam theories are used for cross springs because lateral deflection and longitudinal stress affect each other.

A fully integrated magnetically actuated micromachined relay has been successfully fabricated and tested by Taylor and co-workers (Taylor and Allen, 1997; Taylor, Brand and Allen, 1998). This fully integrated micro relay design has several advantages such as minimum size and it does not require any assembly of coils to an actuator plate, thus allowing maximum advantages of batch fabrication.

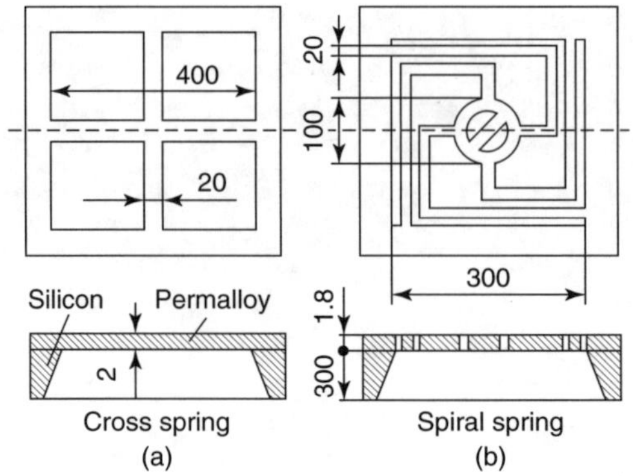

Figure 3.41 Schematic diagram of springs used in micro relays: (a) cross springs; (b) spiral springs (dimensions are in micrometers). Reproduced from H. Hosaka, H. Kuwano and K. Yanagisawa, 1993, 'Electromagnetic micro relays: concepts and fundamental characteristics', in *Proceedings of IEEE Symposium on Microelectromechanical Systems MEMS '93*, IEEE, Piscataway, NJ, USA: 12–15, by permission of IEEE, © 1993 IEEE

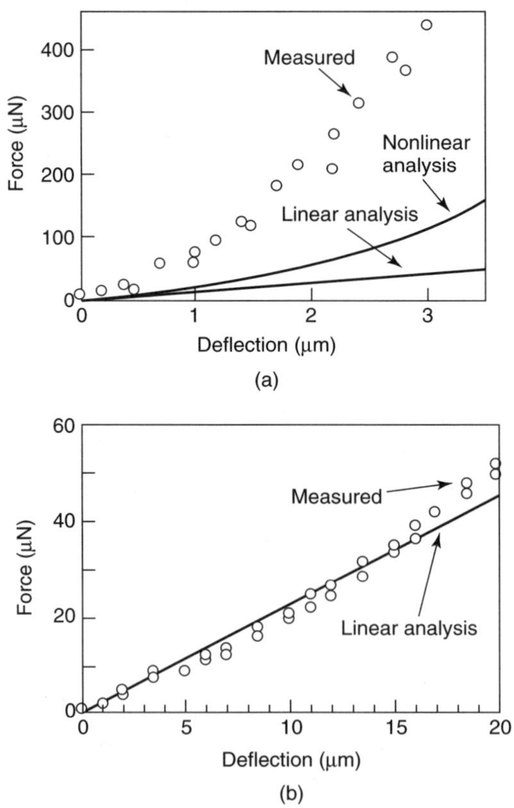

Figure 3.42 Measured and computed static deflection of (a) cross spring; (b) spiral spring. Reproduced from H. Hosaka, H. Kuwano and K. Yanagisawa, 1993, 'Electromagnetic micro relays: concepts and fundamental characteristics', in *Proceedings of IEEE Symposium on Microelectromechanical Systems MEMS '93*, IEEE, Piscataway, NJ, USA: 12–15, by permission of IEEE, © 1993 IEEE

The single-layer coil design reduces the fabrication complexity and eliminates the via connections required in a multilayer coil. The meander nature of the coil design and the elimination of via connections results in a low coil resistance. A schematic diagram of the cantilever type, normally open, micro relay is shown in Figure 3.43. The operation of the micro relay is similar to normally open electromagnetic relays. Here the magnetic flux is distributed rather than additive as in the case of most conventional relays. The device actuates by passing a current of sufficient magnitude through the coil and generating magnetic flux that is concentrated by the lower side of the magnetic cores. This flux generates a force on the upper magnetic plate, resulting in motion of the upper plate toward the electromagnet. When the upper plate moves down toward the electromagnet, it encounters contacts A and B, as shown in Figure 3.43 and the relay will be in the on state. Since the upper plate is conductive, the current can flow from one contact to the other. When the current in the coil is discontinued, the mechanical restoring forces of the upper plate suspension arms are sufficient to pull the upper plate off the contacts and the relay is in the off state. Measured contact resistance of the micro relay shows that the maximum is 38.6 mΩ and the minimum is 22.4 mΩ. This resistance includes the resistances from the bonding pad, interface resistance. A photograph of the micro relay is shown in Figure 3.44, which is able repeatedly to switch a 1.2 A dc load.

The fabrication of micro relay is based on the standard polyimide mould electroplat ing technique. It consists of integrated planar meander coil and a pair of relay contacts

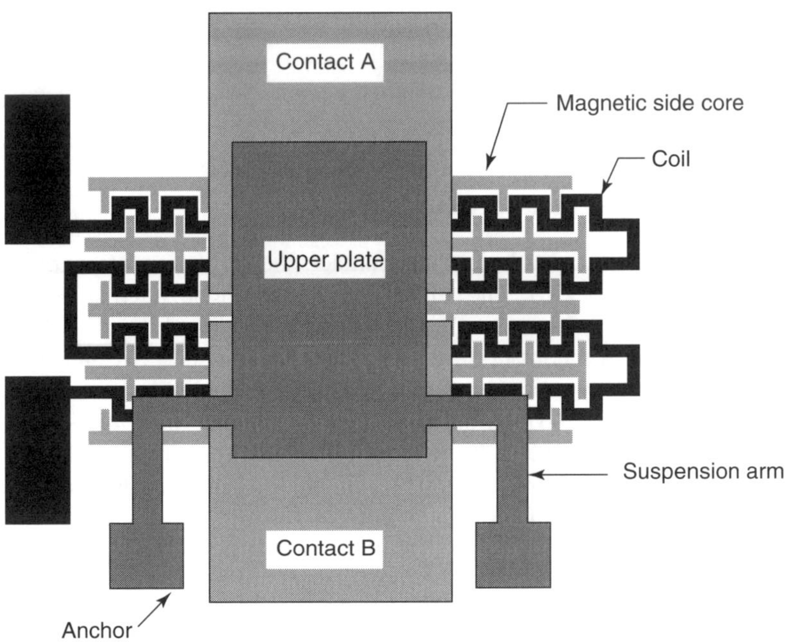

Figure 3.43 Schematic view of a micro relay showing the upper movable plate, side cores and coils. Reproduced from W.P. Taylor, O. Brand and M.G. Allen, 1998, ‘Fully integrated magnetically actuated micromachined relays’, *Journal of Microelectromechanical systems* **7**(2): 181–191, by permission of IEEE, © 1998 IEEE

Figure 3.44 Photograph of the cantilever micro relay shown next to a dime. Reproduced from W.P. Taylor, O. Brand and M.G. Allen, 1998, 'Fully integrated magnetically actuated micromachined relays', *Journal of Microelectromechanical systems* **7**(2): 181–191, by permission of IEEE, © 1998 IEEE

positioned above the coil. A movable magnetic plate is surface-micromachined above the contacts.

3.6.2 Relay contact force and materials

In conventional relays, for a good electrical contact, the on-state contact force should be 20 mN (Peek and Wager, 1955). The advantage of the silicon micro relay is the clean environment of the electrical contacts inside the micromachined device. The absence of any insulating films on the contact surface reduces the on-state resistance and enables stable contact with a very low contact force. Recent investigations show that a minimum force of about 0.1 mN up to 0.6 mN, depending on the contact material, is sufficient for a stable contact (Hosaka *et al.*, 1994). If the required contact resistance is allowed to exceed 1 Ω, even a contact force as low as 10 μN will be sufficient (Majumdar *et al.*, 1997). The knowledge of contact characteristics at contact force in the domain of millinewtons or below, which is generated by microactuators, is still poor. The precision technique developed by Schimakt (1998) enables the measurement of electric contact forces varying continuously in the range from 0.1 mN to 10 mN. It is found that $AuNi_5$ and rhodium (Rh) are contact materials well suited for micro relays, whereas gold is not appropriate because of its high adherence. Forces as low as 0.6 mN provide stable and reliable contacts. The contact resistance of $AuNi_5$ and Rh is found to be 100 mΩ and 1 Ω, respectively.

The experimentally determined characteristics of the contact materials are summarized in Table 3.6. It can be seen that pure gold exhibits a significant disproportion of contact

Table 3.6 Experimentally determined characteristics of old, $AuNi_5$ and rhodium contact material

Parameter	Material		
	Au	$AuNi_5$	Rh
F_{min} (mN)	<0.1	0.3	0.6
$R(F_{min})$ (mΩ)	<30	<100	<1000
F_{adh} (mN)	2.7	0.3	<0.1

Note: F_{min}, minimum force for stable contact; $R(F_{min})$; resistance at F_{min}; F_{adh}, minimum adherence force.
Source: Schimakt, 1998.

and lift-off force, which would result in an oversized actuator. Therefore gold is not an appropriate contact material for a micro relay.

3.7 DYNAMICS OF THE SWITCH OPERATION

The cantilever consists of a thin strip of metal and dielectric that is fixed at one end and suspended over a free space. If the thin metal and the dielectric is fixed at both ends and suspended in middle, the structure is a bridge. In a diaphragm, the thin membrane of metal and dielectric is fixed around its periphery and suspended at the middle. In all these configurations, the structures are suspended over a bottom metal electrode so that a capacitor is formed. When the bias voltage is applied between contacts, charge distributes in such a way that an electrostatic force occurs between them, independent of the voltage polarity. This force pulls the top electrode towards the bottom one creating an opposing tensile force as the structure is bent. When the applied force reaches a certain threshold value, the tensile force no longer balances the electrostatic force and the cantilever abruptly falls to the bottom contact. Figure 3.45 presents a schematic diagram

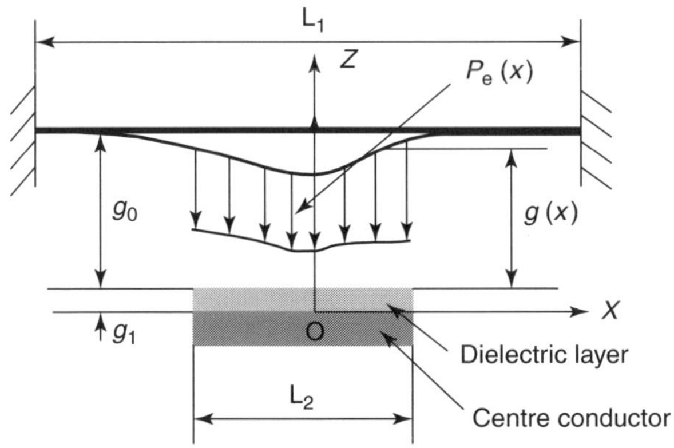

Figure 3.45 Schematic diagram of RF MEMS capacitive switch structure modeled as a fixed-fixed beam. Note: L_1, L_2, g_0, g_1 and g_x are as defined here; $p_e(x)$ is the equivalent load on the cantilever as a result of the applied electric field

of the capacitive bridge RF MEMS switch modeled as a fixed-fixed beam. This model gives an approximate solution for the nonlinear differential equation, which is difficult to solve in general.

3.7.1 Switching time and dynamic response

3.7.1.1 Switching speed

In a parallel plate electrostatic actuation mechanism, the electrostatic pull-in is a well-known instability because it is sharp and sudden. The opposing tensile force cannot balance the electrostatic force for a long time and the switch snaps down once the threshold voltage is reached. The time required to close the contacts depends on the applied voltage, the effective stiffness of the beam and the gap between the contacts. Three-dimensional finite element method (FEM) simulations to examine the geometrical effects of structures are computationally time-consuming and are limited to static analysis (Chan, Kan and Dutton, 1997). The mechanical characteristics, including effective stiffness constant and static contact deformation, were studied using a static model, and a dynamic model was used for calculation of switching speed and frequency of operation (Huang *et al.*, 2001). The capacitive micromachined bridge structure can be modeled as an elastic beam with electrostatic force and squeeze-film damping force, which can be written as (Huang *et al.*, 2001):

$$\rho A\frac{\partial^2 z}{\partial t^2}+\frac{\partial^2}{\partial x^2}\left(\tilde{E}I\frac{\partial^2 z}{\partial x^2}\right)-T_{\mathrm{b}}\frac{\partial^2 z}{\partial x^2}=F_{\mathrm{e}}+F_{\mathrm{d}} \tag{3.15}$$

where the electrostatic force, F_{e}, can be written as

$$F_{\mathrm{e}}=-\frac{\varepsilon_0 V^2 w}{2h^2}\left(1+0.65\frac{h}{w}\right)\left[\delta\left(x+\frac{L_2}{2}\right)-\delta\left(x-\frac{L_2}{2}\right)\right],\quad |x|\le\frac{L_1}{2} \tag{3.16}$$

and

$$h(x,t)=g(x,t)+\frac{g_1}{\varepsilon_{\mathrm{r}}} \tag{3.17}$$

$$z(x,t)=h_0-h(x,t) \tag{3.18}$$

with the step function $\delta(x)$ given by

$$\delta(x)=\begin{cases}1, & x>0\\ 0, & x<0\end{cases} \tag{3.19}$$

The damping force, F_{d}, which is assumed proportional to the velocity of the beam at that point, can be written as

$$F_{\mathrm{d}}=K_{\mathrm{s}}w\eta\frac{L_1^2}{g^3(x,t)}\frac{\partial z}{\partial t} \tag{3.20}$$

The viscosity, η, can be written as

$$\eta=\eta_0\left[1+9.638\left(\frac{p_0\Lambda_0}{p_\ell g_0}\right)^{1.159}\right]^{-1} \tag{3.21}$$

where K_s is the flow coefficient (equal to 0.085), p_0 is the ambient pressure (101 396.16 N m^{-2}), g_0 is the initial gap between the cantilever and the lower plate, η_0 is the air viscosity (1.82×10^{-5} kg m^{-1} s^{-1}), Λ_0 is (0.064 μm) the mean free path of air under standard atmospheric conditions and p_ℓ is the pressure at which the viscosity is calculated.

Equation (3.15) can be reduced to an ordinary differential equation in terms of the first-order approximations of the model shape function. The displacement of the beam can be approximated as

$$z(x,t) = \alpha(t)\phi(x) \tag{3.22}$$

where $\alpha(t)$ is a scaling factor for the mode shape and $\phi(x)$ is the first mode shape function. Equation (3.15) is then reduced to

$$M\frac{\mathrm{d}^2\alpha}{\mathrm{d}t^2} + D\frac{\mathrm{d}\alpha}{\mathrm{d}t} + K\alpha = F \tag{3.23}$$

in which M, D, K and F are defined as follows:

$$M = \int_{-L_1/2}^{L_1/2} \rho A\phi^2(x)\,\mathrm{d}x \tag{3.24}$$

$$D = \int_{-L_1/2}^{L_1/2} d\phi^2(x)\,\mathrm{d}x \tag{3.25}$$

$$K = \int_{-L_1/2}^{L_1/2} \tilde{E}I\phi_{xx}^2(x)\,\mathrm{d}x - \int_{-L_1/2}^{L_1/2} T_\mathrm{b}\phi(x)\phi_{xx}(x)\,\mathrm{d}x \tag{3.26}$$

$$F = \int_{-L_1/2}^{L_1/2} -\frac{\varepsilon_0 V^2 w}{2h^2}\left(1 + 0.65\frac{h}{w}\right)\phi(x)\,\mathrm{d}x \tag{3.27}$$

$\tilde{E}$ is the effective Young's modulus of the beam, w is its width and $|x| \le L_1/2$ (x is the direction of the X axis). I is equal to $wt^3/12$, T_b is equal to $\sigma' wt$, where σ' is the effective residual stress and t is the thickness of the bridge; ε_0 is the free space permittivity and V is the applied voltage. The gap parameter $h = h(x)$ can be written as

$$h(x) = g(x) + \frac{g_1}{\varepsilon_\mathrm{r}} \tag{3.28}$$

where $g(x)$ is the gap between the bridge and the dielectric layer, g_0 is the initial gap, g_1 is the thickness of the dielectric layer with permittivity ε_1. The effective Young's modulus depends on the width of the beam. In general, a beam is considered to be wide if the width of the beam, w, is greater than $5t$, where t is the thickness of the beam, and is considered narrow when w is less than 5t. Therefore, the effective modulus can be written as

$$\tilde{E} = \begin{cases} \dfrac{E}{1-v^2}, & w > 5t \\ E, & w < 5t \end{cases} \tag{3.29}$$

The effective residual stress σ' for the present case is $\sigma(1 - v)$.

The model shape function of the vibration system obtained from Equation (3.15) can be written as

$$\phi(x) = -\frac{\cos(mL_1/2)\cosh(nx) - \cosh(nL_1/2)\cos(mx)}{\cosh(nL_1/2) - \cos(mL_1/2)} \tag{3.30}$$

where m and n are given by

$$m = \left[\frac{-a + (a^2 + 4b)^{1/2}}{2}\right]^{1/2} \tag{3.31}$$

and

$$n = \left[\frac{a + (a^2 + 4b)^{1/2}}{2}\right]^{1/2} \tag{3.32}$$

where

$$a = \frac{T_b}{\tilde{E}I} \tag{3.33}$$

and

$$b = \frac{\rho A}{\tilde{E}I}p^2 \tag{3.34}$$

and p is the natural frequency of the structure satisfying the eigenvalue equation:

$$m\sin\left(\frac{mL_1}{2}\right)\cosh\left(\frac{nL_1}{2}\right) + n\cos\left(\frac{mL_1}{2}\right)\sinh\left(\frac{nL_1}{2}\right) = 0 \tag{3.35}$$

The natural frequency of the bridge can be written as

$$f = \frac{p}{2\pi} \tag{3.36}$$

The switching time can be obtained by solving Equation (3.23).

3.7.2 Threshold voltage

The equivalent load on the cantilever due to the applied electric field can be written as (Huang *et al.*, 2001):

$$P_e(x) = \frac{\varepsilon_0 V^2}{2h^2}\left(1 + 0.65\frac{h}{w}\right) \tag{3.37}$$

The total force acting upward on the bridge is

$$F(h) = K_{eff}(h_0 - h) - P_e \tag{3.38}$$

here

$$h_0 = g_0 + \left(\frac{g_0}{\varepsilon_r}\right) \tag{3.39}$$

When the applied voltage is about to reach the threshold value, the bridge is in static equilibrium and $F(h) = 0$.

When $\partial F/\partial h < 0$, the static equilibrium is stable until reaching the collapse condition by changing the gap between the electrodes. When voltage increases, the gap decreases and the bridge become unstable. The threshold voltage, which depends on the gap between the electrodes, the material in between them and the effective stiffness constant of the beam, can be written as

$$V_c = \left\langle 8K_{\text{eff}}h_0^3 \left\{ 27\varepsilon_0 \left[1 + 0.42 \left(\frac{h_0}{w} \right) \right] \right\}^{-1} \right\rangle^{1/2} \tag{3.40}$$

and

$$h_c = \frac{3}{2}h_0 \tag{3.41}$$

K_{eff} can be approximated from the equivalent configuration of a rigid body suspended on a lumped linear spring with spring constant K_{eff}. However, for a bridge stature, when it starts deflecting, the gap at different points will be different. The gap at the location of maximum deflection is taken as the reference point for the computation of the effective stiffness constant. It can be approximated as a partially uniform distributed load P over the centre of the structure as

$$K_{\text{eff}} = \frac{P}{d_{\max}} \tag{3.42}$$

For a uniformly loaded beam,

$$P(x) = P\left[\delta\left(x + \frac{L_2}{2} \right) - \delta\left(x - \frac{L_2}{2} \right) \right] \tag{3.43}$$

where L_1 and L_2 are the length of the bridge and the electrode, respectively.

The residual stress constant can be written as

$$K_{\text{eff}} = \frac{32\tilde{E}t^3}{L_2(2L_1^3 - 2L_1L_2^2 + L_2^3)} + \frac{8\sigma' t}{2L_1L_2 - L_2^2} \tag{3.44}$$

The switching time is more difficult to predict because it depends on the time required for the bridge to drop from the threshold state to the bottom contact under the effect of the electrostatic force. Since the electrostatic force increases as the gap closes, the switch-down time is shorter than the actual estimate. Typical structures have the up to down time as roughly 1 μs, while the switching from down state to up state is much slower, taking roughly 10 μs. It is this longer time that is usually quoted as the limitation of RF MEMS switching speed.

When an ac signal with frequencies much less that the natural frequency of the microswitch is applied, the membrane follows the ac waveform with nearly the same response as dc. At frequencies much higher than the natural frequency, the membrane no longer follows the ac waveform, instead responding only to the root mean square (rms) voltage between the electrodes. This makes MEMS switches very linear to the high-frequency signal. When signals of two different frequencies are applied to the switch through the RF line there is practically no mixing or intermodulation between the two signals. This is quite unlike the case of solid-state switches, where the inherent nonlinearity is a serious problem.

As we can expect, some of these devices have displayed RF-induced switching. This occurs roughly when the rms voltage becomes large enough to close the switch by itself with no assistance from the dc bias, depending on the RF power level, switch type and physical characteristics.

3.8 MEMS SWITCH DESIGN, MODELING AND EVALUATION

One of the significant challenges limiting design and commercialization of RF MEMS products is RF systems developers' own computer-aided design (CAD) tools. The challenge is to bring the integrated designs together with numerous analog and mixed signal microelectronics blocks and MEMS components on a single chip or assembled in an integrated package. Silicon micromachined structures, despite using IC processing techniques, are no less complex than structures like bridges. Success in fabricating and utilizing them is tremendously complicated by their small size and impossibility of using many traditional mechanical diagnostic methods. Although the recent system-on-chip (SoC) methodologies have triggered efforts in top-to-bottom integration, MEMS design is a different discipline that traditionally has required tedious manual design techniques and in-depth understanding of the fabrication procedures. The lack of common platforms for sharing the design information increases the likelihood of serious design flaws and adds additional iterations. For example, the automatic transferring of the information derived from the 3D field solvers on MEMS structures should be compatible with state-of-art IC design tools.

MEMSCAP's MEMS Comm™ Component library (www.memscap.com) available for the popular Agilent Technologies Advanced Design System (ADS) design platform is an effort towards the realization of this challenge. These libraries provide different views such as functional views with S-parameters, electronic views with Spice model (www.pspice.com), implemented on various CAD platforms enabling the design on functional RF MEMS components. The designs of these components are achieved using full-wave electromagnetic simulation based on the finite element method (FEM). The FEM analysis allows modeling of the complete structure, including the different layers and properties of the material with a parameterized description of the components.

Attempts have been made to develop fully integrated FEM-based packages to model the behavior and fabrication process dependency of MEMS devices, which has to consider the physics of the components and geometry effects due to etching, along with material dependence. Mechanical design plays a crucial role in the electronic integration of MEMS devices. The mechanical design can influence the device as well as system performance, although the design techniques may vary based on specific MEMS applications. Conventional MEMS design starts with either a two-dimensional (2D) layout and later adds the parts based on the fabrication process to generate the 3D model, or a 3D model can be directly created using a 3D design tool, which should be compatible with the fabrication process. The Coventor's Designer (www.coventor.com/rf) software, meshed solid models for physical analysis can automatically be created from the 2D layout. Later FEM analysis can be used to mechanical optimization of the design. The analysis can be done

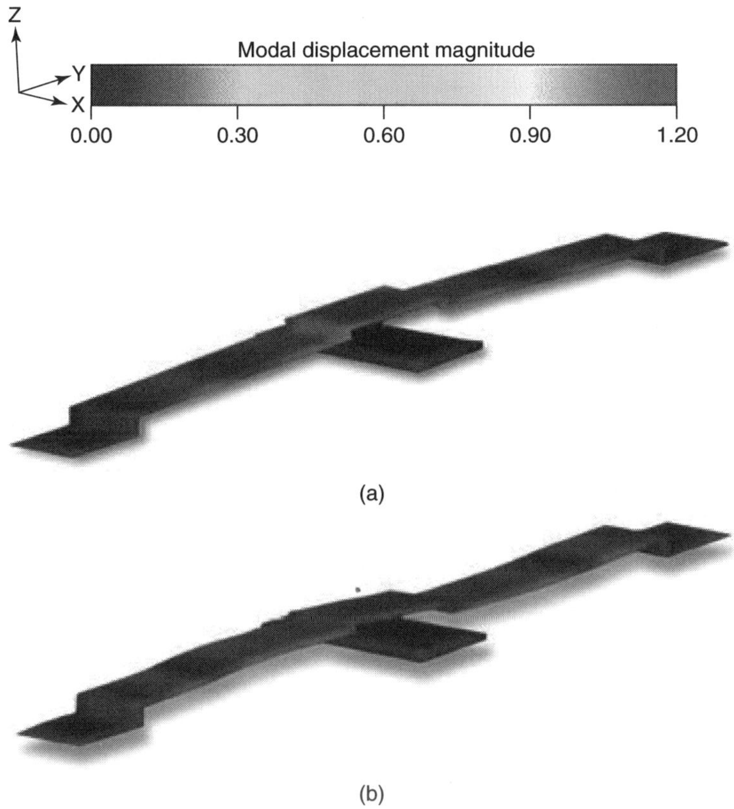

Figure 3.46 Model of (a) undistorted and (b) distorted switch. Reproduced with permission from www.coventor.com

using Analyzer (www.coventor.com/rf). Figure 3.46 highlights the effect of the impact of physical distortion on the performance of a MEMS device. The distortion, which is shown by the shading difference, is caused by the residual stress. The impact of distortion on the RF performance will vary based on the application. However, it is important to consider when designing the switch. Coventor's design methodology provides the ability to analyze distorted and undistorted devices. The software also provides the ability to use a structured custom design approach, starting with the system level of MEMS devices to evaluate their behavior and to converge to the optimized design. The interaction of these MEMS devices with the surrounding electronic circuitry is also important and the software provides the ability to analyze the environmental effects, signal conditioning and packaging. This evaluation will help to refine the MEMS design as well as the design of the control circuitry in the subsystem.

3.8.1 Electromechanical finite element analysis

In-depth finite element analysis of the design of a cantilever switches (Hyman *et al.*, 1999) were performed using Ansys (www.ansys.com) with the geometry and element mesh

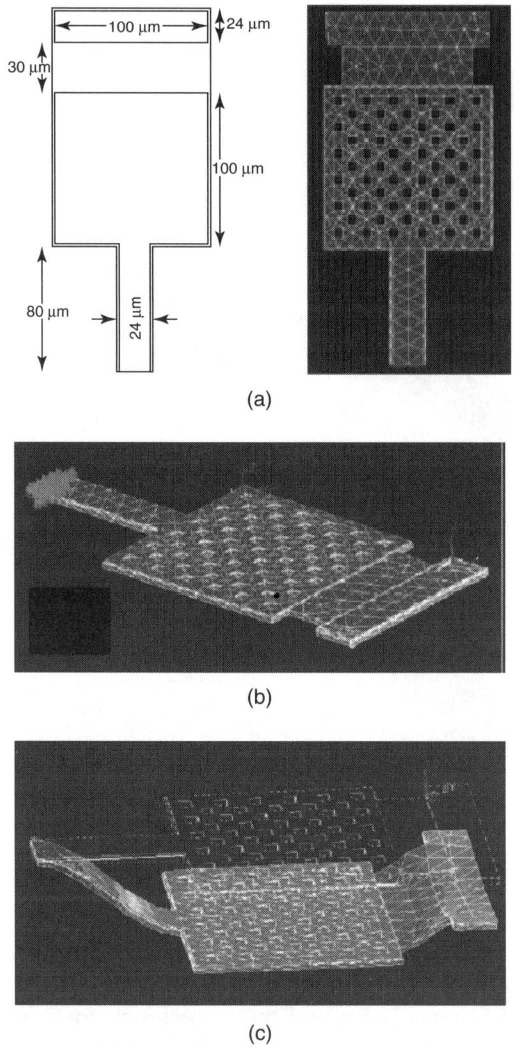

Figure 3.47 (a) Geometry, dimension and meshing of finite element method (FEM) of MEMS switch. The gold layer is 1 μm thick and the silicon nitride layer is 1 μm thick. (b) Isometric view of the FEM mesh of the MEMS switch showing the application of electrostatic pressure over area of bias electrode. (c) deflection of MEMS switch under fully actuated conditions. Reproduced from D. Hyman, J. Lam, B. Warneke, A. Schmitz, T.Y. Hsu, J. Brown, J. Schaffner, A. Waltson, R.Y. Loo, M. Mehregany and J. Lee, 1999a, 'Surface micromachined RF MEMS switches on GaAs substrates', *International Journal of RF and Computer Aided Engineering* **9**: 348–61, by permission of Wiley, © 1999 Wiley

shown in Figure 3.47(a). The gold layer is 1.0 μm thick and is suspended below and overlapped by silicon nitride layers 0.8 μm thick. The magnitude of the pressure applied is related to the applied voltage, which increases as the electrode deflects downwards. The anchor point of the armature is fixed in space with no translation or rotation. Figures 3.47(b) and 3.47(c) show the fully fixed conditions for these nodes.

It can be concluded from the modeling that the armature geometry dominates the switch response and actuation voltages, whereas the contact dimple with its suspended contact beam dominates the contact mechanics.

3.8.2 RF design

In all RF switch designs, the actuation mechanism is important for its characterization. Any switch is assumed to be binary and digital in the sense that it can lie in one of only two possible actuation states. In the on state, the switch is configured to connect to the input port of a system to the output port, whereas in the off state, the switch is configured to disconnect the ports. The number of poles is defined as the number of input terminals or input ports to the switch, and the number of throws is the number of output terminals or output ports. The possible RF measurements to be done on any switch are: (a) insertion loss in the on state; (b) isolation in the off state; and (c) return loss in both states.

3.8.2.1 Transmission lines

Consider a transmission line of characteristic impedance Z_0 feeding a different line of impedance Z_1 as shown in Figure 3.48. It is assumed that there is no reflection from the load, the input impedance seen by the feed line Z_1, so that the reflection coefficient Γ, which is defined as the amplitude of the reflected voltage wave normalized to the amplitude of the incident voltage wave, is

$$\Gamma = \frac{V_0^-}{V_0^+} = \frac{Z_1 - Z_0}{Z_1 + Z_0} \tag{3.45}$$

However, it can be seen that not all the incident wave is reflected: some of it is transmitted to the output port with voltage amplitude given by the transmission coefficient, T.

The transmission coefficient can be written as

$$T = 1 + \Gamma = 1 + \frac{Z_1 - Z_0}{Z_1 + Z_0} = \frac{2Z_1}{Z_1 + Z_0} \tag{3.46}$$

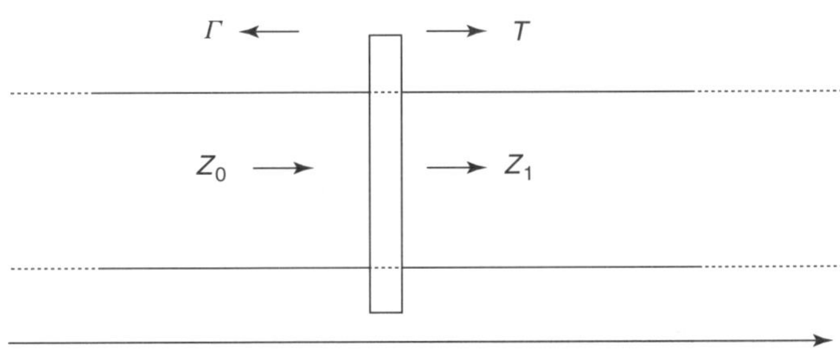

Figure 3.48 Reflection and transmission at the switch connected to two transmission lines with different characteristic impedances. Reproduced from D. Hyman, J. Lam, B. Warneke, A. Schmitz, T.Y. Hsu, J. Brown, J. Schaffner, A. Waltson, R.Y. Loo, M. Mehregany and J. Lee, 1999a, 'Surface micromachined RF MEMS switches on GaAs Substrates', *International Journal of RF and Computer Aided Engineering* **9**: 348–61, by permission of Wiley, © 1999 Wiley

The transmission coefficient between two points in a circuit is often expressed in decibels as the *insertion loss, IL*:

$$IL = -20 \log |T| \, \text{dB} \tag{3.47}$$

When load is mismatched in a transmission line, not all the available power is delivered to the load. This loss is called *return loss* (*RL*) and is defined (in dB) as

$$RL = -20 \log |\Gamma| \, \text{dB} \tag{3.48}$$

so that matched load ($\Gamma = 0$) has a return loss of ∞ dB (no reflected power) while a total reflection ($\Gamma = 1$) has a return loss of 0 dB (all incident power is reflected). As $|\Gamma|$ increases, the ratio of voltage amplitudes V_{max} and V_{min} also changes and its ratio also increases. The standing wave ratio (SWR) is a measure of mismatch of a line and is defined as

$$\text{SWR} = \frac{V_{max}}{V_{min}} = \frac{1 + |\Gamma|}{1 - |\Gamma|} \tag{3.49}$$

3.8.2.2 Microwave considerations

The microwave parameters that should be optimized for any RF switch are the insertion loss, isolation, switching frequency and the return loss. The insertion loss is mainly due to the mismatch between the characteristic impedances of the line and the switch. The contact resistance and the beam metallization loss will also contribute to the insertion loss.

One of the principal requirements of RF MEMS switch design is the design of a transmission line structure that has to be a circuit element in a microwave integrated circuit (MIC). The structure, which is a planar configuration, has the property that its characteristic impedance is determined by the dimensions in a single plane. For example, the impedance of a transmission line on a microwave substrate can be controlled by the width of the line. The common structures of planar transmission lines are shown in Figure 3.49.

The microstrip line is the most commonly used MIC transmission line because of its advantages such as small size, low cost, no cutoff frequency, ease of active device

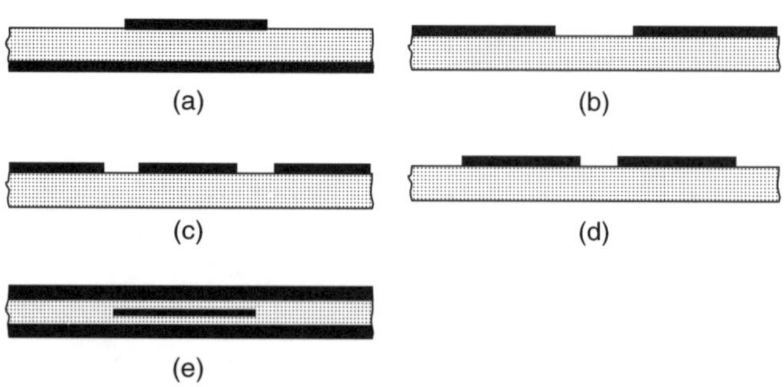

Figure 3.49 Different configurations of planar transmission lines used in microwave integrated circuits: (a) microstrip line; (b) slotline; (c) co-planar waveguide; (d) co-planar strips; (e) stripline configuration

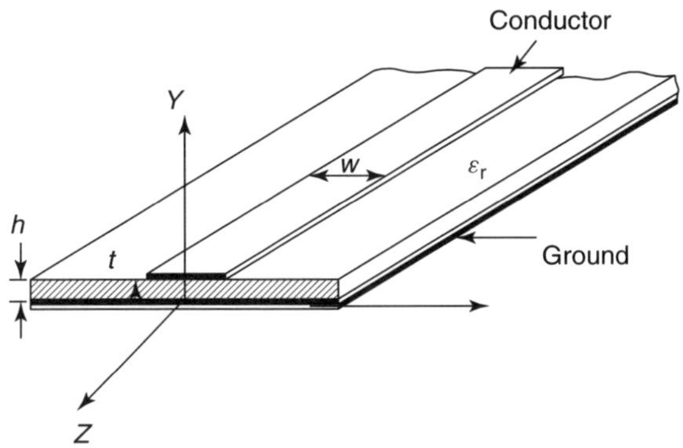

Figure 3.50 Schematic diagram of microstrip transmission line

integration, use of photolithographic method for circuit production, good repeatability and reproduceability, ease of mass production and compatibility with monolithic circuits. Monolithic circuits are MICs on a GaAs substrate with active and passive devices on the same chip. Compared with a rectangular waveguide, the disadvantages of microstrip lines are their higher loss, low power handling capability and greater temperature instability.

Figure 3.50 shows the schematic diagram of the microstrip line. A thin conducting strip of width W is etched on top of a grounded dielectric substrate with thickness h and relative permittivity ε_r. In general, two types of substrates are used: soft substrates and hard substrates. Soft substrates are flexible, cheap and are easy to fabricate. However, they have high thermal expansion coefficients. Some typical soft substrates are RT Duroid 5870 ($\varepsilon_r = 2.3$), RT Duroid 5880 ($\varepsilon_r = 2.2$) and RT Duriod 6010 ($\varepsilon_r = 10.5$) (RT Duroid is the trademark of Rogers Corporation, Chandler, AZ). Hard substrates quartz ($\varepsilon_r = 3.8$), alumina ($\varepsilon_r = 9.7$), sapphire ($\varepsilon_r = 11.7$) and GaAs ($\varepsilon_r = 12.3$) have better reliability and lower thermal expansion coefficients but are more expensive and nonflexible. The most important parameters in microstrip circuit design are W, h and the substrate dielectric constant ε_r.

3.8.2.3 Design equations

A microstrip is a two-conductor transmission line that can be considered to have evolved conceptually from a two-wire transmission line. Microstrip lines differ considerably from other transmission lines. Compared with a stripline, the microstrip structure is open on the top. This configuration makes a microstrip very convenient for use in MICs where discrete lumped devices (active or passive) can be mounted in the circuit. Also because of the planar nature, impedance matching as well as small tuning can be incorporated after the fabrication of the circuit. However, the presence of the dielectric–air interface modifies the mode of propagation in a microstrip and this open structure causes complications in analysis and design. Unlike stripline, where all fields are contained within a homogeneous dielectric medium, microstrip has most of its field lines in the dielectric region, concentrated between the strip conductor and the ground plane and some fraction in air above the substrate. Such a combination cannot support pure transverse electromagnetic (TEM) fields, since the phase velocity of TEM fields in air region would be c and that in the dielectric region would be $c/\sqrt{\varepsilon_r}$. In most of the practical applications, however, the dielectric substrate is electrically very thin ($h \ll \lambda$) and the fields are quasi-TEM. Thus a good approximation for the propagation constant, phase velocity and characteristic impedance can be obtained from the static or quasi-static approximations. The phase velocity and propagation constant can be expressed as

$$v_p = \frac{c}{\sqrt{\varepsilon_e}} \tag{3.50}$$

$$\beta = k_0\sqrt{\varepsilon_e} \tag{3.51}$$

where ε_e is the effective dielectric constant of the microstrip line. Since field lines are in the dielectric region as well as in air, the effective dielectric constant satisfies the relation

$$1 < \varepsilon_e < \varepsilon_r \tag{3.52}$$

and is dependent on the substrate thickens h and conductor width W.

The effective dielectric constant can be interpreted as the dielectric constant of a homogeneous medium that replaces the air and the dielectric regions of the microstrip. The effective dielectric constant of a microstrip line is given by

$$\varepsilon_e = \frac{\varepsilon_r + 1}{2} + \frac{\varepsilon_r - 1}{2} F\left(\frac{W}{h}\right) \tag{3.53}$$

$$\mathrm{F}(W/h) = \begin{cases} \left(1 + \dfrac{12h}{W}\right)^{-1} + 0.04\left(1 - \dfrac{W}{h}\right)^2, & \text{for } \dfrac{W}{h} \le 1 \\ \left(1 + \dfrac{12h}{W}\right)^{-1}, & \text{for } \dfrac{W}{h} \ge 1 \end{cases} \tag{3.54}$$

The characteristic impedance of the microstrip line is given by

$$Z_0 = \begin{cases} \dfrac{60}{\sqrt{\varepsilon_e}} \ln\left(\dfrac{8h}{W} + \dfrac{W}{4h}\right), & \text{for } \dfrac{W}{h} \le 1 \\ 120\pi \left\{\sqrt{\varepsilon_e}\left[\dfrac{W}{h} + 1.393 + 0.667\ln\left(\dfrac{W}{h} + 1.444\right)\right]\right\}^{-1}, & \text{for } \dfrac{W}{h} \ge 1 \end{cases} \tag{3.55}$$

For a given characteristic impedance Z_0 and dielectric constant ε_r, the W/h ratio can be determined from

$$\frac{W}{h} = \begin{cases} \dfrac{8e^A}{e^{2A} - 2}, & \text{for } \dfrac{W}{h} < 2 \\ \dfrac{2}{\pi}\left\{B - 1 - \ln(2B - 1) + \dfrac{\varepsilon_r - 1}{2\varepsilon_r}\left[\ln(B - 1) + 0.39 - \dfrac{0.61}{\varepsilon_r}\right]\right\}, & \text{for } \dfrac{W}{h} > 2 \end{cases} \tag{3.56}$$

where

$$A = \frac{Z_0}{60}\left(\frac{\varepsilon_r + 1}{2}\right)^{1/2} + \frac{\varepsilon_r - 1}{\varepsilon_r + 1}\left(0.23 + \frac{0.11}{\varepsilon_r}\right) \tag{3.57}$$

and

$$B = \frac{377\pi}{2Z_0\sqrt{\varepsilon_r}} \tag{3.58}$$

It can be seen from the equation that the impedance value decreases when the strip-to-height ratio (W/h) of the substrate is increased because an increase in W (or a decrease in h) will increase the line capacitance. These expressions provide accuracy better than 1%. A more accurate expression for the characteristic impedance Z_0 of a microstrip for $t = 0$ and $\varepsilon_r = 1$ is given by (Gupta *et al.*, 1996)

$$Z_0 = 60 \ln \left\{ \frac{f(x)}{x} + \left[1 + \left(\frac{2}{x} \right)^2 \right]^{1/2} \right\} \tag{3.59}$$

where

$$f(x) = 6 + (2\pi - 6) \exp \left[- \left(\frac{30.666}{x} \right)^{0.7528} \right] \tag{3.60}$$

and $x = W/h$. The accuracy of this expression is better that 0.01% for $x \leq 1$ and 0.03 percent for $x \leq 1000$. The effective dielectric constant can be expressed as

$$\varepsilon_e = \frac{\varepsilon_r + 1}{2} + \frac{\varepsilon_r - 1}{2} \left(1 + \frac{10}{x} \right)^{-a(x)b(\varepsilon_e)} \tag{3.61}$$

$$a(x) = 1 + \frac{1}{49} \ln \left(\frac{x^4 + (x/52)^2}{x^4 + 0.432} \right) + \frac{1}{18.7} \ln \left[1 + \left(\frac{x}{18.1} \right)^3 \right] \tag{3.62}$$

$$b(\varepsilon_e) = 0.454 \left(\frac{\varepsilon_r - 0.9}{\varepsilon_r + 3} \right)^{0.053} \tag{3.63}$$

The above expressions are based on the assumption that the thickness of the strip conductor is negligible. However, in practice, the metallic strip has a finite thickness t that affects the performance.

3.8.2.4 Effect of strip thickness on microwave properties

Many practical designs involve loss or attenuation due to the use of good, but not perfect, conductors. It is widely known that at higher frequencies the current is confined almost entirely to a very thin sheet at the surface of a conductor called its skin depth. It is assumed that the thickness of the conductor is very much greater that the depth of penetration, so that there is no reflection from the back surface of the conductor. In other words the surface resistance of a flat conductor at any frequency is equal to the dc resistance of a thickness δ of the same conductor. However, when the thickness of the metallic conductor is not much different compared with the depth of penetration, reflection of the wave occurs at the back surface of the conductor and as a result the transmission line may have very high insertion loss. The skin depth can be defined as

$$\delta = \left(\frac{2}{\omega\mu\sigma} \right)^{1/2} \tag{3.64}$$

Table 3.7 Comparison of skin depth for various conductors at 1 GHz

Metal	Conductivity ($S\,m^{-1}$)	Skin depth (m)
Aluminum	3.816×10^7	2.57×10^{-6}
Copper	5.813×10^7	2.08×10^{-6}
Gold	4.098×10^7	2.48×10^{-6}
Platinum	9.52×10^6	5.15×10^{-6}
Silver	6.173×10^7	2.02×10^{-6}

where $\omega = 2\pi f$ is the radiation frequency and μ and σ are, respectively, the permeability and conductivity of the medium. The amplitude of the fields in the conductor decays by an amount 1/e or 36.8% after traveling a distance of one skin depth. For a good conductor (silver or gold) at microwave frequencies, the distance is very small. Hence it can be easily seen that only a thin metallization of a good conductor is necessary for low-loss microwave components or transmission lines. Calculated skin depths of typical conductors used in present microfabrication technology are given in Table 3.7 for a typical frequency of 1 GHz. These results show that most of the current flow in a good conductor occurs in an extremely thin region near the surface of the conductor.

Many investigators have reported the effect of the thickness of the strip on Z_0 of a microstrip line (Gupta *et al.*, 1996). Simple and accurate formulas for impedance with finite strip thickness can be written as

$$Z_0 = \frac{60}{\sqrt{\varepsilon_e}} \ln\left(\frac{8h}{W_e} + 0.25\frac{W_e}{h}\right), \quad \frac{W}{h} \le 1 \tag{3.65}$$

and

$$Z_0 = \frac{120\pi}{\sqrt{\varepsilon_e}} \left\{\frac{W_e}{h} + 1.393 + 0.667\ln\left(\frac{W_e}{h} + 1.444\right)\right\}^{-1}, \quad \frac{W}{h} \ge 1 \tag{3.66}$$

where

$$\frac{W_e}{h} = \frac{W}{h} + \frac{1.25}{\pi}\frac{t}{h}\left(1 + \ln\frac{4\pi}{t}\right), \quad \frac{W}{h} \le \frac{1}{2\pi} \tag{3.67}$$

$$\frac{W_e}{h} = \frac{W}{h} + \frac{1.25}{\pi}\frac{t}{h}\left(1 + \ln\frac{2h}{t}\right), \quad \frac{W}{h} \ge \frac{1}{2\pi} \tag{3.68}$$

and

$$\varepsilon_e = \frac{\varepsilon_r + 1}{2} + \frac{\varepsilon_r - 1}{2} F\left(\frac{W}{h}\right) - C \tag{3.69}$$

where

$$C = \left(\frac{\varepsilon_r - 1}{4.6}\right)\left(\frac{t}{h}\right)\left(\frac{W}{h}\right)^{-1/2} \tag{3.70}$$

It can be seen that the effect of thickness on Z_0 and ε_e is significant for small values of W/h. However, the effect of strip thickness is significant on conductor loss in the microstrip line. At low frequencies, the insertion loss is the contribution of resistive loss of the signal line, which includes the resistance of the line and contact resistances. At higher frequencies, however, the insertion loss can be attributed to both the resistive loss and the skin depth effect. For frequencies below 4 GHz, the skin-depth effect is much less significant than the resistive loss of the signal lines. The resistive loss can be minimized using thicker signal lines.

3.8.2.5 Power-handling capability

It is well known that, in general, the microstrip lines are suitable only for low-power applications. Although microstrip circuits are not well suited for high-power applications as, like waveguides of coaxial lines, they could possibly be used for several medium-power applications. For example, a 50 Ω microstrip line on a 25-mm thick alumina substrate can handle a few kilowatts of power. The power-handling capacity of the microstrip circuit is limited by the heating because of the ohmic and dielectric losses and by dielectric breakdown. An increase in temperature due to dielectric as well as conductor losses limit the average power, while the breakdown between the strip conductor and ground plane limits its ability to handle peak power.

3.8.2.6 Microstrip losses

RF attenuation in a microstrip structure is caused by two components: conductor loss and dielectric loss. The magnetic loss component will also be present for a magnetic substrate. For a conductor, the surface impedance Z_s(equal to $R + jX$) has a real part R (surface resistance per unit length), which is equal to the imaginary part X. That is,

$$R = X = \omega L \tag{3.71}$$

where L is the inductance per unit length. The inductance L of the microstrip structure can be expressed in terms of the characteristic impedance for the microstrip with the substrate replaced by air and is given as (Chan, Kan and Dutton, 1997)

$$L = \frac{Z_{0\text{air}}}{c} \tag{3.72}$$

where c is the velocity of electromagnetic waves in free space. As a general rule it can be stated that the thickness of the conductors should be greater than about four times the skin depth. The effect of the thickness of the transmission line has been well studied (Garg *et al.*, 1975; Gupta *et al.*, 1996; Horton *et al.*, 1971; Schneider, 1969; Simpson and Tesng, 1976; See also www.coventor.com/rf) and it is observed that the conductor losses are reduced by about 9% when the conductor thickness is $\pi/2$ times the skin depth.

The attenuation constant for a microstrip line due to dielectric loss can be written as

$$\alpha_d = 27.3 \frac{\varepsilon_r}{\sqrt{\varepsilon_e}} \frac{\varepsilon_e - 1}{\varepsilon_r - 1} \frac{(\tan\delta)}{\lambda_0} \tag{3.73}$$

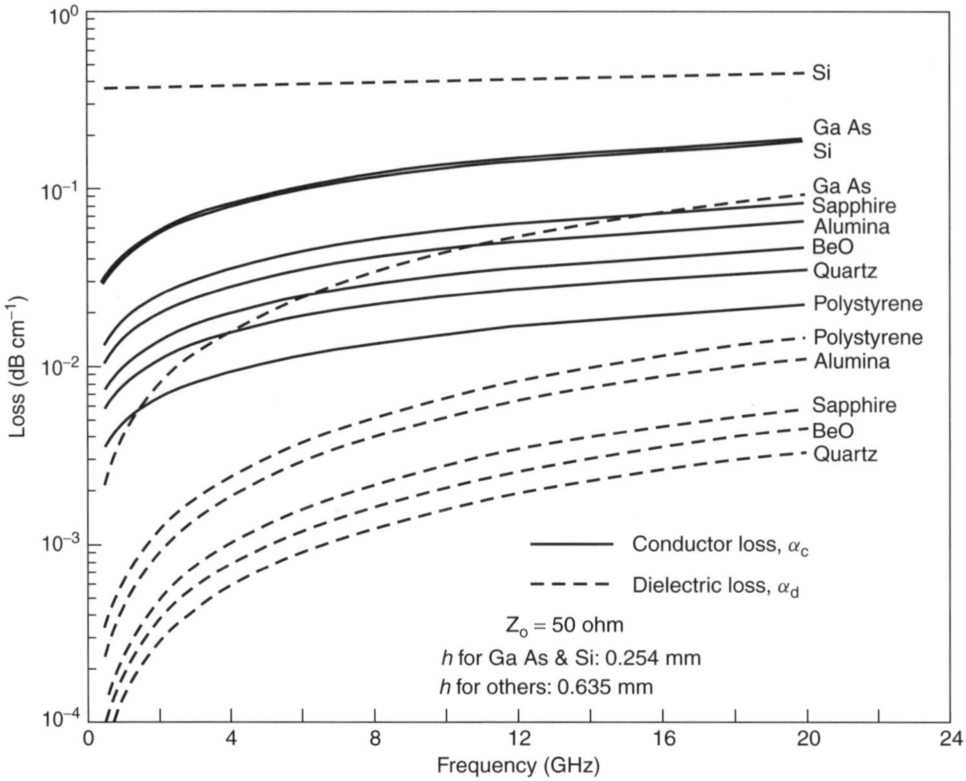

Figure 3.51 Conductor and dielectric losses as a function of frequency for 50-Ω microstrip lines on common substrates. Reprinted with permission from *Microstrip Lines and Slot Lines*, by K.C. Gupta, R. Garg, I. Bahl and P. Bhartia, Artech House Inc., Norwood, MA, © 1996 Artech House Inc.

where $(\tan\delta)$ is the loss tangent of the dielectric substrate. For a microstrip line on alumina substrate the dielectric loss α_d is negligible compared with total loss α. However, for microstrip lines on semiconductor substrates such as silicon, the loss factor is dominant. A 50-Ω microstrip line on a silicon substrate with dielectric constant $\varepsilon_r = 11.7$ and a resistivity of 10^3 Ω cm has a dielectric loss of the order of 0.36 dB cm^{-1}, while the conductor loss is about 0.19 dB cm^{-1} (Chan, Kan and Dutton, 1997). Conductor and dielectric losses per unit length for a 50-Ω microstrip line on various common substrates have been calculated and are presented in Figure 3.51.

3.8.2.7 Sections of lines

The basic building blocks of any RF MEMS circuits are the sections of transmission lines or microstrip lines. When the size of the microstrip section is reduced to dimensions much smaller than the wavelength, it can be used as a lumped element. Examples of lumped microstrip elements are interdigital capacitors, spiral inductors, thin-film resistors, metal–insulator–metal (MIM) capacitors, via holes and airbridges. Microstrip sections in lumped and distributed forms are commonly used in passive and active MEMS circuits. Examples of circuits include MEMS filters (Chapter 5), impedance transformers, couplers, delay lines and phase shifters (Chapter 6). The MEMS inductors and capacitors are discussed in Chapter 4.

3.8.2.8 Switch insertion loss modeling

The electromagnetic scattering computation done by Hyman *et al.*, 1999 using two-dimensional software packages including HP EEsof/Touchstone (http://eesof.tm.agilent.com/) concluded that spatially complex structures such as MEMS devices cannot be accurately handled in two-dimensional programs because of the interaction of the suspended elements with substrate properties. The insertion loss and isolation are modeled using a 3D scattering program, IE3D (Hyman *et al.*, 1999). The components of the switch near the transmission line were included in this model. The structure is modeled as a switch in closed position, with all metals 1-μm thick gold, GaAs substrate and the dielectric constants of the silicon nitride films ignored because of their small thickness. The width of the transmission line is 25 μm. The input and output transmission lines are 50-Ω microstrip lines of 60-mm width. The contact dimples are represented by filled vias with varying resistance due to impact of contact resistance on the insertion loss, as shown in Figure 3.52. The isolation of the switch is similarly calculated by replacing the contact with air gap and is shown in Figure 3.53.

It can be concluded from the modeling results that the contact closure force contributes a significant portion of insertion loss at all frequencies and it can be seen that if these contact force levels are well maintained, the impact on insertion loss should be constant. The presence of unexpected contamination or other contact obstruction would have

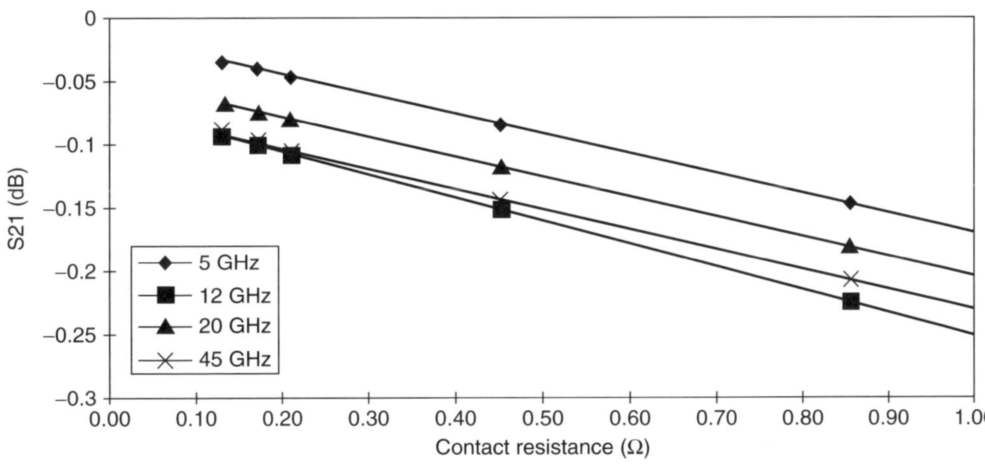

Figure 3.52 Insertion loss for different frequencies as a function of contact resistance. Reproduced from D. Hyman, J. Lam, B. Warneke, A. Schmitz, T.Y. Hsu, J. Brown, J. Schaffner, A. Waltson, R.Y. Loo, M. Mehregany and J. Lee, 1999a, 'Surface micromachined RF MEMS switches on GaAs Substrates', *International Journal of RF and Computer Aided Engineering* **9**: 348–61, by permission of Wiley, © 1999 Wiley

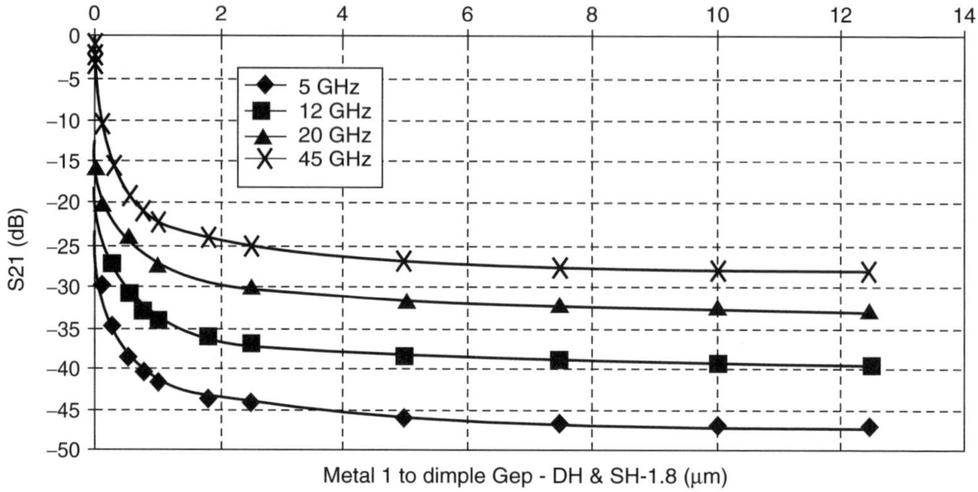

Figure 3.53 Modeled isolation as a function of gap height for different frequencies. Reproduced from D. Hyman, J. Lam, B. Warneke, A. Schmitz, T.Y. Hsu, J. Brown, J. Schaffner, A. Waltson, R.Y. Loo, M. Mehregany and J. Lee, 1999a, 'Surface micromachined RF MEMS switches on GaAs Substrates', *International Journal of RF and Computer Aided Engineering* **9**: 348–61, by permission of Wiley, © 1999 Wiley

debilitating results on switch insertion loss. A 0.5-Ω increase in contact resistance would result in a 0.1-dB increase in loss.

3.8.2.9 Other types of microstrip lines

There are several other types of microstrip lines being used in MEMS and MICs. These include multilayered microstrip, thin-film microstrip, inverted and suspended microstrip. Detailed discussions about these structures are available in Gupta *et al.*, (1996).

3.9 MEMS SWITCH DESIGN CONSIDERATIONS

The main challenge to the MEMS switch designer is in achieving the manufacturer's specified isolation. In general, the isolation levels of <40 dB is not a big issue. However, when the isolation of the order of 50–60 dB is required, a great deal of attention should be paid to circuit layout, particularly in the design of bends and corners. The co-planar waveguide (CPW) transmission lines are commonly used because CPW lines with narrow gaps allow narrower transmission lines, thus reducing the layout constraints and increasing track density. It also reduces radiation and coupling effects due to the presence of an upper ground plane. In addition to careful layout of the RF lines, the low-frequency control lines should be adequately decoupled off-chip using shunt capacitors placed as close to the control line pins as possible. Since the use of MEMS switches can reduce the number of dc lines in a circuit, MEMS switches are preferred over FET and MMIC switches where a high degree of switching is required in a constrained space.

3.10 CONCLUSIONS

The most widely used switching elements at present in RF and millimeter wave integrated circuits (MMICs) are FETs and PIN diode switches. The large insertion loss in the on state and poor isolation in the off state is a problem for many of these semiconductor switches and efforts are being made to develop micromechanical switches for mechanically switching RF signals. The current developments in MEMS technology have made possible the design and fabrication of micromechanical switches as a new switching element in RF circuits. MEMS-based switches with low insertion loss, negligible power consumption, good isolation and higher linearity is achieved by electrostatic actuation mechanisms.

The potential of integrating micromechanical switch structures with silicon microelectronics has shown promise in competing with conventional and integrated circuit switch technology. The advantages of MEMS technology include fast response, low power consumption, integration with microelectronics and high off-to-on state resistance. Over the past couple of years, a significant increase in publication of MEMS switches has highlighted its development profile. Applications of these devices range from automatic test equipment (ATE) where very stringent requirements are needed for isolation and current leakage, to communication systems, where switches with high functionality such as smart switches are required. The basic configuration of the switch is the cantilever, bridge and membrane, which are electrostatically or magnetically actuated. Electrostatic actuation is the preferred method because of the simple processing requirements for integration with IC technology. The target specifications of actuation voltages of these switches range from 5 to 100 V. Table 3.8 presents the state-of-art comparison of RF MEMS switches along with the operating parameters. It can be seen that the insertion loss and isolation are related to the capacitance of the switch when it is on and off.

MEMS switches are generally fabricated using the surface micromachining technique, where sacrificial layers such as oxides are etched to release the switch structure. The major challenge for MEMS switches is their integration with IC fabrications such as CMOS complementary metal oxide semiconductor and GaAs. A number of successful attempts such as polysilicon switches integrated with MOSFETs in which the contacts were made of doped polysilicon for the integration have been reported. However, recent study, Hyman and Mehregany (1999), shows that the use of gold for switch contacts reduces the on state contact resistance and inhibits alien films. Since use of gold appears to be unavoidable, the present aim is the successful fabrication of IC-compatible MEMS switches using gold, because it is not silicon compatible.

An alternative approach to this problem is the use of the electroplating technique, in which suitable metals for contact finish are deposited after CMOS processing and MEMS switch fabrication. The advantage is that this makes the entire process IC compatible with the exception of some final backend plating processes.

The integration of passive components on chips is becoming increasingly important for the communication and wireless industry. Integration, either on-chip or onto the substrate, is seen as providing a solution to the ever-increasing passive component count and the demand for smaller, more reliable systems, especially for handheld products. In the case of magnetic components such as inductors and transformers, the integration will certainly require the use of MEMS technology. The integration of larger value inductances and isolation transformers requires the deposition of magnetic materials and the use of electroplating techniques for the deposition of conducting and/or magnetic layers.

Table 3.8 Comparison of RF MEMS switches

Device	Actuation voltage (V)	On/off speed (μs)	Insertion loss (dB)	Isolation (dB)	Reference
Capacitive: membrane	70				Petersen, 1979
Electrostatic: rotating microswitch	80–200		0.5 at dc to 45 GHz	35 at dc to 45 GHz	Larson, Hackett and Melendes, 1991
Electrostatic: micromachine on GaAs	28	30	0.1 at 4 GHz	50 at 4 GHz	Yao and Chang, 1995
Capacitive: membrane	30	60	0.3–0.5 at 10 GHz	15 at 10 GHz	Goldsmith *et al.*, 1996; Randall *et al.*, 1996
Electrostatic: membrane	35–45		0.25 at 20 GHz	18 at 20 GHz	Yao *et al.*, 1997
Electroplated nickel					Zavaracky, Majumdar and McGruer, 1997
Electrostatic: electroplated materials	100				Zavaracky *et al.*, 1999
Capacitive: serpentine and cantilever springs	14–16			>30 at 40 GHz	Pacheco, Nguyen and Katehi, 1998
Capacitive: bridge	10–23			40 at 21–40 GHz	Barker and Rebeiz, 1998
Capacitive: shunt metal dielectric sandwich	30–50	80–110	0.25 at 35 GHz	35 at 35 GHz	Goldsmith *et al.*, 1998
Cantilever: one end fixed	30	20	<0.2 at dc to 40 GHz	>50 at <2 GHz	Hyman *et al.*, 1999a, 1999b
Hinged: cantilever	14–17		0.5 at 40 GHz	27, at 0.25–40 GHz	Shen and Feng 1999
Capacitive membrane	50	<6	0.14 at 20 GHz 0.25 at 35 GHz	15 at 10 GHz 35 at 35 GHz	Yao *et al.*, 1999
Electrostatic: narrow beams on multichip module	15	2			Hirata *et al.*, 1999

Silicon bulk micromachining on glass: single and double hump	56		0.2 at 30 GHz	>13 at 30 GHz	Suzuki *et al.*, 1999
Electrostatic: resonant tuned cross switch	15–20		0.6 at 22–38 GHz	50	Muldavin and Rebeiz, 1999
Spring folded suspensors with meanders	9	48		26 at 40 GHz	Pacheco, Katehi and Nguyen 2000
Shunt: single membrane, double shunt	15–25		0.2	35 at 10 GHz	Muldavin and Rebeiz, 2000
Capacitive	16–33	10	0.3 at 21 GHz		Ulm *et al.*, 2000
Electrostatic: semi-insulating GaAs, microactuator and CPW	26	10 000	0.2 at10 GHz	17 at 10 GHz	Chang and Chang, 2000
Capacitive: electroplated strontium titanate oxide	8	600	0.8 at 10 GHz	42 at 10 GHz	Park *et al.*, 2000, 2001
Electrostatic: torsional spring	10	10			Plotz *et al.*, 2001
Shunt: on silicon	30		0.4 ± 0.1 at 90 GHz	20 at 80–110 GHz	Rizk *et al.*, 2001
	<5	50	0.05–0.4 at dc to 12 GHz	25–40 at dc to 12 GHz	Campbell, 2001
Capacitive: SPDT X and *K* band			0.95 at 7 GHz 0.69 at 20 GHz	40	Pacheco, Peroulis and Katehi, 2001
Series-shunt absorptive fixed-fixed beam; with dimple	30–35		0.5 at dc to 26 GHz	40 at 5 GHz, 35 at 10 GHz, 25 at 26 GHz	Tan and Rebeiz, 2001

Note: CPW, co-planar waveguide; SPDT, single-phase double-throw.

REFERENCES

Agilent Technologies (2002); http://eesof.tm.agilent.com/about/ HP EES of Design Suite v 61.

Ahn, C.H., Allen, M.G., 1993, 'A fully integrated micromagnetic actuator with a multilevel meander magnetic core', in *IEEE Solid State Sensor and Actuators Workshop, 1993*, IEEE, Piscataway, NJ, USA: 14–18.

Ahn, C.H., Kim, Y.J., Allen, M.G., 1993, 'A planar variable reluctance magnetic micromotor with fully integrated stator and wrapped coils', in *Proceedings of IEEE Microelectromechanical Systems Workshop, 1993*, IEEE, Piscataway, NJ, USA: 1–6.

Alekseev, E., Pavlidis, D., 1998, '77 GHz high isolation coplanar transmit–receive switch using InGaAs/InP PIN diodes', in *Proceedings of IEEE GaAs Integrated Circuit Symposium*, IEEE, Piscataway, NJ, USA: 177–180.

Andricos, C., Bahl, I.J., Griffin, E.L., 1985, 'C-band 6-bit MMIC phase shifter', *IEEE Transactions on Microwave Theory and Techniques* **33**: 1591–1596.

Ayasli, Y., 1982, 'Microwave switching with GaAs FETs', *Microwave Journal* **25**(November): 61–74.

Ayasli, Y., 1982, 'Microwave switching with GaAs FETs', *Microwave Journal* **25**: 61–74.

Ayasli, Y., Platzker, A., Vorhaus, J., Raynolds, L.D., 1982, 'A monolithic single-chip X-band four-bit phase shifter', *IEEE Transactions on Microwave Theory and Techniques* **30**: 2201–2206.

Ayasli, Y., Mozzi, R., Raynolds, L.D., Tsukii, T., 1983, '6 to 19 GHz GaAs FET transmit receive switch', in *Proceedings of the IEEE Galium Arsenide Integrated Circuit Symposium*, IEEE, Piscataway, NJ, USA: 106–108.

Bahl, I.J., Bhartia, P., 1988, *Microwave Solid State Circuit Design*, John Wiley, New York.

Barker, N.S., Rebeiz, G.M., 1998, 'Distributed MEMS true-time delay phase shifter and wide-band switches', *Transactions on IEEE Microwave Theory and Techniques* **46**(11): 1881–1890.

Bozler, C., Drangmeister, R., Duffy, S., Gouker, M., Knecht, J., 2000, 'MEMS microswitch arrays for reconfigurable distributed microwave components', in *Proceedings of IEEE Microwave Theory and Techniques Symposium, 2000*, IEEE, Piscataway, NJ, USA: 153–156.

Brauchler, F., Robertson, S., East, J., Katehi, L.P.B., 1996, 'W-band finite ground plane coplanar (FGC) line circuit elements', in *Proceedings of the IEEE MTT-S Symposium 1996*, IEEE, Piscataway, NJ, USA: 1845–1848.

Browne, J., 1989, 'Switches perform high-frequency signal routing', *Microwaves and RF* (July): 125–132.

Brown, E.R., 1998, 'RF-MEMS switches for reconfigurable integrated circuits', *IEEE Transactions on Microwave Theory and Techniques* **46**(11): 1868–1880.

Bryzek, J., Petersen, K., McCulley, W., 1994, 'Micromachines on the march', *IEEE Spectrum*: **31**(5): 20–31.

Busch-Vishnic, I.J., 1992, 'The case for magnetically driven microactuators', *Sensors and Actuators A* **33**: 207–220.

Campbell, T., 2001, 'MEMS switch technology approaches the ideal switch', *Journal of Applied Microwave and Wireless* **13**(5): 100–107.

Caverly, R.H., 1993, 'Distortion in arsenide MESFET switches', *IEEE Transactions on Microwave Theory and Techniques* **41**: 1323–1328.

Chang, K., 1990, *Handbook of Microwave and Optical Components; Volume 2: Microwave Solid-state Components*, John Wiley, Chichester, Sussex.

Chang, K., 1994, *Microwave Solid-state Circuits and Applications*, John Wiley, Chichester, Sussex.

Chang, C., Chang, P., 2000, 'Innovative micromachined microwave switch with very low insertion loss', *Sensors and Actuators A* **79**: 71–75.

Chan, E.K., Kan, E.C., Dutton, R.W., 1997, 'Nonlinear dynamic modeling of micromachined microwave switches', in *Proceedings of IEEE MTT-S Symposium, 1997*, IEEE, Piscataway, NJ, USA: 1511–1514.

Coutrot, A.L., Defour-Gergam, E., Martincic, E., Gilles, J.P., Grandchamp, J.P., Quemper, J.M., Bosseboeuf, A., Alves, F., Ahamada, B., 2001, 'Electromagnetic micro-device realized by electrochemical way', *Sensors and Actuators A* **91**: 80–84.

de los Santos, H.J., Kao, Y.-H., Caigoy, A.L., Dirmars, D., 1997, 'Microwave and mechanical considerations in the design of MEM switches for aerospace applications', *Proceedings of IEEE Aerospace Conference Aspen, Co* **3**: 235–254.

Drake, J., Jerman, H., Lutze, B., Stuber, M., 1995, 'An electrostatically actuated micro-relay', in *Proceedings of Transducers '95 Eurosensors IX*, IEEE, Piscataway, NJ, USA: 380–383.

Fullin, E., Gobert, J., Tilmans, H.A.C., Bergqvist, J., 1998, 'A new basic technology for magnetic microactuators', in *Proceedings of IEEE Symposium on MEMS '98*, IEEE, Piscataway, NJ, USA: 143–147.

Garg, R., Gupta, K.C., Sharan, R., 1975, 'Optimum thickness of metal in waveguiding structures, ground planes and reflectors', *International Journal of Electronics* **39**: 525–527.

Goldsmith, C., Lin, H.T., Powers, B., Wu, W.R., Norvell, B., 1995, 'Micromachined membrane switches for microwave applications', in *Proceedings of IEEE Microwave Theory and Techniques Symposium 1995*, IEEE, Piscataway, NJ, USA: 91–94.

Goldsmith, C., Randall, J., Eshelman, S., Lin, T.H., Denniston, D., Chen, S., Norvell, B., 1996, 'Characteristics of micromachined switches at microwave frequencies', in *Proceedings of IEEE Microwave Theory and Techniques Symposium, 1996*, IEEE, Piscataway, NJ, USA: 1141–1144.

Goldsmith, C.L., Yao, Z., Eshelman, S., Denniston, D., 1998, 'Performance of low-loss MEMS capacitive switches', *IEEE MW and Guided Wave Letters* **8**(8): 269–271.

Goldsmith, C., Ehmke, J., Malczewski, A., Pillans, B., Eshelman, S., Yao, Z., Brank, J., Eberly, M., 2001, 'Lifetime characterization of capacitive RF MEMS switches', in *Proceedings of IEEE MTT-S, 2001*, IEEE, Piscataway, NJ, USA: 227–230.

Gopinath, A., Rankin, J.B., 1985, 'GaAs FET RF switches', *IEEE Transactions on Electron Devices* **32**: 1272–1278.

Gretillat, M.A., Yang, Y.J., Hung, E.S., Rabinovich, V., Ananthasuresh, G.K., de Rooji, N.F., Senturia, S.D., 1997, 'Nonlinear electromechanical behavior of an electrostatic microrelay', in *Proceedings of Transducers '97*, IEEE, Piscataway, NJ, USA: 1141–1144.

Guckel, H., Christenson, T.R., Skrobis, K.J., Jung, T.S., Klen, J., Hartojo, K.V., Widjaja, I., 1993, 'A first functional current excited planar rotational magnetic micromotor', in *Proceedings of IEEE Microelectromechanical Systems Workshop, 1993*, IEEE, Piscataway, NJ, USA: 7–11.

Gupta, K.C., Garg, R., Bahl, I., Bhartia, P., 1996, *Microstrip Lines and Slot Lines*, Artech House, Boston, MA.

Hashimoto, E., Tanaka, H., Suzuki, Y., Uensishi, Y., Watabe, A., 1994, 'Thermally controlled magnetic actuator (TCMA) using thermo sensitive magnetic materials', in *Proceedings of IEEE Microelectromechanical Systems Workshop, 1994*, IEEE, Piscataway, NJ, USA: 108–113.

Hashimoto, E., Uensishi, Y., Watabe, A., 1995, 'Thermally controlled magnetization microrelay', in *Proceedings of 8th International Conference on Solid State Sensors and Actuators and Eurosensors IX, Transducers, 1995*, IEEE, Piscataway, NJ, USA: 361–364.

Hiltmann, K.M., Schmidt, B., Sandmaier, H., Lang, W., 1997, 'Development of micromachined switches with increased reliability', *Proceedings of Transducers '97*: 1157–1160.

Hirata, A., Machida, K., Kyuragi, H., Maeda, M., 1999, 'A micromechanical switch as the logic element for circuits in multi chip module on Si (MCM-Si)', in *Proceedings of IEEE Symposium*, IEEE, Piscataway, NJ, USA: 582–587.

Horton, R., Easter, B., Gopinath, A., 1971, 'Variation of microstrip losses with thickness of the strip', *Electronics Letters* **7**: 490–491.

Hosaka, H., Kuwano, H., Yanagisawa, K., 1993, 'Electromagnetic microrelays: concepts and fundamental characteristics', in *Proceedings of IEEE Symposium on Microelectromechanical Systems, MEMS '93*, IEEE, Piscataway, NJ, USA: 12–15.

Hosaka, H., Kuwano, H., Yanagisawa, K., 1994, 'Electromagnetic microrelays: concepts and fundamental characteristics', *Sensors and Actuators A* **40**: 41–47.

Huang, J.M., Liew, K.M., Wong, C.H., Rajendran, S., Tan, M.J., Liu, A.Q., 2001, 'Mechanical design and optimization of capacitive micromachined switch', *Sensors and Actuators A* **93**: 273–285.

Hyman, D., Lam, J., Warneke, B., Schmitz, A., Hsu, T.Y., Brown, J., Schaffner, J., Waltson, A., Loo, R.Y., Mehregany, M., Lee, J., 1999a, 'Surface micromachined RF MEMS switches on GaAs Substrates', *International Journal of RF and Computer Aided Engineering* **9**: 348–361.

Hyman, D., Mehregany, M., 1999, 'Contact physics of gold microcontacts for MEMS switches', *IEEE Transactions on Components and Packaging Technology* **22**(3): 357–364.

Hyman, D., Schmit, A., Warneke, B., Hsu, T.Y., Lam, J., Brown, J., Schaffner, J., Walston, A., Lo, R.Y., Tangonan, G.I., Mehregany, M., Lee, J., 1999b, 'GaAs Compatible surface micromachined RF switches', *Electron Letters* **35**(3): 224–226.

Ioth, T., 1993, 'Electromagnetic ultra miniature motor', *Micro Machine* **6**(1): 110–112 [in Japanese].

Kobayashi, K.W., Oki, A.K., Umemoto, D.K., Claxton, S.K.Z., Streit, D.C., 1993, 'Monolithic variable gain amplifiers, attenuators and switches', *IEEE Transactions on Microwave Theory and Techniques* **41**: 2295–2302.

Larson, L.E., Hackett, R.H., Melendes, M.A., Lohr, R.F., 1991, 'Microactuators for GaAs based microwave integrated circuits', in *IEEE Transducers '91 Conference on Solid State Sensors and Actuators, 1991*, IEEE, Piscataway, NJ, USA: 743–746.

Larson, L.E., Hackett, R.H., Melendes, M.A., 1991, 'Micromachined microwave actuator (MIMAC) technology – A new tuning approach for microwave integrated circuits', in *Proceedings of IEEE Microwave and MM wave Monolithic circuits Symposium*, IEEE, Piscataway, NJ, USA: 27–30.

Lau, G.L., Dunn, D.L., Chen, J.C., Pao, C.K., Wang, W.C., 1988, 'A high performance V-band monolithic FET transmit/receive switch', in *Proceedings of IEEE Microwave and Millimeter Wave Monolithic Circuits Symposium*, IEEE, Piscataway, NJ, USA: 99–101.

Losee, F., 1997, *RF Systems, Components, and Circuits Handbook*, Artech House, Boston, MA.

Majumdar, S., McGruer, N.E., Zavracky, P.M., Adams, G.G., 1997, 'Measurement and modeling of surface micromachined, electrostatically actuated microswitches', *Transducers '97*: 1145–1148.

McNie, M.E., King, D.O., Brunson, K.M., Jones, A.R.D., 2000, 'Design and modeling of polysilicon microswitches', in *IEEE Seminar on Microengineering, Modelling and Design*, IEEE, Piscataway, NJ, USA: 6/1–6/4.

Muldavin, J.B., Rebeiz, G.M., 1999, '30 GHz tuned MEMS switches', in *Proceedings of IEEE MTT-S Symposium, 1999*, IEEE, Piscataway, NJ, USA: 1511–1514.

Muldavin, J.B., Rebeiz, G.M., 2000, 'High isolation inductively tuned X-band MEMS shunt switches', in *Proceedings of IEEE MTT-S Symposium, 2000*, IEEE, Piscataway, NJ, USA: 169–172.

Muldavin, J.B., Rebeiz, G., 2000a, 'High isolation CPW MEMS shunt switches, part I: modeling', *IEEE Transactions on Microwave Theory and Techniques* **48**(6): 1045–1052.

Muldavin, J.B., Rebeiz, G., 2000b, 'High isolation CPW MEMS shunt switches, part II: design', *IEEE Transactions on Microwave Theory and Techniques* **48**(6): 1053–1056.

Ota, Y., Sakukara, M., Fujimoto, K., Yamamoto, S., Fujimoto, H., 1995, 'High isolation and low insertion loss switch IC using GaAs MESFETs', *IEEE Transactions Microwave Theory and Techniques* **43**: 2175–2177.

Pacheco, S., Nguyen, C.T.-C., Katehi, L.P.B., 1998, 'Micromechanical electrostatic K-band switch', in *Proceedings of IEEE MTT-S*, IEEE, Piscataway, NJ, USA: 1569–1572.

Pacheco, S.P., Katehi, L.P.B., Nguyen, C.T.-C., 2000, 'Design of low actuation voltage RF MEMS switch', in *Proceedings of 2000 IEEE MTT-S*, IEEE, Piscataway, NJ, USA: 165–168.

Pacheco, S.P., Peroulis, D., Katehi, L.P.B., 2001, 'MEMS single-pole double-throw (SPDT) X and K-band switching circuits', *IEEE MTT-S International Microwave Symposium Digest,* IEEE, Piscataway, NJ, USA, **1**, pp. 321–324.

Park, J.Y., Kim, G.H., Chung, K.W., Bu, J.U., 2000, 'Electroplated RF MEMS capacitive switches', in *Proceedings of IEEE International Conference on MEMS, MEMS 2000*, IEEE, Piscataway, NJ, USA: 639–644.

Park, J.Y., Kim, G.H., Chung, K.W., Bu, J.U., 2001, 'Monolithically integrated micromachined RF MEMS capacitive switches', *Sensors and Actuators A* **89**: 88–94.

Peek, R.L., Wager, H.N., 1955, *Switching Relay Design*, D van Nostand, Princeton, NJ.

Petersen, K.E., 1978, 'Dynamic micromechanics on silicon: techniques and devices', *IEEE Transactions on Electron Devices* **ED-25**(10): 1241–1250.

Petersen, K.E., 1979, 'Micromechanical membrane switches on silicon', *IBM Journal of Research and Development* **23**(4): 376–385.

Poltz, F., Michaelis, S., Aigner, R., Timme, H.J., Binder, C., Noe, R., 2001, 'A low voltage torsional actuator for application in RF microswitches', *Sensors and Actuators A* **92**(1–3), 312–317.

Putnam, J., Fukuda, M., Staecker, P., Yun, Y.H., 1994, 'A 94 GHz monolithic switch with a vertical PIN diode structure', in *Proceedings of IEEE GaAs Symposium*, IEEE, Piscataway, NJ, USA: 333–336.

Randall, J.N., Goldsmith, C., Denniston, D., Lin, T.-H., 1996, 'Fabrication of micromechanical switches for routing radio frequency signals', *Journal of Vacuum Science Technology B* **14**(6): 3692–3696.

Rizk, J., Tan, G.L., Muldavin, J., Rebeiz, G.M., 2001, 'High isolation W-band MEMS Switches', *IEEE Microwave and Wireless Components Letters* **11**(1): 10–12.

Sadler, D.J., Liakopoulos, T.M., Ahn, C.H., 2000, 'A universal electromagnetic micromotor using magnetic interconnection concepts', *Journal of Micromechanical Systems* **9**(4): 460–468.

Saffer, S., Simon, J., Kim, C.J., 1996, 'Mercury contact switching with gap-closing microcantilever', *Proceedings of SPIE 1996* **2882**: 204–209.

Schiele, I., Huber, J., Hillerich, B., Kozlowski, F., 1998, 'Surface micromachined electrostatic microrelay', *Sensors and Actuators A* **66**: 345–354.

Schimakt, J., 1998, 'Contact materials for microrelays', in *Proceedings of MEMS '98*, IEEE, Piscataway, NJ, USA: 190–194.

Schindler, M.J., Morris, A., 1987, 'DC to 40 GHz and 20 to 40 GHz MMIC SPDT switches', *IEEE Transactions on Microwave Theory and Techniques* **35**: 1486–1493.

Schindler, M.J., Miller, M.E., 1988, 'A 3-bit K/Ka band MMIC phase shifter', in *Proceedings of IEEE Microwave and Millimeter Wave Monolithic Circuits Symposium*, IEEE, Piscataway, NJ, USA: 95–98.

Schneider, M.V., 1969, 'Dielectric loss in integrated microwave circuits', *Bell System Technical Journal* **48**: 2325–2332.

Shen, S.-C., Feng, M., 1999, 'Low actuation voltage RF MEMS switches with signal frequencies from 0.25 to 40 GHz', in *Proceedings of IEEE International Electron Devices Meeting*, IEEE, Piscataway, NJ, USA: 689–692.

Simon, J., Safer, S., Kim, C.J., 1997, 'A liquid filled microrelay with a moving mercury drop', *Journal of Micromechanical Systems* **6**(3): 208–216.

Simpson, T.L., Tesng, B., 1976, 'Dielectric losses in microstrip lines', *IEEE Transactions on Microwave Theory and Techniques* **24**: 106–108.

Slobodnik, A.J., Webster, R.T., Roberts, G.A., Scalzie, G.J., 1989, 'Millimeter wave GaAs switch FET modeling', *Microwave Journal* **32**(8), 93–104.

Sokolov, V., Geddes, J.J., Contolatis, A., Bauhahn, P.E., Chao, C., 1983, 'A Ka-band GaAs monolithic phase shifter', *IEEE Transactions on Microwave Theory and Techniques* **31**: 1077–1082.

Sovero, E.A., Mihailovich, R., Deakin, D.S., Higgins, J.A., Yao, J.J., DeNatale, J.F., Hong, J.H., 1999, 'Monolithic GaAs PHEMT MMICs integrated with high performance MEMS microrelays', in *Proceedings of IEEE MTT-S Symposium, 1999*, IEEE, Piscataway, NJ, USA: 257–260.

Street, A.M., 2000, 'RF switch design', *IEEE Training Course 2000: How to Design RF circuits*, 2000/027: 4/1-4/7.

Suzuki, K., Chen, S., Marumoto, T., Ara, Y., Iwata, R., 1999, 'A micromachined RF microswitch applicable to phased-array antennas', in *Proceedings of IEEE MTT-S Symposium*, IEEE, Piscataway, NJ, USA: 1923–1926.

Tangonan, G., Loo, R., Schaffner, J., Lee, J.J., 1999, 'Microwave photonic applications of MEMS technology', in *IEEE Proceedings of International Topical Meetings on Microwave Photonics, MWP '99*, IEEE, Piscataway, NJ, USA: 109–112.

Tan, G.L., Rebeiz, G.M., 2001, 'DC-26 GHz MEMS series-shunt absorptive switches', in *Proceedings of IEEE Microwave Symposium, Volume 1*, IEEE, Piscataway, NJ, USA: 325–328.

Taylor, W.P., Allen, M.G., 1997, 'Integrated magnetic microrelays: normally open, normally closed and multi-pole devices', in *Proceedings of Transducers '97*, IEEE, Piscataway, NJ, USA: 1149–1152.

Taylor, W.P., Brand, O., Allen, M.G., 1998, 'Fully integrated magnetically actuated micromachined relays', *Journal of Microelectromechanical Systems* **7**(2): 181–191.

Tilmans, H.A.C., Fullin, E., Ziad, H., Van de Peer, M.D.J., Kesters, J., Van Geffen, E., 1999, 'A fully packaged electromagnetic microrelay', in *Proceedings of IEEE International Conference on Microelectromechanical Systems, MEMS '99*, IEEE, Piscataway, NJ, USA: 25–30.

Timoshenko, S., Krieger, S.W., 1987, *Theory of Plates and Shells*, McGraw Hill, New York.

Ulm, R.B., Walter, T., Fiedler, R.M., Voigtlender, K., Kasper, E., 2000, 'K-band capacitive MEMS switches', *Topical Meeting on Silicon Monolithic Integrated Circuits in RF Systems 2000*, IEEE, Piscataway, NJ, USA: 119–122.

Wright, J.A., Tai, Y.C., 1999, 'Magnetostatic MEMS relays for the miniaturization of brushless DC motor controllers', in *Proceedings of IEEE Symposium on MEMS '99*, IEEE, Piscataway, NJ, USA: 594–599.

Yao, J.J., Chang, M.F., 1995, 'A surface micromachined miniature switch for telecommunications applications with signal frequencies from DC up to 4 GHz', *Transducers '95–Eurosensors* **IX**: 384–387.

Yao, J.J., 2000, 'RF MEMS from a device perspective', *Journal Micromechanics and Microengineering* **10**: R9–R38.

Yao, Z.J., Chen, S., Eshelman, S., Denniston, D., Goldsmith, C., 1999, 'Micromachined low-loss microwave switches', *Journal of Microelectromechanical Systems* **8**(2): 129–134.

Yao, Z.J., Goldsmith, C., Eshelman, S., Denniston, D., 1997, 'Micromachined RF signal switching devices on high resisitvity silicon substrates', in *Proceedings of Microelectromechanical Systems ASME 1997*, ASME, Fairfield, NJ, USA: 69–73.

Zavracky, P.M., Morrison, R.H., 1984, 'Electrically actuated micromechanical switches with hysteresis', in *Technology Digest IEEE Solid-state Sensors Conference 1984*, IEEE, Piscataway, NJ, USA: 50–51.

Zavracky, P.M., Majumdar, S., McGruer, N., 1997, 'Micromechanical switches fabricated using nickel surface micromachining', *Journal of Microelectromechanical Systems* **6**(1): 3–9.

Zavracky, P.M., McGruer, N.E., Morrison, R.H., Potter, D., 1999, 'Microswitches and microrelays with a view toward microwave applications', *International Journal of RF and Microwave Computer Aided Engineering* **9**: 338–347.

4
MEMS inductors and capacitors

4.1 INTRODUCTION

The widespread use of microwave integrated circuit (MIC) technology along with miniaturization as a result of MEMS have led to the development of RF and microwave circuit elements whose dimensions are much smaller than their wavelength. Micromachined devices are found to be useful today in several emergent applications such as wireless communications, global positioning systems, military and consumer electronics and automotive sensors. High-performance low-power RF transceivers are the integral part of many of these devices. There are still many functions that cannot be implemented using conventional integrated circuit (IC) technology, in particular, components with high Qs (exceeding 30), which are required for high-frequency selectivity in communication systems. The Qs of planar spiral inductors and junction diode capacitors are only of the order of low 10s at higher frequencies and hence alternative off-chip technologies including inductors and tunable capacitors are often used for high Q applications. A low-noise RF voltage-controlled oscillator requires a high Q device because the phase noise of the oscillator is proportional to $1/Q_T^2$, where Q_T is the overall quality factor.

The use of bulk and surface acoustic wave (SAW) filters and tunable tank circuits in cellular applications imposes a significant bottleneck against the miniaturization of transceivers. There has been a recent demand for fully integrated planar inductors and capacitors for the realization of MEMS and monolithic microwave integrated circuits (MMICs). Conventional inductive components are inherently three-dimensional (3D) nature and the implementation of these components in planar shape is quite challenging. Small size and weight, low power consumption, mass production, reliability and reproducibility are some of the numerous advantages of integration of MICs with MEMS. Although there are several technical hurdles remaining to be solved, in the near future the entire RF subsystem incorporating the receiver front end, filters, switches, antennas, IF sections along with amplifiers have to be integrated into a 'single radio chip'.

This chapter introduces various micromachined and MEMS inductors and capacitors along with their design parameters. The efforts to enhance the quality factor of these inductors are discussed in detail. Selecting the right technology for the design of an inductor depends on what a designer is looking for. Each design procedure has its own advantages and disadvantages.

4.2 MEMS/MICROMACHINED PASSIVE ELEMENTS: PROS AND CONS

Passive elements such as inductors, capacitors and transformers play a critical role in today's wireless and high-speed digital applications. Solid-state components such as p–n junctions and Schottky diodes often suffer from small tuning ratio (typically less than 30%), low Q (less than 10), high resistive losses and low self-resonant frequency. The recent efforts in microelectronics and MEMS have shown promising results in realization of high-performance passive components for RF applications. The miniaturization, reliability, ease of assembly and compactness of IC fabrication technology are the factors that paved the way for embedding these passive components directly into the substrate and RF IC technology. When such devices are fabricated on conductive substrates such as silicon, this gives the freedom of low-cost silicon IC fabrication over GaAs IC fabrication and hence the potential for integration with the baseband circuits at the cost of increase in losses and decrease in quality factor. Efforts are being made to improve the Q by multiplayer techniques and leaving a space between substrate and the conducting lines.

Magnetic components operating at high frequencies rapidly increase eddy current and hysteresis loss in the magnetic cores as the operating frequency increases. The micromachining techniques provide several approaches for the miniaturization of inductors operating at high frequencies. Magnetic cores and conductors with desired thickness and width with good dimensional control could be realized using MEMS technology.

A wire in general in any RF circuit has three important electrical characteristics: resistance, capacitance and inductance, which will together add delay in signal. Also, the chances of generating noise due to inductive as well as capacitive coupling between wires are high. Therefore efforts have to be made to follow a well-defined model for the design and implementation of any wires in an RF circuit because a single turn of wire or a strip can drastically change the performance of RF circuits. There is no simple closed-form models existing for computing on-chip inductance, and extracting the inductance is not straightforward.

At present, the miniaturization of radio receivers is challenging job because of unavailability of numerous off-chip, frequency selective, passive, miniature components, which are conventionally operating via a mechanical means in radio receivers. Recent development of MEMS passive RF components retains the need for tunable properties in these radio systems with an order of magnitude smaller in size.

4.3 MEMS INDUCTORS

An inductor is a circuit component as shown in Figure 4.1(a), which is capable of producing voltage across its terminals in response to a changing current flowing through it. As in the case of capacitors, which store electric energy, inductors store magnetic energy. Voltage is generated as a result of magnetic induction. The time-varying magnetic field due to current in an inductor induces electromotive force. Inductors are usually coils of wires, circular or spiral in shape, in which windings are necessary to enhance the flux linkage and hence a large inductance in a small area. In general, the wires in a circuit affect the devices in three ways: (a) the wire capacitance adds load to the driving circuits; (b) signal may be delayed due to wire resistance, capacitance and inductance and (c) adds noise due to inductive and capacitive coupling in different wires. Any change in

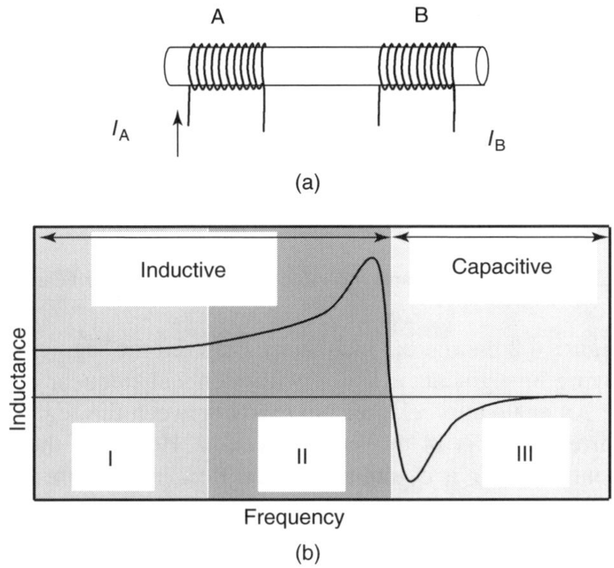

Figure 4.1 (a) Self-inductance and mutual inductance due to a change in current; (b) typical operational regions of an inductor. Note: I_A and I_B, current in coils A and B, respectively

the current of a circuit induces a change in the magnetic field. Faraday's law states that a changing magnetic field induces an electric field and, from Lenz's law, this induced electric filed always opposes further changes in current.

An ideal passive component shows constant values for all frequencies with constant phase. However, every nonideal component exhibits changes in value with frequency as shown as an example for an inductor in Figure 4.1(b). The region I is the useful operational region whereas in transitional region II the inductance value becomes negative when frequency increases. This is the first self-resonance frequency of the inductor. Avoiding operation of an inductor in this region is important since beyond this point the element becomes capacitive and the quality factor is practically zero.

4.3.1 Self-inductance and mutual inductance

As shown in Figure 4.1(a), the change in current in coil A will produce a change in magnetic field in A, which induces a voltage in coil B. The electromotive force induced in circuit B when the current in circuit A changes is proportional to the rate of change of linkage of flux set up by the current in A. The persistent generation of voltages, which oppose the change in magnetic field, is the operating principle of a transformer. The change in current in one coil that affects the current and voltage in the second coil is called the *mutual inductance*. This generation of electromotive force (emf) is explained by Faraday's law and its direction is always opposed to the change in the magnetic field produced by the coupled coil. The change in current induces in each coil an emf due to change in flux linkage of its own turns with its self-produced magnetic field.

In the case of a spiral inductor, the total inductance is equal to the sum of the self-inductances of the straight segments plus the mutual inductance between the segments.

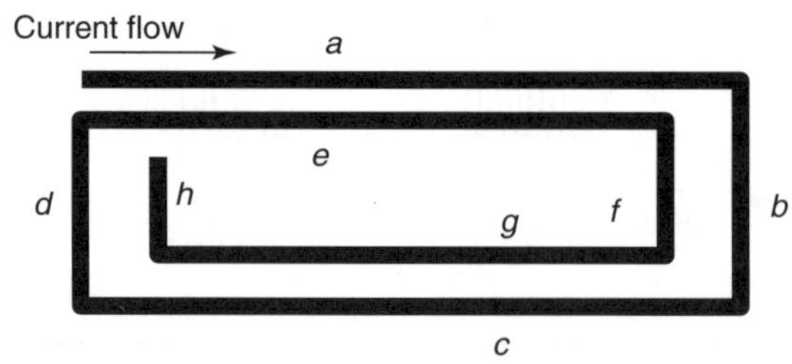

Figure 4.2 Spiral inductor and the effect of current flow in nearby segments

For example, in Figure 4.2 the mutual inductance between the segments a and e is caused by the current flowing in segments a and e, with identical frequency and phase, and can be added together. An analogous relationship exists between the segments $b-f$, $c-g$ and $d-h$, in which current flow is in the same direction. However, the mutual inductance between the segments a and g is due to the current flow in opposite directions. The same relationship exists between the segment pairs $a-c$, $e-g$, $e-c$, $d-f$, $d-b$, $h-f$ and $h-b$. Therefore the inductance for a coil or part of a coil of any shape is

$$L_{\mathrm{T}} = L_0 + \sum M \tag{4.1}$$

where L_{T} is the total inductance, L_0 is the sum of the self-inductances of all the straight segments and $\sum M$ is the sum of all mutual inductances.

A few possible geometries of planar inductive elements for RF applications are shown in Figure 4.3. They can generally be classified as strip inductors or spiral inductors. Straight sections of wires or strips are used for low inductance values typically less than 10 nH while spiral (circular rectangular) have higher Qs and can provide higher inductance values (Grover, 1946). The presence of a ground plane also affects the inductance. The inductance decreases when the ground plane is brought nearer to the conducting line. Planar inductors are made essentially with a single-layer metallization scheme, in which a conducting layer is etched on a dielectric substrate. The finite conductivity of the metal layer and the loss in dielectric substrate can introduce losses in inductors. A metal layer of thickness at least 3 or 4 times the skin depth can reduce the conductive losses. All segments in an inductor depend on the mutual coupling between various lines so that it can achieve a high inductance in a small area. This inevitable area limitation of monolithic integrated circuits prevents further improvements in Q because the Q of an inductor is roughly proportional to its physical area. In order to be this as a lumped element, the total line length in any multisegment component should be a small fraction of the operating wavelength and therefore exhibit negligible phase shift.

The quality factor Q of an inductor can be written as

$$Q = \frac{\omega L}{R} \tag{4.2}$$

where L is the inductance, $\omega = 2\pi f$, f is the operating frequency and R is the resistance. In the case of an inductor folded into a meander line as shown in Figure 4.3(a), the negative mutual inductance associated with the coupling between adjacent tracks must be considered. However, if the line widths are much less than the thickness of the dielectric

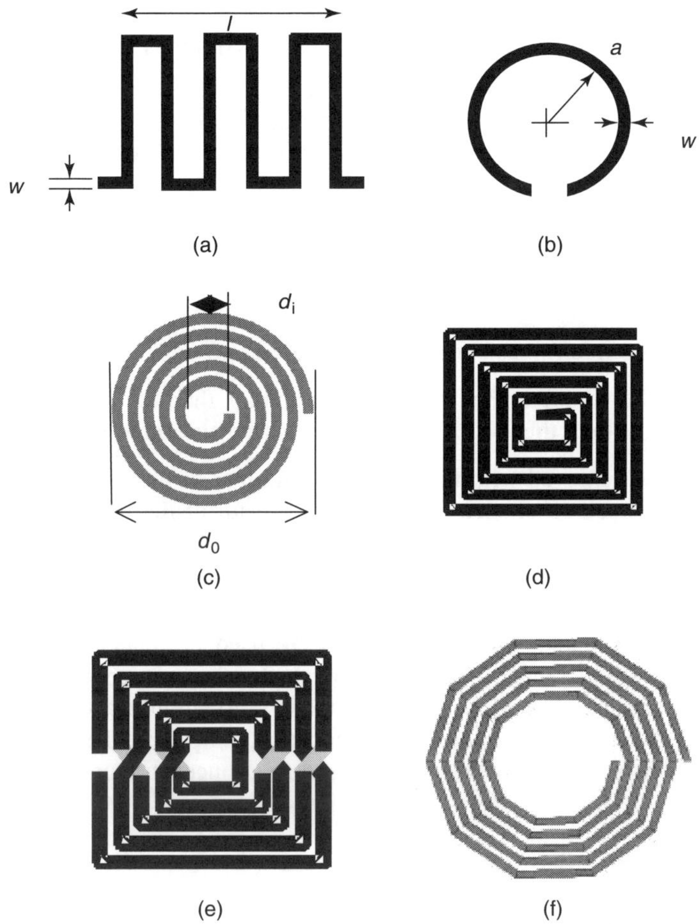

Figure 4.3 Schematic diagram of common planar inductors: (a) meander; (b) loop; (c) circular spiral; (d) square spiral; (e) symmetric spiral; (f) polygon spiral

sheet, the mutual coupling between the lines considered is negligible (Grover, 1946). The inductance of a straight line or strip can be written as

$$L = 2l\left[\ln\left(\frac{l}{w+t}\right) + 0.22\frac{w+t}{l} + 1.19\right] \tag{4.3}$$

where L is the segment inductance in nanohenries, l, w and t are the segment length, width and thickness, respectively, in centimeters. The strip inductors are good in the range of 0.5–4 nH. Higher inductances can be achieved using spiral inductors. The inductance of a single loop in nanohenries is given by (Bahl and Bhartia, 1998)

$$L = 4\pi a\left[\ln\left(\frac{8\pi a}{w}\right) - 2\right] \tag{4.4}$$

where a is the radius and w is the width of the strip, both in centimeters. For a higher number of turns, the approximate expression can be written as

$$L = \frac{394a^2N^2}{8a + 11c} \tag{4.5}$$

where $a = (d_0 + d_i)/4$, and $c = (d_0 - d_i)/2$, with d_0 and d_i as shown in Figure 4.3(c), in centimeters; N is the number of turns.

The important considerations in the design of planar inductors can be summarized as (Greenhouse, 1974; Konishi, 1991):

- The separation between lines should be as small as possible.
- The circular spiral inductor has a shorter conductor than square spiral and the Q is about 10% higher than that of a square spiral having the same value of d_0.
- Higher Q can be achieved with increasing the number of turns per unit area; however, it also lowers the self-resonance frequency as a result of the increase in capacitance.
- To avoid parasitic effects, the maximum diameter of the inductor should be less than $\lambda/30$.

The spiral, rectangular or circular inductors are used in general as reactive elements in filters, couplers, dividers and matching elements and impedance transformers in MICs. Inductors are integrated with MICs using standard fabrication procedures and the innermost turn is connected using either wire bonding or flip-chip methods. In monolithic MICs, instead of wire bond, air bridges are used to connect the inner coil to the circuit.

4.3.2 Micromachined inductors

Two general categories of components, called distributed and lumped elements, are used MICs. The most common distributed circuits are microstrip line, slot line and co-planar waveguides (CPWs). The lumped elements are truly lumped and have no variation in L or C with frequency or any variation of phase over the element. Micromachined planar inductors and capacitors are used as passive elements in MICs. In general, these lumped elements show low loss and wide bandwidth for frequencies below 12 GHz. For planar inductors, the parasitic capacitance between the inductor and the ground plane is a problem (Chi and Rebeiz, 1994, 1995). These parasitics both lower the Q of the inductors and create a self-resonance frequency that limits the maximum frequency of operation, making the devices insufficient for communication applications. Attempts to fabricate large-value spiral inductors on silicon in the early 1960s (Warner, 1965) led to the conclusion that the parasitic capacitances of these structures cause self-resonance, which would limit its use for high-frequency applications. The large parasitic capacitance of a planar interdigitated capacitor, which is directly related to the fabrication process, also affects the performance as a true lumped element. However, later, the measured unloaded Q of inductors on high resistive silicon (3000–7000 Ω cm) shows that the Qs are comparable to those fabricated on GaAs and quartz (Park *et al.*, 1997a, 1997b, 1997c; Reyes *et al.*, 1995). It is also demonstrated that a silicon-on-sapphire (SoS) technology (Johnson *et al.*, 1996) can provide inductors with high self-resonant frequencies and good quality factors.

4.3.2.1 Meander inductors

One of the MEMS inductors easy to fabricate is the meander type. However, owing to the negative turn-to-turn nature of the mutual inductance, it suffers the drawback of a very low inductance value. One of the major difficulties in achieving magnetic microactuators (MMAs) is in fabricating three-dimensional solenoid-type coils using planar fabrication process. The interest behind the realization of microactuators has led to the development of hybrid techniques such as to place magnetic components into planar coils (Wanger and Beneckle, 1991), to introduce external magnetic fields onto integrated high-permeability moving parts (Guckel, 1991) and dry etching technique to fabricate thin-film meander inductors with 7 μm coil spacing (Yamaguchi *et al.*, 1990). In a meander actuator (Ahn and Allen, 1993) fabricated using surface micromachining technique, the low-resistance meander conductors in a single plane were interwoven with multilevel meander magnetic cores as shown in Figure 4.4. To achieve an inductive component in a standard solenoid, the conducting wires have to be wrapped around a magnetic core. Such a solenoid can be fabricated using multilevel metal interconnects to wrap around a magnetic material, as shown in Figure 4.4. This inductor structure has 26 turns with a size of 0.9 mm × 4 mm and has a measured inductance of 0.2 μH at 100 kHz. Since the structure has a relatively short and planar conductor, it reduces the total resistance of the coil. This multilevel topology offers advantages in developing magnetic microactuators such as integrated magnetic recording heads on the same substrate with an integrated circuit.

The meander inductor can be modeled as shown in Figure 4.5. Figure 4.5(b) shows the model including the direction of the magnetic flux. The magnetic flux density at the center of each meander coil can be calculated by evaluating the magnetic field at the center, which is generated by the current flowing through all meander conductors as shown in Figure 4.5(a). The inductance due to both self and mutual flux linkage, assuming the material remains magnetically linear, can be written as (Ahn and Allen, 1998)

$$L = \frac{\sum \Lambda}{I} \tag{4.6}$$

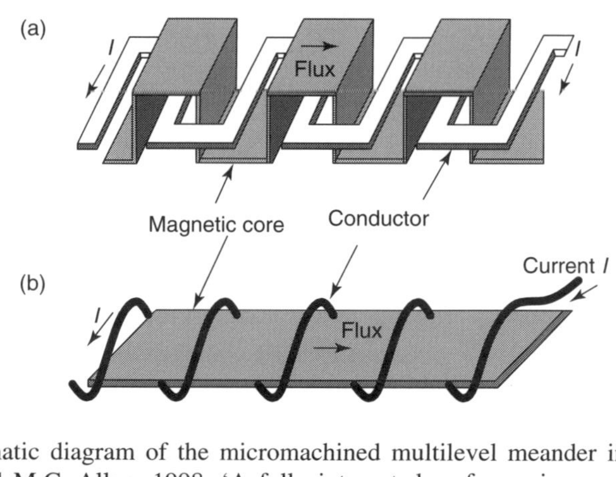

Figure 4.4 Schematic diagram of the micromachined multilevel meander inductor. Reproduced from C.H. Ahn and M.G. Allen, 1998, 'A fully integrated surface micromachined microactuator with a multilevel meander magnetic core', *Journal of Microelectromechanical Systems* **2**(1): 15–22, by permission of IEEE, © 1998 IEEE

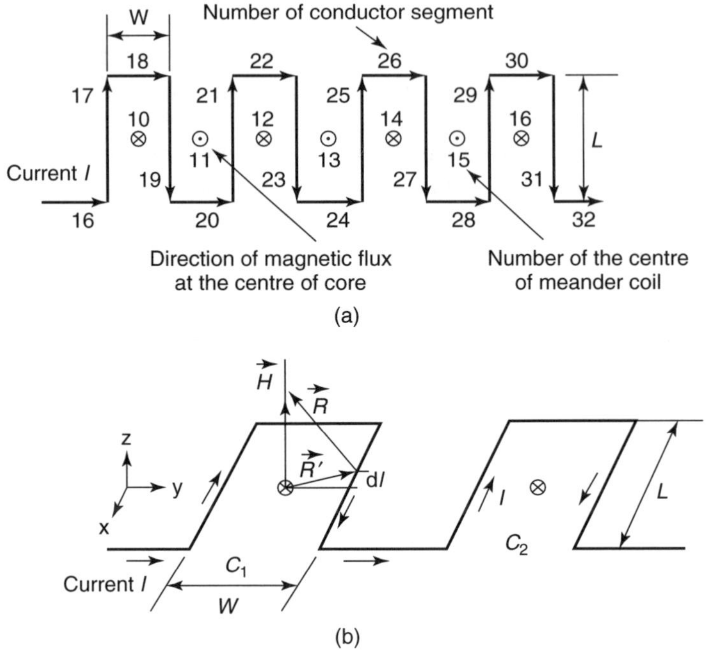

Figure 4.5 Model of the meander inductor: (a) coordinates for calculation of flux using Boit–Savart law; (b) direction of magnetic flux. Reproduced from C.H. Ahn and M.G. Allen, 1998, 'Micromachined planar inductors on silicon wafers for MEMS applications', *IEEE Transactions on Industrial Electronics* **45**(6): 866–876, by permission of IEEE, © 1998 IEEE

where $\sum \Lambda$ denotes the flux linkage. This flux linkage is in between the closed multilevel meander magnetic circuit and the flux generated from the current flowing through all meander conductor elements.

The Q of the inductor can be written as

$$Q = \frac{\omega L}{R} = \frac{w\mu_0\mu_r N A_c A_w}{2(w+l)\rho l_c} \tag{4.7}$$

where A_c is the cross-sectional area of the magnetic core, l_c is the length of the core, μ_0 and μ_r are the permeability of free space and relative permeability, respectively, A_w is the cross-sectional area of the conductor, $2(w+l)$ is the length of one meander coil turn, ρ is the resistivity of the metallic conductor, N is the number of turns of the coil and ω is the frequency of operation. It can be seen from Equation (4.7) that the introduction of a thin-film magnetic core in micromachined inductors greatly improves the inductance as well as the Q factor. Figure 4.6 shows a scanning electron micrograph of the fabricated inductor.

4.3.2.2 Spiral inductors

IC-compatible thin-film inductors have applications in the on-chip generation of magnetic fields in magnetic microactuators. Owing to the relatively low inductance values of these

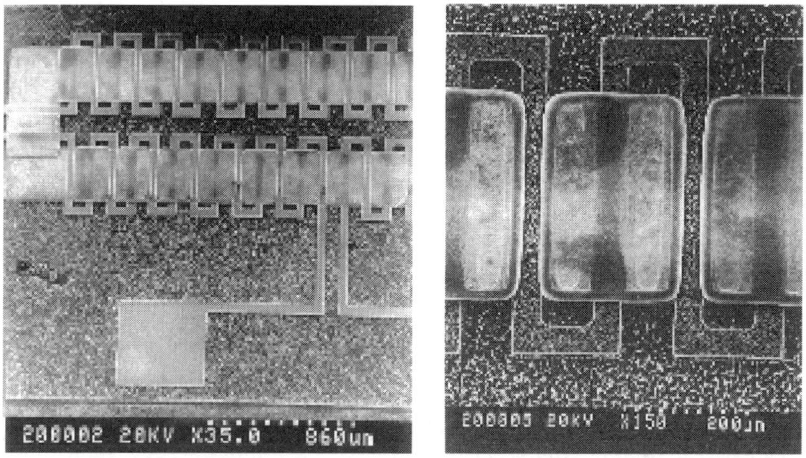

(a) (b)

Figure 4.6 Scanning electron micrograph of the fabricated toroidal-meander inductor: (a) half of the inductor and (b) detailed view. Reproduced from C.H. Ahn and M.G. Allen, 1998, 'Micromachined planar inductors on silicon wafers for MEMS applications', *IEEE Transactions on Industrial Electronics* **45**(6): 866–876, by permission of IEEE, © 1998 IEEE

inductors, most of these structures can be operated in a very high frequency regime as passive components. Also, owing to the geometrical characteristics of the spiral inductors developed on insulated substrates (Burghartz *et al.*, 1998; Olivei, 1969), on magnetic substrates (Rodrigues *et al.*, 1980), the flux generated is spread throughout the surface of the substrate. It is difficult to guide the magnetic flux towards the point of interest without using a core with magnetic permeability. The inductors for micropower applications should have closed magnetic circuits with a high-permeability material. Also the resistance of the coils should be as small as possible to reduce the power consumption. The micromachined planar inductor developed by Ahn and Allen (1993, 1994) has overcome these problems. Figure 4.7 shows a schematic diagram of the micromachined inductor, which has a closed magnetic circuit and thick conductor line. The central magnetic core is an electroplated high-permeability Ni–Fe structure, which concentrates the flux and completely encapsulates the windings. The magnetic flux generated from the spiral coil is confined by the closed magnetic circuits, which ensures better magnetic flux linkage between the spiral coils and the magnetic circuits. This results in maximum inductance along with minimum electromagnetic field interference.

A simplified model using double spiral coils for the inductance calculation of the outer core is shown in Figure 4.7(b). The total inductance is the sum of the inductance for path 1 and path 2, and the internal self-inductance of the coil can be written as

$$L = L_1 + L_2 + L_\mathrm{i} \tag{4.8}$$

where the inductance for path 1 is L_1, for path 2 is L_2 and for the self-inductance of the coil is L_i. The fabricated geometries of Figure 4.7(b) are: length $a = 1346\,\mu\mathrm{m}$, dimension of the center magnetic core $c = 508\,\mu\mathrm{m}$, thickness $d = 30\,\mu\mathrm{m}$, $N = 18.5$, $t_\mathrm{m} = 8\,\mu\mathrm{m}$ and conductor width $w = 12.5\,\mu\mathrm{m}$. The evaluated values are $L_1 = 14.5\,\mu\mathrm{H}$, $L_2 = 10.2\,\mu\mathrm{H}$,

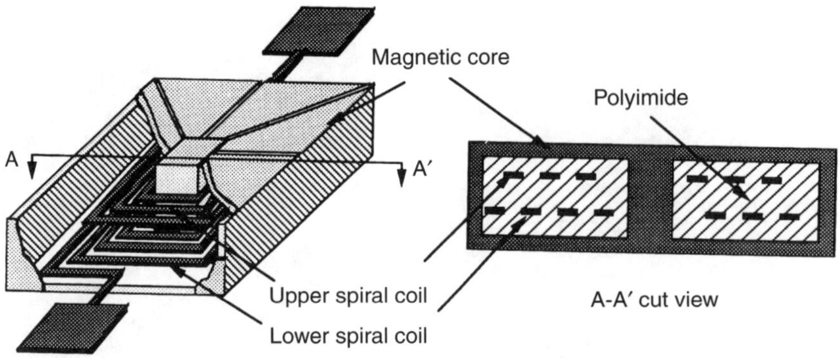

Figure 4.7 Schematic diagram of a micromachined spiral inductor. Reproduced from C.H. Ahn and M.G. Allen, 1993, 'A planar micromachined spiral inductor for integrated magnetic microactuator applications', *Journal of Micromechanics Microengineering* **3**: 37–44, by permission of the Institute of Physics

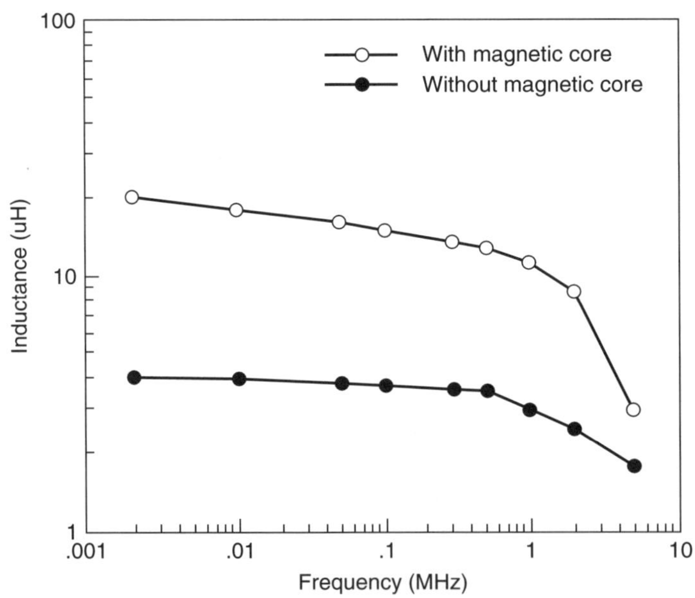

Figure 4.8 Measured inductance with and without magnetic core. Reproduced from C.H. Ahn and M.G. Allen, 1993, 'A planar micromachined spiral inductor for integrated magnetic microactuator applications', *Journal of Micromechanics Microengineering* **3**: 37–44, by permission of the Institute of Physics

$L_i = 0.01\,\mu H$ and $L = 24.71\,\mu H$. A relative permeability of 800 is assumed for the calculation.

For a device of 36 turns and a size of $3\,mm \times 3\,mm$, an inductance of $20\,\mu H$ was measured at 10 kHz. Figure 4.8 presents the effect of magnetic core on the performance of the inductor. It can be seen that the magnetic core has increased the inductance by a factor of 4 to 5 compared with the structure without the magnetic core. The inductance falls off above 3 MHz because of the decreasing permeability of the Ni–Fe permalloy at higher frequencies.

4.3.2.3 Solenoid inductors

The conventional toroidal inductor is fabricated by wrapping conducting wires around a magnetic core. However, there is tremendous difficulty in realizing this 3D structure in a planar shape. Fabrication of a coil wrapped around a core has been more difficult in conventional IC fabricating techniques than has the fabrication of meander or spiral shapes. Even though meander inductors are easy to fabricate, they suffer from a low value of inductance because of the negative mutual inductance. Spiral inductors require a lead wire, which connects from the inside end of the coil to the outside. This introduces dominant stray capacitance. Attempts were made to realize a planar solenoid inductor by manually wrapping coils around a magnetic thin film (Soohoo, 1979) as a hybrid fashion. Kawahito *et al.*, (1991) presented a design of integrated inductor by wrapping coils around an air core on a silicon substrate. However, the electrical parameters of these inductors are not suitable for micromagnetic devices and applications. A toroidal inductor in planar shape is fabricated as a bar-type micromachined inductor on silicon, as shown in Figure 4.9. It could be possible to achieve a closed (toroidal) magnetic circuit using this structure, minimizing the flux leakage. The bar-type inductor is constructed by a 25 μm thick nickel–iron permalloy magnetic core wrapped with 30 μm thick multilevel copper conductor lines. The inductor has a size of 4 mm × 1.0 mm with a thickness of 110 μm having 33 turns of multilevel coils.

The wrapping coils are constructed using metal interconnect via contacts. The via contacts may produce high contact resistance and, in order to achieve high inductance value, more turns of the solenoid coils are required which in turn increases the number of via contacts and hence increases the contact resistance. Electroplating the conductor lines and vias solves the problem because electroplated metal contacts usually have low contact resistance. The width of the conductor line and the bar core are 80 and 300 μm respectively. The measured dc resistance of the conductor line was 0.3 Ω.

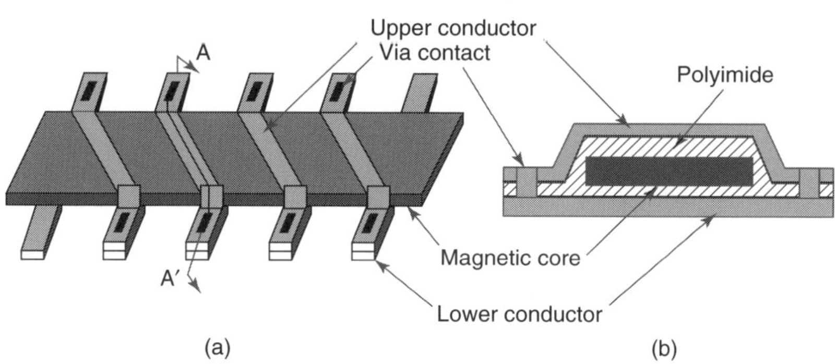

Figure 4.9 Schematic diagram of a solenoid-type inductor: (a) schematic view; (b) cross-sectional view at AA′. Reproduced from C.H. Ahn and M.G. Allen, 1998, 'Micromachined planar inductors on silicon wafers for MEMS applications', *IEEE Transactions on Industrial Electronics* **45**(6): 866–876, by permission of IEEE, © 1998 IEEE

The calculation of the inductance of a solenoid inductor is analogous to the toroidal inductor and can be written as (Ahn and Allen, 1998)

$$L = \frac{\mu_0 \mu_r N^2 A_c}{l_c} \tag{4.9}$$

where A_c is the cross-sectional area, l_c is the length of the closed magnetic core and μ_0 and μ_r are the free space permittivity and relative permittivity, respectively. The design geometry for Figure 4.9 is $A_c = 330\,\mu\text{m} \times 20\,\mu\text{m}$, $l_c = 9000\,\mu\text{m}$, $N = 33$ turns and $\mu_r = 800$. The calculated inductance is $0.729\,\mu\text{H}$.

The Q of the inductor can be written as

$$Q = \frac{\omega L}{R} = \frac{w \mu_0 \mu_r N A_c A_w}{2W \rho l_c} \tag{4.10}$$

where A_w is the cross-sectional area of the conductor, $2W$ is the length of the coil per turn and ρ is the resistivity of the metal conductor. It can be concluded that the inductance as well as Q of the solenoid-type inductor and meander-type inductor are proportional to μ_r.

4.3.3 Effect of inductor layout

The key parameters for the design of inductors involve the outer dimensions, width and spacing of the metal tracks, thickness of the metal, number of turns and the substrate material. The characteristics of the micro and nano dimensional materials used for the fabrication of MEMS inductors differ significantly from the from their bulk properties. The inductance as well as its quality factor can be fine tuned by the proper selection of the above parameters.

4.3.3.1 Effect of line spacing

Figure 4.10 presents the measured and calculated values of inductance arising from changing the spacing between the conductors for meander and spiral inductors. When line spacing decreases, it is observed that the inductance of the spiral coil increases whereas that of the meander coil decreases. This is because of spiral coil has positive mutual inductance and meander coil has negative mutual inductance.

4.3.3.2 Effect of line width

The series resistance of a planar coil is related to the sheet resistance of the metal strip, which is inversely proportional to the width of the strip. The series resistance of the coil also affects the magnetically induced losses, which depends on the time derivative of the magnetic flux through the metal strip since the magnetic flux is related to the flow of current. This loss usually increases with increase in frequency as well as the strip width. Hence there will be an optimum strip width, which minimizes the series resistance and maximizes the Q factor.

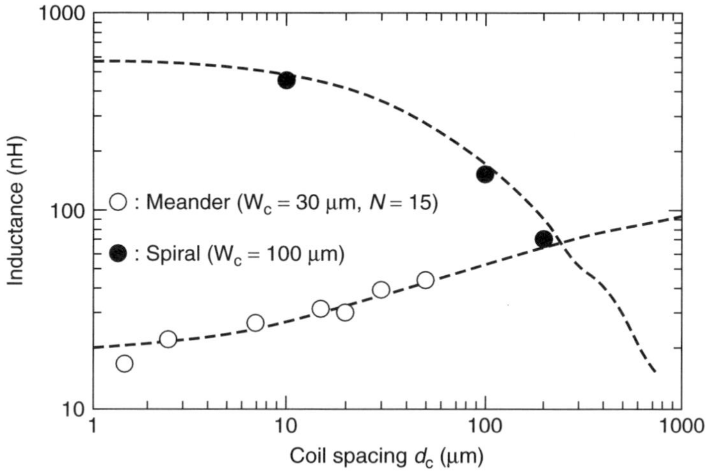

Figure 4.10 Change in inductance arising from changes in spacing between the conductors for meander and spiral inductors. Reproduced from M. Yamaguchi, M. Mastumo, H. Ohzeki and K.I. Arai, 1991, 'Analysis of the inductance and the stray capacitance of the dry-etched micro inductors', *IEEE Transactions on Magnetics* **27**(6): 5274–5276, by permission of IEEE, © 1991 IEEE

It is well known that large values of planar inductors can be fabricated using wider conducting strips on a substrate. When the inductance of the spiral is made larger, the capacitance between the turns and the ground also increases, leading to a decrease in resonant frequency. The stray capacitance between the conductor lines of a meander coil can be written as (Yamaguchi *et al.*, 1991)

$$C_{cc} = \frac{\varepsilon_0 \varepsilon_r K (1 - k^2)^{1/2}}{K(k)\,(2N - 1)} \tag{4.11}$$

where

$$k = \left(1 + \frac{2W_c}{d_c}\right)^{1} \tag{4.12}$$

$K(k)$ is the complete elliptic integral of the first kind, W_c is the width of the line, d_c is the distance between the lines and N is the number of turns. The value of stray capacitance is calculated for different widths of meander inductor and is shown in Figure 4.11.

It can be seen from the figure that the stray capacitance between the parallel conductors remains less than 100 fF when the line spacing is close to 1 μm. In the case of a spiral coil, the dominant capacitance is from the line connecting the center conductor and the outer side of the spiral coil. Also, the stray capacitance between the ground plane and the conductive elements can play a major role in inductance measurement. Efforts should be made to eliminate or minimize the stray capacitance in planar inductors. There is a trade-off between the minimum stray capacitance and the high inductance values, and these factors must be taken into consideration while designing inductors.

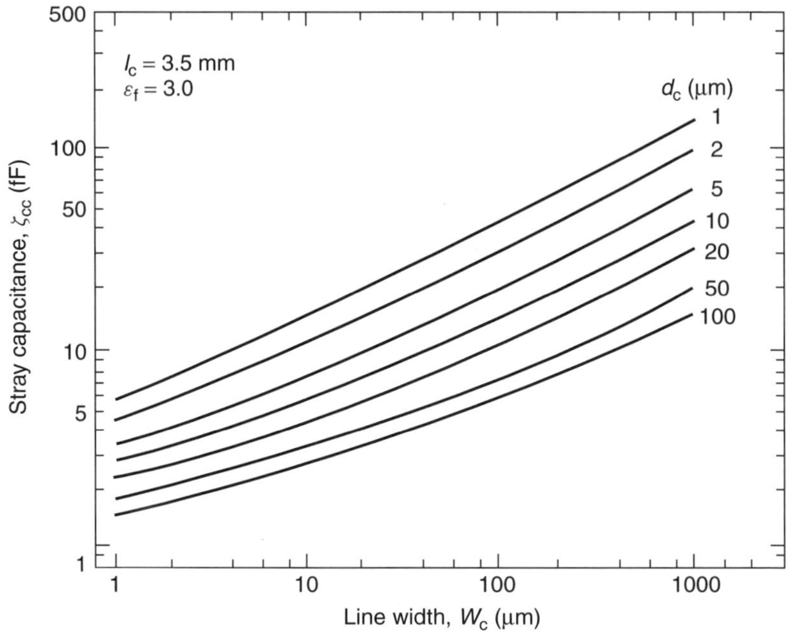

Figure 4.11 Computed stray capacitance due to change in line width W_c. Reproduced from M. Yamaguchi, M. Mastumo, H. Ohzeki and K.I. Arai, 1991, 'Analysis of the inductance and the stray capacitance of the dry-etched micro inductors', *IEEE Transactions on Magnetics* **27**(6): 5274–5276, by permission of IEEE, © 1991 IEEE

4.3.3.3 Effect of magnetic cores

The magnetic cores of the micromachined inductors have to have high saturation flux in order to obtain high saturation current; these cores should have high permeability to obtain high inductance values; also it should have high resistance to reduce the eddy current losses. The study of the effect of two different magnetic cores – permalloy ($Ni_{0.80}$, $Fe_{0.20}$) and orthonol ($Ni_{0.50}$, $Fe_{0.50}$) – on micromachined inductors on silicon at low frequencies shows that the permalloy inductor has slightly higher inductance compared with the orthonol core inductor (Park and Allen, 1996, 2000). The dc saturation current of the orthonol inductor is much higher than the permalloy inductor. This is important for high-power applications since the dc saturation current is proportional to the saturation flux density and high saturation flux density is more important than permeability.

4.3.3.4 Effect of number of turns on inductance and quality factor

There exists a trade-off between the inductance and quality factor when increasing the number of turns of a spiral inductor. Figure 4.12 shows the variation of inductance and Q due to the change in number of turns for 8 planar spiral inductors of the same dimensions except the number of turns (Koutsoyannopoulos and Papananos, 2000). It can be seen from the figure that, when the number of turns varies from 3 to 8, the inductance increases

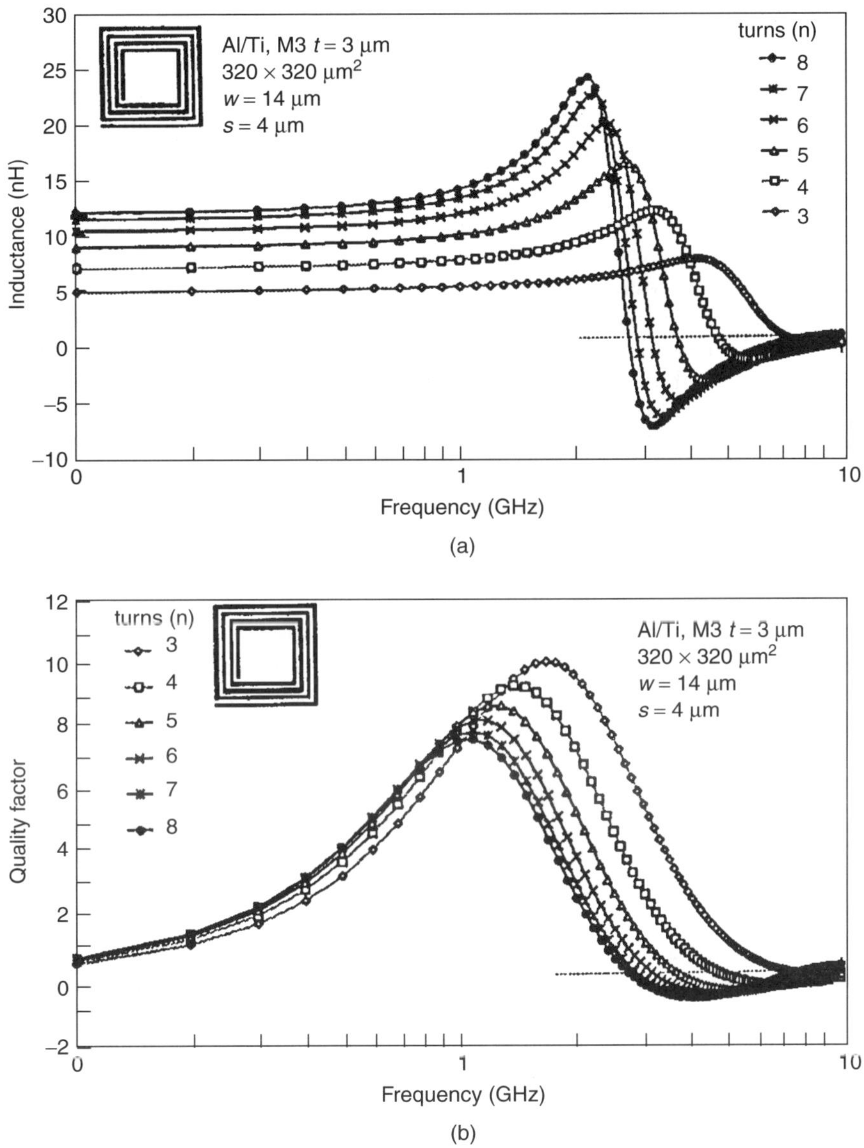

Figure 4.12 Effect of number of turns on (a) the inductance value and (b) Q factor. Reproduced from Y.K. Koutsoyannopoulos and Y. Papananos, 2000, 'Systematic analysis and modeling of integrated inductors and transformers in RF IC design', *IEEE Transactions on Circuits and Systems II* **47**(8): 699–713, by permission of IEEE, © 2000 IEEE

while Q decreases. As the area of the inductor increases due to increase in number of turns, the capacitance between the turns also increases; thus reducing the Q.

Hence it is understood that for a design engineer significant inductor performance can be obtained with the proper selection of number of turns, line width and spacing along with the proper selection of substrate.

4.3.4 Reduction of stray capacitance of planar inductors

Large values of planar inductors can be fabricated using many spirals of metal on silicon. When the inductance of the spiral is made larger, the capacitance between the turns and the ground increases, leading to a decrease in resonant frequency. Several attempts have been made to reduce these unwanted effects out of which one is integration with high resistive silicon (Koutsoyannopoulos and Papananos, 2000; Lu *et al.*, 2000). This has improved the loss factor of the inductor; however, the coupling capacitor remains unchanged. This characteristic problem can be overcome by making the substrate under the passive component an insulator by selectively etching out the silicon leaving the inductor on a suspended oxide layer (Chang, Abidi and Gaitan, 1993; Lopez-Villegas *et al.*, 1997; Ribas *et al.*, 2000). Suspended membranes (Sun, Tauritz and Baets, 1999; Sun *et al.*, 1996a, 1996b), fabricating the devices on thick dielectric substrates (Dahlmann and Yeatman, 2000; Lubecke *et al.*, 2000), self-assembling variable inductors (Case, 1997; Yoon *et al.*, 1999) and multilevel metallization (Ashby *et al.*, 1996; Burghartz, Soyuer and Jenkins, 1996; Burghartz *et al.*, 1995) are some of the approaches to reduce the parasitic effect in planar inductors. The measurement results of spiral inductors on suspended membranes shows that inductors with larger dimensions and deeper etching yield higher Qs (Lu *et al.*, 2000; Sun, Tauritz and Baets, 1999). The ratio of the outer dimension d to the average etching depth h is inversely proportional to the Q.

Fabrication of an inductor starts with depositing a $SiO_2/Si_3N_4/SiO_2$ layer on a silicon substrate using thermal oxidation and high-temperature chemical vapor deposition. This can also be done on GaAs substrates using techniques such as plasma-enhanced chemical vapor deposition (PECVD) on a substrate. This layer acts as a supporting layer once the silicon substrate is etched underneath the membrane. Planar inductor is defined on this layer using standard photolithography, evaporation of gold and formation of the air bridge. Finally, a deep opening is created by etching the substrate and leaving the dielectric membrane. The etchant for silicon wafer is KOH or EDP (ethylene–diamine–pyrocatechol) and for GaAs wafers is an $H_2SO_4/H_2O_2/H_2O$ based solution or dry etching using a reactive ion etching (RIE) machine.

Figure 4.13 shows a 3D electromagnetic simulation of the resonant frequency of a 100-nH inductor, which increases from 800 MHz to 3 GHz after removal of the underlying substrate. The inductor was designed with 20 turns of square spiral of 4-μm wide lines, and 4-μm spacing results in an outer dimension of 440 μm. The inductor exhibits a 14-fold reduction in parasitic capacitance, which increases the resonant frequency without changing the inductance value and associated series resistance.

The parasitic capacitance of a planar inductor is reduced by fabricating it in a small dielectric membrane of 1.2 to 1.4 μm thick, as shown in Figure 4.14.

The thin membrane has to be designed such that it is mechanically stable and will not affect the RF propagation properties. Planar inductors were fabricated on 1.2-mm thick dielectric membrane using micromachining technique. A 1.0-μm thick microstrip line is gold plated with air-bridge dimensions 250 μm × 40 μm. The measured and modeled reactance of the inductors on thin dielectric membrane (L_{1M}, L_{2M}) and thick silicon substrate (L_{1S}, L_{2S}) are shown in Figure 4.15. The resonant frequency of 1.2 nH and 1.7 nH planar inductors are 22 GHz and 17 GHz for thick silicon substrate, while the resonant frequencies are changed to 70 and 50 GHz, respectively, for the membrane inductors. Since the geometries are identical for both silicon and membrane inductors, the inductance L_s and the resistance R_s are not changed as a result of change in substrate.

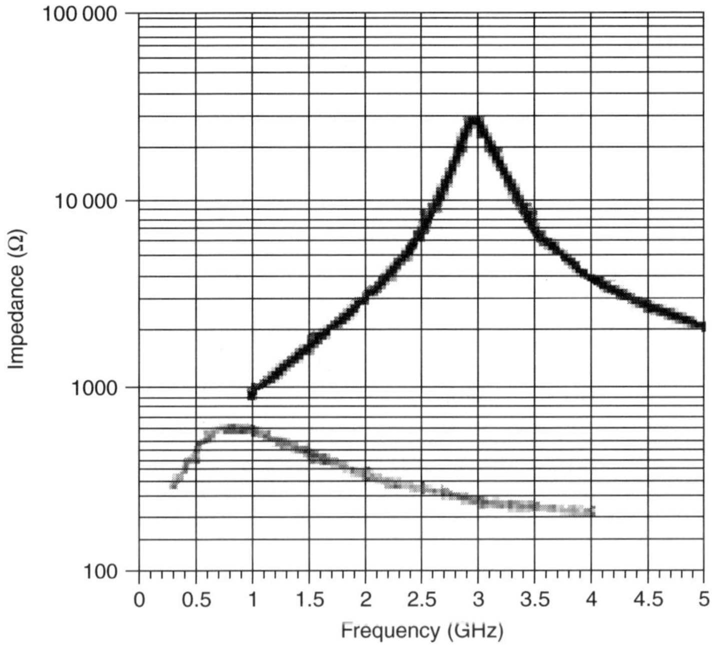

Figure 4.13 Simulated change in resonant frequency of 100 nH inductor with (gray) and without (black) underlying substrate. Reproduced from J.Y.-C. Chang, A.A. Abidi and M. Gaitan, 1993, 'Large suspended inductors on silicon and their use in a 2 mm CMOS RF amplifier', *IEEE Electron Device Letters* **14**(5): 246–248, by permission of IEEE, © 1993 IEEE

The success in bulk reproduction and integration of discrete components on silicon has generated an interest in fabrication of MEMS devices on silicon. However, the problems associated with the silicon technologies at RF frequencies have forced the development of alternative materials for RF applications. GaAs possesses many interesting properties such as it can be used as a substrate for optoelectronics and MMIC applications, high mobility of electron devices and higher piezoresistive values than silicon (Hjort, Solderkvist and Schweitz, 1994). GaAs bulk micromachining using 0.2-μm HEMT (high electron mobility transistor) MMIC technology has led to the development of free-standing structures by removing the walls of GaAs substrate (Ribas *et al.*, 1997). Micromachining approach also reduces losses and parasitic effects by suspending microwave devices such as co-planar waveguides, Lange-couplers and inductors (Chi and Rebeiz, 1997; Milanovic *et al.*, 1997). Devices such as suspended GaAs/AlGaAs mesa-shaped structures, triangular prism shaped structures and free-standing structures containing only metal and intermetallic layers are realized on GaAs substrate. It is observed that the suspended microstrip lines show higher characteristic impedance and phase velocity. The planar inductors also have a higher *Q*-factor, self-resonant frequency and input impedance when fabricated on micromachined GaAs substrate. The advantage of fabricating devices on semi-insulating GaAs is that the thick gold metallization allows high-frequency transmission lines and components. However, the major limitation is the parasitic capacitance caused by the wires and the air bridges. The reduction of capacitance of the air-bridge line is quite significant in improving the self-resonant frequency of a spiral inductor. As shown in Figure 4.16, an

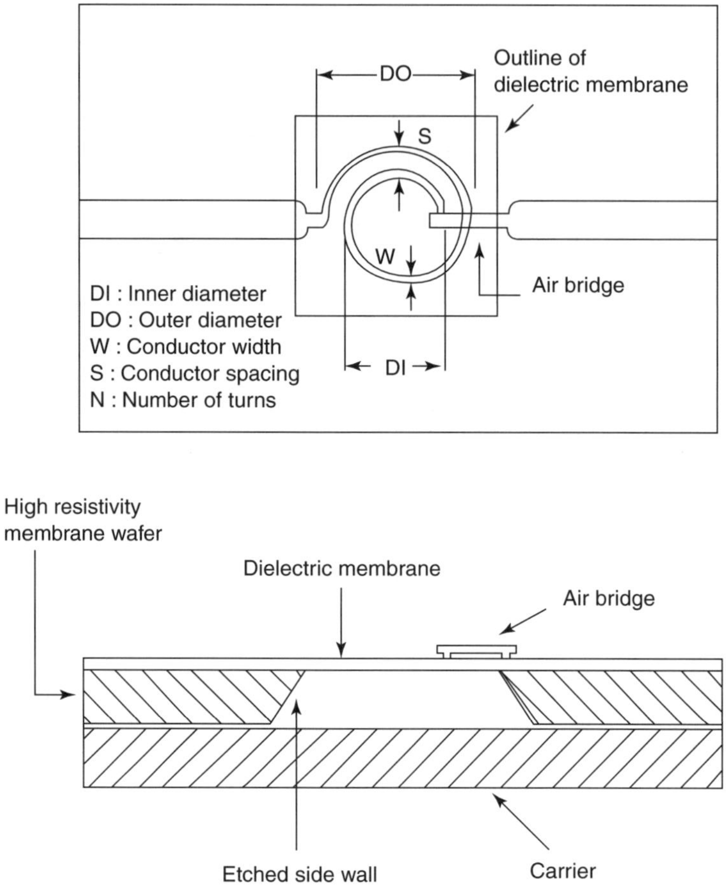

Figure 4.14 Schematic diagrams of the planar inductor and the membrane outline. Reproduced from C.-Y. Chi and G.M. Rebeiz, 1995, 'Planar microwave and millimeter wave lumped elements and coupled line filters using micromachining technique', *IEEE Transactions on Microwave Theory and Techniques* **43**(4): 730–738, by permission of IEEE, © 1995 IEEE

improvement of approximately 150% in Q factor and 130% in self-resonant frequency is observed by removing only 5 μm of the GaAs substrate below the device.

4.3.5 Approaches for improving the quality factor

The quality factor Q is one of the important figure-of-merits of inductors. Higher values of inductors are necessary for designing RF circuits with low insertion loss, low noise, high gain and good frequency selectivity. In general, the Q value is inversely proportional to the finite resistance of the metal layer. The series resistance becomes a complex function at high frequencies and the losses in inductors increases as a result of the induced currents and dielectric losses. The quality factor of spiral inductors can be improved by layout optimization by making the inner spiral line width narrower than the width of the outer spiral line (Bahl, 1999; Lopez-Villegas *et al.*, 2000). The layout optimization minimizes the

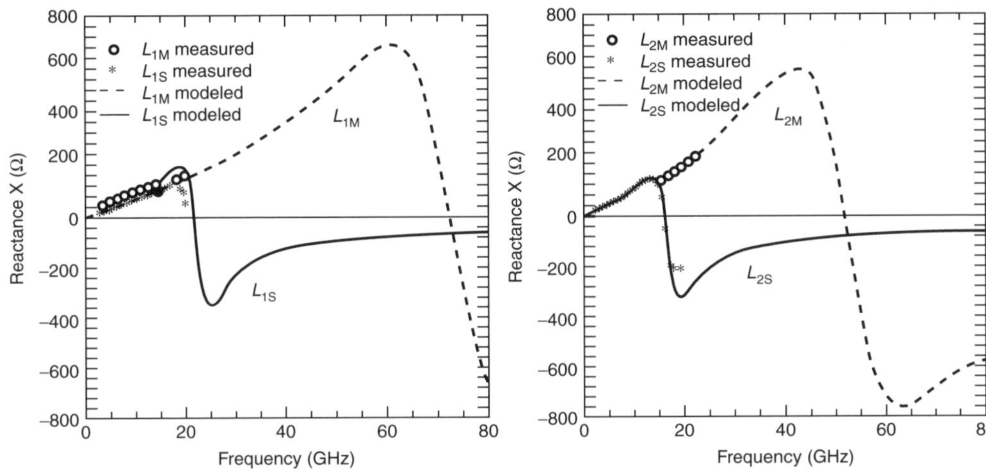

Figure 4.15 Measured and modeled reactance for inductors on silicon (L_{1S}, L_{2S}) and on membrane (L_{1M}, L_{2M}). Reproduced from R. Rodrigues, J.M. Dishman, F.D. Dickens and E.W. Whelan, 1980, 'Modeling of two-dimensional spiral inductors', *IEEE Transactions Components, Hybrids, Manufacturing Technology* **5**: 535–541, by permission of IEEE, © 1980 IEEE

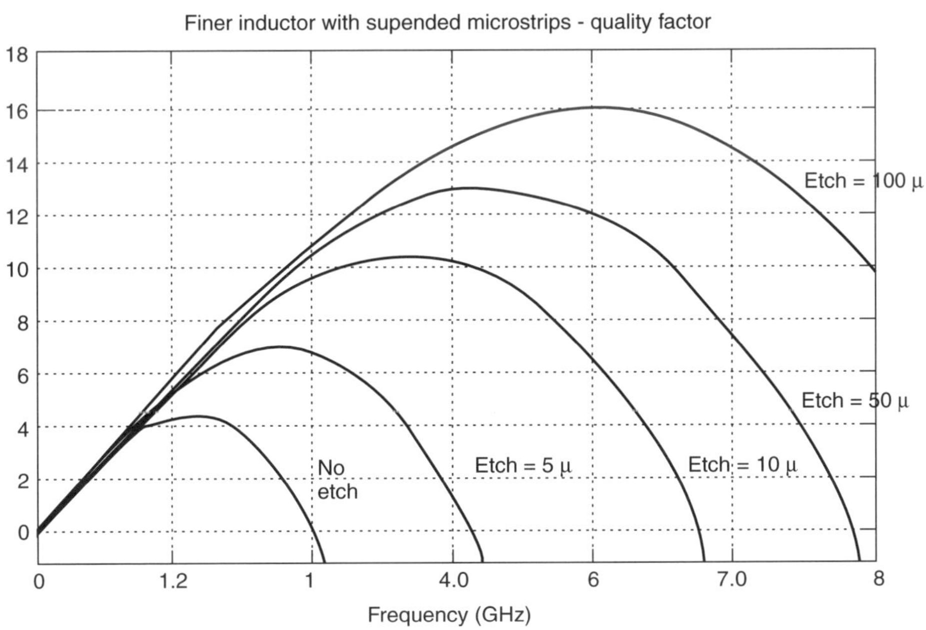

Figure 4.16 The change in Q for a suspended planar inductor for different etch depth. Reproduced from R.P. Ribas, N. Bennouri, J.M. Karam and B. Courtois, 1997, 'GaAs MEMS design using 0.2 μm HEMT MMIC technology', in *Proceedings of the 19th Annual IEEE Gallium Arsenide Integrated Circuit Symposium*, IEEE, Piscataway, NJ, USA: 127–130, by permission of IEEE, © 1997 IEEE

series resistance of a coil by taking into account its ohmic losses due to conduction current and magnetically induced losses due to eddy current. Three-dimensional helical inductors fabricated on a multilayer ceramic-based multichip module (MCM-C) technology (Sutono *et al.*, 1999), double rectangular (Shin *et al.*, 1999), self-assembling structures (Dahlmann and Yeatman, 2000; Fan *et al.*, 1998), planar spiral inductors with separately suspended strips (Ribas *et al.*, 2000), selectively removing the substrate and suspending the inductors in air (Chang, Abidi and Gaitan, 1993; Lee *et al.*, 2000) are also found to be improving the Q of the inductor. Experimental measurements show that the Q is increased by 22% by changing the layout (Bahl, 1999). These techniques are explained in the following sections.

4.3.5.1 Effect of air gap: solenoid inductors

It can be seen that conventional IC fabrication techniques are suitable for the fabrication on meander type and spiral inductors. The low value of inductance associated with meander inductors and the problems in size as well as the direction of flux in spiral inductors can be solved by using solenoid inductors. However, fabrication of a conducting coil wrapped around a core has been found to be more difficult because of the limitations in microfabrication techniques. A fully integrated solenoid inductor with air core and electroplated copper coil (Kim and Allen, 1998) using micromachining and polymer/metal multilayer processing techniques has been found to reduce stray capacitance. It is also observed that by giving an air gap between the coils and the substrate the effect of dielectric substrate is reduced. Figure 4.17 presents a schematic diagram of an air core solenoid inductor with air gap.

The inductor is fabricated on an alumina substrate using polymer/metal multiplayer processing and surface micromachining techniques. A sacrificial layer introduces the air gap between the coils and the substrate. The support structure between the substrate and the coil is defined using conventional photolithography and wet etching techniques.

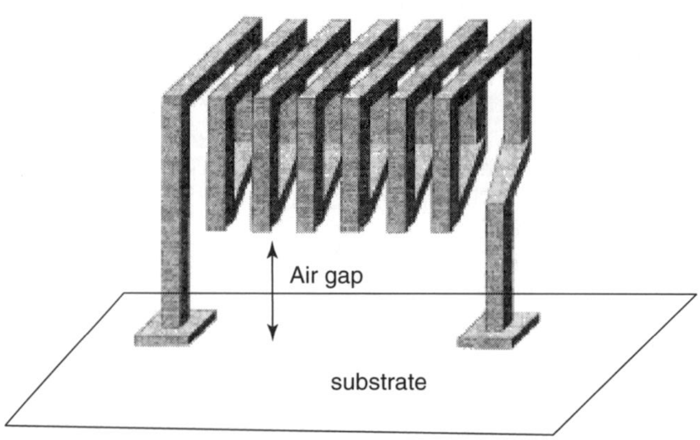

Figure 4.17 Schematic diagram of a solenoid inductor with an air gap. Reproduced from Y.J. Kim and M.G. Allen, 1998, 'Surface micromachined solenoid inductors for high frequency applications', *IEEE Transactions on Components, Packaging and Manufacturing Technology, Part C* **21**(1): 26–33, by permission of IEEE, © 1998 IEEE

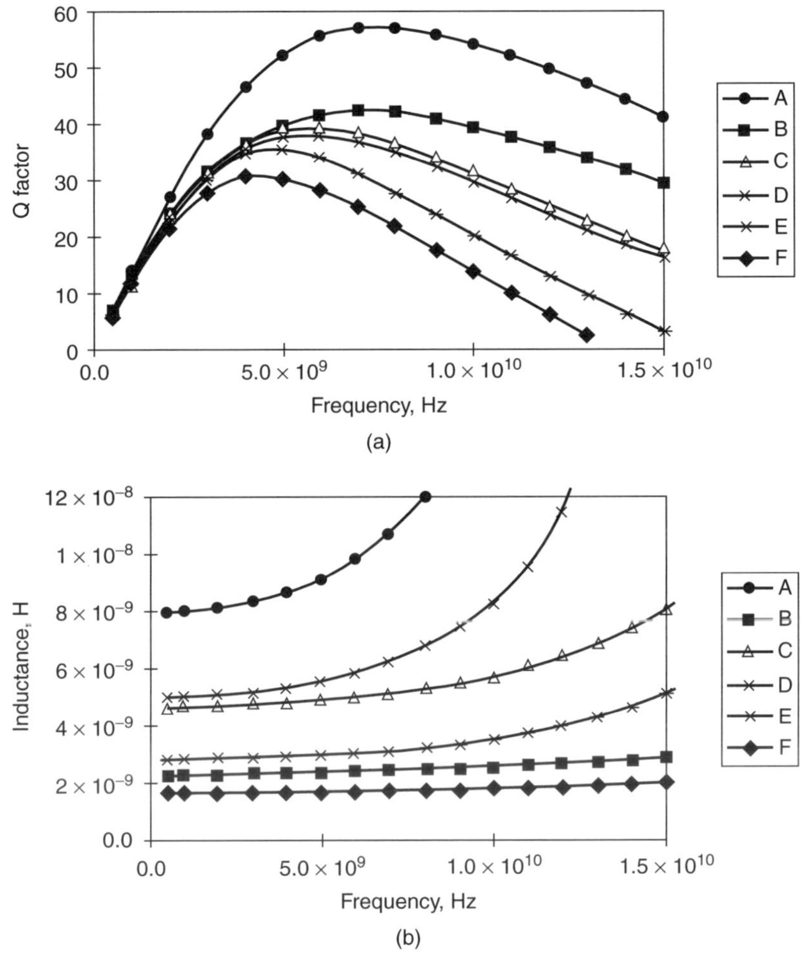

Figure 4.18 (a) Measured Q and (b) inductance for the inductors shown in Table 4.1. Reproduced from Y.J. Kim and M.G. Allen, 1998, 'Surface micromachined solenoid inductors for high frequency applications', *IEEE Transactions on Components, Packaging and Manufacturing Technology, Part C* **21**(1): 26–33, by permission of IEEE, © 1998 IEEE

Table 4.1 Details of inductors A–F cited in Figure 4.18

	A	B	C	D	E	F
Number of Turns	6	10	20	10	20	20
Core width (μm)	200	200	200	400	300	400

The measured Q factor and the inductance as functions of frequency are presented in Figure 4.18 for the list of inductors fabricated as shown in Table 4.1. The inductor's resistance values vary from 0.32 to 1 Ω, with stray capacitance from 13 to 30 pF.

The stray capacitance of a solenoid inductor can be determined by considering only the conductor-to-conductor capacitance as shown in the equivalent circuit of Figure 4.19.

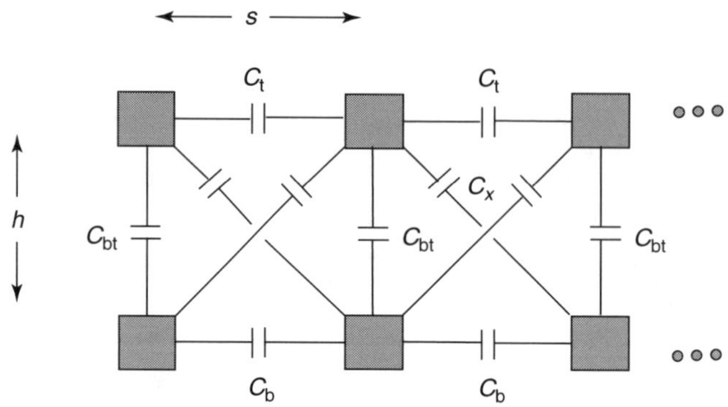

Figure 4.19 Equivalent circuit for the calculation of stray capacitance, C, between conductors of a solenoid

In this figure, C_t is the capacitance between two top conductors, C_b is that between two bottom conductors, C_{bt} is that between the top and bottom conductors and C_x is that between two diagonally placed lines. The capacitance can be approximated by neglecting the fringing fields and can be written as

$$C_t = C_b = \frac{\varepsilon A}{d} = \frac{\varepsilon(wb)}{s} \tag{4.13}$$

$$C_{bt} = \frac{\varepsilon(wa)}{h} \tag{4.14}$$

$$C_x = \frac{\varepsilon w(a^2 + b^2)^{1/2}}{(s^2 + h^2)^{1/2}} \tag{4.15}$$

where ε is the dielectric constant of air, a and b are the width and height of the conductor line, respectively, w is the length of the conductor line, h is the vertical spacing between the top and bottom of the conductor lines and s is the spacing between the conductors.

Figure 4.20 presents the effect of the vertical spacing h between the top and bottom of the coils on stray capacitance. The stray capacitance is calculated by substituting the computed values from Equations (4.13)–(4.15) in a circuit simulation program. The effect of line spacing between two adjacent conductor lines is shown in Figure 4.21. The total stray capacitance is found to be decreasing sharply when the line spacing changes from 10 to 30 μm. Even though the line spacing reduces the stray capacitance, it also increases the inductor size and decreases the inductance.

Figure 4.22 presents the effect of air gap on quality factor of the inductor. The stray capacitance has been reduced from 25.1 fF with no air gap to 17.7 fF with introduction of 20 μm air gap between the coils and the substrate. The inductors with air gap have higher Q and self-resonance frequency than one without any air gap.

4.3.5.2 Effect of air gap: spiral inductors

A surface micromachined air core spiral inductor suspended approximately 60 μm above the silicon substrate is found to reduce the effect of substrate proximity on the performance

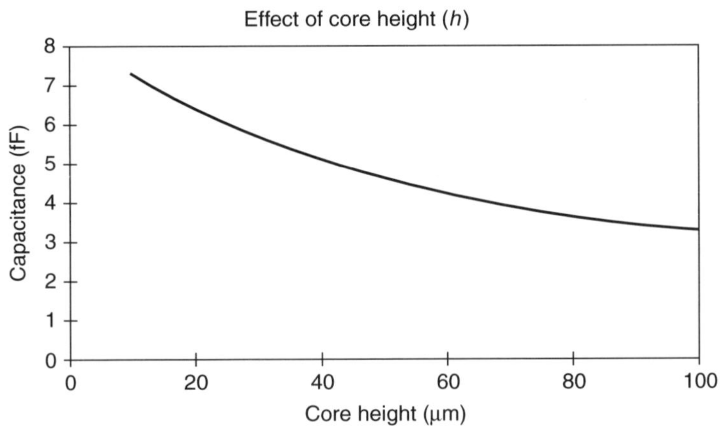

Figure 4.20 The change in stray capacitance due to the change in spacing between the top and bottom conductors, for constant line spacing. Reproduced from Y.J. Kim and M.G. Allen, 1998, 'Surface micromachined solenoid inductors for high frequency applications', *IEEE Transactions on Components, Packaging and Manufacturing Technology, Part C* **21**(1): 26–33, by permission of IEEE, © 1998 IEEE

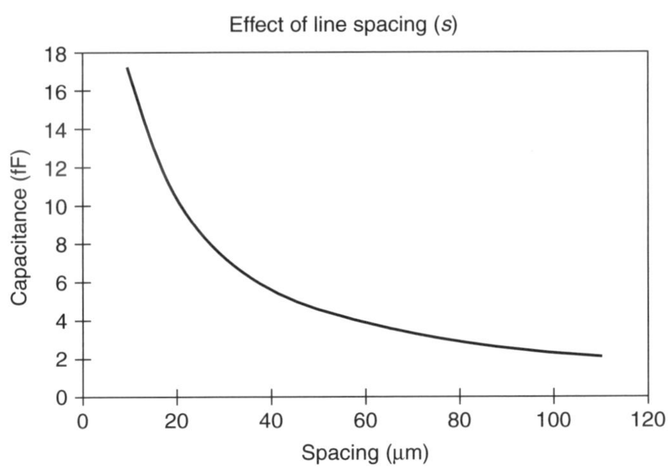

Figure 4.21 Change in stray capacitance for different conductor spacing, assuming constant h and b. Reproduced from Y.J. Kim and M.G. Allen, 1998, 'Surface micromachined solenoid inductors for high frequency applications', *IEEE Transactions on Components, Packaging and Manufacturing Technology, Part C* **21**(1): 26–33, by permission of IEEE, © 1998 IEEE

of the inductor (Park and Allen, 1998; Ronkainen *et al.*, 1997). This can lead to higher Q value and self-resonant frequency by reducing the substrate losses and parasitic effects. Figure 4.23 shows the schematic diagram of the air core spiral inductor fabricated with a large air gap between the conductor lines and substrate. As it differs from the traditional air bridge approach for air gap, in this design the plated copper metal supports help to maintain the air gap. Thick conductor lines, which increases the cross-sectional area, reduces the conductor resistance.

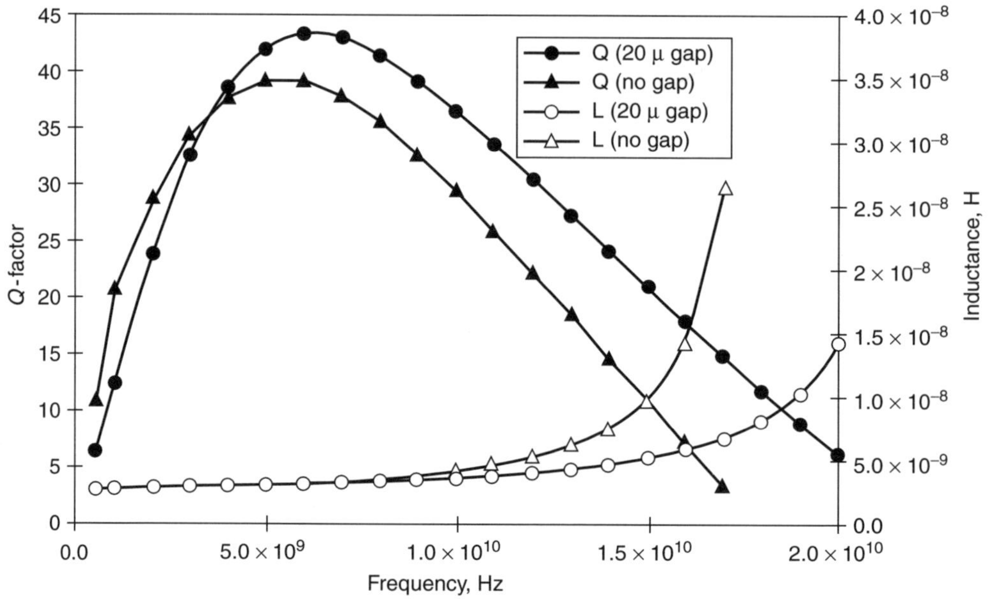

Figure 4.22 Effect of air gap on Q factor. Reproduced from Y.J. Kim and M.G. Allen, 1998, 'Surface micromachined solenoid inductors for high frequency applications', *IEEE Transactions on Components, Packaging and Manufacturing Technology, Part C* **21**(1): 26–33, by permission of IEEE, © 1998 IEEE

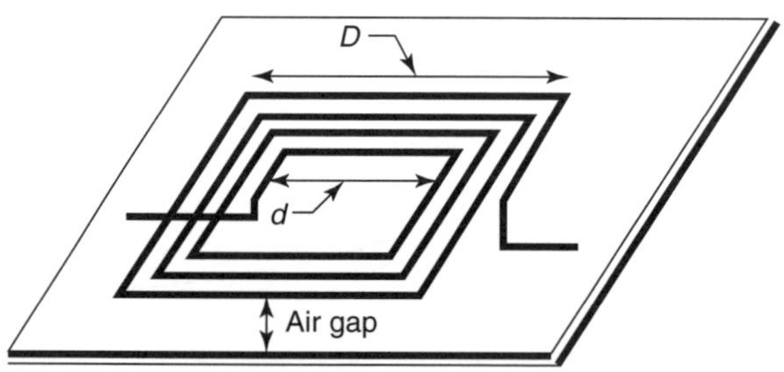

Figure 4.23 Schematic diagram of the spiral air core inductor with air gap to reduce the stray capacitance. Reproduced from J.Y. Park and M.G. Allen, 1999, 'Packaging-compatible high Q microinductors and microfilters for wireless applications', *IEEE Transactions on Advanced Packaging* **22**(2): 207–213, by permission of IEEE, © 1999 IEEE

The fabricated inductors have inductances ranging from 15 to 40 nH, and Q factors from 45 to 50 at frequencies of 0.9 to 2.5 GHz. Figure 4.24 shows the comparison of inductance as a function of frequency for different inductors whose dimensions are shown in Table 4.2.

4.3.5.3 Effect of substrate resistivity

Substrate resistivity seriously affects the resonant frequency and maximum Q value of the inductor. Owing to the relatively low resistivity of silicon substrate, the losses are the most

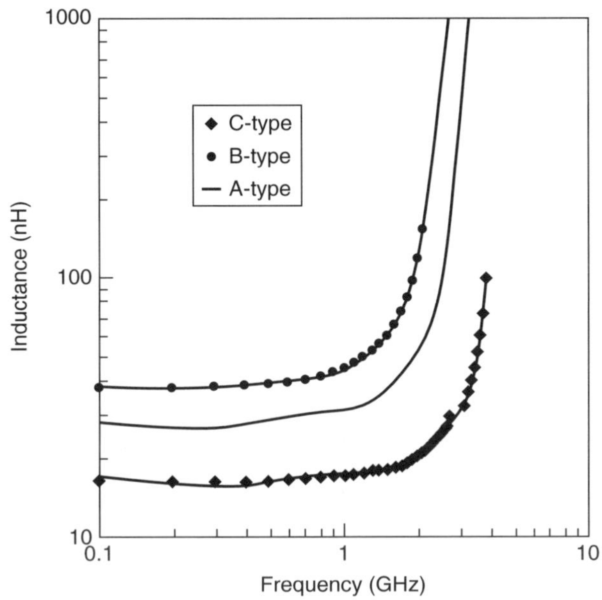

Figure 4.24 Comparison of inductance of spiral air core inductors suspended from substrate. For definitions of inductors A–C, Table 4.2. Reproduced from J.Y. Park and M.G. Allen, 1999, 'Packaging-compatible high Q microinductors and microfilters for wireless applications', *IEEE Transactions on Advanced Packaging* **22**(2): 207–213, by permission of IEEE, © 1999 IEEE

Table 4.2 Dimensions of the fabricated inductors A–C, as cited in Figure 4.24

Inductor	Dimensions of inductor (mm)	Inductor lines (μm)			Core dimensions (mm)	Number of turns
		width	thickness	spacing		
A	1.28 × 1.28	40	9	40	0.8 × 0.8	3.5
B	1.3 × 1.3	40	9	40	0.5 × 0.5	3.5
C	1.03 × 1.03	40	9	40	0.5 × 0.5	3.5

important factor degrading the performance of an inductor. This is mainly because of two reasons: one is the capacitive coupling which allows the flow of conduction current not only through the metal strips but also through the silicon substrate. The other reason is the inductive coupling, which induces current loops and associated losses by the penetration of magnetic field through the substrate. These losses affect the Q of the inductor, and the change in Q for different resistivity of silicon is shown in Figure 4.25.

4.3.5.4 Effect of width of the strip

The Q factors of different 20-nH inductors are plotted as a function of the width of metal strips, as shown in Figure 4.26. All results are simulated using HP Momentum planar solver (Lopez-Villegas *et al.*, 2000). It can be clearly seen that, although the Q factor is optimized for a given frequency and width, better results can be obtained if a different

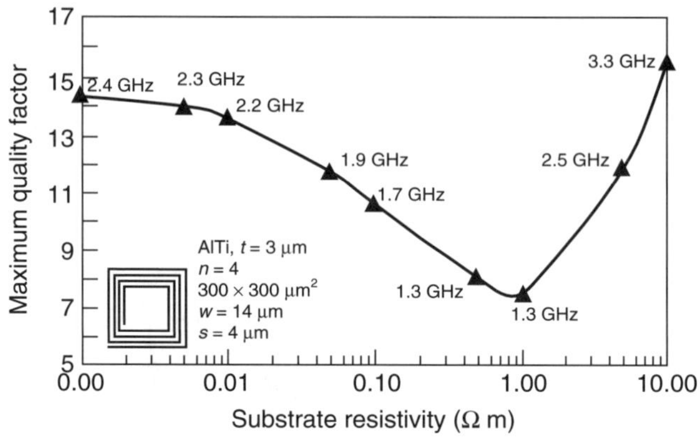

Figure 4.25 Change in quality factor due to change in resistivity of silicon substrate. Reproduced from Y.K. Koutsoyannopoulos and Y. Papananos, 2000, 'Systematic analysis and modeling of integrated inductors and transformers in RF IC design', *IEEE Transactions on Circuits and Systems II* **47**(8): 699–713, by permission of IEEE, © 2000 IEEE

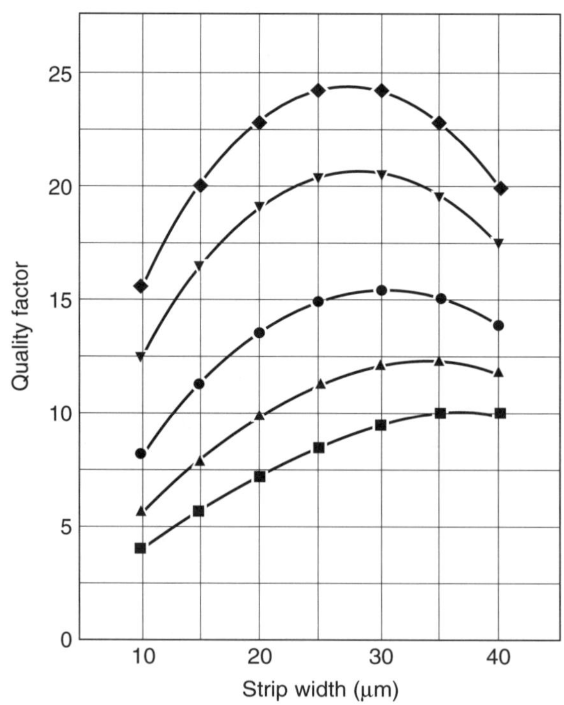

Figure 4.26 Change in Q due to change in strip width for 20-nH inductors for different frequencies: ■, 7 GHz; ▲, 1 GHz; •, 1.5 GHz; ▼, 2.5 GHz; and ◆, 3.5 GHz. Reproduced from I.J. Bahl, 1999, 'Improved quality factor spiral inductors on GaAs substrates', *IEEE Microwave and Guided Wave Letters* **9**(10): 398–400, by permission of IEEE, © 1999 IEEE

width is used in each turn of the coil. Since the maximum field is inside the inductor, the losses in inner strips can be optimized using narrow strips. Also, as the ohmic losses are predominant at the outer turns, it can be minimized by using wider strips. The magnetic field distribution will change as a result of the difference in widths and hence a layout optimization must be performed for better results using variable strip widths.

4.3.5.5 Effect of thickness of the metallization

The Q value of spiral inductors can be enhanced by increasing the conductor thickness since it reduces the series resistance. The most important parameter for improving the inductor properties at frequencies below 3 GHz is the series resistance (Park and Allen, 1999; Parisot *et al.*, 1984; Ronkainen *et al.*, 1997). Even though the thick metallization is a nonstandard IC processing, it could be achieved after the standard IC processing for the rest of the circuit. Techniques such as electroplating for fabrication of 6-μm thick metal layers are possible. It is experimentally observed that enhancing the thickness of the metallization from 4.5 μm to 9 μm and placing inductors on top of a 10-μm polyimide layer improves the Q factor by 93%, compared with standard inductors fabricated in silicon substrate (Bahl, 2001). Results of a double level of metallization using via holes to increase the thickness of metallization (Q-enhanced) are shown in Figure 4.27. The Q-enhanced inductors [Figure 4.27(a)] display lower dc resistance compared with corresponding normal inductors.

Figure 4.28 shows the change in maximum Q values and ac and dc resistances of the fabricated inductors as a function of metal thickness. The thicker metallic structures are realized using the standard silicon technology by connecting multiple metal layers with dense via arrays (Burghartz, Soyuer and Jenkins, 1996). The inductors are fabricated with one (M3), two (M2/M3 or M3/M4) and three (M2/M3/M4) metal levels. The thickness of the metal layers M1, M2, M3 are about 1 μm and that of M4 is about 2 μm. It can be

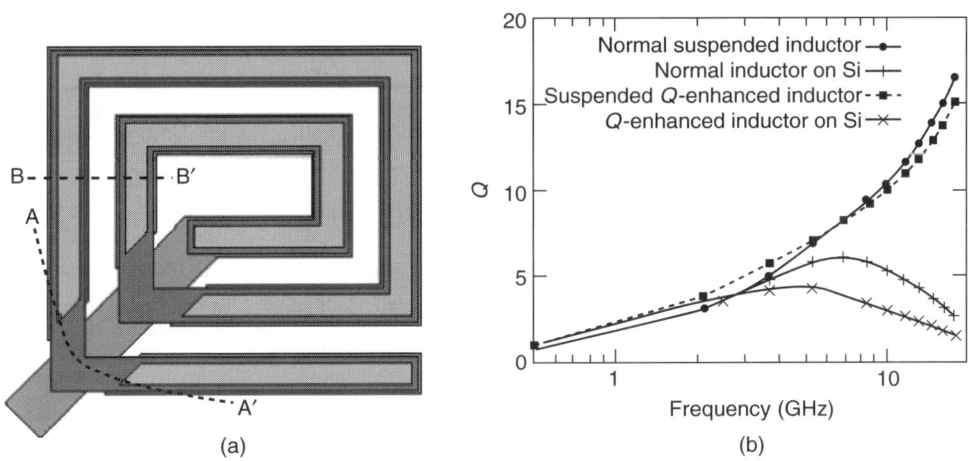

Figure 4.27 (a) Schematic diagram of a Q-enhance inductor; (b) measured results of normal and Q-enhanced inductors. Reproduced from Y. Sun, J.L. Tauritz and R.G.F. Baets, 1999, 'Micromachined RF passive components and their applications in MMICs', *International Journal of RF and Microwave CAE* **9**: 310–25, © Wiley (1999), by permission of Wiley

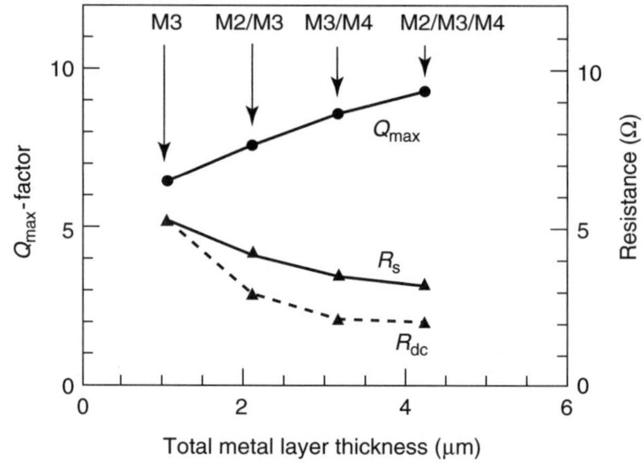

Figure 4.28 Measured change in Q factor and ac and dc resistances due to change in thickness. Reproduced from J.N. Burghartz, M. Soyuer and K.A. Jenkins, 1996, 'Microwave inductors and capacitors in standard multilevel interconnect silicon technology', *IEEE Transactions on Microwave Theory and Techniques* **44**(1): 100–104, by permission of IEEE, © 1996 IEEE

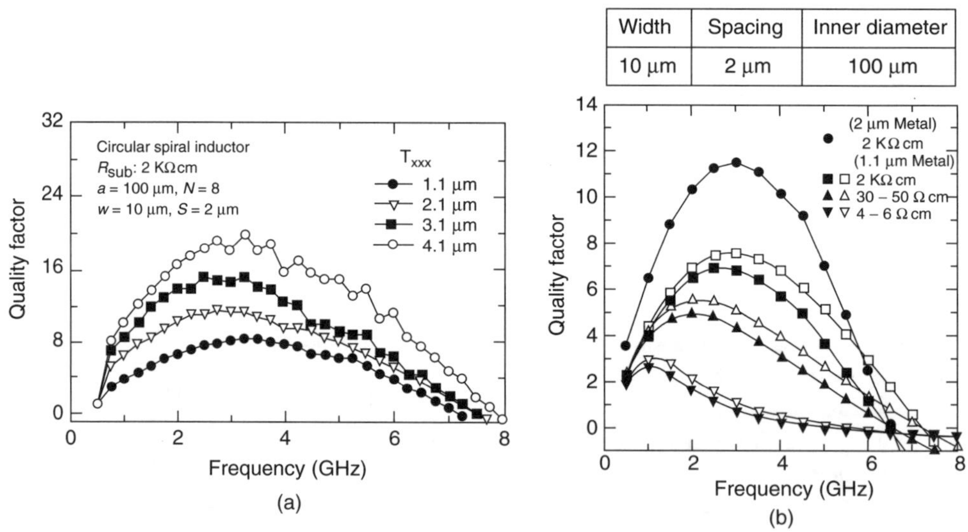

Figure 4.29 Change in Q of an inductor for (a) different metal thickness; (b) for substrates with different resistivity. Reproduced from M. Park, C.S. Kin, J.M. Park, H.K. Yu and K.S. Nam, 1997b, 'High Q microwave inductors in CMOS double metal technology and its substrate bias effects for 2 GHz RF IC application', in *Proceedings of IEDM 97*, IEEE, Washington, DC: 59–62, by permission of IEEE, © 1997 IEEE

seen from the figure that the Q value was not as high as one would expect because of the difficulties in fabricating continuous sandwiched metallic structures. Figure 4.29(a) shows the measured Q for circular spiral inductors fabricated in 2 kΩ cm silicon substrate for different metal thickness (Park *et al.*, 1997b). It is observed that with increasing metal thickness from 1.1 μm to 4.1 μm, the Q increases up to 20.1 at 3.25 GHz. The measured Q

for rectangular and spiral inductors of eight turns fabricated on different silicon substrates is shown in Figure 4.29(b).

4.3.6 Folded inductors

It is obvious from Section 4.3.5 that the parasitic capacitance is reduced when the current-carrying conductors are separated from the lossy substrate. This could be done by removal of the substrate underneath the inductor or giving an air gap between the coils and substrate. A serious drawback to this approach is the limitation in maximum separation that can be achieved between the coils and the substrate. Also there are concerns about substrate etching in terms of process compatibility and the mechanical stability of the substrate related to the relatively large openings in it. The difference in fabrication process can make it difficult to integrate the RF components monolithically on the same substrate. Therefore a unified fabrication technique for the passive components is highly preferred.

The solder surface-tension self-assembly technique (Dahlmann and Yeatman, 2000) allows separation between coil and substrate along with the possibility of rotating the coil into a plane perpendicular to the substrate. The meander and spiral inductors fabricated on low-resistivity silicon substrate shows improved Q factor from 4 to 20 when it is decoupled from the substrate. Planar copper inductors are fabricated with solder pads being placed between anchored and released portions of the device. Heating the substrate melts the solder pads and drives the hinges by the surface-tension force, which rotates the structures out of plane. Solder pads are then resolidified by cooling the substrate. Figure 4.30(a) shows a spiral inductor after self-assembly. Figure 4.30(b) shows the effect of separation of coil from the substrate through self-assembly technique for different folding angles between the coil and the substrate. There is a significant increase in Q

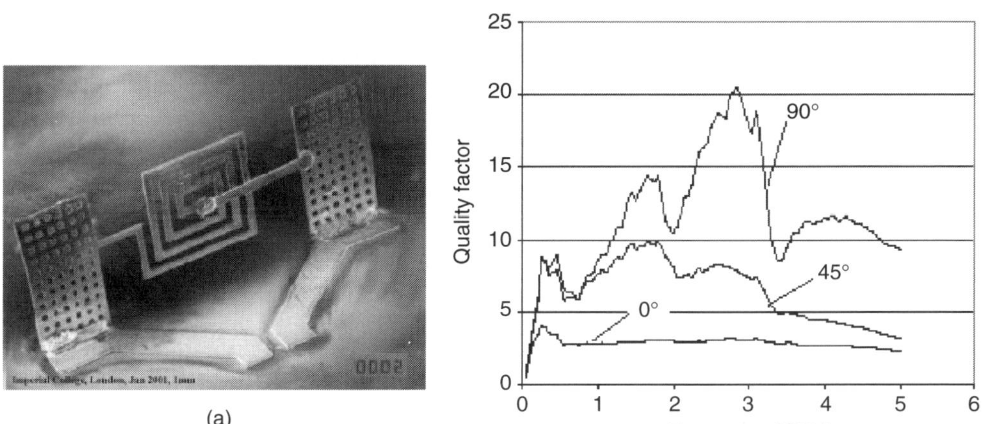

(a) (b)

Figure 4.30 (a) Three-turn spiral folded inductor after self-assembly; and (b) change in Q against frequency for different angles between coil and substrate. All devices are $4\frac{1}{2}$-turn meander inductors ($L = 2\,\text{nH}$). Reproduced from G.W. Dahlmann and E.M. Yeatman, 2000, 'High Q microwave inductors on silicon by surface tension self-assembly', *Electronics Letters* **36**(20): 1707–1708, by permission of IEEE, © IEEE 2000

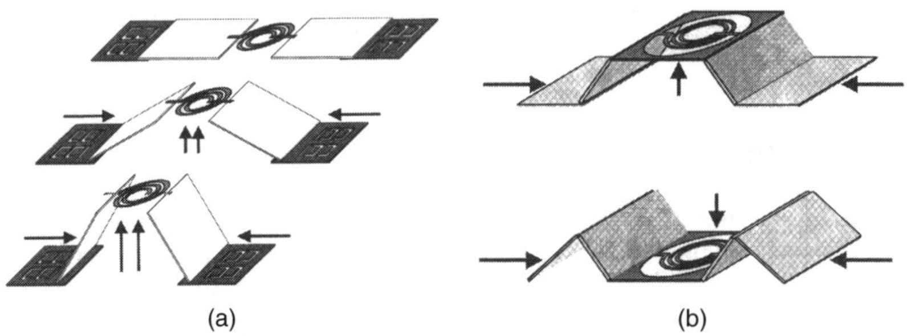

Figure 4.31 (a) Schematic diagram of the MESA micro-elevator by self-assembly structure; (b) the center platform can move upward or downward. Reproduced from L. Fan, R.T. Chen, A. Nepolsa and M.C. Wu, 1998, 'Universal MEMS platforms for passive RF components: suspended inductors and variable capacitors', in *Proceedings of 11th Annual International Workshop on MEMS '98*, IEEE, Washington, DC: 29–33, by permission of IEEE, © 1998 IEEE

and the frequency of maximum Q for the self-assembly, because it mainly reduces the capacitance between the coil and the substrate. It is observed that an inductor that is not folded but released from the substrate has a Q value of only 4 at 0.5 GHz. However, the inductor standing upright on the substrate achieves a Q of 21 at 3 GHz.

Another approach to separate the conducting coils from the lossy substrate is the micro-elevator by self-assembly (MESA) structure proposed by Fan *et al.* (1998). It is demonstrated that adjustable platforms as high as 250 μm and with a size of 5 mm x 5 mm are possible to fabricate for the applications of passive components such as inductors and variable capacitors. The height of the MESA structure is determined by the length of the side support, as shown in Figure 4.31. The 3D MESA structure is fabricated using three-polysilicon layer surface micromachining at the Microelectronics Center of North Carolina (MCNC). The spiral inductor is patterned on the center platform. The microactuators move toward each other to lift the inductor vertically and suspend it above the substrate. The microactuators pull or push the microhinges to both ends of the support arms, translating the lateral movement to a vertical motion in order to vary the vertical gap spacing.

4.3.7 Modeling and design issues of planar inductors

The key parameters in the design of inductors involve the outer dimensions, width and spacing of the metal tracks, thickness of the metal, number of turns of the spiral and the substrate material. The characteristics of the micro and nano dimensional materials used in the fabrication of MEMS inductors differ significantly from its bulk properties. Hence, 2D and 3D field modeling such as FEM of MEMS devices are important. A very accurate numerical solution can be obtained by using a 3D FEM such as MagNet (Mohan *et al.*, 1999), however, it is computationally intense and time-consuming. It can be used as a tool for design verification rather than a design tool. Modeling strategies are presented by combining 2D and 3D FEM modeling (Driesen *et al.*, 1999), thermal and mechanical characteristics (Ribas *et al.*, 2000), full geometry description with frequency dependence (Sieiro *et al.*, 2001), parallel coupled and single transmission line models (Cahana, 1983; Koutsoyannopoulos and Papananos, 2000; Long and Copeland, 1997), method of lines analysis (Schmuckle, 1993), classic circuit and network analysis techniques (Niknejad and Meyes, 1998), design equations based on Bryan's equations (Li, 1996) and, physical modeling (Yue and Wong, 2000). Mohan *et al.* (1999) presented simple expressions for the calculation of inductance of square, hexagonal, octagonal and circular spiral shapes and compared the results with measured values.

A planar meander inductor is easy to fabricate using IC fabrication techniques. However, it suffers negative mutual inductance and hence a very low inductance value. The spiral inductors can offer relatively high inductance values because of their planar nature. The spiral inductors require a lead connection between the inside end of the coil to the outside, which causes a high stray capacitance. The planar spiral coil flux, which is perpendicular in direction, can interfere with the underlying circuits in MCM modules. The solenoid-type inductors can address some of these problems, however, with complex fabrication procedures.

A planar inductor can be modeled as shown in Figure 4.32 as the lumped-element equivalent circuit. L_s is the spiral inductance, C_s is the fringing capacitance and R_s is the wire series resistance. The oxide capacitance between the metallic spiral and the silicon substrate is modeled as C_{oxide}; silicon capacitance and resistance are C_{si} and R_{si}, respectively. The value of L_s is the sum of the self-inductance L_i of each straight segment and the mutual inductance M_{ij} between the ith and jth elements. Equation (4.1) can be written as (Ribas *et al.*, 2000)

$$L_s = \sum_{i=1}^{N} L_i + \sum_{i=1}^{N}\sum_{j=1}^{N} M_{ij} \tag{4.16}$$

The value of M_{ij} depends on the length of the strip, l and the separation between the strips d and can be written as

$$M_{ij} = \frac{\mu_0 l}{2\pi}\left\langle \ln\left\{\frac{l}{D} + \left[1 + \left(\frac{l}{D}\right)^2\right]^{1/2}\right\} - \left[1 + \left(\frac{l}{D}\right)^2 + \frac{D}{l}\right]^{1/2}\right\rangle \tag{4.17}$$

where

$$\ln(D) = \ln(d) - \left[\frac{1}{12}\left(\frac{w}{d}\right)^2 + \frac{1}{60}\left(\frac{w}{d}\right)^4 + \frac{1}{168}\left(\frac{w}{d}\right)^6 + \cdots\right] \tag{4.18}$$

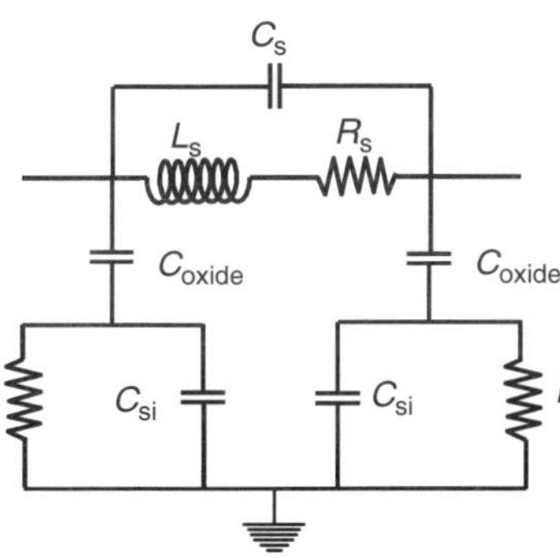

Figure 4.32 Lumped-element equivalent circuit of a planar inductor

For a planar inductor, several different definitions of Q factors have been used in the literature since there is no unique definition of Q factors available. Q is defined in general based on the ratio of energy stored to energy dissipated in any device per cycle. The inductors reactance is inductive when the frequency is much less that its resonance frequency and the Q is defined for such cases as

$$Q = \frac{\omega L}{R} \tag{4.19}$$

where R is the series resistance. When the inductor is used as a resonant component, the Q is defined as the ratio of the resonance frequency, $f_{\text{resonance}}$, to the 3-dB bandwidth (3-dB BW):

$$Q = \frac{f_{\text{resonance}}}{\text{3-dB BW}} \tag{4.20}$$

However, in RF and microwave circuits, the inductors are commonly used well below their self-resonance frequency. In such cases the Q factor is the effective $Q_{\text{eff.}}$, which can be defined as the ratio of the imaginary part to the real part of the one-port input impedance with the other side grounded. This can be written as

$$Q_{\text{eff.}} = \frac{\omega L_s}{R_s}\left[1 - \frac{R_s^2(C_s + C_0)}{L_s} - \varpi^2 L_s(C_s + C_0)\right] \tag{4.21}$$

At its self-resonance this Q becomes zero because $\text{Im}(Z_{\text{in}})$ is zero and $\text{Re}(Z_{\text{in}})$ is maximum. The inductive reactance and parasitic capacitive reactance become equal. Beyond this resonant frequency, the inductor becomes capacitive.

The series resistance R_s can be derived from a free-standing microstrip line equation with width w, thickness t, conductivity σ and permeability μ:

$$R_s = \frac{l}{wt\sigma}\left[\frac{0.431x_w}{1 + 0.041(w/t)^{1.19}} + \frac{1.1147 + 1.2868x_w}{1.2296 + 1.287x_w^3} + 0.0035\left(\frac{w}{t} - 1\right)^{1.8}\right] \tag{4.22}$$

for $x_w > 2.5$ and

$$R_s = \frac{l}{wt\sigma}\left[1 + 0.0122x_w^{3+0.01x_w^2}\right], \qquad \text{for } x_w < 2.5 \tag{4.23}$$

The series capacitance C_s accounts for the fringing capacitance between the spiral wires and the wires connecting outside. C_0 represents the substrate capacitance between the strips and the ground plane.

The figure-of-merit of an inductor (FMI) is defined as (Bahl, 2001):

$$\text{FMI} = \frac{Q_{\text{eff.}} f_{\text{resonance}}}{\text{inductor area}} \tag{4.24}$$

and the design aim is to achieve the highest FMI value.

4.3.8 Variable inductors

A programmable inductor network can be implemented by controlling various turns of a multi-turn inductor. This can be done by digitally controlled micro relays, which can be integrated with the inductors by combining surface and bulk micromachining techniques. Using four electrostatic micro relays, 16 different inductors whose values ranged from 2.5 nH to 324.8 nH were obtained (Zhou, Sun and Carr, 1997, 1999). Using a phase-shifting network with coupled RF and drive coils, an electrically tunable inductor demonstrated a measured tunability of 100% and Q of 2000 (Pehlke, Burstein and Chang, 1997). Lubecke *et al.* (2000) developed inductors with variable inductance values using 3D self-assembling structures. These variable inductors use interlayer stress to bend out of plane of the substrate in which it is fabricated. The inductors are separated out from the substrate to minimize the parasitic losses and use thermally controlled intermember positioning to alter the inductance. The inductors with Q greater than 13 have been demonstrated to have a continuous inductance variation greater than 18%, and, for Q values greater than 20, the inductance variations are greater than 30%.

4.3.9 Polymer-based inductors

The deformation characteristics of a polymer to create a sacrificial core have been utilized to fabricate on-chip 3D air core micro-inductors (Chomnqwang and Lee, 2001). The inductor is fabricated on silicon substrate with an electroplated base. A 15-μm SU8 resist with electroplated copper has been used as a polymeric mold for the bottom conductor. UV lithography is used to pattern 40-μm thick temporary core, which on curing deformed to a bell shape. Gold film is sputtered on top of this sacrificial polymer and is developed using electrodeposition of photoresist. The inductor has been formed with bottom conductors on the substrate while the top conductors have an arch-like structure with an air bridge.

4.4 MEMS CAPACITORS

There are many broadband applications with specific design requirements in which the capacitor controls critical electrical parameters. They include low-noise amplifiers, harmonic frequency generators and frequency controllers. Many of the modern wireless systems place stringent requirements on high-quality, stable, low-phase noise with wide-tuning-range voltage-controlled oscillators (VCOs). The tuning range of these VCOs must be large enough to cover the entire frequency band of interest. The electronically tunable capacitors are the key elements in such VCOs. The difficulties in implementing high-Q on-chip variable capacitors make them often designed as an external component in many circuits. Components such as band select, channel select and tuning elements of VCOs are still external to the chip because inductors and tunable capacitors (varactors) with a high quality factor are not available in standard silicon processes. The Q factor of the conventional varactors made in silicon or gallium arsenide p–n or Schottky-barrier junction varactors used for the current tuning application is not adequate for low-phase noise devices. However, the performance of recent micromachined capacitors is promising and silicon micromachined devices and electronics on a single chip may have many applications in the near future. MEMS allow precise positioning and repositioning of suspended membranes and cantilevers with a very small voltage, which can be integrated into RF systems for tuning and switching applications. Compared with solid-state varactors, the MEMS tunable capacitors have the advantage of low loss and greater tuning range.

Before designing a capacitive element for broadband application, the designer must carefully consider the magnitude of its impedance, insertion loss, the capacitor's parasitic element, equivalent series resistance (ESR), device linearity in response to RF and the quality factor (Q) for the entire frequency of interest. The magnitude of both real and imaginary parts of the capacitor's impedance can be seen on a Smith chart using an impedance analyzer. At the series resonant frequency, the net reactance of the capacitor is zero. The impedance of the device will then be equal to a small ESR value, typically 100 mΩ at 1 GHz for a high Q ceramic capacitor, which can be seen on the S_{11} Smith chart.

As a general design rule, large value capacitors are selected for a broadband design requirement since the performance satisfies the low-frequency region. At this low frequency, the impedance will be low enough to provide a good through path to the RF signal. The operating conditions of a capacitor can be determined from the insertion loss (S_{21}) measurements. The excessive magnitude of S_{21} of a given capacitor makes them unusable for a given application. A capacitor quality factor can be defined as

$$Q = \frac{|X_C - X_L|}{\text{ESR}} \tag{4.25}$$

where $|X_C - X_L|$ is the net reactance. The capacitor ESR should be taken into account over the entire frequency band of interest. It can be seen from Equation (4.25) that the larger the resistance the smaller the Q and the greater resistive loss for the device. Also, the inductance associated with the tunable capacitor will resonate at a frequency known as the electrical self-resonance for the capacitor. The capacitor becomes unusable beyond the self-resonance frequency because the inductance dominates the total device impedance. Therefore, the inductance associated with a capacitor needs to be kept as low as possible so that the self-resonance should be much higher than the signal frequencies for which the tunable capacitor is designed.

Recent efforts in microelectronics and MEMS show promising results in realization of wide-band tunable capacitors for RF applications. MEMS capacitors tune their capacitance by adjusting the device's physical parameters and dimensions via electromechanical means: electrostatic or thermal. The capacitance of the capacitor with two electrodes of area A, separated by a gap d, can be written as

$$C = \frac{\varepsilon A}{d} \tag{4.26}$$

neglecting the fringing fields. Here, ε is the dielectric constant of the medium. It is clear from Equation (4.26) that the three physical parameters, ε, A and d, can be varied to realize a tunable capacitor. Thus the MEMS tunable capacitors can be classified based on their tuning schemes such as gap tuning, area tuning and dielectric tuning.

4.4.1 MEMS gap-tuning capacitors

4.4.1.1 Electrostatic tuning

The tunable capacitor can be made by one of the electrodes suspended on the top of a fixed electrode. The suspended electrode, supported by micromachined springs, is movable in the vertical direction normal to the substrate. The gap between the movable and the fixed electrodes can be adjusted electrostatically by applying a tuning voltage, resulting in a change in its capacitance. The electrostatic actuation is preferred over the other actuation mechanisms because of its low power consumption. Young and Boser (1996; Young *et al.*, 2001) presented the design of a micromachined gap-tuning capacitor. Figure 4.33 shows the top view of the capacitor made on silicon substrate. It consists of 1-μm thick aluminum plate suspended in air with four mechanical folded beam suspensions acting as springs. The electrodes are 200 μm by 200 μm with 2 μm by 2 μm holes spaced 10 μm apart to ensure a complete removal of sacrificial material. The initial gap is 1.5 μm with a measured Q of 62. The capacitance varied from 2.11 to 2.46 pF when applied voltage changed from 0 to 5.5 V, which corresponds to a tuning range of 16%.

When an electric field is applied to a parallel plate system, the movable plate starts moving towards the fixed plate as a result of electrostatic force. This force is distributed along the length of the movable plate and, when the threshold bias voltage is reached, the plate snaps down to the bottom plate and the applied voltage no longer controls the beam. The equilibrium between the electrostatic attracting force and the force at the supports holds only for a deflection smaller than one-third of the initial gap between them. This limits the tunability of MEMS capacitors to two-thirds of the initial gap, which restricts the theoretical limit to 50% to any electrostatically actuated parallel plate system. The

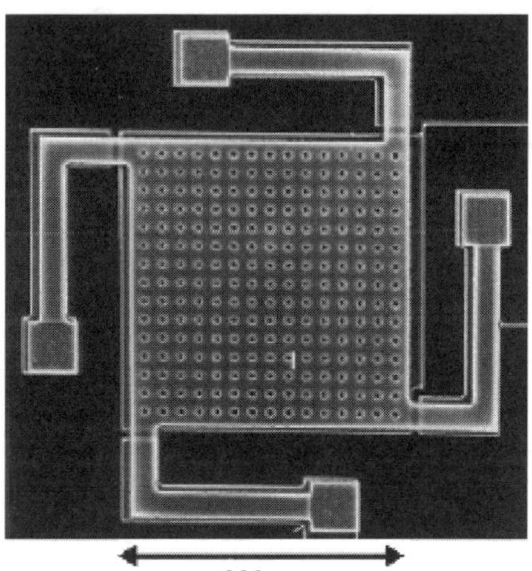

Figure 4.33 Top view of a micromachined variable capacitor. Reproduced from D.J. Young and B.E. Bover, 1996, 'A micromachined variable capacitor for monolithic low-noise VCOs', in *Proceedings of the International Conference on Solid-state sensors and Actuators*, IEEE, Washington, DC: 86–89, by permission of IEEE, © 1996 IEEE

maximum tuning range R_{max} can be written as

$$\left\{\varepsilon A\left[\left(1-\frac{1}{3}\right)^{x_0}\right]^{-1}-\frac{\varepsilon A}{x_0}\right\}\left(\frac{\varepsilon A}{x_0}\right)^{-1} \tag{4.27}$$

The theoretical limitation in tunability of a two-plate system can be overcome by the redesign of the top movable plate, demonstrated by Zou and co-workers (Zou, Liu and Schutt-Aine, 2001; Zou *et al.*, 2000) or a three-plate system (Dec and Suyama, 1997, 1998a, 1998b, 2000). Dec and Suyama (1997) demonstrated that the tuning range can be changed from 50% to 100% by using a micromachined three-plate tunable capacitor as shown in Figure 3.34(b). The device was designed in a multi-user MEMS process (MUMPS) available at MCNC. The principle of a two-plate varactor consists of a fixed plate and a suspended plate as shown in Figure 4.34(a). The top plate is suspended by a spring with spring constant k while the bottom plate is fixed. The dc voltage applied to the top plate causes an electrostatic force, which moves the suspended plate closer to the bottom plate and thus increases its capacitance. The capacitor shows a normal capacitance of 0.57 pF with a 0.75-μm spacing between Poly1 and Poly2 and a maximum capacitance of 0.85 pF when 3.3 V is applied. The suspended plate with Poly2 and gold has a mass of 0.8 mg. A total of 141 holes were made in an area of 230 μm × 230 μm to help the etching of oxide layer between the top and bottom plate.

The three-plate micromachined variable capacitor consists of three layers of polysilicon; one suspended plate and two fixed plates, as shown in Figure 4.35. The top and bottom plates are fixed while the middle plate is suspended by a spring arrangement. Separate tuning voltages, V_1 and V_2, can be applied to C_D and C_P. The application of V_1 to C_D increases its capacitance and that of V_2 will decrease C_D. The maximum tuning range for capacitor C_D with respect to voltage V_1 and for C_P with respect to voltage V_2 still remains 50%.

However, the voltage V_2 reduces the tuning range of C_D by approximately 50%, which results in a theoretical tuning range of 100%. Even though polysilicon is less conductive compared with aluminum, the superior mechanical properties make it preferable for

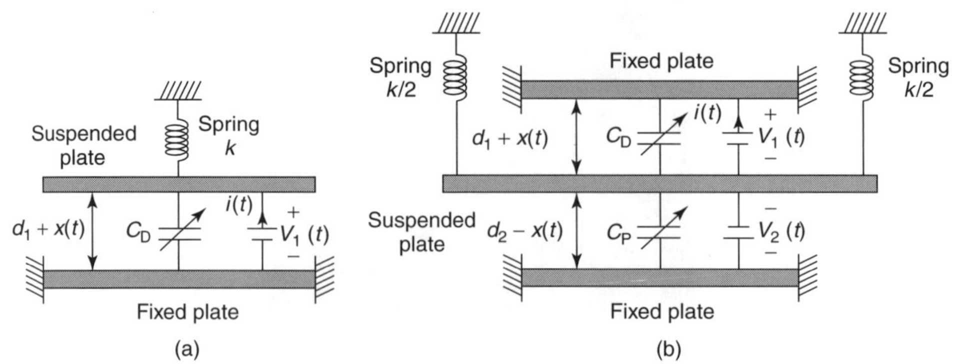

Figure 4.34 Principle of an electromechanically tunable parallel plate capacitor: (a) with two plates; (b) with three plates. Reproduced from A. Dec and K. Suyama, 1998b, 'Micromachined electromechanically tunable capacitors and their applications to RF IC's', *IEEE Transactions on Microwave Theory and Techniques* **46**(12): 2587–2596, by permission of IEEE, © 1998 IEEE

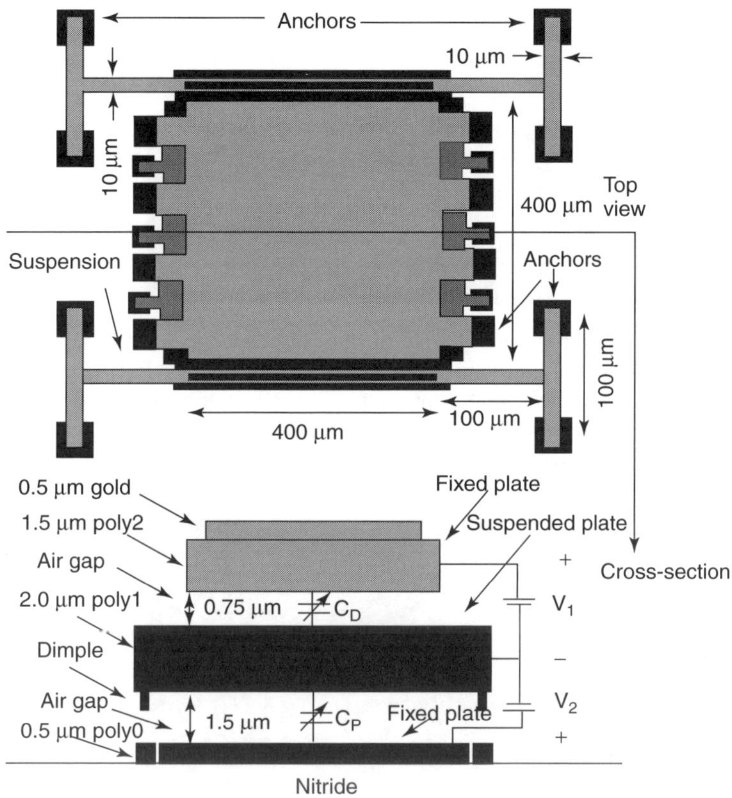

Figure 4.35 Top and cross-sectional views of three-plate varactor. Reproduced from A. Dec and K. Suyama, 1998b, 'Micromachined electromechanically tunable capacitors and their applications to RF IC's', *IEEE Transactions on Microwave Theory and Techniques* **46**(12): 2587–2596, by permission of IEEE, © 1998 IEEE

capacitor applications. The 1.9-pF tunable capacitor is designed with an air gap of 0.75 μm with an area of 398 μm × 398 μm. The capacitance can be changed to 2.84 pF with a bias voltage of $V_1 = 3.3\ V$ and $V_2 = 0$. The quality factor of the device is measured as 15.4 at 1 GHz and a tuning range of 1.87 : 1.

Variable capacitors with gap tuning can also be realized using a surface-micromachined microelevator by self-assembly (MESA) structure (Fan *et al.*, 1998). The schematic diagram of the MESA structure is shown in Figure 4.36. The 250 × 250 μm^2 polysilicon plate is raised above the substrate by four 300-μm long side supports, which are controlled by microactuators. The capacitance is changed from 500 fF to 20 fF when the suspended electrode is raised by 250-μm. The main drawback of this system is that the fine tuning is difficult because the change in capacitance is not linear with the displacement of the top plate.

It has already been established that selective etching and removal of silicon substrate supporting the passive component is an efficient method to improve the performance of a passive component. The metal–insulator–metal (MIM) capacitor fabricated on a suspended membrane shows that the best Q of a 2.6 pF capacitor exceeds 100 at 2 GHz, while the same capacitor fabricated directly on silicon has a Q less than 10 (Sun, Tauritz

Figure 4.36 Scanning electron micrograph of the suspended capacitor supported by MESA micro-elevator by self-assembly structure. Reproduced from L. Fan, R.T. Chen, A. Nepolsa and M.C. Wu, 1998, 'Universal MEMS platforms for passive RF components: suspended inductors and variable capacitors', in *Proceedings of 11th Annual International Workshop on MEMS '98*, IEEE, Washington, DC: 29–33, by permission of IEEE, © 1998 IEEE

Figure 4.37 Photograph of the suspended MIM (metal–insulator–metal) capacitor. Reproduced from Y. Sun, J.L. Tauritz and R.G.F. Baets, 1999, 'Micromachined RF passive components and their applications in MMICs', *International Journal of RF and Microwave CAE* **9**: 310–325, © Wiley (1999), by permission of Wiley

and Baets, 1999). The photograph of the MIM capacitor is shown in Figure 4.37; this device can be actuated electrostatically.

The capacitors are fabricated on standard p-type 3.5 Ω cm silicon ⟨1 0 0⟩ with first (M1) and second (M2) metal layers 0.6 μm and 1.4 μm, respectively. Si_3N_4 is deposited by the low-pressure chemical-vapor deposition method. High-value small-area MIM capacitors are realized by depositing an Al_2O_3 layer in between the two metals. Passive components are patterned after two metallizations and the underlying silicon is selectively removed later using KOH wet etching.

A variable MEMS capacitor using electrostatic actuation with digital control (Hoivik *et al.*, 2001) shows a tuning ratio of 4 : 1 and a measured Q of 140 at 750 MHz. Electrostatic actuation is preferred over thermal actuation for a variable capacitor because of higher

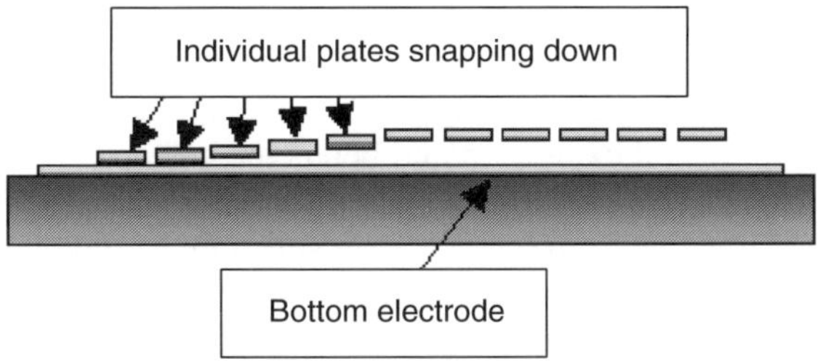

Figure 4.38 Schematic diagram of the capacitor plate arrangement. Reproduced from N, Hoivik, M.A. Michalicek, Y.C. Lee, K.C. Gupta and V.M. Bright, 2001, 'Digitally controllable variable high-Q MEMS capacitor for RF applications', in *Proceedings of IEEE MTT-S Symposium, May 2001, Volume 3*, IEEE, Washington, DC: 2115–2118, by permission of IEEE, © 2001 IEEE

actuation speed and greater deflection stability. A linear capacitance variation due to change in voltage is achieved because the individual capacitor plates are connected to the bond pads by beam fixtures of different widths as shown in Figure 4.38. The top plate is an array of 30 plates of equal area. When the actuation voltage is applied, the top capacitor plate moves towards the bottom plate in a cascading manner, depending on the stiffness of the individual beam.

Devices such as micromachined frequency variable impedance tuners (Jung *et al.*, 2001) and double stub tuners (Lang *et al.*, 2001) use micromachined capacitors for a wide tuning range.

4.4.1.2 Electro-thermal tuning

It is known that the tunability of electrostatic MEMS capacitors is limited to two-thirds of the initial gap, which restricts the theoretical limit to 50% to any electrostatically actuated parallel plate system. This tuning limit due to the pull-in effect can be overcome by using thermal actuators (Feng *et al.*, 2000a, 2000b; Harsh *et al.*, 1999; Wu *et al.*, 1998). The thermal actuator to control the gap is driven by the principle of differential thermal expansions of the thick/thin polysilicon arms, as shown in Figure 4.39. The variable capacitor driven by a vertical electrothermal actuator (Feng *et al.*, 2001; Harsh *et al.*, 2000; Young and Boser, 1996) shows a Q factor of 300 at 0.1 pF at 10 GHz. The electrothermal actuator drives the top plate of a parallel plate capacitor.

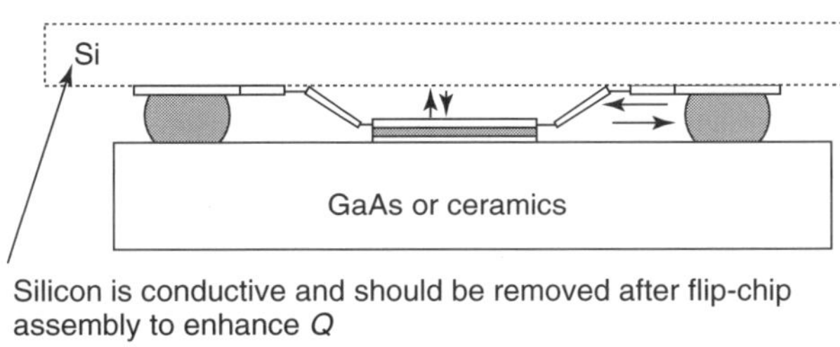

Figure 4.39 Flip-chip assembly of silicon-based MEMS. Reproduced from K.F. Harsh, B. Su, W. Zhang, V.M. Bright and Y.C. Lee, 2000, 'The realization and design considerations of flip-chip integrated MEMS tunable capacitor', *Sensors and Actuators A: Physical* **80**: 108–118, with permission from Elsevier Science, © 2000 Elsevier Science

The vertical displacement of the electrothermal actuator is achieved by the thermal mismatch resulting from different temperature distributions. The electrothermal actuator is driven by the differential thermal heating of the wide and narrow polysilicon arms. The narrow arm exhibits significant expansion and bends the actuator, thus controlling the gap and the capacitance. The device performs well at millimeter wave frequencies owing to the flip-chip assembly and transfer from lossy silicon substrate to low-loss alumina substrate.

The silicon substrate, which is standardized for many MEMS processes, is not acceptable for RF applications without any post-process modifications. One such post-process is the packaging demonstrated by Harsh *et al.* (1999) to flip-chip bond on unreleased MEMS into a ceramic substrate. This technique is compatible with optimal integration of MEMS into existing circuitry or fabrication of new surface micromachined MEMS devices. Figure 4.39 shows the flip-chip assembly of MEMS device to a GaAs or ceramic substrate. The lossy silicon substrate is removed during the sacrificial release.

A schematic diagram of the capacitor with electrothermal actuators is shown in Figure 4.40. Figure 4.41 shows a photograph of the MEMS tunable capacitor on a ceramic substrate, which uses electrothermal actuators to control the gap between the plates. Experimental results show that a capacitance variation of 2.7 : 1, with a bias voltage varying from 0 to 2.5 V, is possible using the above configuration. Figures 4.42 and 4.43 show the measured properties of the MEMS capacitor.

The electrothermal actuation of tuning of a capacitor is an attractive approach because it removes the theoretical tuning limit of 50% for any electrostatically actuated system. However, electrothermal actuation is slow and it consumes power, typically 10 mA at 3 V. Gap-tuning capacitors in general suffer another drawback of low RF power handling capability due to a small electrode gap. When the capacitor is tuned for higher capacitance, the electrodes move closer and closer; RF breakdown in the air gap is thus more likely to occur.

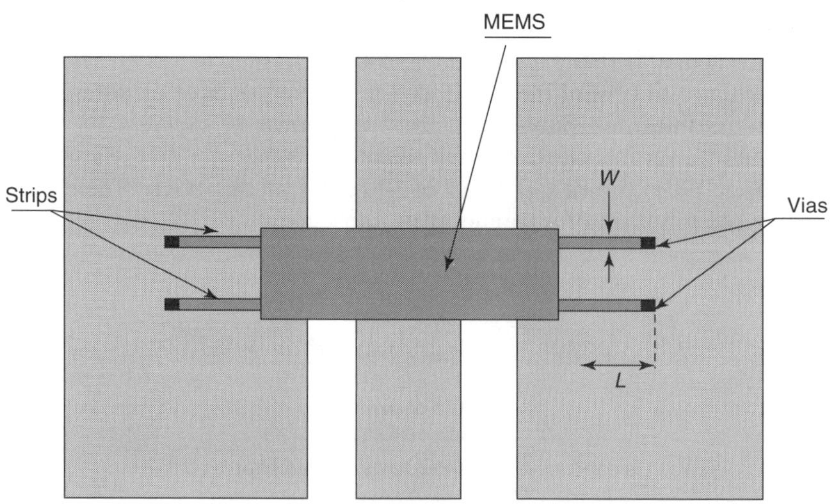

Figure 4.40 Schematic diagram of the shunt capacitor with electrothermal actuators. Reproduced from Z. Feng, H. Zhang, K.C. Gupta, W. Zhang, V.M. Bright and Y.C. Lee, 2001, 'MEMS-based series and shunt variable capacitors for microwave and millimeter-wave frequencies', *Sensors and Actuators A* **91**: 256–265, with permission from Elsevier Science, © 2001 Elsevier Science

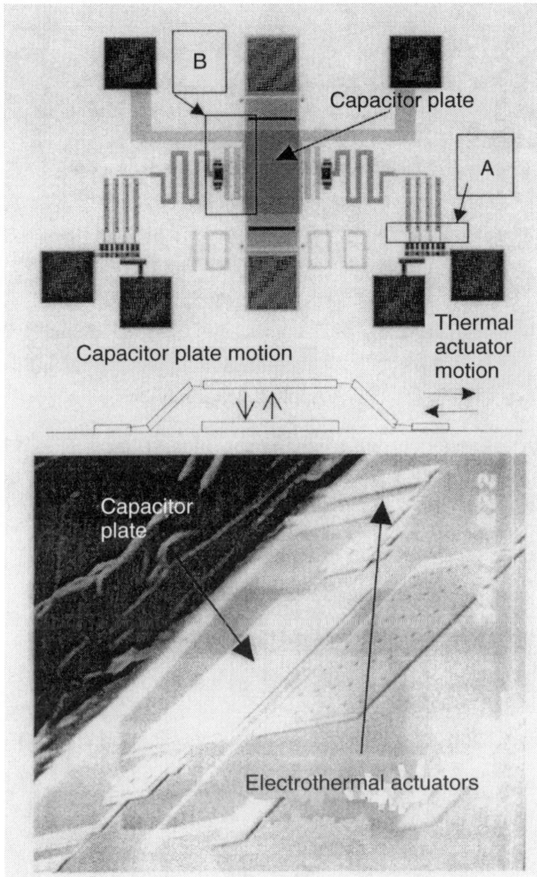

Figure 4.41 MEMS tunable capacitor with electrothermal actuators. Reproduced from K.F. Harsh, B. Su, W. Zhang, V.M. Bright and Y.C. Lee, 2000, 'The realization and design considerations of flip-chip integrated MEMS tunable capacitor', *Sensors and Actuators A: Physical* **80**: 108–118, with permission from Elsevier Science, © 2000 Elsevier Science

4.4.1.3 Piezoelectric-actuator tuning

The gap between the electrodes of a tunable capacitor can be changed using a piezoelectric actuator. Park *et al.* (2001) presented a MEMS tunable capacitor in a CPW transmission line circuit using integrated PZT (lead zirconate titanate) actuator. It has a C_{max}/C_{min} ratio of 3.1 to 1 for a bias voltage of 6 V. The MEMS capacitor integrating with piezoelectric actuator has advantages such as low driving voltages and linear tuning of capacitance. The PZT actuators are fabricated on silicon substrate and are diced and bonded to transmission lines on a quartz substrate using flip-chip bonding technology, as shown in Figure 4.44. The bias voltages applied to the control pad move the PZT actuator vertically onto the dielectric layer on top of the fixed electrodes. The variation of the gap between top and bottom electrodes results in a change in device capacitance. The CPW transmission line and control electrodes with Pt/Cu/Ar metals are fabricated on quartz substrate using the lift-off process. Dielectric layer is then deposited followed by selectively wet etching to

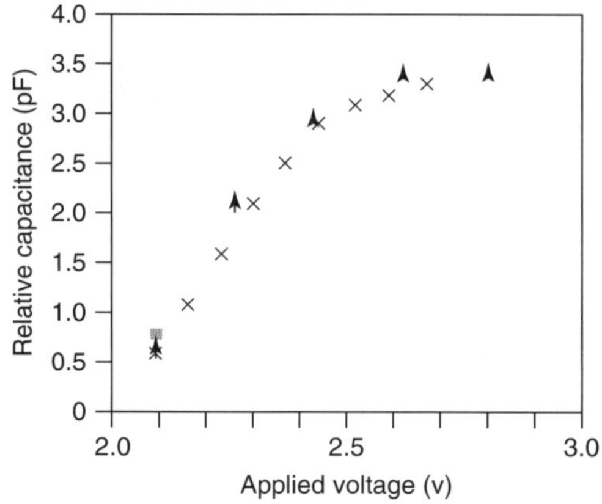

Figure 4.42 Change in capacitance due to change in bias voltage. Reproduced from K.F. Harsh, B. Su, W. Zhang, V.M. Bright and Y.C. Lee, 2000, 'The realization and design considerations of flip-chip integrated MEMS tunable capacitor', *Sensors and Actuators A: Physical* **80**: 108–118, with permission from Elsevier Science, © 2000 Elsevier Science

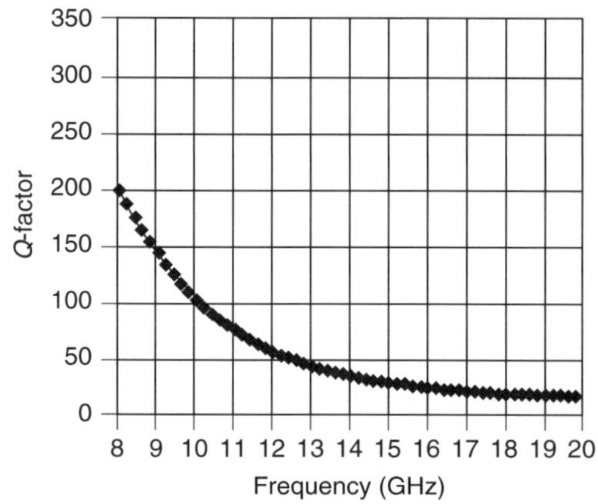

Figure 4.43 Plot of Q against frequency for a 0.1-pF capacitor measured using HP 8519C Network Analyzer. Reproduced from K.F. Harsh, B. Su, W. Zhang, V.M. Bright and Y.C. Lee, 2000, 'The realization and design considerations of flip-chip integrated MEMS tunable capacitor', *Sensors and Actuators A: Physical* **80**: 108–118, with permission from Elsevier Science, © 2000 Elsevier Science

form a parallel plate capacitor. Gold is electroplated on the top of the transmission line, ground plane and the control electrodes. Figure 4.44 shows the fabricated PZT actuator, which is used as the movable top electrode for the tunable capacitor. The capacitor shows a quality factor of 210 at 1 GHz.

4.4.2 MEMS area-tuning capacitors

Area tuning is preferred over gap tuning for MEMS capacitors because there is no theoretical tuning limit for area-tuning capacitors. One of the simple methods of area

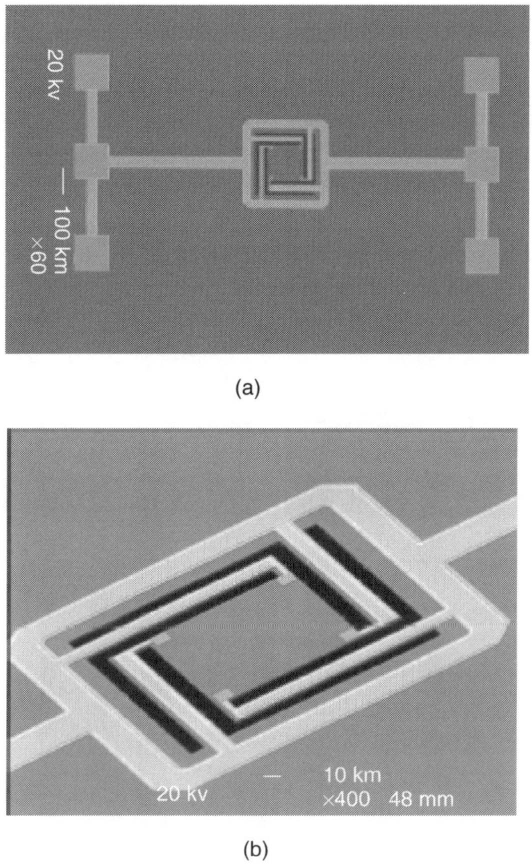

Figure 4.44 Scan electron micrograph picture of (a) fabricated PZT (lead zirconate titanate) actuator on a silicon substrate using silicon bulk micromachining; (b) close-up view. Reproduced from J.Y. Park, Y.J. Yee, H.J. Nam and J.U. Bu, 2001, 'Micromachined RF MEMS tunable capacitors using piezoelectric actuators', in *Proceedings of IEEE MTT-S Symposium*, May 2001, Volume 3, IEEE, Washington, DC: 2111–2114, by permission of IEEE, © 2001 IEEE

tuning is using a comb structure. In a comb structure, only the supporting spring design and the length of the comb practically limits its tuning range. Larson *et al.* (1991) designed a micromachined electrostatically controlled variable positioning device with metal conductors. An electrostatic micromotor is used to slide the fingers so that it can change the overlapping area between the capacitor electrodes. This device is ideal for MEMS capacitive switches, variable capacitors and tuning stubs at RF frequencies from 2 to 45 GHz. The micromachined variable capacitor shows a change in capacitance from 0.035 pF to 0.1 pF with bias voltages ranging from 80 to 200 V.

In an interdigitated comb structure, one set of comb is stationary and the other one is movable. When a tuning voltage is applied in between these structures, the electrostatic force between the fingers laterally actuates and slides the finger so that it can change the overlapping area, while the gap between them remains unchanged. The micromachined interdigitated comb structure shown in Figure 4.45 clearly demonstrated that a silicon

Figure 4.45 Scanning electron micrograph images of interdigitated comb tunable capacitor at three different magnifications. Reproduced from J.J. Yao, S. Park and J. DeNatale, 1998, 'High tuning ratio MEMS based tunable capacitors for RF communications applications' in *Proceedings of solid-state sensors and Actuators Workshop*, IEEE, Washington, DC: 124–127, by permission of IEEE, © 1998 IEEE

MEMS capacitor has a continuous tuning range of at least 200% or 3 : 1 tuning ratio (Yao, Park and DeNatale, 1998).

Figure 4.46 presents a series of images showing a portion of tunable capacitor electrode for different actuation voltages. It is clear from the figure that the change in overlapping area due to the actuation voltage changes its capacitance, as plotted in Figure 4.47. The

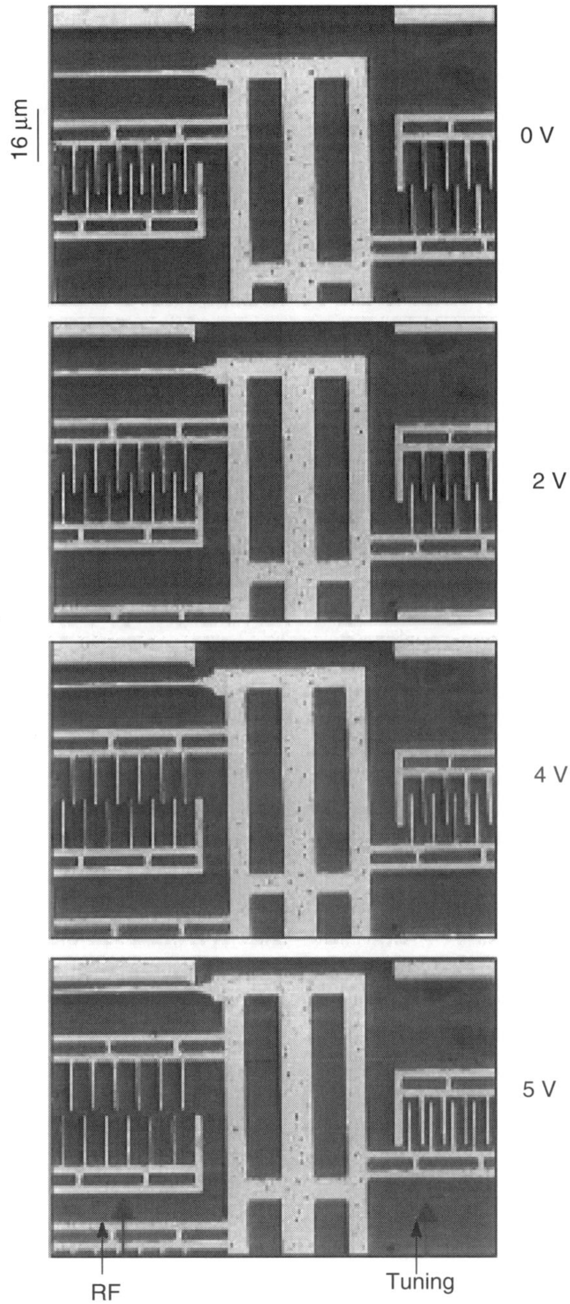

Figure 4.46 Series of images showing a MEMS tunable capacitor with a tuning voltage of 0 to 5-V. Reproduced from J.J. Yao, S. Park and J. DeNatale, 1998, 'High tuning ratio MEMS based tunable capacitors for RF communications applications' in *Proceedings of solid-state sensors and Actuators Workshop*, IEEE, Washington, DC: 124–127, by permission of IEEE, © 1998 IEEE

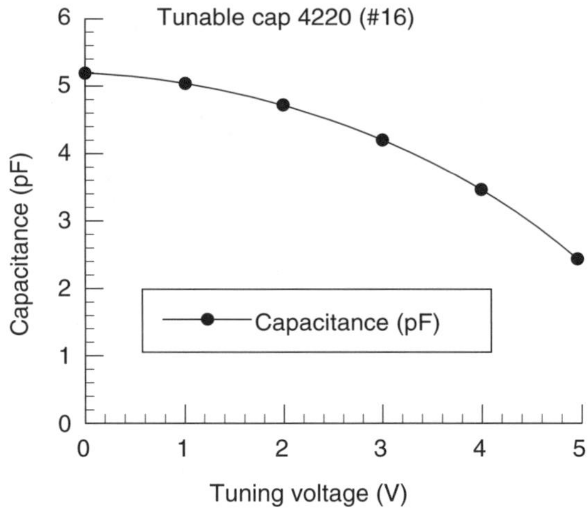

Figure 4.47 Change in capacitance with tuning voltage for the device shown in Figure 4.46. Reproduced from J.J. Yao, S. Park and J. DeNatale, 1998, 'High tuning ratio MEMS based tunable capacitors for RF communications applications', in *Proceedings of Solid-state Sensors and Actuators Workshop*, IEEE, Washington, DC: 124–127, by permission of IEEE, © 1998 IEEE

device has a total capacitance of 5.19 pF at 0 V and 2.48 pF with a tuning voltage of 5 V. It has a quality factor of 34 at 500 MHz and 3 to 1 tuning ration for bias voltages 2 to 14 V.

4.4.3 Dielectric tunable capacitors

High-Q tunable MEMS capacitors are possible by changing the material properties in between the conducting plates of the capacitor. It is clear from Equation (4.26) that the change in dielectric constant can directly translate to change in capacitance of a parallel plate capacitor. The dielectric tunable capacitor can be implemented in planar interdigital (Jose *et al.*, 2001; Kirchoefer *et al.*, 1998) or in coupled (Lu *et al.*, 2000; Van Keuls *et al.*, 1999; Varadan *et al.*, 1995) lines so that a dc bias voltage can change its electrical properties. Voltage-tunable ferroelectric thin films such as $Ba_xSr_{1-x}TiO_3$ is deposited on $LaAlO_3$ or MgO substrates using laser vapor deposition with proper stoichiometric ratios necessary for the desired film. An interdigital capacitor is fabricated on top of it by photolithography and the lift-off process. A schematic diagram of a microwave interdigitated finger capacitor fabricated on ferroelectric material is shown in Figure 4.48. The device shows 3.4 : 1 tuning range for a frequency band of 50 MHz to 20 GHz for the change in bias voltage ranging from 1 to 40. This bias voltage can be further reduced to smaller voltages by designing the interdigital fingers on silicon and then depositing the tunable ferroelectric thin film on top of it (Jose *et al.*, 2001).

These voltage tunable capacitors find applications in MEMS phase shifter designs (Lu *et al.*, 2000; Van Keuls *et al.*, 1999; Varadan *et al.*, 1995).

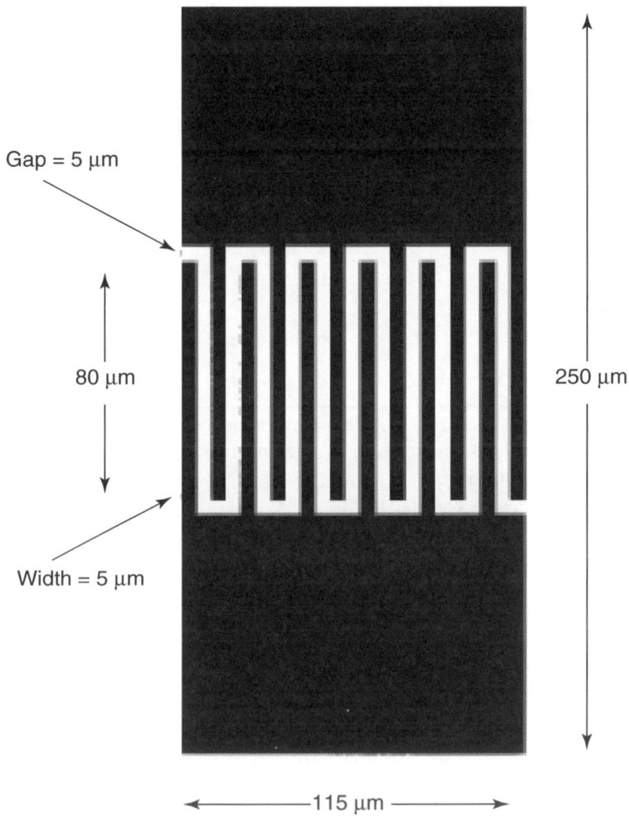

Figure 4.48 Schematic diagram and dimensions of the interdigital capacitor. Reproduced from S.W. Kirchoefer, J.M. Pond, A.C. Carter, W. Change, K.K. Agarwal, J.S. Horwitz and D.B. Chrisey, 1998, 'Microwave properties of $Sr_{0.5}Ba_{0.5}TiO_3$ thin film interdigitated capacitors', *Microwave and Optical Technology Letters* **18**(3): 168–171, by permission of IEEE, © 1998 IEEE

4.5 CONCLUSIONS

Even though silicon ICs are now operating in the gigahertz frequency range and modern bipolar, CMOS (complementary metal oxide semiconductor) and BiCMOS (bipolar complementary metal oxide semiconductor) processes provide high-frequency silicon RF-ICs to compete with GaAs in low gigahertz frequency regime, the lack of high-quality passive components on silicon has made it a poor choice for high-frequency microwave circuits. Also, the lossy silicon substrate makes the design of high-Q reactive components in silicon difficult. Despite this difficulty, the low cost of silicon IC fabrication techniques over the GaAs IC has the potential for integration of micromachined RF MEMS components with RF circuits, which makes silicon one of the choices.

Since the introduction of micromachined inductors and capacitors, many authors have reported higher performance on silicon substrate, using advances in the processing

Table 4.3 Integrated inductor performance trends

			Q_{max}	L	f_{res}
Conductor thickness (i)	↑	⇒	↗	–	–
Conductor sheet resistance (R_{sh})	↑	⇒	↘	–	–
Insulator thickness (h)	↑	⇒	↗	–	↗
Substrate resistivity (p)	↑	⇒	↘↗	–	↗
Area	↑	⇒	↘	↗	↘
Number of turns (n)	↑	⇒	↘	↗	↘
Track width (w)	↑	⇒	↗	↘	↘
Multilayer inductor, extra layer	↑	⇒	↘	↗	↘

Note: ↗, increase; ↘, decrease; – , almost constant; ↘↗, exhibits minimum; Q_{max}, maximum quality factor; L, inductance; f_{res}, resonant frequency; ↑, when the parameter increases; ⇒, the result is.
Source: Koutsoyannopoulos and Papananos, 2000.

technology. This includes removal of the substrate behind the components and use of floating inductors to isolate the inductor from the lossy substrate, higher conductivity metal layers to reduce the loss of the inductor, use of multimetal layers to either shunt inductors to reduce loss or to reduce the area of the inductors and low-loss substrates to reduce losses in the substrate at high frequency.

From a series of design analysis it has been shown that increasing the number of turns or total area of an inductor cannot linearly increase its Q value and there is a trade-off between the inductance and the quality factor. The results are summarized in Table 4.3 aiming to help designers to make the right decision towards the optimization of inductors. At low frequencies, the larger inductors offer higher Qs because of lower series resistance at frequency less than 1 GHz. When frequency increases, the substrate effects dominate and the smaller inductors show higher Q.

Q can also be improved by fabricating the inductor by separating the inductor from silicon with a thick oxide layer. Since increasing the thickness of the oxide layer suppresses the substrate effects, the Q will improve by increasing the oxide thickness. But as frequency increases, the oxide capacitance is effectively short-circuited. This can lead to dominate the substrate effects and eventually reduce the Q.

A significant improvement in Q can be obtained by increasing the metal thickness up to a certain level because the thinner metal spirals suffer severe skin effects. This is because the current flow is at the bottom of the conductor and a thicker metal is ineffective in lowering the series resistance.

Stray capacitance of an inductor determines the self-resonant frequency and its operational frequency range. Controlling the stray capacitance is very important for a practical design.

Little, though, has been written on the modeling, analysis and optimization of these structures. This is because the characteristics of the micro and nano dimensional materials used on fabrication of these MEMS components differ significantly from its bulk properties. Most past researchers have used measurement results of previously built inductors to construct their models. While this technique is most practical, it does not allow the possibility of optimization, nor does it allow the circuit designer freedom to choose

Table 4.4 Evolution of MEMS inductors

Device description	Inductance	Maximum Q	Frequency	Remarks	References
Planar spiral with magnetic core	20 000 nH		0.000010 GHz	Ni/Fe core	Ahn and Allen, 1993
Suspended inductors on silicon	100 nH		3.0 GHz	2 μm CMOS	Chang, Abidi and Gaitan, 1993
Toroidal meander with magnetic core, 30 turns	200 nH		0.01 GHz	$\mu_r = 500$	Ahn and Allen, 1994
Micromachined spiral suspended on membrane	1.2 and 1.7 nH		73 and 54 GHz		Chi and Rebeiz, 1994, 1995
Spiral and meander on different substrate				Comparable with HR silicon	Reyes *et al.*, 1995
Micromachining of HR silicon suspended on membrane	1.2 nH 1.7 nH		70 GHz 50 GHz		Chi and Rebeiz, 1995
Inductors on different core 30 turns on	6700 nH 5700 nH		<10 MHz	Permalloy Orthonol	Park and Allen, 2000
Spiral on silicon and silicon-on-sapphire	4.0 nH	11.9	13.9 GHz		Johnson *et al.*, 1996
Rectangular spiral using high-speed complementary bipolar	34 nH	12	<10 GHz		Ashby *et al.*, 1996
Spiral inductors using standard 0.8 μm BiCMOS	4.5 nH	10	<20 GHz	Multilayer metal	Burghartz, Soyuer and Jenkins, 1996
Spiral using copper damascene interconnect of HR silicon	1.4 nH	30	5.8 GHz		Burghartz *et al.*, 1997
Micro variable inductor	2.5–324.8 nH	1.7	1.9 GHz	Actuation 20 V	Zhou, Sun and Carr, 1997, 1999
Planar spiral inductors 0.8 μm BiCMOS technology		16	4 GHz	4 μm metal 4 μm oxide	Ronkainen *et al.*, 1997
Spiral inductors, CMOS	12.98 nH	11.5	3 GHz		Park *et al.*, 1997c
Micromachined semiencapsulated	1600 nH		<1MHz	Ni–Fe core	Sadler *et al.*, 1997

(*continued overleaf*)

Table 4.4 *(continued)*

Device description	Inductance	Maximum Q	Frequency	Remarks	References
Micromachined spiral on silicon	0.5–100 nH	40	<15		Burghartz *et al.*, 1998
	4.88 nH				Niknejad and Meyer, 1998
Si/SiGe HBT technology on thick polyimide dielectric	0.5–15 nH	22	10 GHz		Laney *et al.*, 1998
Spiral, 36 turns	20 μH	0.25	10 kHz		Ahn and Allen, 1998
Solenoid, 33 turns	0.4 μH	1.5	10 kHz		
Meander, 30 turns	0.2 μH	1.0	10 kHz		
12.5 turns spiral using MESA technique	24 nH		6.6 GHz	Change in SRF	Fan *et al.*, 1998
Spiral free-standing GaAs HEMT technology	4.8 nH	16	15 GHz		Ribas *et al.*, 1998
Micromachined air-core solenoid with air gap	1–20 nH	7–60			Kim and Allen, 1998
Toroidal planar with different magnetic core	10 000 nH		<0.001 GHz		Liakopoulos and Ahn, 1999
Spiral surface micromachined suspended air core	15–40 nH	40–50	0.9–2.5 GHz		Park and Allen, 1999
Surface micromachined solenoid on silicon with insulator layer	Silicon: 2.67 nH Glass: 2.3 nH	16.7 25.1	2.4 GHz 8.4 GHz		Yoon *et al.*, 1999
3D helical, multilayer ceramic-based MCM-C	9.6 nH	93	1.15 GHz		Sutono *et al.*, 1999
Spiral suspended Q and L enhanced	1–3 nH	6–25	6–18 GHz		Sun, Tauritz and Baets, 1999

Micromachined planar spiral, membrane supported	12 nH	20	7 GHz		Ribas *et al.*, 2000
Self-assembling MEMS variable and fixed	0.67–0.82 nH	13	15 GHz		Lubecke *et al.*, 2000
Spiral inductors in CMOS suspended in air	9 nH	5.88	1.5 GHz		Lee *et al.*, 2000
Spiral on 10 Ω-cm silicon with 4.5-μm oxide	8 nH	5	2 GHz	Physical model	Yue and Wong, 2000
Micromachined spiral with etch depth 20 μm	1.8 nH	20.2	2–40 GHz		Lu *et al.*, 2000
3D air core-using sacrificial polymer	1.6–1.8 μH	40	0.01–10 MHz	Bell-shaped	Chomanwang and Lee, 2001
Copper meander using self-assembly	1.5–2.5 nH	20	0.5–3 GHz		Dahlmann and Yeatman, 2000
Micromachined six turn solenoid on 0.24 μm CMOS	2.6 nH	21	4.5 GHz		Chen *et al.*, 2001
Spiral on GaAs 9-μm metal on 10-μm polyimide	0.3–4.5 nH	41.5	>40 GHz		Bahl, 2001

Note: BICMOS, bipolar complementary metal oxide semiconductor; CMOS, complementary metal oxide semiconductor; HBT, heterojunction bipolar transistor; HEMT, high electron mobility transistor; HR, high resistivity; MCM-C, multilayer ceramic-based multichip; MESA, micro-elevator by self-assembly; SRF, self resonant frequency; L, inductance; Q, quality factor; μ_r, relative permeability.

Table 4.5 Evolution of MEMS capacitors

Device description	Capacitance (pF)	Maximum Q	Frequency (GHz)	Remarks	References
Micromachined variable positioning device	0.035–0.1			80–200 V	Larson *et al.*, 1991
MEMS tunable with electrostatic actuation	2.11	62	1	Shunt Capacities in parallel	Young and Boser, 1996
Interdigitated comb	5.19	34	0.5	200% tuning, 2–14 V	Yao, Park and DeNatale, 1998
Micromachined two-plate and three-plate assembly	4.0–4.4		1	0–0.8 V tunable	Dec and Suyama, 1998a, 1998b
	2.05	20	1		
	1.9				
Interdigitated fingers on ferroelectric thin film	4		0.05 to 20	3.4 : 1, tuning 1–40 V	Kirchoefer *et al.*, 1998
MEMS tunable capacitor with electrothermal actuators		256	1	90%	Wu *et al.*, 1998
MIM on suspended membrane	2.6	100	2		Sun, Tauritz and Baets, 1999
	0.1	300	10	Tunable	Feng *et al.*, 2000a
MEMS flip-chip transfer process	0.1	100	10		Harsh *et al.*, 2000
		1050	1		
MEMS tunable with integrated PZT actuator		210	1	3.1 : 1, 0–6 V	Park *et al.*, 2001
Electrostatic digitally controllable MEMS capacitor		140	0.75	4 : 1, tuning 0–30 V	Hoivik *et al.*, 2001
MEMS suspended top plate and two bottom plates		30	5	69.8% at 1 MHz	Zhou, Liu and Schutt-Aine, 2001
Tunable ferroelectric thin film on interdigital fingers	40		0.148	80% tunable at 6 V	Jose *et al.*, 2001

Note: MIM, metal–insulator–metal; PZT, lead zirconate titanate.

parameters such as inductance, resistance, capacitance, and Q freely. We presented some of the simple and accurate design tools for the design of passive components.

Finally, Tables 4.4 and 4.5 present developmental milestones in the state-of-the-art of MEMS inductors and capacitors.

REFERENCES

Ahn, C.H., Allen, M.G., 1993, 'A fully integrated surface micromachined microactuator with a multilevel meander magnetic core', *Journal of Microelectromechanical Systems* **2**(1): 15–22.

Ahn, C.H., Allen, M.G., 1993, 'A planar micromachined spiral inductor for integrated magnetic microactuator applications', *Journal of Micromechanics and Microengineering* **3**: 37–44.

Ahn, C.H., Allen, M.G., 1994, 'A new toroidal-meander type integrated inductor with a multilevel meander magnetic core', *IEEE Transactions on Magnetics* **30**(1): 73–79.

Ahn, C.H., Allen, M.G., 1998, 'Micromachined planar inductors on silicon wafers for MEMS applications', *IEEE Transactions on Industrial Electronics* **45**(6): 866–876.

Ashby, K.B., Koullias, I.A., Finley, W.C., Bastek, J.J., Moinian, S., 1996, 'High Q inductors for wireless applications in a complementary silicon bipolar process', *IEEE Journal of Solid State Circuits* **31**(1): 4–9.

Bahl, I.J., Bhartia, P., 1998, *Microwave Solid-state Circuit Design*, John Wiley, New York.

Bahl, I.J., 1999, 'Improved quality factor spiral inductors on GaAs substrates', *IEEE Microwave and Guided Wave Letters* **9**(10): 398–400.

Bahl, I.J., 2001, 'High performance inductors', *IEEE Transactions on Microwave Theory and Techniques* **49**(4): 654–664.

Burghartz, J.N., Soyeur, M., Jenkins, K.A., Hulvey, M.D., 1995, 'High Q inductors in standard silicon interconnect technology and its application to an integrated RF power amplifier', in *Proceedings of IEDM '95*, IEEE, Washington, DC: 1015–1017.

Burghartz, J.N., Soyuer, M., Jenkins, K.A., 1996, 'Microwave inductors and capacitors in standard multilevel interconnect silicon technology', *IEEE Transactions on Microwave Theory and Techniques* **44**(1): 100–104.

Burgharz, J.N., Edelstein, D.C., Jenkins, K.A., Kwark, Y.H., 1997, 'Spiral inductors and transmission lines in silicon technology using copper-damascene interconnects and low-loss substrates', *IEEE Transactions on Microwave Theory and Techniques* **45**(10): 1961–1968.

Burghartz, J.N., Edelstein, D.C., Soyuer, M., Ainspan, H.A., Jenkins, K.A., 1998, 'RF circuit design aspects of spiral inductors on silicon', *IEEE Journal of Solid-state Circuits* **33**(12): 2028–2034.

Cahana, D., 1983, 'A new transmission line approach for designing spiral microstrip inductors for microwave integrated circuits', in *Proceedings of IEEE MTT-S Symposium, 1983*, IEEE, Washington, DC: 245–247.

Case, M., 1997, 'SiGe MMICs and flip-chip MICs for low cost microwave systems', *Microwave Journal* **40**: 264–276.

Chang, J.Y.-C., Abidi, A.A., Gaitan, M., 1993, 'Large suspended inductors on silicon and their use in a 2 μm CMOS RF amplifier', *IEEE Electron Device Letters* **14**(5): 246–248.

Chen, E.Y., Yoon, Y.K., Laskar, J., Allen, M.G., 2001, 'A 2.4 GHz integrated CMOS power amplifier with micromachined inductors', in *Proceedings of IEEE MTTS Symposium 2001*, IEEE, Washington, DC: Vol. 1, pp. 523–526.

Chi, C.-Y., Rebeiz, G.M., 1994, 'Planar millimeter wave microstrip, lumped elements using micromachining techniques', in *Proceedings of IEEE MTT-S Symposium*, IEEE, Washington, DC: 657–660.

Chi, C.-Y., Rebeiz, G.M., 1995, 'Planar microwave and millimeter wave lumped elements and coupled line filters using micromachining technique', *Transactions on IEEE Microwave Theory and Techniques* **43**(4): 730–738.

Chi, C.-Y., Rebeiz, G.M., 1997, 'Design of Lange-couplers and single sideband mixers using micromachining techniques', *IEEE Transactions on Microwave Theory and Techniques* **45**: 291–294.

Chomnawang, N., Lee, J.-B., 2001, 'On-chip 3D air core micro-inductor for high frequency applications using deformation of sacrificial polymer', *Proceedings of SPIE Smart Electronics and MEMS* **4334**: 54–62.

Dahlmann, G.W., Yeatman, E.M., 2000, 'High *Q* microwave inductors on silicon by surface tension self-assembly', *Electronics Letters* **36**(20): 1707–1708.

Dec, A., Suyama, K., 1997, 'RF micromachined varactors with wide tuning range', *Electronics Letters* **33**(11): 922–924.

Dec, A., Suyama, K., 1998a, 'RF micromachined varactors with wide tuning range', in *Proceedings of IEEE Radio Frequency Integrated Circuit Symposium*, IEEE, Washington, DC: 309–312.

Dec, A., Suyama, K., 1998b, 'Micromachined electro-mechanically tunable capacitors and their applications to RF ICs', *IEEE Transactions on Microwave Theory and Techniques* **46**(12): 2587–2596.

Dec, A., Suyama, K., 2000, 'Microwave MEMS-based voltage controlled oscillators', *IEEE Transactions on Microwave Theory and Techniques* **48**(11): 1943–1949.

Driesen, J., Ruythooren, W., Belmans, R., DeBoeck, J., Celis, J.-P., Hameyer, K., 1999, 'Electric and magnetic FEM modeling strategies for microinductors', *IEEE Transactions on Magnetics* **35**(5): 3577–3579.

Fan, L., Chen, R.T., Nespola, A., Wu, M.C., 1998, 'Universal MEMS platforms for passive RF components: suspended inductors and variable capacitors', in *Proceedings of 11th Annual International Workshop on MEMS '98*, IEEE, Washington, DC: 29–33.

Feng, Z., Zhang, W., Su, B., Harsh, K.F., Gupta, K.C., Bright, V.M., Lee, Y.C., 2000a, 'Design and modeling of RF MEMS tunable capacitors using electro-thermal actuators', in *Proceedings of 1999 IEEE MTT-S Symposium*, IEEE, Washington, DC: 1507–1510.

Feng, Z., Zhang, H., Zhang, W., Su, B., Gupta, K.C., Bright, V.M., Lee, Y.C., 2000b, 'MEMS based variable capacitors for millimeter wave applications', in *Proceedings of Solid-state Sensors and Actuators Workshop, South Carolina, June 2000*, IEEE, Washington, DC: 255–258.

Feng, Z., Zhang, H., Gupta, K.C., Zhang, W., Bright, V.M., Lee, Y.C., 2001, 'MEMS-based series and shunt variable capacitors for microwave and millimeter-wave frequencies', *Sensors and Actuators A: Physical* **91**: 256–265.

Greenhouse, H.M., 1974, 'Design of planar rectangular microelectronic inductors', *IEEE Transactions on Parts, Hybrids and Packaging* **10**(2): 101–109.

Grover, F.W., 1946, *Inductance Calculations*, Van Nostrand, New York.

Guckel, H., Skrobis, K., 1992, 'Metal micromechanisms via deep X-ray lithography, electroplating and assembly', *Journal of Micromechanics and Microengineering* **2**(4): 225–228.

Harsh, K.F., Zhang, W., Bright, V.M., Lee, Y.B., 1999, 'Flip-chip assembly for Si-based RF MEMS', in *Proceedings of 20th IEEE International Symposium on Microelectromechanical Systems*, IEEE, Washington, DC: 273–278.

Harsh, K.F., Su, B., Zhang, W., Bright, V.M., Lee, Y.C., 2000, 'The realization and design considerations of flip-chip integrated MEMS tunable capacitor', *Sensors and Actuators A: Physical* **80**: 108–118.

Hjort, K., Solderkvist, J., Schweitz, J.-A., 1994, 'Gallium arsenide as a mechanical material', *Journal of Micromechanics and Microengineering* **4**: 1–13.

Hoivik, N., Michalicek, M.A., Lee, Y.C., Gupta, K.C., Bright, V.M., 2001, 'Digitally controllable variable high-*Q* MEMS capacitor for RF applications', in *Proceedings of IEEE MTT-S Symposium*, May 2001, Volume 3, IEEE, Washington, DC: 2115–2118.

Johnson, R.A., Chang, C.E., Asbeck, P.M., Wood, M.E., Garcia, G.A., Lagnado, I., 1996, 'Comparison of microwave inductors fabricated on silicon-on-sapphire and bulk silicon', *IEEE Microwave and Guided Wave Letters* **6**(9): 323–325.

Jose, K.A., Yoon, H., Vinoy, K.J., Sharma, P., Varadan, V.K., Varadan, V.V., 2001, 'Low voltage tunable capacitors for RF MEM filters and antenna applications', in *Proceedings of IEEE*

Antennae and Propagation Society International Symposium, July 2001, Boston MA, IEEE, Washington, DC: 670–673.

Jung, S., Kang, K., Park, J.Y., Chung, K.W., Kim, Y.K., Kwon, Y., 2001, 'Micromachined frequency variable impedance tuners using resonant unit cell', in *Proceedings of IEEE MTT-S Symposium*, May 2001, Volume 1, IEEE, Washington, DC: 333–336.

Kawahito, S., Sasaki, Y., Ashiki, M., Nakamura, T., 1991, 'Micromachined solenoid for high sensitive magnetic sensors', in *Proceedings of Transducers 1991*: IEEE, Piscataway, NJ, USA: 1077–1080.

Kim, Y.J., Allen, M.G., 1998, 'Surface micromachined solenoid inductors for high frequency applications', *IEEE Transactions on Components, Packaging and Manufacturing Technology, Part C* **21**(1): 26–33.

Kirchoefer, S.W., Pond, J.M., Carter, A.C., Change, W., Agarwal, K.K., Horwitz, J.S., Chrisey, D.B., 1998, 'Microwave properties of $Sr_{0.5}Ba_{0.5}TiO_3$ thin film interdigitated capacitors', *Microwave and Optical Technology Letters* **18**(3): 168–171.

Konishi, Y., (Ed.),1991, *Microwave Integrated Circuits*, Marcel Dekker, New York.

Koutsoyannopoulos, Y.K., Papananos, Y., 2000, 'Systematic analysis and modeling of integrated inductors and transformers in RF IC design', *IEEE Transactions on Circuits and Systems II* **47**(8): 699–713.

Laney, D.C., Larson, L.E., Malinowski, J., Harame, D., Subbanna, S., Volant, R., Case, M., Chan, P., 1998, 'Low-loss microwave transmission lines and inductors implemented in a Si/SiGe HBT process', in *Proceedings of IEEE BCTM 1998*, IEEE, Washington, DC: 101–104.

Lang, K.L., Papapolymerou, J., Goldsmith, C.L., Malezewski, A., Kleber, J., 2001, 'A reconfigurable double-stub tuner using MEMS devices', in *Proceedings of IEEE MTT-S Symposium*, May 2001, Volume 1, IEEE, Washington, DC: 337–340.

Larson, L.E., Hackett, R.H., Lelendes, M.A., Lohr, R.S., 1991, 'Micromachined microwave actuator (MIMAC) technology – a new tuning approach for microwave integrated circuits', in *Proceedings of IEEE Microwave and Millimeter Wave Monolithic Circuits Symposium*, IEEE, Washington, DC: 27–30.

Lee, C.-Y., Kao, Y.-H., Luo, J.-Y., Chang, K.-M., 2000, 'The enhanced Q spiral inductors with MEMS technology for RF applications', in *Proceedings of Asia Pacific Microwave Conference* IEEE, Piscataway, NJ, USA: 2000: 1326–1329.

Li, P., 1996, 'A new closed form formula for inductance calculation in microstrip line spiral inductor design', in *Proceedings of IEEE 5th Topical Meeting on Electrical Performance of Electronic Packaging*, 1996, IEEE, Washington, DC: 58–60.

Liakopoulos, T.M., Ahn, C.H., 1999, '3-D microfabricated toroidal planar inductors with different magnetic core schemes for MEMS and power electronic applications', *IEEE Transactions on Magnetics* **35**(5): 3679–3681.

Long, J.R., Copeland, M.A., 1997, 'The modeling, characterization and design of monolithic inductors for silicon RF ICs', *IEEE Journal of Solid-state Circuits* **32**(3): 357–369.

Lopez-Villegas, J.M., Samitier, J., Bausells, J., Merlos, A., Cane, C., Knochel, R., 1997, 'Study of integrated RF passive components performed using CMOS and Si micromachining technologies', *Journal of Micromechanics and Microengineering* **7**: 162–164.

Lopez-Villegas, J.M., Samitier, J., Cane, C., Losantos, P., Bausells, J., 2000, 'Improvement of quality factor of RF integrated inductors by layout optimization', *IEEE Transactions on Microwave Theory and Techniques* **48**(1): 76–83.

Lu, L.-H., Ponchak, G.E., Bhattacharya, P., Katehi, L.P.B., 2000, 'High Q X-band and K-band micromachined spiral inductors for use in S-based integrated circuits', in *Proceedings of the Symposium on Silicon MICs in RF Systems*, IEEE, Piscataway, NJ, USA: 108–112.

Lu, Y., Nagra, A.S., Erker, E.G., Periaswamy, P., Taylor, T.R., Speck, J., York, R.A., 2000, '$BaSrTiO_3$ interdigitated capacitors for distributed phase shifter applications', *IEEE Microwave and Guided Wave Letters* **10**(11): 448–450.

Lubecke, V.M., Barber, B., Chan, E., Lopez, D., Gammel, P., 2000, 'Self-assembling MEMS variable and fixed RF inductors', in *Proceedings of 12th Asia Pacific Microwave Conference December 2000*, IEEE, Piscataway, NJ, USA: 7–10.

Milanovic, V., Gaitan, M., Bowen, E.D., Zaghloul, M.E., 1997, 'Micromachined microwave transmission lines in CMOS technology', *IEEE Transactions on Microwave Theory and Techniques* **45**: 630–635.

Mohan, S.S., Hershenson, M., Boyd, S.P., Lee, T.H., 1999, 'Simple accurate expressions for planar spiral inductors', *IEEE Journal of Solid-state Circuits* **34**(10): 1419–1424.

Niknejad, A.M., Meyer, R.G., 1998, 'Analysis, design and optimization of spiral inductors and transformers for Si RF ICs', *IEEE Journal of Solid-state Circuits* **33**(10): 1470–1481.

Niknejad, A.M., 2000, *Inductors and Transformers for Si RF ICs*, Kluwer Academic, Boston, MA.

Olivei, A., 1969, 'Optimized miniature thin film planar inductors compatible with integrated circuits', *IEEE Transactions Part, Material Packaging* **5**: 71–88.

Parisot, M., Archamnault, Y., Pavlidis, D., Magarshack, J., 1984, 'Highly accurate design of spiral inductors for MMICs with small size and high cut-off frequency characteristics', in *Proceedings of IEEE MTT-S symposium*, 1984, IEEE, Washington, DC: 106–110.

Park, J.Y., Allen, M.G., 1996, 'A comparison of micromachined inductors with different magnetic core materials', *Proceedings of IEEE Electronic Components and Technology Conference*, IEEE, Washington, DC: 375–381.

Park, M., Kin, C.S., Park, J.M., Yu, H.K., Nam, K.S., 1997, 'High *Q* microwave inductors in CMOS double metal technology and its substrate bias effects for 2 GHz RF IC application', in *Proceedings of IEDM 97*, IEEE, Washington, DC: 59–62.

Park, M., Lee, S., Yu, H.K., Koo, J.G., Nam, K.S., 1997c, 'High *Q* CMOS compatible microwave inductors using double metal interconnection silicon technology', *IEEE Microwave and Guided Wave Letters* **7**(2): 45–47.

Park, M., Lee, S., Yu, H.K., Nam, K.S., 1997, 'Optimization of high *Q* CMOS compatible microwave inductors using silicon CMOS technology', in *Proceedings of 1997 IEEE Radio frequency Integrated Circuit Symposium*, IEEE, Washington, DC: 181–84.

Park, J.Y., Allen, M.G., 1999, 'Packaging-compatible high *Q* microinductors and microfilters for wireless applications', *IEEE Transactions on Advanced Packaging* **22**(2): 207–213.

Park, J.Y., Allen, M.G., 2000, 'Integrated electroplated micromachined magnetic devices using low temperature fabrication process', *IEEE Transactions on Electronics Packaging and Manufacturing* **23**(1): 48–55.

Park, J.Y., Yee, Y.J., Nam, H.J., Bu, J.U., 2001, 'Micromachined RF MEMS tunable capacitors using piezoelectric actuators', in *Proceedings of IEEE MTT-S Symposium*, May 2001, Volume 3, IEEE, Washington, DC: 2111–2114.

Pehlke, D.R., Burstein, A., Chang, M.F., 1997, 'Extremely high *Q* tunable inductors for Si-based RF integrated circuit applications', in *Proceedings of International Electronics Devices Meeting IEDM*, 1997, IEEE, Washington, DC: 63–66.

Reyes, A.C., El-Ghazaly, S.M., Dorn, S.J., Dydyk, M., Schroder, D.K., Patterson, H., 1995, 'Coplanar waveguides and microwave inductors on silicon substrate', *IEEE Transactions on Microwave Theory and Techniques* **43**(9): 2016–2022.

Ribas, R.P., Bennouri, N., Karam, J.M., Courtois, B., 1997, 'GaAs MEMS design using 0.2 μm HEMT MMIC technology', in *Proceedings of the 19th Annual IEEE Gallium Arsenide Integrated Circuit Symposium*, IEEE, Piscataway, NJ, USA: 127–130.

Ribas, R.P., Lescot, J., Leclercq, J.L., Bennouri, N., Karam, J.M., Courtois, B., 1998, 'Micromachined planar spiral inductor in standard GaAs HEMT MMIC technology', *Transactions on IEEE Electron Device Letters* **19**(8): 285–287.

Ribas, R.P., Lescot, J., Leclercq, J.-L., Karam, J.M., Ndagijimana, F., 2000, 'Micromachined microwave planar spiral inductors and transformers', *IEEE Transactions on Microwave Theory and Techniques* **48**(8): 1326–1335.

Rodrigues, R., Dishman, J.M., Dickens, F.D., Whelan, E.W., 1980, 'Modeling of two-dimensional spiral inductors', *IEEE Transactions on Components, Hybrids, and Manufacturing Technology* **5**: 535–541.

Ronkainen, H., Kattelus, H., Tarvainen, E., Riihisaari, T., Andersson, M., Kuivalainen, P., 1997, 'IC compatible planar inductors on silicon', *IEE Proceedings on Circuits, Devices and Systems* **144**(1): 29–35.

Sadler, D.J., Zhang, W., Ahn, C.H., Kim, H.J., Han, S.H., 1997, 'Micromachined semi-encapsulated spiral inductors for microelectromechanical systems (MEMS) applications', *IEEE Transactions on Magnetics* **33**(5): 3319–3321.

Schmuckle, F.J., 1993, 'The method of lines for the analysis of rectangular spiral inductors', *IEEE Transactions on Microwave Theory and Techniques* **41**(6/7): 1183–1186.

Shin, D.-H., Kim, C.S., Jeong, J.-H., Nam, S.-E., Kim, H.J., 1999, 'Fabrication of double rectangular type ReTaN film inductors', *IEEE Transactions on Magnetics* **35**(5): 3511–3513.

Sieiro, J., Lopez-Villegas, J.M., Cabanillas, J., Osorio, J.A., Samitier, J., 2001, 'A complete physical frequency dependent lumped model for RF integrated inductors', in *Proceedings of 2001 IEEE Radio Frequency Integrated Circuit Symposium*, IEEE, Washington, DC: 121–124.

Soohoo, R.F., 1979, 'Magnetic thin film inductors for integrated circuit applications', *IEEE Transactions on Magnetics* **15**: 1803–1805.

Sun, Y., van Vliet, F.E., Tauritz, J.L., Baets, R.G.F., 1996, 'Monolithic narrow band active inductors using suspended membrane passive components on silicon substrate', High Performance Electron Devices for Microwave and Optoelectronic Applications Workshop, EDMO, IEEE, Piscataway, NJ, USA, 127–130.

Sun, Y., van Zeji, H., Tauritz, J.L., Baets, R.G.F., 1996a, 'Suspended membrane inductors capacitors for applications on silicon MMICs', in *Proceedings of IEEE Microwave and Millimeter Wave Monolithic Circuits Symposium*, IEEE, Washington, DC: 99–102.

Sun, Y., Tauritz, J.L., Baets, R.G.F., 1999, 'Micromachined RF passive components and their applications in MMICs', *International Journal of RF and Microwave CAE* **9**: 310–325.

Sutono, A., Pham, A., Lasker, J., Smith, W.R., 1999, 'Development of three-dimensional ceramic-based MCM inductors for hybrid RF/microwave applications', in *Proceedings of 1999 IEEE Radio Frequency Integrated Circuit Symposium*, IEEE, Washington, DC: 175–178.

Van Keuls, F.W., Romanofsky, R.R., Varaljay, N.D., Miranda, F.A., Canedy, C.L., Aggarwal, S., Venkatesan, T., Ramesh, R., 1999, 'A Ku-band Gold/$Ba_xSr_{1-x}TiO_3$/$LaaAlO_3$ conductor thin film ferrielectric microstrip line phase shifter for room-temperature communications applications', *Microwave and Optical Technology Letters* **20**(1): 53–56.

Varadan, V.K., Jose, K.A., Varadan, V.V., Hughes, R., Kelly, J.F., 1995, 'A novel microwave planar phase shifter', *Microwave Journal* (April): 244–254.

Wanger, B., Beneckle, W., 1991, 'Microfabricated actuator with moving permanent magnet', *IEEE Microelectro Mechanical Systems Workshop*, IEEE, Washington, DC: 27–32.

Warner, R.M. (Ed.), 1965, *Integrated Circuits: Design Principles and Fabrication*, McGraw-Hill, New York.

Wu, H.D., Harsh, K.F., Irwin, R.S., Zhang, W., Mickelson, A.R., Lee, Y.C., 1998, 'MEMS designed for tunable capacitors', in *Proceedings of IEEE MTT-S Symposium*, IEEE, Washington, DC: 127–129.

Yamaguchi, M., Mastumo, M., Ohzeki, H., Arai, K.I., 1990, 'Fabrication and basic characteristics of dry-etched micro inductors', *IEEE Transactions on Magnetics* **26**(5): 2014–2016.

Yamaguchi, M., Mastumo, M., Ohzeki, H., Arai, K.I., 1991, 'Analysis of the inductance and the stray capacitance of the dry-etched micro inductors', *IEEE Transactions on Magnetics* **27**(6): 5274–5275.

Yao, J.J., Park, S., DeNatale, J., 1998, 'High tuning ratio MEMS based tunable capacitors for RF communications applications', in *Proceedings of Solid-state Sensors and Actuators Workshop*, IEEE, Washington, DC: 124–127.

Yoon, J.-B., Kim, B.-K., Han, C.-H., Yoon, E., Kim, C.-K., 1999, 'Surface micromachined solenoid on Si and on-glass inductors for RF applications', *IEEE Electron Device Letters* **20**(9): 487–489.

Young, D.J., Boser, B.E., 1996, 'A micromachined variable capacitor for monolithic low-noise VCOs', in *Proceedings of International Conference on Solid-state Sensors and Actuators*, IEEE, Washington, DC: 86–89.

Young, D.J., Boser, B.E., Malba, V., Bernhardt, A.F., 2001, 'A micromachined RF low noise voltage controlled oscillator for wireless communications', *International Journal of RF and Microwave CAE* **11**: 285–300.

Yue, C.P., Wong, S.S., 2000, 'Physical modeling of spiral inductors on silicon', *IEEE Transactions on Electron Devices* **47**(3): 560–568.

Zhou, S., Sun, X.-Q., Carr, W.N., 1997, 'A micro variable inductor chip using MEMS relays', in *Proceedings of Transducers '97, International Conference on Sensors and Actuators*, Chicago, June 1997, IEEE, Piscataway, NJ, USA: 1137–1140.

Zhou, S., Sun, X.-Q., Carr, W.N., 1999, 'A monolithic variable inductor network using microrelays with combined thermal and electrostatic actuation', *Journal of Micromechanics and Microengineering* **9**: 45–50.

Zou, J., Liu, C., Schutt-Aine, J., Chen, J., Kang, S.M., 2000, 'Development of a wide tuning range MEMS tunable capacitor for wireless communication systems', in *Proceedings of IEDM-00*, IEEE, Washington, DC: 403–406.

Zou, J., Liu, C., Schutt-Aine, J.E., 2001, 'Development of a wide tuning range two parallel plate tunable capacitor for integrated wireless communication systems', *International Journal of RF and Microwave CAE* **11**: 322–329.

5

Micromachined RF filters

5.1 INTRODUCTION

An area that has got significant attention, and remains technically challenging, is miniaturization of the well-proven mechanical filters. These filters involve a form of mechanical wave propagation at some stage between their input and output terminals. These mechanical waves are often vibrations. We include filters operating with acoustic waves also under this class, since acoustic waves may be considered as a form of mechanical waves. Some filters are only fabricated with micromachining techniques and do not involve mechanics for their operation. The common element in all these filters is their fabrication approach, micromachining. Hence we propose to call all these filters together micromachined filters.

Several different types of filters are used in communication equipment. These are generally classified based on the frequency band they transmit as low pass, high pass, band pass and band stop filter. However, in most communication systems the primary need is for band pass filters with an extremely narrow pass band and rapid roll off on either side of the pass band. Some of these performance parameters of filters are described below.

The most important characteristic of a filter is the *insertion loss*. This is the ratio of the signal delivered at the output side to that supplied to the filter at its input. Obviously, the design goal is to minimize this quantity within its pass band. The pass band characteristics of a filter are generally expressed as its *quality factor* (Q-factor). The quality factor is a measure of the energy stored in the system to the energy dissipated per cycle. In terms of the frequency band this can be expressed more conveniently as

$$Q = \frac{f_0}{\Delta f} \tag{5.1}$$

where f_0 is the center frequency and $\Delta f = f_2 - f_1$ (Figure 5.1). The Q-factor therefore gives a good picture of the effectiveness of the filter *vis-à-vis* its frequency response. There are few other quantities often prescribed for filters. *Roll off* is the rate at which the transfer response of the filter changes from pass band to stop band. *Stop band rejection* is the signal transmitted through the system at frequencies beyond the pass band. This is expressed in decibels, often relative to the minimum insertion loss.

A typical modern personal communication system is shown in Figure 5.2. These systems are designed to handle many communication channels operating simultaneously. The selection between these channels is accomplished by using band pass filters. In an effort to pack more channels within the limited spectrum available, these filters should

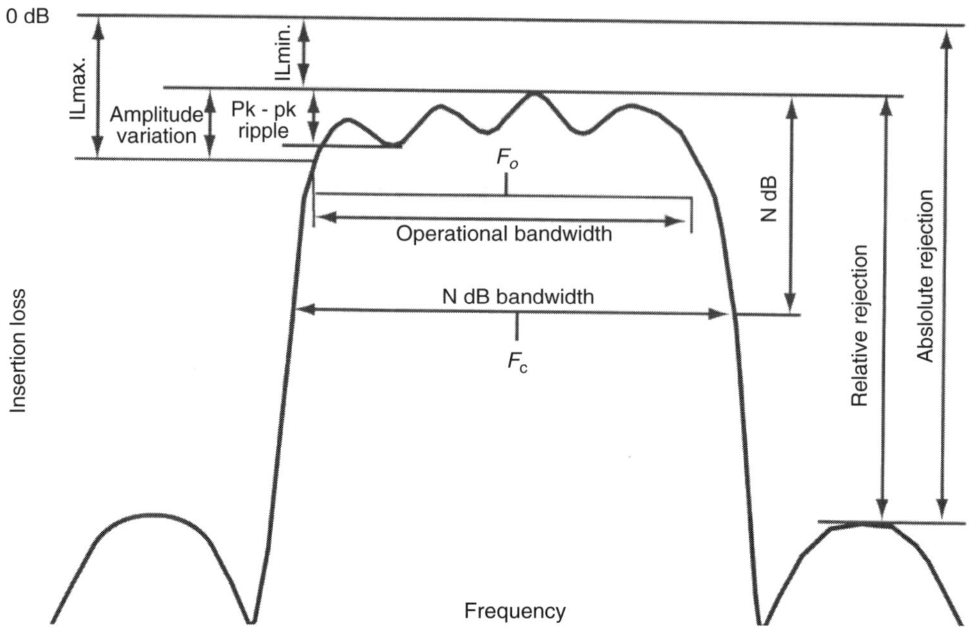

Figure 5.1 Parameters for characterizing bandpass filters

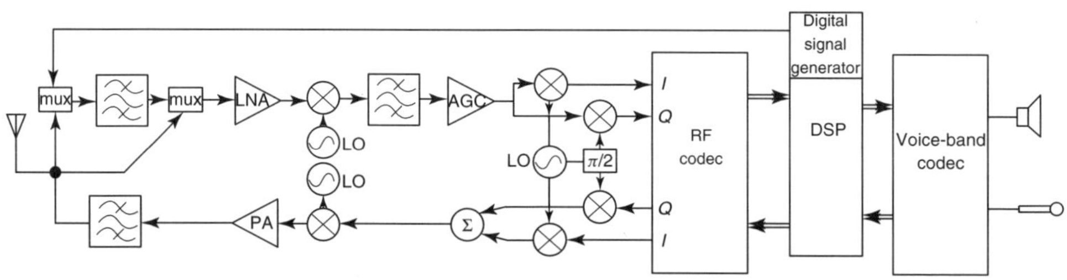

Figure 5.2 Block schematic of a wireless transceiver: mux, multiplexer; LNA, low noise amplifier; LO, local oscillator; PA, power amplifier; AGC, automatic gain control

be designed such that they have a uniform pass band with very low insertion loss, rapid roll off, and high out-of-band rejection ratio.

At various frequency regimes these objectives are accomplished with various types of filter implementations. Conceptually, the simplest of these is a design involving lumped inductors and capacitors. However, in practice, this configuration is limited by attainable Q factors. Another approach that caught the attention is the use of digital filters that makes use of recent advancement in high-speed processors and modern digital signal processing algorithms. However, this approach is limited by the maximum sampling frequency. This is especially true at higher frequencies. Alternative schemes are therefore required.

From the early days of telecommunications various forms of electromechanical filters have been used to obtain desirable characteristics such as a high quality factor (Q). These use two electromechanical transducers, one at the input side and another at the output side and a form of mechanical transmission line connecting them. These mechanical and electromechanical components generally have a strong resonance, which results in the excellent Q of such filters. Modeling of these systems is generally done by using their equivalent circuits, which can be translated to the electrical domain by simple transformations for design and optimization. These transducers are assumed conservative or lossless for simplicity in the analysis. Approaches for obtaining the equivalent circuits for several useful electromechanical transducers have already been presented in Chapter 1.

In Section 5.2 the approaches taken for modeling components of mechanical filters are discussed. Models for resonator components such as bars, wires and membranes, which are responsible for the sharp resonance characteristics of these systems, are presented. These mechanical models form the basis for their electrical equivalent circuits that can be developed making use of electromechanical analogies. The characteristics of various mechanical transmission and coupling components also play a key role in the filter performance and are also discussed in Section 5.2.

High-Q filters using resonating mechanical components are widely used in many communication systems and radars for frequencies in the kilohertz range. As the frequency is increased the size of these devices becomes smaller, almost become infeasible to fabricate. These are therefore not amenable for mass production, thereby increasing their cost. One of the key factors in popularizing modern telecommunication equipment is the low cost. Hence alternative design and fabrication schemes need be evolved to replace these mechanical filters.

The principles behind mechanical filters can, however, be translated to smaller devices operative at relatively higher frequencies. These micro devices can be used for frequencies up to 10s of megahertz and can have Q in 1000s with proper packaging (Bannon, Clark and Nguyen, 2000). The electrical energy is converted into a form of mechanical energy such as vibration at the electromechanical transducer at the input side of the filter. This mechanical energy is propagated towards another electromechanical transducer which converts it back into the electrical domain, using a form of mechanical transmission line. These mechanical and electromechanical components are generally frequency sensitive. Their mechanical resonance determines the frequency band of operation of the device. Design of these microelectromechanical filters is presented in Section 5.3. It may be mentioned here that in this chapter terms such as micromechanical filters, RF-MEMS filters or microelectromechanical filters are used interchangeably to represent these devices. It is estimated that this approach can be extended to cover most of very-high-frequency (VHF) band by a few small design modifications and improvements in fabrication accuracies.

Current fabrication limitations restrict the extension of micromechanical filters for frequencies above ~100 MHz and planar distributed filters below gigahertz frequencies. Filters and resonators based on the principles of surface acoustic waves (SAWs) can bridge this gap, and provide high-Q devices for frequencies up to 2 GHz (Campbell, 1983). These devices consist of an interdigitated transducer (IDT) which launches surface acoustic waves on a piezoelectric substrate. Another equivalent transducer at the output side can be used to convert this acoustic signal back to the electrical form. These devices are highly frequency dependent and hence are utilized in filter design. The design of IDTs and the wave propagation involving surface acoustic waves are discussed in Section 5.4. These planar devices can be accurately fabricated with modern microfabrication facilities. The limitations of these devices, basically driven by the fabrication constraints, are also presented. For yet higher frequency applications, filters based on bulk acoustic waves are used. Preliminary ideas of these are introduced in Section 5.5. It may be mentioned that both these types of filters, relying on forms of mechanical wave propagation, are the most significant contributors leading to the proliferation of miniaturized modern communication equipment.

At microwave and millimeter wave frequencies distributed components are used extensively to realize filters. However, the Q factors obtained by conventional implementations of these devices are limited by parasitics. Several micromachining techniques have been effectively used to minimize these effects. Planar filters on thin dielectric membrane show low loss and are suitable for low-cost, compact, high-performance MMICs. These filter designs do not involve principles of mechanics, but are discussed in Section 5.6. A brief summary of the chapter is presented in Section 5.7.

5.2 MODELING OF MECHANICAL FILTERS

A brief comparative study of various electromechanical actuation schemes has been discussed in the previous section. These transducers behave as resonators and are analyzed in terms of electrical and mechanical characteristics. Important mechanical properties in the context of filter action are the resonant frequency and the Q factor of these elements. Several of such resonators are joined together using coupling elements such as wires, to improve the filter performance. The number of these resonators plays a key role in determining the shape factor of the filter performance, while their resonant frequency is the center frequency of the band pass filter. The filter bandwidth is reduced by increasing the equivalent mass of the resonators, or by increasing the compliance of the coupling wires.

Although the form of these components is significantly different in microdevices, an understanding of their operational mechanism would aid successful design of such devices. Approaches to the mathematical modeling of these components are studied in this section.

5.2.1 Modeling of resonators

Mechanical properties of resonators depend on their shape, type of materials used and the relevant modes of vibration. Several classical references are available for the resonance modes of vibration of bars, rods, thin plates and disks useful in filter design. The analysis is relatively simple when these bodies have one of the dimensions very much different from the other two. Longitudinal, torsional, flexural or radial modes are easily obtained for these bodies. The wave motion is relatively complex for thick bodies. Analysis is further involved for complex shapes such as dumbbell resonators and tuning forks (Johnson, 1983). We do not intend to cover these complex shapes as they are beyond the scope of the present objective.

A simplified analysis, the results of which are presented in this section, makes use of several assumptions. These include (Johnson, 1983):

- vibrations are of small amplitude, and the stress–strain relationship is linear;
- there are no internal losses and no external damping of vibrations by air resistance, etc.;
- effects of external gravity and magnetic forces can be neglected.

The following steps are involved in the analysis of resonators (Johnson, 1983):

1. Start with the differential equations of wave motion within the resonator. These are second or fourth order in space coordinates and second order in time.
2. To eliminate the time dependency, sinusoidal excitations are assumed, and phasor notation is used.
3. Solutions to these equations are expressed in terms of trigonometric, hyperbolic or Bessel functions.
4. The boundary conditions are mathematically represented. These are then substituted into the solutions for displacement to eliminate constants. The frequency equation is obtained for different modes.
5. Substituting these in the original differential equation, one can obtain a relationship between the wave number and frequency.
6. Using this relation, and the frequency equation, the resonant frequencies for different modes can be obtained.
7. The equivalent mass is defined as an equivalent lumped mass placed at a specified location on the resonator that matches with the kinetic energy of the distributed parameter element vibrating at a given mode and resonant frequency.

These result in dynamic properties of various resonator shapes discussed below. Since a detailed step-by-step derivation in each case would be digressive, only the starting differential equation, the frequency equation, the expressions for resonant frequency and equivalent mass are given here.

5.2.1.1 Longitudinal mode rod resonator

The schematic for the longitudinal mode vibration in a slender rod is shown in Figure 5.3. The material of the rod has a density ρ and Young's modulus E. The wave equation in this case is

$$\frac{E}{\rho}\frac{\partial^2 u}{\partial x^2} = \frac{\partial^2 u}{\partial t^2} \tag{5.2}$$

where u is as illustrated in Figure 5.3. After substituting for the time dependency using phasors, its solution can be expressed in terms of trigonometric functions. Assuming both ends of the bar are free, the frequency equation is:

$$k_n l = n\pi, \quad n = 1, 2, 3, \ldots \tag{5.3}$$

From the wave equation we obtain the relationship between propagation constant and frequency, ω:

$$k = \omega\left(\frac{\rho}{E}\right)^{1/2} \tag{5.4}$$

This leads to the resonant frequencies as

$$f_n = \frac{n}{2l}\left(\frac{E}{\rho}\right)^{1/2} \tag{5.5}$$

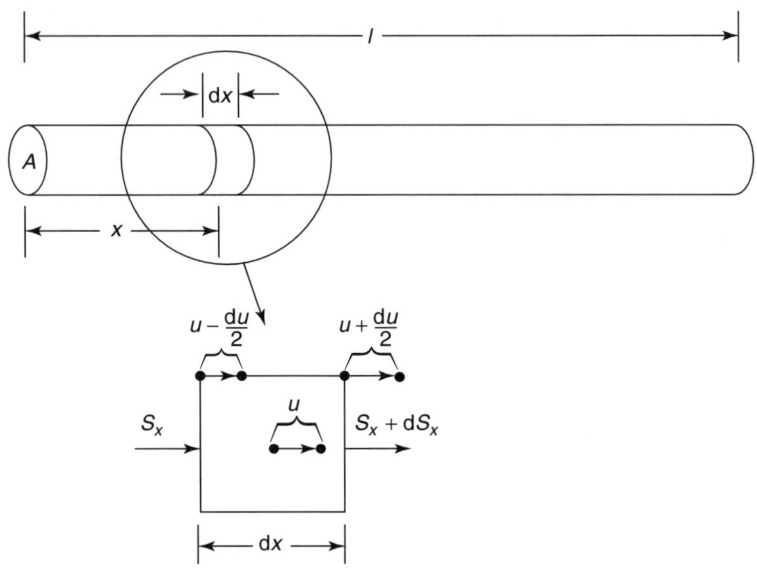

Figure 5.3 Longitudinal mode resonator. Reproduced from R.A. Johnson, 1983, *Mechanical Filters in Electronics*, Wiley Interscience, New York, by permission of Wiley, © 1983 Wiley

The displacement $u_n(x)$ for the nth mode is

$$u_n(x) = A \cos \frac{n\pi x}{l} \tag{5.6}$$

where A is a constant.

Furthermore, the equivalent mass is obtained as

$$M_{\mathrm{eq}\,x} = \frac{1}{V_0} \int_0^l (V_0 \cos k_n x)^2 \rho A \,\mathrm{d}x \tag{5.7}$$

where V_0 is the velocity at $x = 0$.

When the radius of the rod is greater than a tenth of a wavelength, the radial variation in displacement is taken into account to obtain the generalized resonant frequency:

$$f_n = \frac{n}{2l} \left(\frac{E}{\rho}\right)^{1/2} \left[1 - \left(\frac{n\mu\pi a}{2l}\right)^2\right] \tag{5.8}$$

5.2.1.2 Torsional mode rod resonator

The schematic for a rod vibrating in the torsional mode is shown in Figure 5.4. The wave equation in this case is the same as the longitudinal mode of the rod; however, we obtain angular displacement and resonant frequency in terms of shear modulus G as

$$\theta = \theta_0 \cos \frac{n\pi x}{l} \tag{5.9}$$

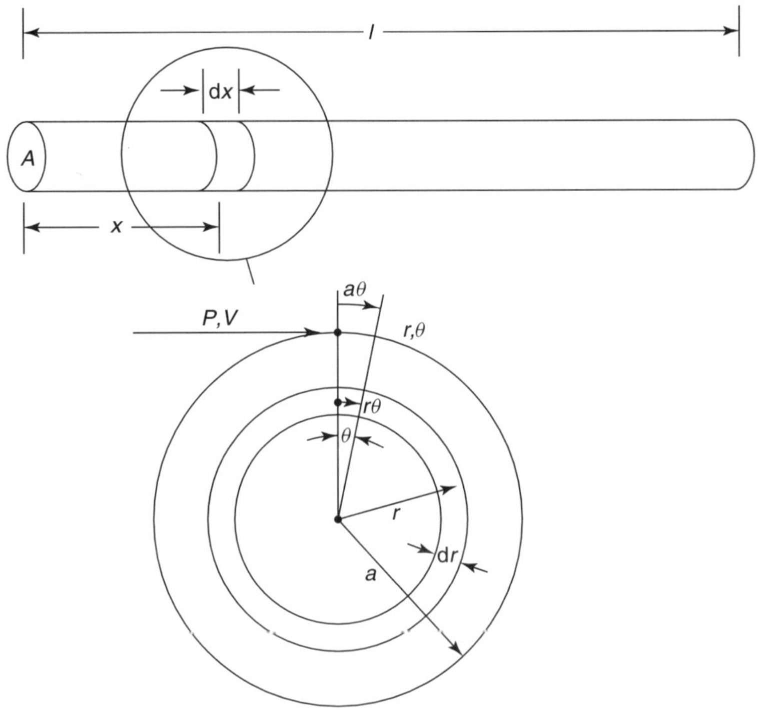

Figure 5.4 Torsional modes on a rod. Reproduced from R.A. Johnson, 1983, *Mechanical Filters in Electronics*, Wiley Interscience, New York, by permission of Wiley, © 1983 Wiley

$$f_n = \frac{n}{2l}\left(\frac{G}{\rho}\right)^{1/2} \tag{5.10}$$

where θ_0 is the angular displacement at the ends of the rod. The equivalent mass in this case is

$$M_{\text{eq}}|_{r,l} = \pi\rho l a^4\left[4r^2\cos^2\left(\frac{n\pi x}{l}\right)\right]^{-1} \tag{5.11}$$

5.2.1.3 Flexural mode bar resonator

A schematic of the fundamental flexural mode on a bar or a rod is shown in Figure 5.5. The wave equation for this is

$$\frac{\partial^4 u}{\partial x^4} = \frac{\rho A}{EI}\frac{\partial^2 u}{\partial t^2} \tag{5.12}$$

where the bending moment of inertia I of a rectangular bar of width w and thickness t is

$$I = \frac{wt^3}{12} \tag{5.13}$$

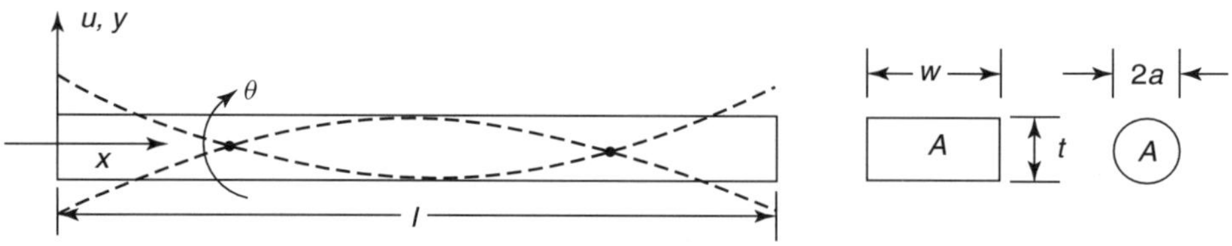

Figure 5.5 Flexure modes on a rod. Reproduced from R.A. Johnson, 1983, *Mechanical Filters in Electronics*, Wiley Interscience, New York, by permission of Wiley, © 1983 Wiley

For a circular rod of radius a it is

$$I = \frac{\pi a^4}{4} \tag{5.14}$$

The frequency equation for this bar, when its ends are both free, or both clamped, can be obtained as

$$\cos kl = \frac{1}{\cosh kl} \tag{5.15}$$

The roots of this are: $k_{11} = 4.73$, $k_{21} = 7.853$, $k_{31} = 10.996$, $k_{41} = 14.137, \ldots$.

The dispersion relationship between propagation constant and frequency is:

$$k^4 = \frac{\rho A}{EI}\omega^2 \tag{5.16}$$

This results in the resonant frequency for this mode as:

$$f_n = \frac{(k_n l)^2}{2\pi l^2}\left(\frac{EI}{\rho A}\right)^{1/2} \tag{5.17}$$

The equivalent mass is

$$M_{\text{eq}x} = \rho Al\left\{4\left[\frac{u_n(x)}{u_n(0)}\right]\right\}^{-1} \tag{5.18}$$

5.2.1.4 Flexural disk resonators

A schematic showing the flexural mode in a circular disk resonator is shown in Figure 5.6. The wave equation in rectangular coordinates is

$$\frac{\partial^4 u}{\partial x^4} + \frac{\partial^4 u}{\partial y^4} + 2\frac{\partial^4 u}{\partial x^2 \partial y^2} = \frac{1}{c^2}\frac{\partial^2 u}{\partial t^2} \tag{5.19}$$

It is much easier to solve for this equation in the cylindrical coordinate system as:

$$u(r, \theta) = [AJ_n(kr) + BI_n(kr)]\cos n\theta \tag{5.20}$$

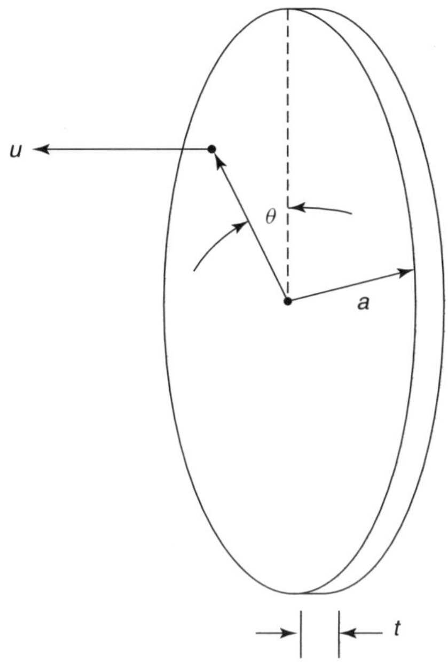

Figure 5.6 Flexure modes on a disk. Reproduced from R.A. Johnson, 1983, *Mechanical Filters in Electronics*, Wiley Interscience, New York, by permission of Wiley, © 1983 Wiley

The frequency equation is obtained as

$$2(\mu - 1) + \left[ka\frac{J_0(ka)}{J_1(ka)} + \frac{I_0(ka)}{I_1(ka)}\right] = 0 \tag{5.21}$$

The dispersion relation for this case is

$$k^2 = \frac{\omega}{c} \tag{5.22}$$

The resonant frequency is then obtained as

$$f_s = (ka)_s^2 \frac{t}{\pi d^2}\left[\frac{E}{3\rho(1-\mu^2)}\right]^{1/2} \tag{5.23}$$

the index s corresponding to nodal circular modes of flexural vibration. The displacement is

$$u(r)_s = A\left[I_0(kr) - \frac{I_1(kr)_s}{J_0(kr)_s} J_0\left(\frac{r}{a}(ka)_s\right)\right] \tag{5.24}$$

The equivalent mass for $s = 1$ is given by

$$M_{\text{eq}}\Big|_{\substack{r \\ s=1}} = 0.247 M_{\text{static}}\left[\frac{u(r)}{u(0)}\right]^{-2} \tag{5.25}$$

5.2.1.5 Thick disks and plates

The differential equation representing the vibration of a thick circular plate is given by

$$\frac{\partial^4 u}{\partial x^4} + \frac{\partial^4 u}{\partial y^4} + 2\frac{\partial^4 u}{\partial x^2 \partial y^2} = \frac{\rho h}{D}\frac{\partial^2 u}{\partial t^2} \tag{5.26}$$

where ρ is the density of the material, h is the thickness of the plate and D its diameter.
The dispersion relationship in this case can be obtained as

$$k^4 = \frac{\rho h}{D}\omega^2 \tag{5.27}$$

The resonant frequencies are given by

$$f_n = \frac{1}{2\pi}\left(\frac{D}{\rho h}\right)^{1/2}\frac{\alpha_n^2}{a^2} \tag{5.28}$$

where α_n are the roots of the characteristic equation ($\alpha_n = k_n a$):

$$J_0(ka)I'_0(ka) - I_0(ka)J'_0(ka) = 0 \tag{5.29}$$

The displacement is given by

$$u = J_0\left(\frac{\alpha_n r}{a}\right)\left[\frac{1}{J_0(\alpha_n)}\right]^{-1} - I_0\left(\frac{\alpha_n r}{a}\right)\left[\frac{1}{I_0(\alpha_n)}\right] \tag{5.30}$$

5.2.1.6 Circular and rectangular membranes

For a circular membrane of radius a the two-dimensional propagation equation in the polar coordinates, with origin at the center of the membrane, is:

$$\frac{\partial^2 u}{\partial r^2} + \frac{1}{r}\frac{\partial u}{\partial r} + \frac{1}{r^2}\frac{\partial u}{\partial \phi} - \frac{\rho''}{F'}\frac{\partial^2 u}{\partial t^2} = 0 \tag{5.31}$$

where ρ'' is mass per unit area of the membrane, and F' is the tension at its edges. The dispersion relation in this case is

$$k = \omega\left(\frac{\rho''}{F'}\right)^{1/2} \tag{5.32}$$

We can obtain the solution in the spatial domain in the form:

$$u(r, \phi) = A\cos m\phi\, J_m\left(\frac{\alpha_{m,n} r}{a}\right) \tag{5.33}$$

where A is an arbitrary constant and $\alpha_{m,n}$ are the values of α such that $J_m(\alpha a) = 0$, meeting the boundary condition. Correspondingly, the resonant frequencies are given by:

$$f_{m,n} = \frac{\alpha_{m,n}}{2\pi a}\left(\frac{F'}{\rho''}\right)^{1/2} \tag{5.34}$$

For the first five resonant frequencies, the mode coefficients in the above equation are $\alpha_{01} = 2.406$, $\alpha_{02} = 5.52$, $\alpha_{03} = 8.654$, $\alpha_{04} = 11.792$ and $\alpha_{05} = 14.931$.

The differential equations differ in their form, when rectangular membranes are considered. In Cartesian coordinates, the solution can be obtained as

$$u(x, y) = A \sin\frac{m\pi x}{a} \sin\frac{n\pi y}{b} \tag{5.35}$$

where a and b are the dimensions of the membrane in the x and y directions, A is a constant and m and n are positive integers (both are not simultaneously zero, but any one of them can be).

The resonant frequencies are given by:

$$f_{m,n} = \frac{c}{2}\left[\left(\frac{m}{a}\right)^2 + \left(\frac{n}{b}\right)^2\right]^{1/2} \tag{5.36}$$

5.2.2 Mechanical coupling components

Mechanical resonance characteristics of coupling components such as bar, string and beam have been studied in the context of conventional mechanical filters. Although their micro size counterparts may not behave identically as the larger ones, an analysis of these would give an insight into the performance of these systems. For these components, we first endeavor to develop an equivalent circuit model, treating them as an ideal transmission line. As mentioned earlier, obtaining an electrical equivalent circuit model of these simplifies design of the overall filter. These equivalent circuits are developed based on electromechanical mobility analogies. Their wave propagation characteristics for specified boundary conditions are used to obtain their resonance characteristics.

5.2.2.1 Electrical transmission lines

To enable an easy understanding of the equivalent circuit of these components we briefly mention the equivalent circuit of a two-conductor low-loss electrical transmission line. The values of these distributed components shown in Figure 5.7 are specified per unit length of the line. The voltage and current equations of the transmission line based on this model are derived in many textbooks dealing with network synthesis and basic electromagnetics. However, an understanding of the physical features of the line that contribute to the values of these components in the equivalent circuit model would be beneficial.

Permeability of the metal used and the inductance contribution of having different phases for currents in different depths of the line are the primary reasons for the inclusion of the inductance term L. Geometry of the line and the permittivity of the medium between

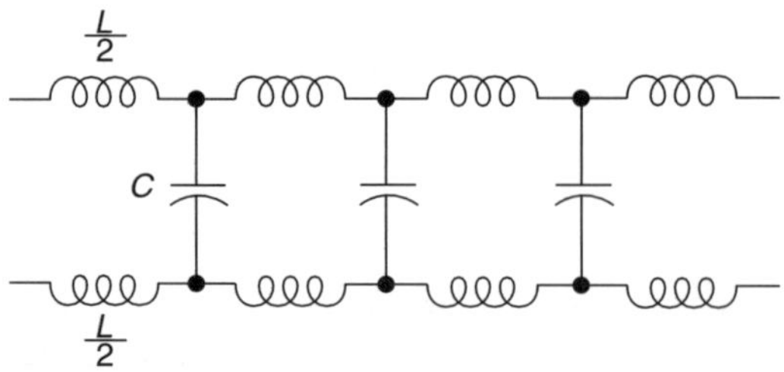

Figure 5.7 Equivalent circuit for a transmission line

the conductors contribute to the capacitance term C. The dielectric and conductance losses of the latter can be account for by adding a conductance term G. Similarly, the conductivity of the metal, geometrical features such as length and cross-sectional area, radiation losses and the skin-depth effect contribute to the resistance term in the equivalent circuit. It is now apparent that if the line is assumed lossless, the resistance and conductance terms vanish, simplifying the analysis significantly.

The governing differential equations for this model are

$$\frac{\mathrm{d}I}{\mathrm{d}z} = -(G + j\omega C)V \tag{5.37}$$

$$\frac{\mathrm{d}V}{\mathrm{d}z} = -(R + j\omega L)I \tag{5.38}$$

Differentiating Equation (5.38) and substituting Equation (3.37), gives

$$\frac{\mathrm{d}^2 V}{\mathrm{d}z^2} = \gamma^2 V \tag{5.39}$$

Similarly, we can also get

$$\frac{\mathrm{d}^2 I}{\mathrm{d}z^2} = \gamma^2 I \tag{5.40}$$

In Equations (5.39) and (5.40), the complex propagation constant is

$$\gamma = [(R + j\omega L)(G + j\omega C)]^{1/2} \tag{5.41}$$

This complex number can be rewritten as:

$$\gamma = \alpha + j\beta \tag{5.42}$$

where α is the attenuation constant and β is the propagation constant in the medium.

The solutions of these homogeneous differential equations can be obtained in the form

$$V = V_1 e^{-\gamma z} + V_2 e^{+\gamma z} \tag{5.43}$$

$$I = I_1 e^{-\gamma z} + I_2 e^{+\gamma z} \tag{5.44}$$

The characteristic impedance of this line is

$$Z_0 = \left(\frac{R + j\omega L}{G + j\omega C}\right)^{1/2} \tag{5.45}$$

If one were to assume an ideal transmission line, the terms contributed by the losses disappear. The propagation constant becomes

$$\beta = \omega(LC)^{1/2} \tag{5.46}$$

The characteristic impedance of such a transmission line is

$$Z_0 = \left(\frac{L}{C}\right)^{1/2} \tag{5.47}$$

The phase velocity for the waves propagating in this line is

$$v = (LC)^{-1/2} \tag{5.48}$$

It is also of interest when the transmission line is of finite length. A short-circuited quarter wavelength behaves as a parallel resonant circuit. To analyze this, the input impedance of a transmission line is defined first as the ratio of Equation (5.43) to Equation (5.44). If this line is terminated in a short circuit, the input impedance Z_{in} is written as:

$$\begin{aligned} Z_{in} &= Z_0 \tanh \gamma l \\ &= Z_0 \frac{\sinh \alpha l \cos \beta l + j \cosh \alpha l \sin \beta l}{\cosh \alpha l \cos \beta l + j \sinh \alpha l \sin \beta l} \end{aligned} \tag{5.49}$$

Applying boundary conditions, the resonant frequency is such that

$$\beta l = \frac{n\pi}{2}, \quad n \text{ is a odd integer} \tag{5.50}$$

The corresponding resonant frequency is

$$f_0 = \frac{nv}{4l} \tag{5.51}$$

where v is the velocity of electromagnetic waves in the medium between the conductors of the transmission line. Using the resonant condition in Equation (5.50), Equation (5.49) reduces to:

$$Z_{in} = Z_0 \frac{\cosh \alpha l}{\sinh \alpha l} = \frac{Z_0}{\tan \alpha l} \approx \frac{Z_0}{\alpha l} \tag{5.52}$$

At this point it is also possible to derive expressions for the Q factor of such a resonating segment. At frequencies close to the resonant frequency f_0,

$$\beta l = \frac{2\pi f}{v} l = \frac{2\pi(f_0 + \delta f)}{v} l = \frac{n\pi}{2} + \frac{2\pi \delta f l}{v} \tag{5.53}$$

Substituting these into Equation (5.49), after simple trigonometric transformations, the input impedance becomes:

$$Z_{\text{in}} = Z_0 \frac{-\sinh \alpha l \sin(2\pi \delta f l/v) + j \cosh \alpha l \cos(2\pi \delta f l/v)}{-\cosh \alpha l \sin(2\pi \delta f l/v) + j \sinh \alpha l \cos(2\pi \delta f l/v)} \tag{5.54}$$

Substituting simplifications for small arguments of these trigonometric functions:

$$Z_{\text{in}} = Z_0 \left(\alpha l + j \frac{2\pi \delta f l}{v} \right)^{-1} \tag{5.55}$$

Comparing this with Equation (5.52), it is clear that if the imaginary term in the denominator of Equation (5.55) is made equal to its real term, the input impedance is half that at the resonant frequency. The corresponding frequency deviation is:

$$\delta f = \frac{\alpha v}{2\pi} = \frac{\alpha f_0}{\beta} \tag{5.56}$$

The corresponding Q is therefore:

$$Q = \frac{f_0}{2\delta f} = \frac{\beta}{2\alpha} \tag{5.57}$$

In the following some of the systems used in mechanical filters are discussed. Parallels will be drawn to the above discussion as and when such analogies arise.

5.2.2.2 Assumptions and theorems for mechanical modeling

For a simple straight forward analysis the present discussion is restricted to homogeneous, isotropic, continuous, elastic, lossless solids. Even for micro size systems, these assumptions are valid if the grain size of the crystalline materials is much smaller than the wavelength. It is also assumed that disturbances that travel along these solids are continuous motions around their rest positions with a relatively small magnitude of variation. The elasticity law defines the normal stress σ_x due to deformation in the direction of propagation x as

$$\sigma_x = E_1 \varepsilon_x \tag{5.58}$$

where E_1 is the longitudinal modulus of elasticity and ε_x is the fractional variation in thickness (strain). The longitudinal modulus of elasticity is

$$E_1 = E \frac{(1-\mu)}{(1+\mu)(1-2\mu)} \tag{5.59}$$

where E is the modulus of elasticity of the material and μ is the Poisson ratio.

The deformations in the transverse directions on a rectangular element make it a parallelogram:

$$\tau_{xy} = \tau_{yx} = G\gamma_{xy} \tag{5.60}$$

where τ_{xy} and τ_{yx} are the tangential stresses on the element, γ_{xy} is the shear angle and G is the shear modulus. For long thin bars, Hooke's and Poisson's laws are also relevant:

$$F = SE\frac{\delta l}{l} \tag{5.61}$$

$$\frac{\delta a}{a} = -\mu\frac{\delta l}{l} \tag{5.62}$$

where F is the force applied, S is the cross-sectional area, $\delta l/l$ is the strain and $\delta a/a$ is the relative variation in the lateral dimension.

5.2.2.3 Longitudinal mode in a solid bar

Consider a long thin solid bar of length l and a uniform cross-sectional area S placed along the x-axis. A small longitudinal deformation traveling in the x direction causes a force $F(x)$ the cross-section at x. The resultant displacement at x is $\xi(x)$. For a section of length $\mathrm{d}x$ on the bar, Newton's law can be applied to equate the force to mass of the section and the acceleration

$$-F(x+\mathrm{d}x) + F(x) = \rho S\mathrm{d}x\frac{\mathrm{d}^2\xi(x)}{\mathrm{d}t^2} \tag{5.63}$$

After rearranging terms, this becomes:

$$\frac{\partial F}{\partial x} = \rho S\frac{\mathrm{d}v}{\mathrm{d}t} \tag{5.64}$$

Hooke's law in Equation (5.61) is used to express the force at a location in terms of displacement as

$$F(x) = SE\frac{\xi(x+\mathrm{d}x) - \xi(x)}{\mathrm{d}x} = SE\frac{\partial\xi(x)}{\partial x} \tag{5.65}$$

Taking the derivative with respect to time and rearranging terms:

$$\frac{\partial v}{\partial x} = \frac{1}{SE}\frac{\mathrm{d}F}{\mathrm{d}t} \tag{5.66}$$

Assuming sinusoidal variations for these disturbances, phasor notation can be employed to remove the time dependencies of Equations (5.64) and (5.66). These then become:

$$\frac{\mathrm{d}\tilde{F}}{\mathrm{d}x} = -j\omega\rho S\tilde{v} \tag{5.67}$$

$$\frac{\mathrm{d}\tilde{v}}{\mathrm{d}x} = -\frac{j\omega}{SE}\tilde{F} \tag{5.68}$$

Comparing Equations (5.67) and (5.68) with the characteristic equations of lossless transmission lines in Equations (5.37) and (5.38), respectively (after making $R = G = 0$), the similarities are striking. Now, using electromechanical mobility analogies, an equivalent circuit representation can be obtained for the transmission line equivalent circuit of the bar.

It would also be of interest to obtain the propagation constant and the velocity of propagation on this transmission line equivalent of the bar. Comparing with Equation (5.48), the velocity of waves in the bar c_b is

$$c_b = \left(\frac{E}{\rho}\right)^{1/2} \tag{5.69}$$

The propagation constant is

$$\beta = \omega\left(\frac{\rho}{E}\right)^{1/2} \tag{5.70}$$

5.2.2.4 Stretched-string transmission line

A stretched string when disturbed about its rest position gets into vibrations with transverse standing waves. In order to simplify the analysis, we consider an ideal flexible string with a constant mass per unit length. We also assume that the disturbance is small and is applied in a direction transverse to its length. At a point on the string, the transverse component of stress shown in Figure 5.8 is

$$F_x = -T\sin\alpha \approx -T\tan\alpha = -T\frac{\mathrm{d}x}{\mathrm{d}z} \tag{5.71}$$

where α is the angle between the length of the wire and the tangent to the displacement in the string. Taking the derivative with respect to time,

$$\frac{\mathrm{d}v_x}{\mathrm{d}z} = -\frac{1}{T}\frac{\mathrm{d}F_x}{\mathrm{d}t} \tag{5.72}$$

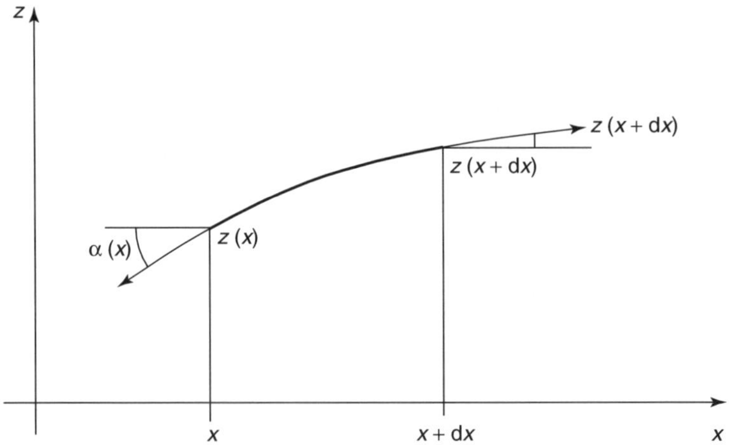

Figure 5.8 Distribution of stress on a stretched string. Reproduced from M. Rossi, 1988, *Acoustics and Electroacoustics*, Artech House, Norwood, MA, by permission of Artech House, © 1988 Artech House

where v_x is the transverse component of velocity. Now, considering a small element of the string of length dx, and making use of Newton's law, components of forces in the transverse direction can be written as:

$$F_x(z) - F_x(z + \mathrm{d}z) = -\frac{\mathrm{d}F_x(z)}{\mathrm{d}z}\mathrm{d}z = \rho'\mathrm{d}z\frac{\mathrm{d}v_x}{\mathrm{d}t} \tag{5.73}$$

where ρ' is the linear mass density of the string. This reduces to

$$\frac{\mathrm{d}F_x(z)}{\mathrm{d}z} = -\rho'\frac{\mathrm{d}v_x}{\mathrm{d}t} \tag{5.74}$$

Using phasors for time dependencies, Equations (5.72) and (5.74) become:

$$\frac{\mathrm{d}\tilde{F}_x}{\mathrm{d}z} = -j\omega\rho'\tilde{v}_x \tag{5.75}$$

$$\frac{\mathrm{d}\tilde{v}_x}{\mathrm{d}z} = -j\omega\frac{1}{T}\tilde{F}_x \tag{5.76}$$

The similarity of these expressions to the voltage and current Equations (5.37) and (5.38) of a transmission line is quite clear. Thus we can make use of the electromechanical analogies to obtain the equivalent circuit model for the string, as shown in Figure 5.9. The velocity of the wave propagation in the string is

$$v = \left(\frac{T}{\rho'}\right)^{1/2} \tag{5.77}$$

5.2.3 General considerations for mechanical filters

Mechanical filters consist of a series of resonators coupled together with some form of coupling elements discussed above. These components critically affect the performance of the filter. For example the number of resonators determines the shape factor for the filter response. Operational frequencies of such resonators decide the center frequency of the filter pass band. The compliance of coupling wires as well as the equivalent mass of the resonator determine bandwidth of the filter. With this background we now proceed to smaller sized versions of mechanical filters.

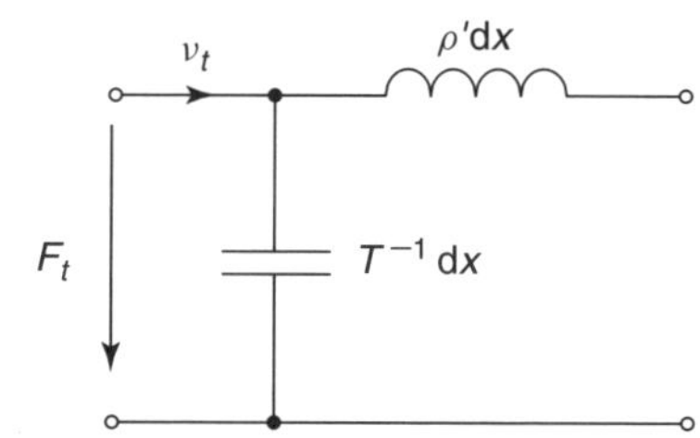

Figure 5.9 Equivalent circuit for a string modeled as a transmission line. Reproduced from M. Rossi, 1988, *Acoustics and Electroacoustics*, Artech House, Norwood, MA, by permission of Artech House, © 1988 Artech House

5.3 MICROMECHANICAL FILTERS

The basic principles described so far for mechanical filters can be used in the design their micro-sized counterparts. However, the accuracy of these formulations is plausible at the micro scale for reasons such as the structural dimensions being not large enough compared with wavelength, nonidealities of boundary conditions and other nonlinear effects. Nonetheless these formulations are definite indicators for understanding the operational principles of these micro devices. Many of the actuation mechanisms discussed earlier in this chapter have not made their way to realizable practical configurations of such micromechanical filters.

The goal is to fabricate devices such as filters so small that they can be integrated into rest of the circuit in a single chip leading to a 'system-on-a-chip'. Conventional filters using crystal oscillators are not amenable to such miniaturization. Hence a larger emphasis has been put on miniaturization of mechanical filters, using standard IC fabrication techniques, so that they can be integrated with other circuits easily.

The performance for micromachined filters is enhanced by using a series of resonator tanks connected together with coupling networks. In general, the number of such tanks used is equal to the (order of filter is the order of its polynomial transfer function). The higher this number the better the frequency selectivity for the filter. But the insertion loss is simultaneously degraded, which can, however, be improved by designing the filter with very large Q factors.

Only electrostatically actuated devices are discussed in this section, although the possibility of using other schemes may not be ruled out. A parallel plate capacitor configuration is common for such large electromechanical filters. However, an additional possibility exists for micro devices, where electrostatic comb drives vibrating in a plane parallel to the substrate can be fabricated. The theory and operation of these devices are presented below.

5.3.1 Electrostatic comb drive

An electrostatically driven parallel plate actuator has a clamped-clamped beam configuration. Although this is amenable for micro fabrication, as derived previously in this chapter, this configuration has nonlinear response characteristics. This nonlinearity can cause frequency instability in the filter operation (Nguyen, 1995). Hence another electrostatically actuated structure is preferred at the microscale. The layout of such a laterally driven electrostatic resonant structure is shown in Figure 5.10. Two resonator configurations are possible with this structure (Tang, Nguyen and Howe, 1989). In the first, a two-port configuration, the structure is driven on one of the comb structures and sensed at the other, for capacitance variations. In the second configuration, both comb structures are used to drive differentially, while sensing is achieved by monitoring shift in impedance at resonance. The folded beam truss suspension has large compliance and is capable of reducing the residual strain in the structural film.

In the two-port configuration, the driving force and the sensitivity of the output are both proportional to the variation of capacitance with lateral displacement $\partial C/\partial x$. The static displacement at the drive port, for an applied drive voltage v_{D}, is (Tang, Nguyen and Howe, 1989):

$$x = \frac{F_x}{k_s} = \frac{1}{2k_s} v_{\mathrm{D}}^2 \frac{\partial C}{\partial x} \tag{5.78}$$

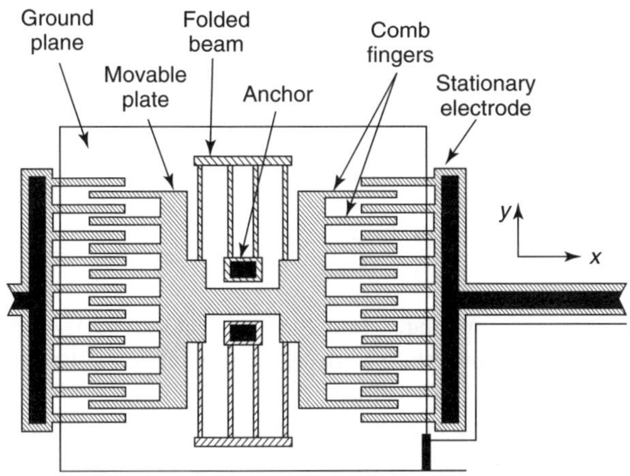

Figure 5.10 Lateral electrostatic comb actuator. Reproduced from W.C. Tang, T.C.H. Nguyen and R.T. Howe, 1989, 'Laterally driven polysilicon resonant microstructures', *Sensors and Actuators* **20**: 25–32, with permission from Elsevier Science, © 1989 Elsevier Science

where F_x is the x component of the electrostatic force and k_s is the spring constant of the system. Assuming the trusses that join the folded beam are rigid, the spring constant is obtained analytically as:

$$k_s = 24\frac{EI}{L^3} = 2Eh\left(\frac{W}{L}\right)^3 \tag{5.79}$$

To ensure stability, the drive voltage consists of an ac voltage of amplitude v_d superimposed on a dc bias V_P such that

$$v_D = V_P + v_d \sin \omega t \tag{5.80}$$

It may be mentioned at this point that realization of a small interelectrode gap is essential to reduce the drive voltage requirements of this actuation mechanism. A fabrication technique using combination of oxidation machining with a suitable post-release positioning has been developed to address this issue (Hirano *et al.*, 1992). Submicron gaps can be achieved by this approach.

Substituting Equation (5.80) in Equation (5.78) and taking the time derivative, we get

$$\frac{\partial x}{\partial t} = \frac{1}{2k_s}\frac{\partial C}{\partial x}\frac{\partial V_D^2}{\partial t} = \frac{1}{2k_s}\frac{\partial C}{\partial x}(2V_P v_d \omega \cos \omega t + v_d^2 \omega \sin 2\omega t) \tag{5.81}$$

For ac voltages much smaller than the dc bias, the second harmonic term on the right-hand side can be neglected. At resonance, the magnitude is multiplied by the quality factor, to get the magnitude of the electromechanical transfer function which relates the phasor displacement X to the phasor drive voltage V_d:

$$\left|\frac{X}{V_d}\right| = \frac{V_P \omega Q}{k_s}\frac{\partial C}{\partial x} \tag{5.82}$$

This shows that since $\partial C/\partial x$ is independent of displacement x, the comb drive has a linear electromechanical transfer function between the displacement and the drive voltage. It may, however, be recalled that this analysis assumes that the amplitude of the ac component of the drive voltage is smaller than the dc bias.

The quality factor for this structure is estimated to be

$$Q = \frac{d}{\mu A_{\mathrm{p}}}(M_{\mathrm{b}}k_{\mathrm{s}})^{1/2} \tag{5.83}$$

where d is the gap between the plates and the substrate, μ is the absolute viscosity of air, A_{p} is the surface area of the plate and M_{b} is the mass of the support beam. In the derivation of Equation (5.83), it is assumed that the primary loss mechanism is Couette flow with a linear velocity profile (Tang, Nguyen and Howe, 1989). This can be improved further by incorporating the Stokes flow of the fluid above, and the damping between the comb fingers (Zhang and Tang, 1994). In practice, however, the value of Q is often controlled by attaching series resistors in the input and output circuits (Nguyen, 1995).

The sensed current i_{s} at the output port is

$$i_{\mathrm{s}} = V_{\mathrm{s}}\frac{\partial C}{\partial x}\frac{\partial x}{\partial t} \tag{5.84}$$

where V_{s} is the bias voltage between the structure and the sense electrode. Substituting Equation (5.81) in Equation (5.84), the magnitude of the transconductance of the resonant structure is given by:

$$\left|\frac{I_{\mathrm{s}}}{V_{\mathrm{d}}}\right| = \frac{V_{\mathrm{P}}V_{\mathrm{s}}\omega Q}{k_{\mathrm{s}}}\left(\frac{\partial C}{\partial x}\right)^2 \tag{5.85}$$

The resonant frequency of the structure is determined by Rayleigh method as:

$$f_{\mathrm{r}} = \frac{1}{2\pi}\left(\frac{k_{\mathrm{s}}}{M_{\mathrm{p}} + 0.3714M_{\mathrm{b}}}\right)^{1/2} = \frac{1}{2\pi}\left[2Eh\left(\frac{W}{L}\right)^3\frac{1}{M_{\mathrm{p}} + 0.3714M_{\mathrm{b}}}\right]^{1/2} \tag{5.86}$$

Their fabrication uses a single mask for most of the critical features; this eases the process design and can potentially reduce cost.

The parasitic capacitive coupling between the input and output ports is minimized by including a grounded planar electrode, which also helps suppress excitation of undesired modes.

5.3.2 Micromechanical filters using comb drives

A number of resonant structures can be coupled together in either series or parallel configuration to obtain high-quality filter characteristics. Schematics for these configurations are shown in Figure 5.11 (Lin *et al.*, 1992). In the series filter, a square truss coupling spring connects the two resonators. In the parallel configuration of band pass filter, the input and output terminals of the resonators are connected in parallel such that the output

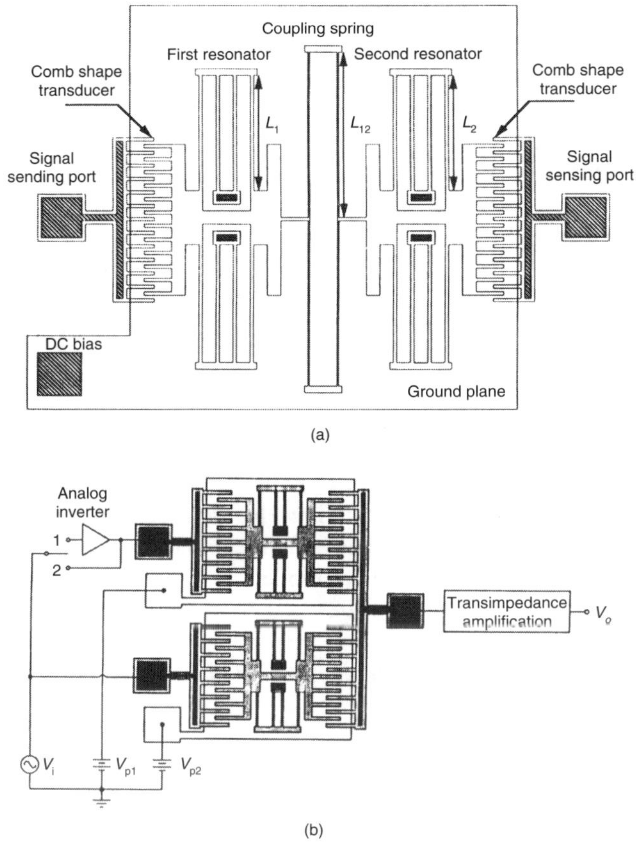

Figure 5.11 (a) Series and (b) parallel combination of resonators. Reproduced from L. Lin, C.T.-C. Nguyen, R.T. Howe, and A.P. Pisano, 1992, 'Micro electromechanical filters for signal processing', in *IEEE Conference on Micro Electro Mechanical Systems '92, February 4–7 1992*, IEEE, Washington, DC, by permission of IEEE, © 1992 IEEE

currents are added up. A notch filter can also be realized in a similar way, by adding the currents in opposite phases.

In this analysis, mass of the coupling beam is assumed negligible. The bandwidth of the filter depends on the ratio of stiffness of the coupling beam (k_{sij}) to that of the resonator beam (k_r).

$$\text{bandwidth} = \frac{f_L}{k_{nij}} \frac{k_{sij}}{k_r} \tag{5.87}$$

where f_L is the filter center frequency and k_{nij} is the normalized coupling coefficient used in filter design. These resonators are designed to have slightly different resonant frequencies such that their difference is related to the Q factor:

$$f_2 - f_1 = \frac{f_1}{Q_1} \tag{5.88}$$

This ensures flat and symmetrical band pass characteristics, provided the individual resonators have identical 3-dB bandwidths and resonance amplitudes. In other words, the difference in frequencies is equal to the 3-dB bandwidth of the filter. To obtain a steep roll-off and flat pass band characteristics, a large number of resonators should be connected in parallel. In terms of the highest and lowest resonant frequencies, f_{u} and f_{L}, respectively, the number of resonators is obtained as

$$N = Q\frac{f_{\mathrm{u}} - f_{\mathrm{L}}}{f_{\mathrm{L}}} \tag{5.89}$$

For such a filter with n resonators coupled in series, the overall transfer function of Equation (5.85) is modified to

$$\frac{I_{\mathrm{out}}}{V_{\mathrm{in}}} = j\omega V_{\mathrm{P,in}} V_{\mathrm{P,out}} \left(\frac{\partial C}{\partial x}\right)_{\mathrm{in}} \left(\frac{\partial C}{\partial x}\right)_{\mathrm{out}} [C_{2n}(j\omega)^{2n} + C_{2n-1}(j\omega)^{2n-1} + \cdots + C_0]^{-1} \tag{5.90}$$

A mechanical model and the corresponding electrical equivalent circuit for the filter configuration are shown in Figure 5.12. The equivalent mass, spring constant and damping constant for ith resonator in the mechanical model are expressed in terms of the physical parameters as (Lin, Howe and Pisano, 1998)

$$M_i = M_{\mathrm{p}i} + 0.3714 M_{\mathrm{b}i} \tag{5.91}$$

$$k_i = 2E_{\mathrm{P}}h\left(\frac{w_i}{L_i}\right)^3 \tag{5.92}$$

$$D_i = \frac{(M_i k_i)^{1/2}}{Q} \tag{5.93}$$

where $M_{\mathrm{p}i}$ is the mass of the plate and $M_{\mathrm{b}i}$ is that of the folded beams at the ith resonator; w_i and L_i are, respectively, the width and length of the folded suspension in the ith resonator; h is the thickness of the polysilicon structures and E_{p} is its Young's modulus. The stiffness of the coupling spring is similarly obtained as:

$$k_{ij} = E_{\mathrm{P}}h\left(\frac{w_{ij}}{L_{ij}}\right)^3 \tag{5.94}$$

Making use of the electromechanical mobility analogy described in Chapter 1, the parameters in the electrical equivalent circuits are related to the above mechanical parameters by (Lin, Howe and Pisano, 1998):

$$L_i = \eta^2 M_i \tag{5.95}$$

$$C_i = \frac{1}{\eta^2 k_i} \tag{5.96}$$

$$R_i = \eta^2 D_i \tag{5.97}$$

$$C_{ij} = \frac{1}{\eta^2 k_{ij}} \tag{5.98}$$

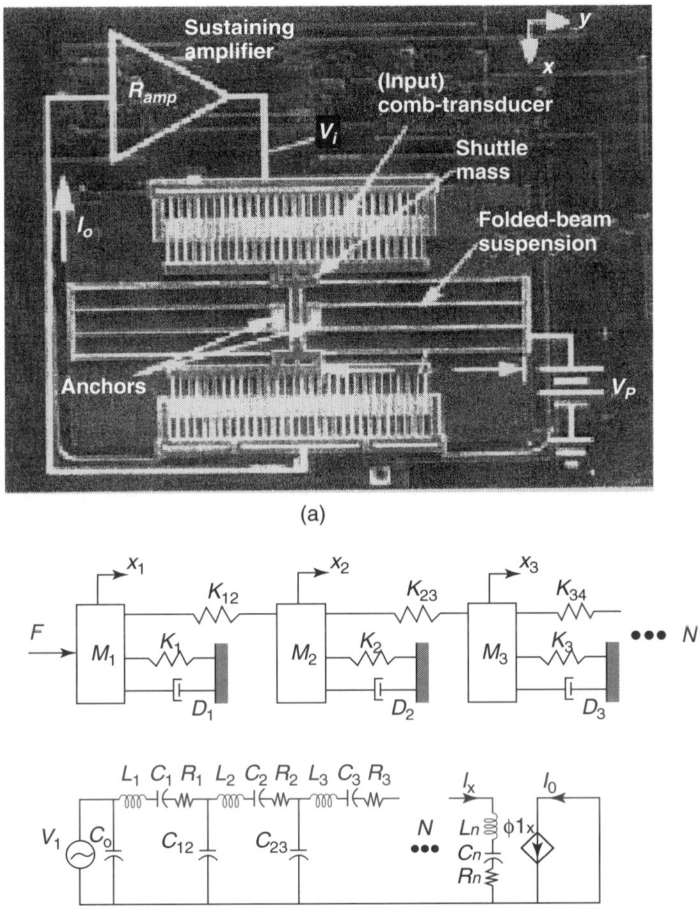

Figure 5.12 (a) Photograph and (b) equivalent circuits of micromechanical filters. Reproduced from K. Wang and C.T.-C. Nguyen, 1997, 'High-order Micromechanical electronic filters', in *Proceedings of 1997 International Microelectromechanical Systems Workshop*, IEEE, Washington, DC: 25–30, by permission of IEEE, © 1997 IEEE

where the transformation parameter η for the filter is defined as

$$\eta = \left[V_{P_i} \left(\frac{\partial C}{\partial x} \right)_i \right]^{-1} \tag{5.99}$$

The amplification factor in the electrical equivalent circuit of Figure 5.12 is theoretically obtained to be:

$$\phi = V_{P,out} \left(\frac{\partial C}{\partial x} \right)_{out} \left(\frac{\partial C}{\partial x} \right)_{in}^{-1} \tag{5.100}$$

The above model assumes the coupling beam is massless. However, it is possible to isolate the effect of its mass on the filter properties by making the beam a quarter wavelength long. This requires that its length and width be chosen such that (Wang and Nguyen, 1997):

$$\sin\alpha \sinh\alpha + \cos\alpha \cosh\alpha = 0 \tag{5.101}$$

$$k_{sij} = \frac{EI\alpha^3}{L_{ij}^3} \frac{\sin\alpha + \sinh\alpha}{\cos\alpha \cosh\alpha - 1} \tag{5.102}$$

where

$$\alpha = \left(\frac{\rho A \omega^2}{EI}\right)^{1/4} \tag{5.103}$$

$$I = \frac{W_{ij} h^3}{12} \tag{5.104}$$

and A is the cross-sectional area of the beam.

It has been found that as the size of the resonator gets smaller at higher frequencies, the above closed-form expressions for its resonant frequency and parameters in the model tend to be inaccurate. In such cases, the distributed matrix technique is used to predict the resonant frequency (Wang and Nguyen, 1997). The effective lumped mass and spring constant are then given as:

$$M_r = \frac{2KE_{\text{total}}}{v_c^2} \tag{5.105}$$

$$k_r = \omega_0^2 m_r \tag{5.106}$$

The total kinetic energy KE_{total} of the system is given by

$$KE_{\text{total}} = \tfrac{1}{2}\omega_0^2 X_0^2 M_0 \tag{5.107}$$

with X_0 the shuttle displacement amplitude and M_0 the effective equivalent mass seen at any point on the resonator:

$$M_0 = M_p + \frac{1}{4}M_t + \frac{12}{35}M_b \tag{5.108}$$

where the subscripts p, t and b correspond to shuttle plate, folding truss and folded beams, respectively. The parameters in the electrical equivalent circuit are obtained from Equations (5.95)–(5.98) by substituting Equations (5.105) and (5.106) for the mass and spring constant. The resonant frequency by this approach is (Nguyen, 1995):

$$f_r = \frac{1}{2\pi}\left[2Eh\left(\frac{W}{L}\right)^3\left(M_p + \frac{1}{4}M_t + \frac{12}{35}M_b\right)^{-1}\right]^{1/2} \tag{5.109}$$

The above equivalent circuit is deemed applicable in most cases. However, it does not take into consideration parasitics due to the presence of substrate. This can be overcome to a great extent by CMOS integration, which may also improve the frequency and phase stability of the system (Nguyen and Howe, 1993).

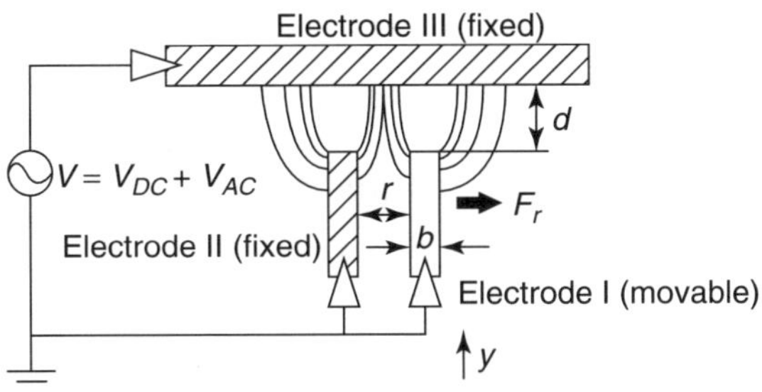

Figure 5.13 Lateral repulsive force. Reproduced from K.B. Lee and Y.-H. Cho, 2001, 'Laterally driven electrostatic repulsive-force microactuators using asymmetric field distribution', *Journal of Microelectromechanical Systems* **10**: 128–136, by permission of IEEE, © 2001 IEEE

The above equivalent circuit approach to modeling of such systems is useful in understanding the behavior of these systems with regard to various structural, material and geometrical parameters. But to predict accurately their mechanical, electromechanical and electrical performance, advanced modeling and simulation software have to be employed. For example, MEMCAD can be used to simulate complex systems such as this from their mask descriptions (Gilbert *et al.*, 1993).

To avoid the problems associated with stiction and deterioration of electrodes during operation, an electrostatic repulsive actuator has been recently developed (Lee and Cho, 2001). The repulsive force in this is generated by the asymmetry of the in-plane electric field (Figure 5.13), causing the movable electrode to slide in the direction shown. The force generated, resonant frequency and quality factor of this configuration are generally derived using finite elements analysis (Lee and Cho, 2001).

5.3.3 Micromechanical filters using electrostatic coupled beam structures

Lateral drive actuators have a linear transfer function between displacement and voltage and hence have significant advantages on filter performance. However, these are relatively large structures. It may be recalled that the resonant frequency of a simple spring mass system is

$$f = \frac{1}{2\pi}\left(\frac{k}{m}\right)^{1/2} \tag{5.110}$$

This shows that in order to increase the resonant frequency, the structure should have equivalent higher spring constant and/or very low mass. Reducing the mass of the relatively large comb structure may not be feasible. An alternative configuration for a high-frequency filter is based on coupled clamped-clamped beam resonators, a schematic of which is shown in Figure 5.14 (Bannon, Clark and Nguyen, 2000). A similar resonator is also developed with free-free beam configuration (Wang, Wong and Nguyen, 2000). Although inherently nonlinear, their operation can be assumed linear for the small signal case. On the actuator side, a voltage applied between an electrode below the beam and a fixed electrode on the substrate causes the beam to move down, by electrostatic attractive

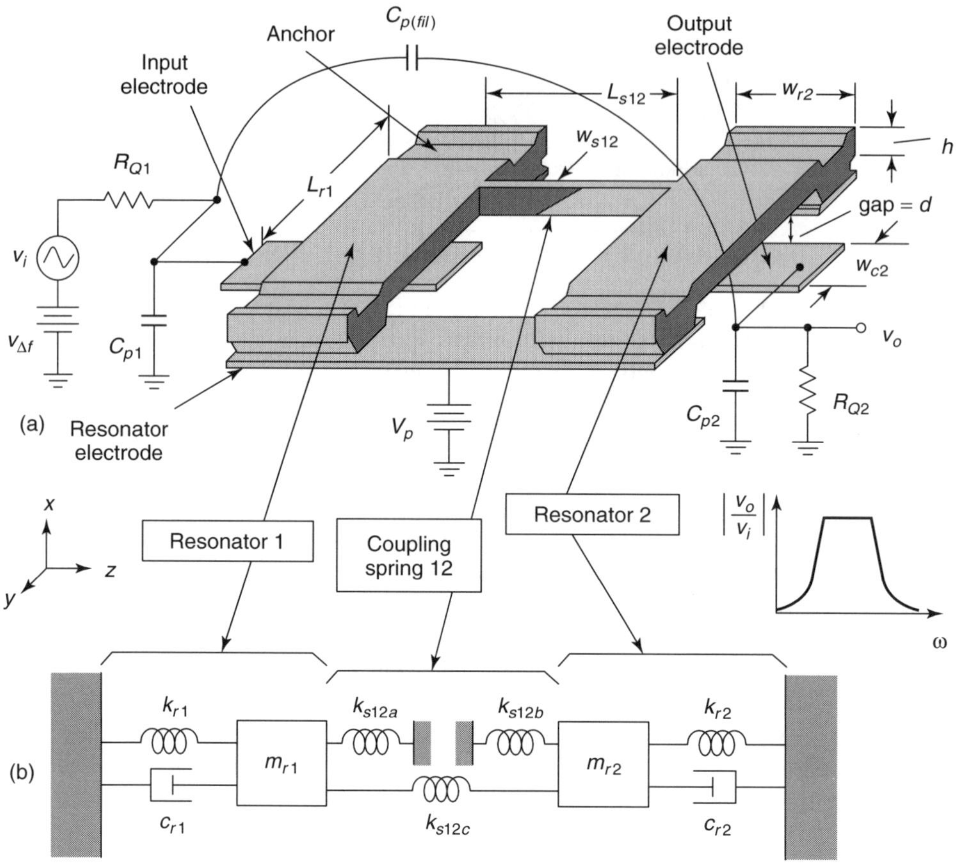

Figure 5.14 Perspective view and equivalent circuit of a resonator with two clamped beams. Reproduced from F.D. Bannon III, J.R. Clark and C.T.-C. Nguyen, 2000, 'High-*Q* HF microelectromechanical filters', *IEEE Journal of Solid-state Circuits* **35**: 512–526, by permission of IEEE, © 2000 IEEE

force. This movement of the beam is coupled towards the next beam, which operates as a capacitive transducer that senses the displacement of the beam.

The dynamic analysis of clamped-clamped beam presented earlier in this chapter is valid for these micro structures as well. The input voltage consists of dc bias V_P and a dynamic ac signal v_d. The resonant frequency can be obtained as (Bannon, Clark and Nguyen, 2000)

$$f_0 = \left[1.03\kappa\left(\frac{E}{\rho}\right)^{1/2}\right][1 - g(d, V_P)]^{1/2} \tag{5.111}$$

where h and L are, respectively, the thickness and length of the resonating beam, the function g models the effect of applied voltage in reducing the effective spring constant of the beam and κ is a scale factor that accommodates surface topography, determined by finite element analysis. It may be recalled that the first term on the right-hand side represents the resonant frequency of the beam without considering the effects of applied bias voltage. This change in stiffness due to an applied voltage is useful for electrically tunable filter characteristics.

When the applied frequency of the ac component of the input matches the resonant frequency of the structure, the beam vibrates such that its displacement at a location is given by:

$$x(y) = \frac{Q}{k_{\text{eff}}(y)} V_{\text{P}} \frac{\partial C}{\partial x} v_{\text{d}} \tag{5.112}$$

At the output side, this displacement in turn, causes a current

$$i_x = (V_{\text{P}} - V_2) \frac{\partial C_2}{\partial x} \frac{\partial x}{\partial t} \tag{5.113}$$

where V_2 is the bias voltage at the output beam and C_2 is the capacitance between the electrodes at this beam and the substrate.

The filter analysis and synthesis is significantly simplified using an electrical equivalent circuit. One such model is obtained through the corresponding mechanical model (with spring mass system), shown in Figure 5.15 (Bannon, Clark and Nguyen, 2000). The parameters used in this model are

$$\eta_{\text{e1}} = \left\{ \int_{L_{\text{e1}}}^{L_{\text{e2}}} \int_{L_{\text{e1}}}^{L_{\text{e2}}} \frac{V_{\text{P}}^2 (\varepsilon_0 W_{\text{r}})^2}{[d(y')d(y)]^2} \frac{k_{\text{re}}}{k_{\text{r}}(y)} \frac{1}{W_{\text{e}}} \frac{X_{\text{mode}}(y)}{X_{\text{mode}}(y')} \, \mathrm{d}y' \, \mathrm{d}y \right\}^{-1} \tag{5.114}$$

This is the turns ratio equivalent at the input side. The value for the output side can be calculated similarly. The index corresponding to the resonator number is omitted in the following formulae for the rest of the parameters:

$$L_x = m_{\text{e}} \tag{5.115}$$

$$C_x = \frac{1}{k_{\text{e}}} \tag{5.116}$$

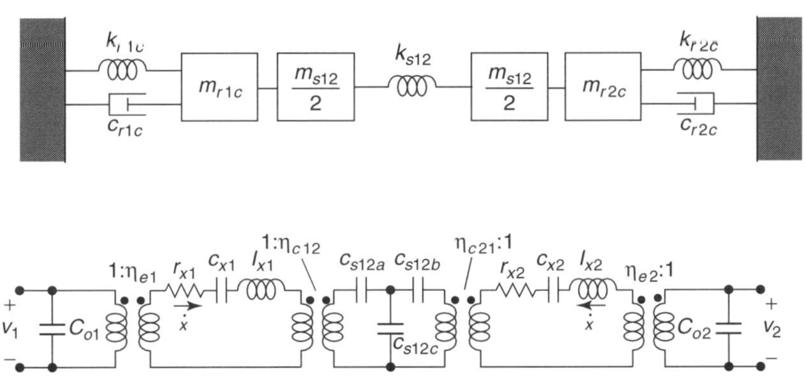

Figure 5.15 Equivalent circuits for the filters in Figure 5.14. Reproduced from F.D. Bannon III, J.R. Clark and C.T.-C. Nguyen, 2000, 'High-Q HF microelectromechanical filters', *IEEE Journal of Solid-state Circuits* **35**: 512–526, by permission of IEEE, © 2000 IEEE

$$R_x = \frac{(k_e m_e)^{1/2}}{Q} \tag{5.117}$$

$$C_0 = \frac{\varepsilon_0 W_r W_e}{d} \tag{5.118}$$

As with lateral drive filters, the maximum attainable quality factor is proportional to the ratio of spring constants of the resonator and the coupling spring. The quality factor of the resulting filter is

$$Q = k_{12} \frac{k_c}{k_{s12}} \tag{5.119}$$

The dynamic spring constant k_c of the beam varies with distance from anchor points. Thus to improve the quality factor of the filter, the coupling beam is not attached at the center of the beams. Instead it is attached at a point closer to the anchor point, where the dynamic spring constant is higher and thus the filter Q factor. As with the previous case, the length of the coupling beam is taken as quarter acoustic wavelength. The coupling in this case is therefore modeled as:

$$\eta_{c12} = \left(\frac{k_{1c}}{k_{1e}} \right)^{1/2} \tag{5.120}$$

$$C_{s12a} = C_{s12b} = \frac{1}{k_{s12a}} \tag{5.121}$$

$$C_{s12c} = \frac{1}{k_{s12c}} \tag{5.122}$$

where the spring constants of the coupling beam are:

$$k_{s12a} = -k_{s12c} = \frac{E I_s (\sin\alpha + \sinh\alpha)}{L_s^3 (\cos\alpha \cosh\alpha - 1)} \tag{5.123}$$

5.4 SURFACE ACOUSTIC WAVE FILTERS

The maximum resonant frequency of mechanical filters and their microsized counterparts discussed so far in this chapter are limited by their minimum feature size. With present-day technology, the maximum operational frequency of these filters is generally limited to tens of megahertz. Compared with the resonant vibrations used in these filters, a different wave motion mechanism, acoustic waves in elastic solids, can be used to extend this limit upwards in frequency. Filters based on surface acoustic waves (SAWs) have a significant, almost monopolistic market share for applications in ultra-high-frequency (UHF) and VHF bands. These filters have several features common with their lower frequency counterparts at high frequency (HF) RF MEMS filters discussed so far in this chapter, and micromachined microwave filters discussed later in this book. Furthermore, the acoustic wave propagation, a key element in their operation, is defined by the laws of mechanics. Hence it would be prudent to include SAW filters as part of this chapter on micromechanical filters.

These filters require specialized piezoelectric substrates that prevent their integration with the circuits in a single chip. However, the goal of miniaturization of systems can definitely be achieved using them. Moreover, novel crystal cuts of semiconductor substrates that allow piezoelectric wave propagation suitable for SAW filters thus having a potential of further system integration are being investigated.

Various design aspects of SAW filters are presented in this section. First, the basic principle of operation of a simple SAW filter is described to provide a preliminary understanding. The surface wave excitation and propagation mechanism on piezoelectric solids are then compared with other types of acoustic waves to understand their significant features. The design of interdigital transducers (IDTs) used in the generation of SAWs is discussed next. Contributing factors to losses in the basic SAW filter structure and methods to overcome some of these are also presented. Several aspects of SAW devices as an enabling technology and their applications and limitations are also discussed. It may, however, be mentioned that several text books are already available on SAW filters and the treatment in this section by no means may be deemed comprehensive. However, it is expected that this section will enlighten the reader with a balanced perspective of various technologies used in RF filter design.

5.4.1 Basics of surface acoustic wave filter operation

Surface acoustic waves were extensively studied by Lord Rayleigh in the nineteenth century, but filters and other devices based on these were not developed until the late 1960s. A historical account of the technology of SAW devices is available in (Morgan, 1998). We however present some preliminary concepts essential to their understanding.

A schematic of the basic structure of a filter based on the principle of surface acoustic waves is shown in Figure 5.16. This consists of a pair of metallic IDTs patterned on a piezoelectric substrate. These IDTs are reciprocal devices, in that identical devices can be used as input and output transducers. Being anisotropic, the crystal structure and orientation of piezoelectric substrates determine the characteristics of the SAW propagation between IDTs.

Several piezoelectric materials have been used as substrates for SAW filters. Important mechanical and electrical properties of some of these are reproduced in Table 5.1. The

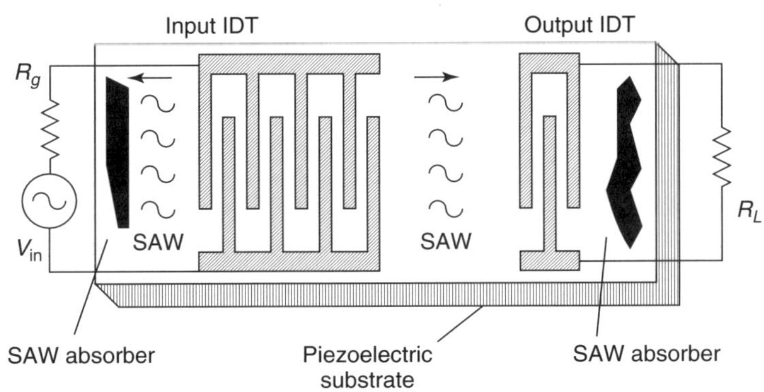

Figure 5.16 Schematic of a surface acoustic wave filter. Reproduced from C. Campbell, 1998, *Surface Acoustic Wave Devices for Mobile and Wireless Communications*, Academic Press, San Diego, CA, by permission of Academic Press, Elsevier

Table 5.1 Properties of some piezoelectric materials used in surface acoustic wave (SAW) filters

Material	Crystal cut	Direction of propagation	Wave velocity	Attenuation at 1 GHz	Coupling factor K_2%	Dielectric constant	Temperature coefficient	Relative bandwidth
Quartz	ST	x	3158	2.6	0.16	4.5	0	4
Lithium niobate	y	z	3488	1.07	4.5	46	94	10
Lithium niobate	128	–	3992	–	–	–	75	–
Lithium tantalite	y	z	3230	1.14	0.9	47	38	23
Gallium arsenide	(1 0 0)	⟨1 1 0⟩	2841	–	0.06	12	35	–
ZnO/AlN/glass	–	–	5840	–	4.3	–	21	–
Langasite ($La_3Ga_5SiO_{14}$)	–	–	2400	–	0.3	–	–	–

choice of the substrate is, however, based on the application and compatibility with the rest of the circuits and fabrication approaches. It may also be noted that material properties depend on crystal cuts and the direction of wave propagation.

When a voltage is applied between the terminals of the input IDT an electric field is created between the adjacent pairs of fingers of this IDT. These fields fringe into the substrate material and cause generation time-varying strains on the substrate, which propagate as acoustic signals. The width and spacing of these fingers are designed in such a way that these acoustic signals generated from each such electrode pair add up in directions normal to the length of these fingers. The IDT shown in Figure 5.16, like several other similar designs, produces acoustic waves propagating in two opposing directions, causing wastage of half the energy after it is converted into acoustic form. Owing to this loss at both the input and the output IDTs, a minimum of 6-dB loss is expected in this basic design of SAW filter. However, this loss is further compounded by the poor electromechanical conversion efficiency and propagation losses on the substrate. Absorbers at the ends of the substrate attenuate acoustic waves, thereby reducing reflections caused at the edges. The acoustic energy reaching the output IDT causes electric signals to be generated at its terminals.

Since the velocity of the acoustic wave in these devices is much smaller (almost five orders of magnitude) than that of electromagnetic waves, the acoustic wavelength in these devices are correspondingly smaller, hence facilitating smaller devices. But as operational frequencies are increased, substrates with higher acoustic velocities are preferred to overcome the fabrication limitations on the smallest feasible feature size. The velocity being a material property is dependent on the type of piezoelectric material used in the design.

5.4.2 Wave propagation in piezoelectric substrates

Piezoelectric substrate forms an important element that influences the performance of the filters. The properties of some such piezoelectric crystals have already been mentioned in Table 5.1. In these materials, an applied mechanical strain produces dielectric polarization, and, conversely, an applied electric field causes mechanical strain. One necessary condition for a crystalline material to be piezoelectric is that the absence of a center of symmetry. Dimensional changes occur only if mechanical force is applied to a symmetric crystal. In piezoelectric crystals, however, the mechanical force causes a shift in the effective centers for the positive and negative charges, forming a dipole moment at the crystal. The relationship between the dipole moment and the mechanical deformation is expressed as constitutive relations:

$$\sigma = cS - eE \tag{5.124}$$

and

$$D = \varepsilon_0 E + eS \tag{5.125}$$

where σ is the mechanical stress, S is the strain, E is electric field, D is flux density, c is the elastic constant, e is the piezoelectric constant and ε_0 is the permittivity of free space. It may be noticed that in the absence of piezoelectricity these relations reduce to Hooke's law and the constitutive relation for dielectric materials, respectively.

As mentioned previously, the electric field on the electrodes of an IDT causes mechanical strains, which travels on the surface substrate as acoustic waves. The displacement has a component parallel to the direction of propagation, and another orthogonal component normal to the surface. These two wave motion components are 90° out of phase in time and need not match in amplitude. Thus the resulting displacement waves are elliptical.

The effectiveness of a piezoelectric material is best expressed in terms of its electromechanical coupling coefficient K^2. By definition this is related to other material parameters used in the above constitutive equations by

$$K^2 = \frac{e^2}{c\varepsilon} \tag{5.126}$$

However, this parameter is often determined experimentally from the relation

$$K^2 = -\frac{2\Delta v}{v} \tag{5.127}$$

where Δv is the reduction in SAW velocity when the surface of the piezoelectric substrate is shorted with a thin electrically conducting film, and v is the unperturbed SAW velocity (Campbell, 1998). The SAW velocity on the substrate depends on its density and elastic and piezoelectric constants.

Surface acoustic waves in their conventional form are called Rayleigh waves. However, there are several other modes of wave propagation on piezoelectric substrates that have been found to be preferable for filter applications. Their use has suggested investigation of newer crystal cuts, and some improvements into the geometry of IDTs.

These newer wave modes include the leaky surface acoustic wave, the shallow bulk wave and the surface transverse wave. Their wave motion is compared schematically in Figure 5.17 with normal SAWs. Leaky SAW has enabled the design of low-loss filters and antenna duplexers for modern wireless transceivers (Campbell, 1998).

5.4.3 Design of interdigital transducers

Design of IDTs is the most critical part in terms of frequency response characteristics. In the basic configuration shown in Figure 5.16, each filter has a pair of IDTs. In its simplest

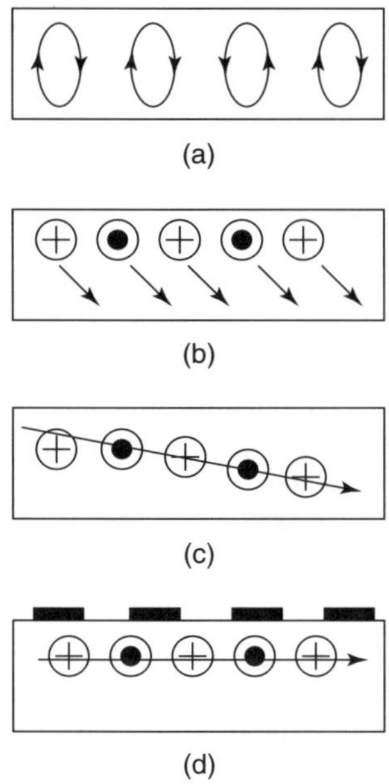

Figure 5.17 Different wave modes: (a) Surface acoustic wave (SAW); (b) leaky SAW; (c) Shallow bulk wave; (d) surface transverse wave. Reproduced from C. Campbell, 1998, *Surface Acoustic Wave Devices for Mobile and Wireless Communications*, Academic Press, San Diego, CA, by permission of Academic Press, Elsevier

form, each IDT consists of an array of metallized finger electrodes, connected by two bus bars.

It may be recalled that the polarity of neighboring fingers are opposite. When a voltage is applied, this causes mechanical strains in opposing directions at adjacent pairs of fingers. These form the crests and dips of the acoustic waveform. The wavelength of acoustic waves on the substrate is given by

$$\lambda = \frac{v}{f} \tag{5.128}$$

The mechanical strains generated at each consecutive pair of fingers are algebraically added in phase if their separation is

$$d = \frac{\lambda}{2} \tag{5.129}$$

The constructive addition of the signals is possible for higher harmonics as well. This fact is used in designing higher-frequency devices with limited fabrication capabilities.

The width of each electrode finger is generally chosen as half the period. Its length determines the acoustic beamwidth and hence is not as significant in this preliminary design. The number of pairs of fingers is, however, critical in choosing the filter bandwidth. The filter design often uses a finite impulse response approach, similar to that used in designing digital filters. The impulse response of the basic IDT is a rectangle. The Fourier transform of a rectangle is a sinc function whose bandwidth in the frequency domain is proportional to the length of the rectangular window in the space domain. As a result, narrow bandwidth requires the IDT to have a large number of fingers.

The transfer function of the filter may be written as

$$H(f) = H_1(f)H_2(f)e^{j\beta d(f)} \tag{5.130}$$

Where H_1 and H_2 are the transfer functions for the input and output IDTs, d is the distance between their effective phase centers and β is the wave propagation constant on the substrate. Near the resonant frequency f_0 the transfer function of an IDT may conveniently be written as:

$$|H_1(f)| \cong N\left|\left\{\sin\left[\frac{N_p\pi(f-f_0)}{f_0}\right]\left[\frac{N_p\pi(f-f_0)}{f_0}\right]^{-1}\right\}\right| \tag{5.131}$$

where N is the number of fingers and $N_p = N/2$ or $(N-1)/2$ for even and odd N, respectively. Being a sinc function in the frequency domain, the sidelobe for this transfer function is approximately 13-dB below the main lobe. To improve the out-of-band rejection characteristics of bandpass filters further, several modifications are attempted to this basic design of IDT. One such approach is to vary the overlap length of fingers, a process known as apodization. The contribution of each finger to the impulse response of the filter is proportional to its length, a fact that draws parallels in digital far infrared (FIR) filter design approaches (Smith, 1995). However, in contrast to digital filters, these operate in real time (except for the finite propagation delay).

The above design approach can only be used as a general guideline. A more comprehensive modeling and analysis is required for improving the filter performance. This rigorous analysis should take into consideration factors such as:

- Electromagnetic cross-talk between IDTs
- Bulk wave modes generated on the substrate
- Internal reflections inside the transducers
- Effects of mass, topography and conductivity of metallization
- Diffraction, attenuation and dispersion in wave propagation
- External impedances such as of the source and load
- Parasitic impedances of metallizations
- Reflections between transducers and triple transit interference.

Several modifications to the simple configuration discussed here have been reported to overcome some of these losses. These modifications have significantly enhanced the performance of SAW filters, making them versatile components in several modern communication systems.

For example, SAW ring filters (Figure 5.18) with insertion loss less than 1 dB are reported in (Dobershtein and Malyukhov, 1997) using reflective multistrip couplers on a

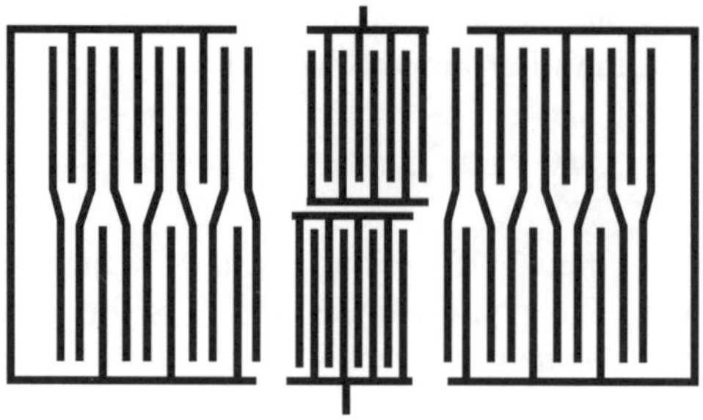

Figure 5.18 Surface acoustic wave ring filter. Reproduced from S.A. Dobershtein and V.A. Malyukhov, 1997, 'SAW ring filters with insertion loss of 1 dB', *IEEE Transactions on Ultrasonics, Ferroelectrics and Frequency Control* **44**: 590–596, by permission of IEEE, © 1997 IEEE

different crystal cut of lithium niobate. Another such modified design uses a single-phase unidirectional transducer (SPUDT), the principles of which are introduced next.

5.4.4 Single-phase unidirectional transducers

The schematic for a simple SPUDT filter is shown in Figure 5.19 (Campbell, 1998). These have acoustic reflectors within the transducer such that the surface waves add constructively in one direction and destructively in the reverse direction. This approach results in a unidirectional transducer, thus eliminating power loss associated with bidirectional IDTs discussed earlier.

Several IDT configurations have been developed for SPUDT filters. The presence of reflector strips within the IDT results in a shift in excitation centers with respect to the reflection centers (Ruppel *et al.*, 1993). The transducer can be designed in such a way that the reflection and regeneration cancel each other out. This can be accomplished over a broad bandwidth. These can also be designed to cancel out triple transit signals with those reflected at the reflector electrodes.

Having lower insertion loss compared with the basic configuration, the performance of a SPUDT filter is rather independent of external matching networks. However, suitable

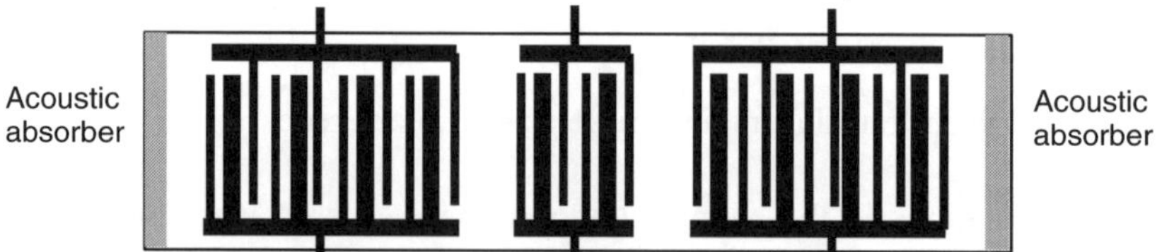

Figure 5.19 Single-phase unidirectional transducer. Reproduced from C. Campbell, 1998, *Surface Acoustic Wave Devices for Mobile and Wireless Communications*, Academic Press, San Diego, CA, by permission of Academic Press

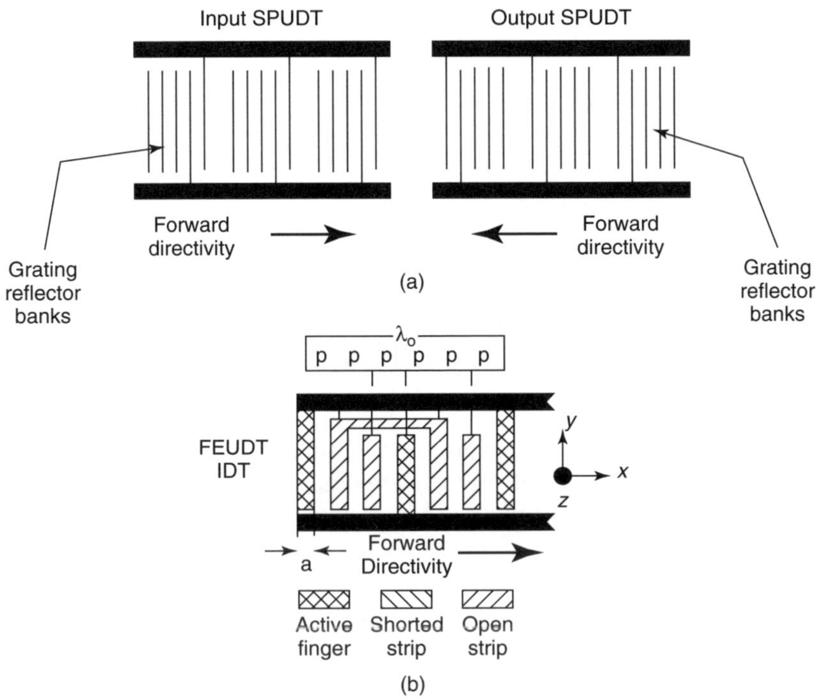

Figure 5.20 Filter configuration with improved characteristics. Reproduced from C. Campbell, 1998, *Surface Acoustic Wave Devices for Mobile and Wireless Communications*, Academic Press, San Diego, CA, p. 181. Note: IDT, interdigital transducers; FEUDT floating electrode unidirectional transducer; SPUDT, single-phase unidirectional transducer

matching networks can reduce the pass band and group delay ripple without affecting the insertion loss.

A third set of metallization, in the form of a grating with reflector, is inserted between the input and output IDTs, along with some modifications to these IDTs themselves, to further improve the filter performance (Figure 5.20). This configuration results in the formation of resonant cavities, which modifies the frequency response of the SPUDT structures resulting in further reduction in insertion loss and size (Gopani, 1998).

5.4.5 Surface acoustic wave devices: capabilities, limitations and applications

Filters based on SAW technology are extensively used in several types of electronic equipment. Advantages of these filters include ruggedness, reliability, linear phase characteristics, small value for the shape factor of frequency response and temperature stability. Semiconductor processing facilities can be used for large-volume production of these devices, a technology that enables repeatability as well as lowering the cost. The maximum and minimum frequency limits of the SAW filter technology is also bounded by the fabrication capabilities: at higher frequencies the wavelength gets very small, the feature size becomes too small to be fabricated reliably, and at lower frequencies the wavelength is so large that the device becomes impractical. Typically, these devices can be used for applications ranging from 10 MHz to 3 GHz.

Proper packaging is essential to maintain the performance of these devices. Acoustic waves in these devices are propagating as surface waves and hence can be perturbed easily by modifications to the substrate surface. This has enabled a large number of resonant sensors for applications as varied as chemical sensors and accelerometers. SAW devices also find application in oscillators, pulse compressors, convolvers, correlators, multiplexers and demultiplexers. In short, SAW devices find widespread use in TV, digital radio, cell phones, satellites, modem, radar, remote control, sensors and ID tags.

5.5 BULK ACOUSTIC WAVE FILTERS

Filters using principles of surface acoustic waves are generally limited in the upper frequency range. These also have higher insertion loss than is acceptable in most instances. An alternate, but similar, approach is to incorporate bulk acoustic waves, using those devices that can be fabricated for higher frequencies.

Similar to SAW-based systems discussed in the previous section, these also use wave propagation though piezoelectric materials. However, in this case, thin films of materials such as lead zircorate titanate (PZT) and ZnO are used. One advantage of using thin-film technology is that a very precise control of thickness is achievable with modern film deposition techniques such as RF sputtering.

A thin-film bulk acoustic wave resonator is shown in Figure 5.21. The thickness of PZT thin film is 0.9 μm. The size of this device is $0.69 \times 0.55\,\text{mm}^2$ and has 47 MHz bandwidth at 1.5 GHz (Misu *et al.*, 1998). Both electrodes of these devices are made free to vibrate to aid resonance. These electrodes also act as reflectors that restrain acoustic waves. The bottom electrode is made free to vibrate by removing the substrate material

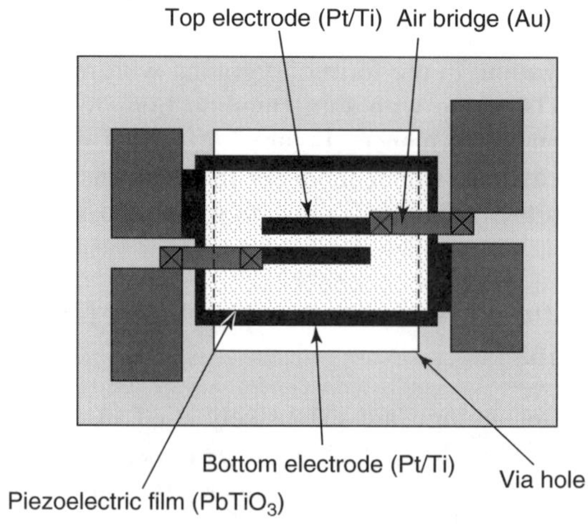

Figure 5.21 Top view of a filter using a thin-film bulk acoustic wave resonator approach. Reproduced from K. Misu, T. Nagatsuka, S. Wadaka, C. Moeda and A. Wadaka, 1998, 'Film bulk acoustic wave filters using lead titanate on silicon substrate; in *IEEE 1998 ultrasonics Symposium*, IEEE, Washington, DC: 1091–1094, by permission of IEEE, © 1998 IEEE

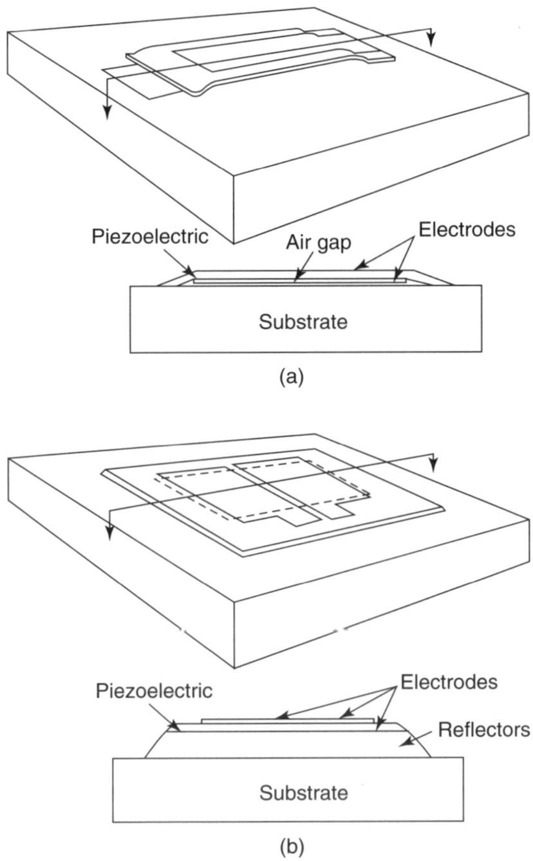

Figure 5.22 (a) Membrane-supported and (b) solidly supported configurations. Reproduced from K.M. Lakin, K.T. McCarron, J. Belsick and J.F. McDonald, 2001, 'Thin film bulk acoustic wave filters technology', in *RAWCON 2001: IEEE Radio and Wireless Conference*, IEEE, Washington, DC: 89–92, by permission of IEEE, © 2001 IEEE

below. These are called membrane-supported structures. In solidly mounted structures, however, a sequence of quarter wavelength sections of material are used to isolate the reflector from the substrate (Lakin *et al.*, 2001). Schematics of these configurations are shown in Figure 5.22.

Individual resonators are arranged in a ladder configuration for filters with better performance. A basic configuration of such filters is shown in Figure 5.23. These sections can be cascaded, and Π or T networks may be used for convenience. The maximum of S_{21} occurs when the series resonance of the series crystal X_a coincides with the parallel resonance of the shunt crystal X_b. The piezoelectric coupling coefficient has a bearing on the bandwidth of the filter (Lakin, Kline and McCarron, 1992). Thickness of piezoelectric layers can be used as the control parameter for the coupling coefficient. ZnO thin films have a better coupling coefficient and hence can improve filter performance (Su *et al.*, 2000). Ladder configurations of filters with this approach have been fabricated for frequencies ranging from 300 MHz to 12 GHz for various applications (Lakin *et al.*, 2001).

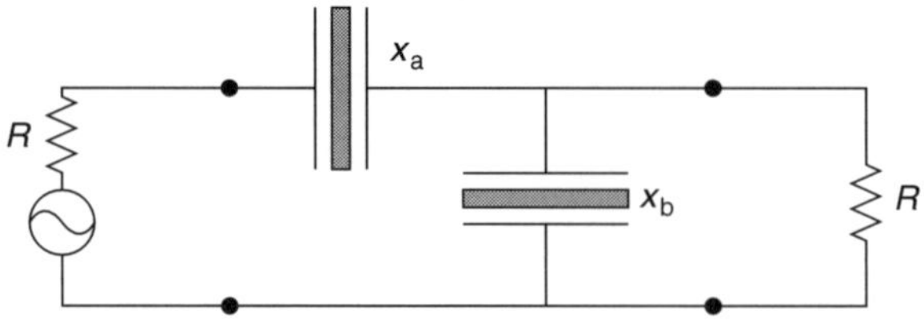

Figure 5.23 Basic ladder configuration for filter design. Reproduced from K.M. Lakin, G.R. Kline and K.T. McCarron, 1992, 'Thin film bulk acoustic wave filters for GPS', in *IEEE 1992 Ultrasonics Symposium*, IEEE, Washington, DC: 471–476, by permission of IEEE, © 1992 IEEE

5.6 MICROMACHINED FILTERS FOR MILLIMETER WAVE FREQUENCIES

Filters for higher frequencies, especially at millimeter wave frequencies devices using principles discussed thus far are not practical. However, fabrication approaches similar to those developed for MEMS can be effectively used in their design.

At microwave frequencies filters are fabricated using distributed component approach (Pozar, 1988). However, as frequencies increase, the size of these components gets smaller, calling for improved precision in the fabrication process. Common substrate materials are either bulky or contribute to added losses in these systems. Membrane microstrip lines and other components fabricated by micromachining techniques offer significant improvement over conventional fabrication approaches.

A coupled line band pass filter for W-band (94.7 GHz) is shown in Figure 5.24 (Rosertson, Katchi and Rebeiz, 1996). The coupled line geometry is made on a membrane-supported transmission line. This filter has an insertion loss of 3.6 dB in the pass band. Conductor loss is the primary contributor to the insertion loss of this device. This has a bandwidth of 6.1% (Figure 5.25).

A similar approach is followed for bandpass filters with low loss for 37 and 60 GHz operation (Blondy *et al.*, 1998). These are fabricated on high-resistivity silicon substrate. The layout is patterned with gold by electroplating after depositing stress-compensated

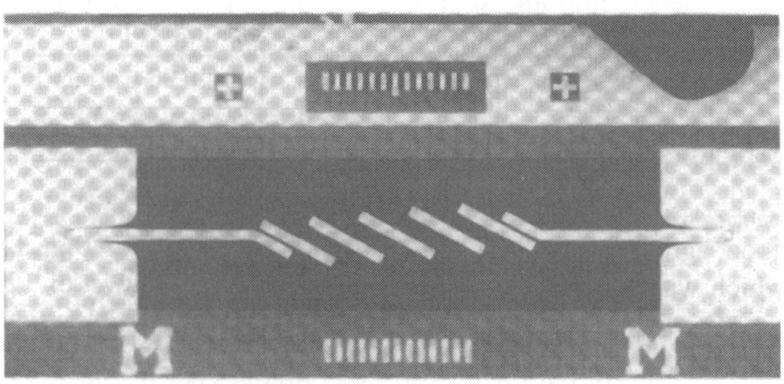

Figure 5.24 Photograph of W-band filter. Reproduced from S.V. Robertson, L.P.B. Katchi and G.M. Reseiz, 1996, 'Micromachined W-band filter', *IEEE Transactions on Microwave Theory and Techniques* **44**: 598–606, by permission of IEEE, © 1996 IEEE

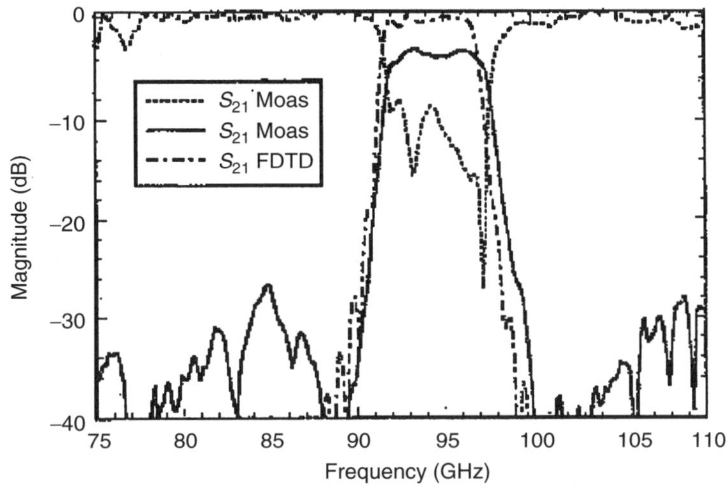

Figure 5.25 Performance of the micromachined bandpass filter pictured in Figure 5.24. Reproduced from S.V. Robertson, L.P.B. Katchi and G.M. Reseiz, 1996, 'Micromachined W-band filter', *IEEE Transactions on Microwave Theory and Techniques* **44**: 598–606, by permission of IEEE, © 1996 IEEE

$SiO_2-SiN_4-SiO_2$ membrane layer on the substrate. The silicon below the membrane is etched out completely to reduce the dielectric losses of the circuit. The lower and upper cavities are formed on separate wafers, by etching and subsequent metallization with gold. The wafers are stacked and bonded together with silver epoxy, as shown in Figure 5.26. Layouts of these filters are shown in Figure 5.27.

In the design for 37 GHz, the input and output lines are coupled capacitively and the hairpin sections are coupled magnetically to obtain two transmission zeros. This configuration is useful for sharp roll-off in filter characteristics. A four-pole filter design for 60 GHz is shown in Figure 5.27(b). Transmission line sections marked 1 and 4 in the layout are $\lambda/2$ sections for the operational frequency. The hairpin sections are magnetically coupled and all others are capacitively coupled in this configuration also. This results in

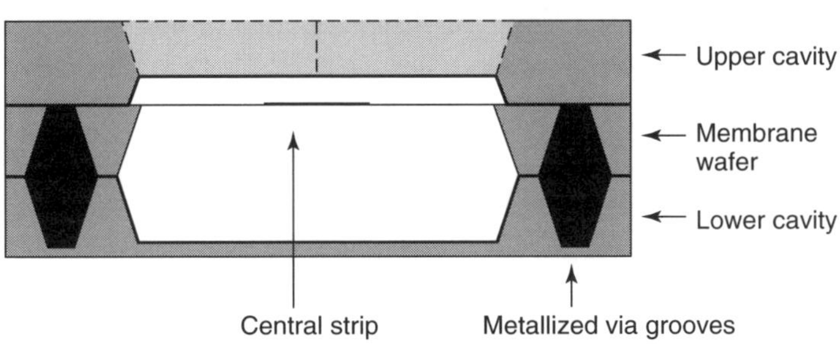

Figure 5.26 Cross-section of the filter structure. Reproduced from P. Blondy, A.R. Brown, D. Cros and G.M. Rebeiz, 1998, 'Low-loss micromachined filters for millimeter-wave communication systems', *IEEE Transactions on Microwave Theory and Techniques* **46**: 2283–2288, by permission of IEEE, © 1998 IEEE

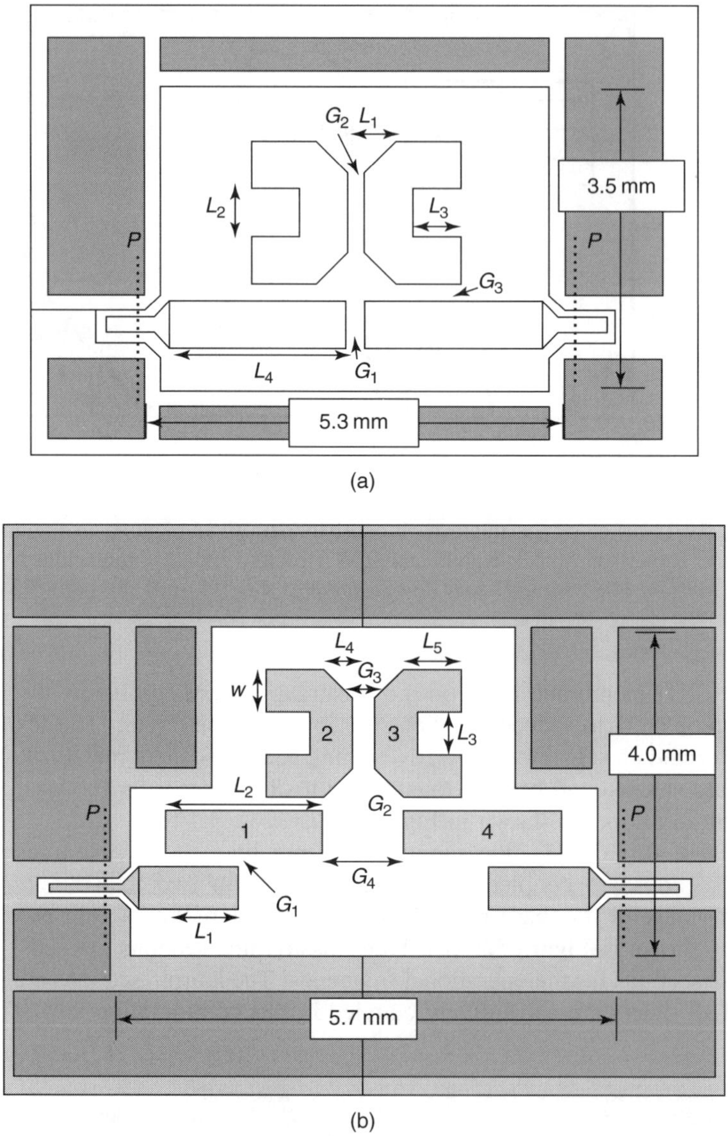

Figure 5.27 Layouts for (a) 37 GHz and (b) 60 GHz band pass filters. Reproduced from P. Blondy, A.R. Brown, D. Cros and G.M. Rebeiz, 1998, 'Low-loss micromachined filters for millimeter-wave communication systems', *IEEE Transactions on Microwave Theory and Techniques* **46**: 2283–2288, by permission of IEEE, © 1998 IEEE

an elliptic response for the filter, thereby having a low insertion loss. The performance of these filters are presented in Figure 5.28.

Yet another filter layout using similar microshield transmission lines is shown in Figure 5.29 (Rebeiz *et al.*, 1997). Series cascaded stubs in CPW configuration are used for filters with low insertion loss. These open-ended stubs are of length $\lambda/4$ for the operational frequency. Insertion loss better than 1 dB can be obtained by this filter design.

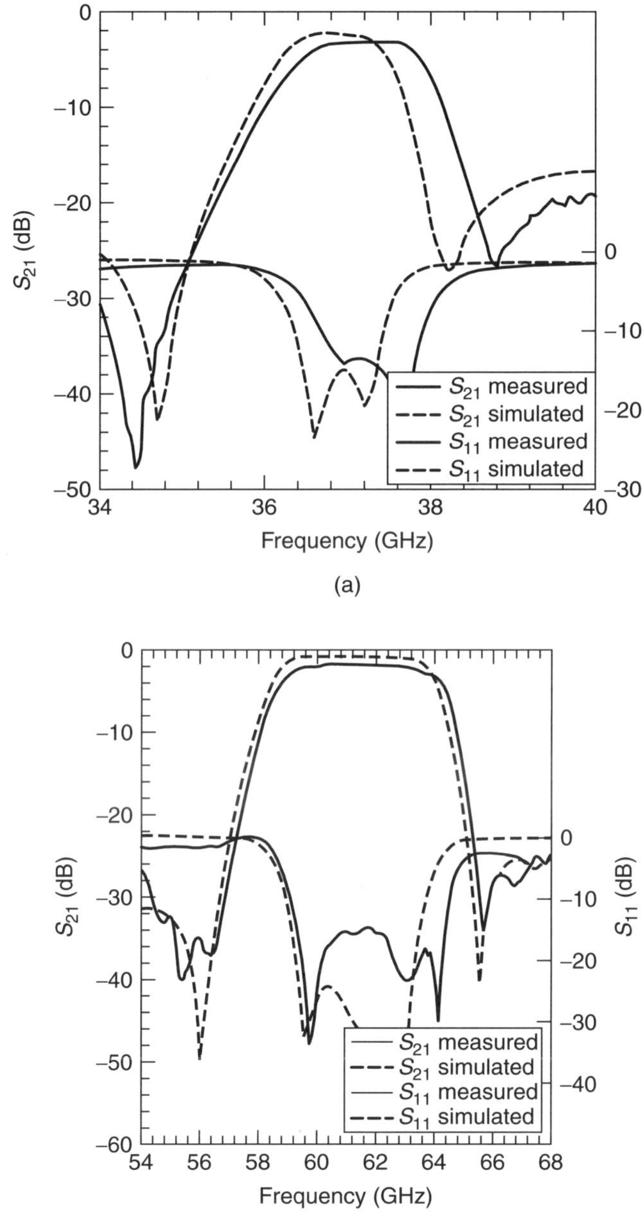

Figure 5.28 Characteristics of filters shown in Figure 5.27: (a) 37 GHz and (b) 60 GHz band pass filters. Reproduced from P. Blondy, A.R. Brown, D. Cros and G.M. Rebeiz, 1998, 'Low-loss micromachined filters for millimeter-wave communication systems', *IEEE Transactions on Microwave Theory and Techniques* **46**: 2283–2288, by permission of IEEE, © 1998 IEEE

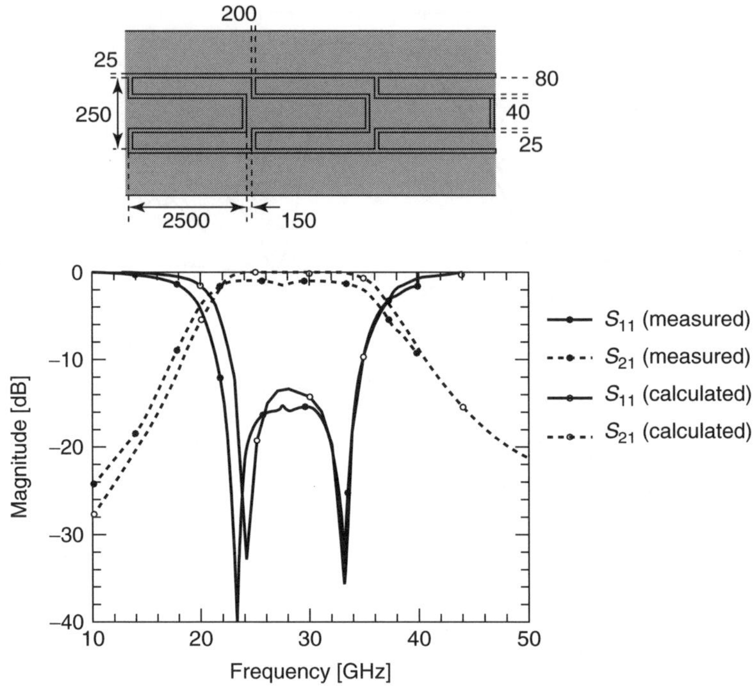

Figure 5.29 A three-section open-ended series stub band pass filter and its response. Reproduced from G.M. Reseiz, L.P.B. Katchi, T.M. Weller, C.-Y. Chi and S.V. Robertson, 1997, 'Micromachined membrane filters for microwave and millimeter-wave applications', *International Journal of Microwave and Millimeter-wave CAE* **9**: 149–166, by permission of Wiley, © 1997 Wiley

5.7 SUMMARY

In this chapter several technologies for designing RF filters in the form of micro scale devices has been presented. Approaches have been presented for the modeling of components of mechanical filters to help lay the foundations for understanding the operational principles of such filters without getting into intricacies of issues related to microfabrication of MEMS filters. Two configurations of micromechanical filters for different frequency regimes have been presented.

Apart from filters with vibratory motion as the mode of energy transmission between input and output terminals, those using forms of acoustic waves have also been discussed in this chapter. SAW filters and filters with bulk acoustic wave resonators have been introduced with this objective.

Filters using such mechanical forms of wave propagation are not convenient at high gigahertz frequencies and for millimeter-wave frequencies. In such cases the existing method of using distributed transmission lines is improvised using micromachining techniques used for other MEMS devices. It is envisaged that the understanding gained from this chapter will inspire newer generation micromachined RF filters for present and future telecommunication equipment.

REFERENCES

Bannon, F.D., Clark, J.R., Nguyen, C.T.C., 2000, 'High-Q HF micro electromechanical filters', *IEEE Journal of Solid-state Circuits* **35**: 512–526.
Bannon III, F.D., Clark, J.R., Nguyen, C.T.-C., 2000, High-Q HF microelectromechanical filters', *IEEE Journal of Solid-state Circuits* **35**: 512–526.
Blondy, P., Brown, A.R., Cros, D., Rebeiz, G.M., 1998, 'Low-loss micromachined filters for millimeter-wave communication systems', *IEEE Transactions on Microwave Theory and Techniques* **46**: 2283–2288.
Campbell, C., 1998, *Surface Acoustic Wave Devices for Mobile and Wireless Communications*, Academic Press, San Diego, CA.
Dobershtein, S.A., Malyukhov, V.A., 1997, 'SAW ring filters with insertion loss of 1 dB', *IEEE Transactions on Ultrasonics, Ferroelectrics and Frequency Control* **44**: 590–596.
Gilbert, J.R., Osterberg, P.M., Harris, R.M., Ouma, D.O., Cai, X., Pfajfer, A., White, J., Senturia, S.D., 1993, 'Implementation of a MEMCAD system for electrostatic and mechanical analysis of complex structures from mask descriptions', in *MEMS '93, Proceedings of IEEE Conference on Micro Electro Mechanical Systems*, IEEE, Washington, DC: 207–212.
Gopani, S., 1998, 'SAW IF filters in mobile communications networks', *Microwave Journal* **41**(11): 5.
Hirano, T., Furuhata, T., Gabriel, K.J., Fujita, H., 1992, 'Design, fabrication, and operation of submicron gap comb drive micro actuators', *Journal of Microelectromechanical Systems* **1**: 52–59.
Johnson, R.A., 1983, *Mechanical Filters in Electronics*, Wiley Interscience, New York.
Lakin, K.M., Kline, G.R., McCarron, K.T., 1992, 'Thin film bulk acoustic wave filters for GPS', in *IEEE 1992 Ultrasonics Symposium*, IEEE, Washington, DC: 471–476.
Lakin, K.M., McCarron, K.T., Belsick, J., McDonald, J.F., 1992, 'Thin film bulk acoustic wave filters technology', in *RAWCON 2001: IEEE Radio and Wireless Conference*, IEEE, Washington, DC: 89–92.
Lee, K.B., Cho, Y.-H., 2001, 'Laterally driven electrostatic repulsive–force microactuators using asymmetric field distribution', *Journal of Microelectromechanical Systems* **10**: 128–136.
Lin, L., Nguyen, C.T.-C., Howe, R.T., Pisano, A.P., 1992, 'Micro electromechanical filters for signal processing', *IEEE Conference on Micro Electro Mechanical Systems '92, Feb. 4–7, 1992*, IEEE, Washington, DC: 226–231.
Lin, L., Howe, R.T., Pisano, A.P., 1998, 'Microelectromechanical filters for signal processing', *Journal of Microelectromechanical Systems* **7**: 286–294.
Misu, K., Nagatsuka, T., Wadaka, S., Maeda, C., Wadaka, A., 1998, 'Film bulk acoustic wave filters using lead titanate on silicon substrate', in *IEEE 1998 Ultrasonics Symposium*, IEEE, Washington, DC: 1091–1094.
Morgan, D.P., 1998, 'History of SAW devices', in *Proceedings of 1998 IEEE International Frequency Control Symposium*, IEEE, Washington, DC: 439–460.
Nguyen, C.T.-C., Howe, R.T., 1993, 'CMOS micromechanical resonator oscillator', in *IEEE International Electron Devices Meeting, 1993. Technical Digest*, IEEE, Washington, DC: 199–202.
Nguyen, C.T.-C., 1995, 'Micromechanical resonators for oscillators and filters', *1995 IEEE Ultrasonics Symposium*, IEEE, Washington, DC: 489–499.
Pozar, D.M., 1998, *Microwave Engineering*, 2nd ed., Wiley, New York.
Rebeiz, G.M., Katehi, L.P.B., Weller, T.M., Chi, C.-Y., Robertson, S.V., 1997, 'Micromachined membrane filters for microwave and millimeter-wave applications', *International Journal of Microwave and Millimeter-wave CAE* **9**: 149–166.
Robertson, S.V., Katehi, L.P.B., Rebeiz, G.M., 1996, 'Micromachined W-band filter', *IEEE Transactions on Microwave Theory and Techniques* **44**: 598–606.
Rossi, M., 1988, *Acoustics and Electroacoustics*, Artech House, Norwood, MA.

Ruppel, C.C.W., Dill, R., Fischerauer, A., Gawlik, A., Machui, J., Muller, F., Reindl, L., Ruile, W., Scholl, G., Schroll, I., Wagner, K.C., 1993, 'SAW devices for consumer communication applications', *IEEE Transactions on Ultrasonics, Ferroelectrics and Frequency Control* **40**: 438–452.

Smith, P.M., 1995, 'Power SAWs: analog signal processing with surface acoustic wave devices', *IEEE Potentials* **14**(5): 18–20.

Su, Q.X., Kirby, P.B., Komuro, E., Whatmore, R.W., 2000, 'Edge supported ZnO thin film bulk acoustic wave resonators and filter design', in *2000 IEEE/EIA International Frequency Control Symposium and Exhibition*, IEEE, Washington, DC: 434–440.

Tang, W.C., Nguyen, T.C.H., Howe, R.T., 1989, 'Laterally driven polysilicon resonant microstructures', *Sensors and Actuators* **20**: 25–32.

Wang, K., Nguyen, C.T.-C., 1997, 'High-order micromechanical electronic filters', in *Proceedings of 1997 IEEE International Microelectromechanical Systems Workshop*, IEEE, Washington, DC: 25–30.

Wang, K., Wong, A.-C., Nguyen, C.T.-C., 2000, 'VHF free-free beam high-Q micromechanical resonator', *Journal of Microelectromechanical Systems* **9**: 347–360.

Zhang, X., Tang, W.C., 1994, 'Viscous air damping in laterally driven microresonators', *MEMS '94, Proceedings, IEEE Workshop on Micro Electro Mechanical Systems*, IEEE, Washington, DC: 199–204.

6
Micromachined phase shifters

6.1 INTRODUCTION

A phase shifter is a two-port network with the provision that the phase difference between output and input signals can be controlled by a control signal, usually dc bias. Phase shifters with low insertion loss, low drive power, continuous tunability and low production cost are the key to the development of lightweight phased array antennas. Modern microwave and millimeter wave phased array antennas are attractive because of their ability to steer wave beams in space without physically moving the antenna elements, which is required for rapid beam steering and beam shaping. This could be achieved by the electronic control of the phase of the signal into the antenna system. An antenna beam can be formed in any desired shape and can be moved without moving the antenna elements. A typical phased array antenna may have several thousand elements fed by a phase shifter for every antenna element, which can steer the resulting array beam to different directions. Therefore low loss, low cost and lightweight phase shifters are important for the design of phased array antennas.

Phase shifters are generally classified as digital, in which the differential phase shift can be changed only a few predetermined discrete values such as 90°, 45°, 22.5°, 11.25°, etc. and analog in which continuous phase variation is possible.

In a phased array antenna, the phase shifters change the effective path length of the transmission line resulting in different phases for each element. Characteristics such as good impedance match, proper power handling capability, low drive power and fast response speed are the required elements for a phase shifter. Until recently, a wide variety of phase shifters have been developed to meet these requirements. For example, electronically variable phase shifters developed in 1957 (Reggia and Spencer, 1957) were an important milestone because it can provide inertialess phase change in a short time, which was not possible until then using mechanical phase shifters. In addition to the ferrite phase shifters, another important type of phase shifter emerged in the mid-1960s categorized as semiconductor phase shifters, using a PIN diode (see Section 3.4.2) as an electronic switch for phase control (White, 1965). Since then, there have been significant advances and developments in electronic phase shifters. With the aid of semiconductor technologies, a new type of semiconductor phase shifter, such as the GaAs field effect transistor (FET) active phase shifters, emerged and these phase shifters can be integrated into monolithic form (Coget *et al.*, 1989; Kato *et al.*, 1992).

In general, the ferrite phase shifters have low insertion loss and can handle significantly higher powers, but are complex in nature and have a high fabrication cost.

While semiconductor phase shifters using PIN diodes or FET are less expensive and smaller in size than ferrites, their application was limited because of high insertion loss. Recently, other types of phase shifters using microelectromechanical system (MEMS) bridges (Barker and Rebeiz, 1998; De Flaviis and Alexopoulos, 1997; Liu *et al*., 2000; Pillans *et al*., 2000) and thin-film ferroelectric materials such as barium strontium titanate (BST; Jose *et al*., 2001; Varadhan *et al*., 1991; 1995) have been investigated to overcome the above limitations.

Recently, considerable attention has been received for phased array antenna design using MEMS components. An introduction of various types of phase shifters are presented in this chapter along with a detailed analysis of recent MEMS phase shifters and ferroelectric thin-film phase shifters including the fabrication of polymer phase shifters using microstereolithography (MSL).

6.2 TYPES OF PHASE SHIFTERS AND THEIR LIMITATIONS

Figure 6.1 shows the principle of a phased array antenna using phase shifters. The direction of the beam can be controlled by changing the relative phase between the individual antenna elements. There are two distinct methods used for the design of phase shifters: (1) ferrite materials in which phase shift is obtained by changing the bias field, and (2) semiconductor devices.

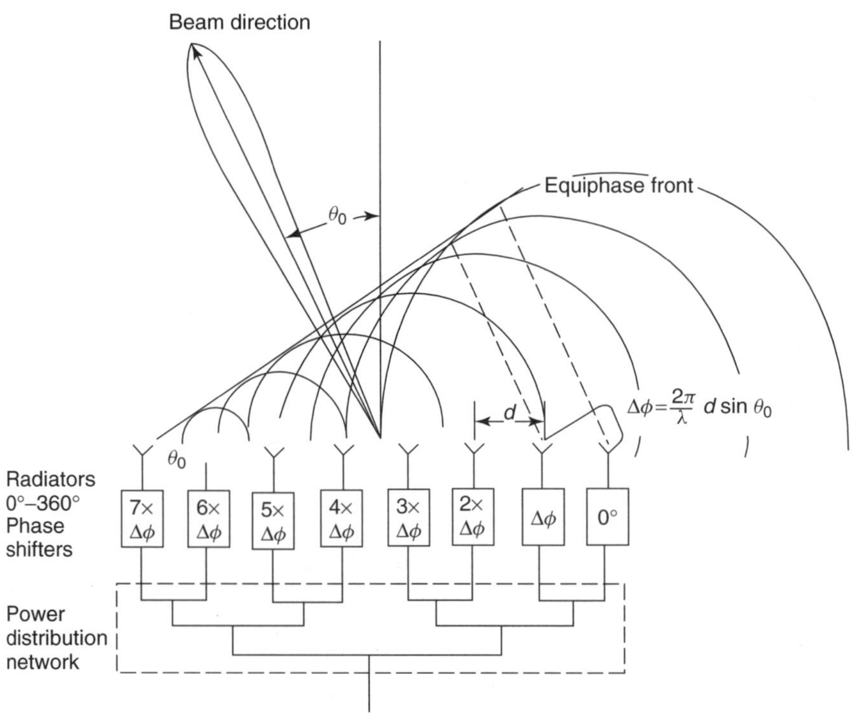

Figure 6.1 Principle of a phased array antenna using phase shifters. Varadan *et al*. 1995

6.2.1 Ferrite phase shifters

The ferrite phase shifter relies on the interaction between the electromagnetic waves and the spinning electrons in a magnetized ferrite. The spinning procession varies with the applied magnetic field, which changes the permeability of the ferrite. Therefore, by electronically controlling the applied magnetic field, it is possible to change the propagation constant of the electromagnetic waves, resulting in a change in their phase. These ferrite phase shifters can be realized in various designs such as waveguides, coaxial lines, striplines and microstriplines operating either in analog or digital mode (Koul and Bhat, 1991).

Ferrite phase shifters have been widely used for phased array antennas due to their lightweight and smaller size, but high cost, complexity and frequency limitations have prevented them to use in many military applications and made them impractical for mobile satellite communication devices.

6.2.2 Semiconductor phase shifters

Semiconductor phase shifters can usually be categorized as PIN diode phase shifters and FET phase shifters depending on which electronic control element is used as an electronic switch.

6.2.2.1 PIN diode phase shifters

PIN diode is a P–N junction that has a very minimally doped or intrinsic region between the p-type and n-type regions. The addition of the intrinsic region results in various characteristics; for example, its conductivity can be controlled in forward bias and the capacitance can be reduced in reverse bias. The PIN diodes are used extensively in microwave circuits for amplitude modulation, attenuation and also make excellent RF switches, phase shifters and limiters.

In phase shifters, the PIN diodes are (employed as electronic switches, by switching the bias current from forward to reverse bias. Figure 6.2 shows a schematic diagram and typical dc voltage–current characteristics of the PIN diode. The intrinsic region controls the on–off state of the diode switch in such a way that under forward bias it lowers the impedance of the diode and under reverse bias it offers very high impedance to the

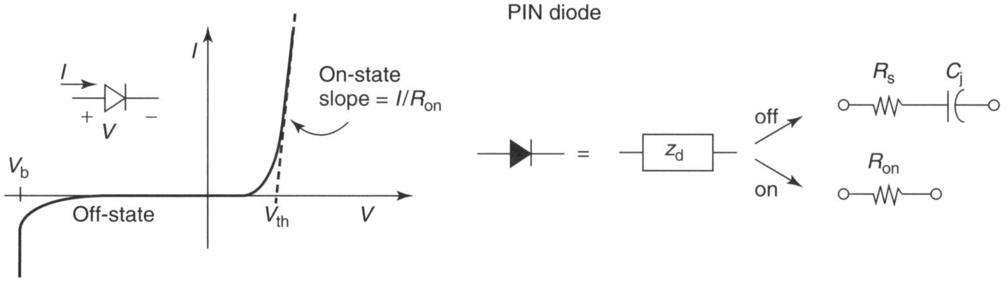

Figure 6.2 (a) Current–voltage (I–V) characteristics of a PIN diode; (b) PIN diode on-state and off-state characteristics (see Figure 3.6)

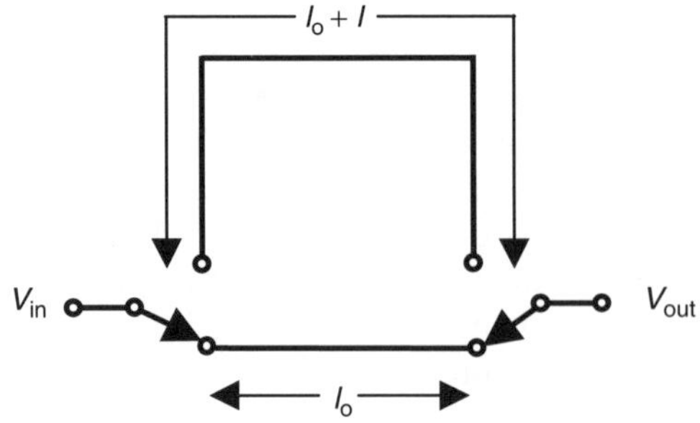

Figure 6.3 Schematic representation of the use of a PIN diode as a switched line phase shifter

diode. Therefore, the PIN diode phase shifters can generate phase shift by switching the signal between two different path lengths l_o and l_{o+l}, as shown in Figure 6.3. The phase shift corresponds to the additional path delay βl, where β is the propagation constant of the medium.

6.2.2.2 Field effect transistor phase shifters

The FET, which is used as a two-terminal switch controlled by the gate bias voltage, provides several advantages compared with PIN diodes. It shows significantly fast switching speed (~ ns), very low dc power consumption and compatibility for monolithic integration (Koul and Bhat, 1991). While the PIN diode phase shifter is a digital phase shifter in nature, the FET phase shifter can be realized in both analog and digital forms.

Since these semiconductor phase shifters with PIN diodes or FET circuits are very expensive and have significant loss at microwave frequencies, there is a need for low-loss, cost-effective components for phase array antennas. From this point of view it is remarkable that two new technologies using thin-film nonlinear dielectrics and MEMS switches have recently been employed for phase shifters. These novel phase shifters show considerably lower insertion loss (<2 dB) for a full phase shift of 360° at microwave frequency ranges. It is believed that both types of phase shifters will find great potential use in phased array antenna design.

6.2.3 Ferroelectric thin-film phase shifters

A thin film of BST has shown great promise for the fabrication of tunable RF and microwave components such as filters and phase shifters. The dielectric permittivity of BST can be tuned via an applied dc field. Detailed analysis of BST thin-film phase shifters is presented in Section 6.4.

6.2.4 Limitations of phase shifters

Even though ferrite phase shifters have low insertion loss and can handle significantly higher power, their cost and complexity are problems yet to be solved. Semiconductor phase shifters using PIN diodes and FETs are less expensive and smaller than ferrites but their application is limited because of the high insertion loss at high frequencies and poor power-handling capability. Inability to obtain a continuous phase variation is a problem for these phase shifters while a continuous phase shift is preferred in modern adaptive antennas and phased array radars.

6.3 MEMS PHASE SHIFTERS

MEMS switches and components demonstrated exceptional performance at RF and millimeter wave frequencies, including low insertion loss, high isolation and low drive power. The MEMS switch can be configured to generate phase shift by switching between two different signal paths, as shown in Figure 6.3, or can be used as a distributed capacitive switch, in which the switch changes the effective capacitance of the transmission line. Combining MEMS technology in a unique design with new dielectric tunable materials can result in lightweight, low-cost large-phased array antennas with significantly reduced production cost. The advantage of using dielectric tunable material is that it can give a continuous change in phase.

6.3.1 Switched delay line phase shifters

MEMS switches (Goldsmith *et al.*, 1995, 1998; Muldavin and Rebeiz, 2000; Yao *et al.*, 1999) have significantly low loss, good isolation at high frequencies, very little dc power consumption and very low intermodulation distortion as compared with semiconductor switches such as FET and PIN diode switches and can be used for a variety of RF applications including phase shifters. Even though the types of MEMS switches can be categorized as a resistive series switch (metal–metal) or a capacitive shunt switch (metal–insulator–metal), the capacitive shunt switch is usually used for high-frequency applications because of its lower actuation voltage and faster switching speed compared with series switches (Isom *et al.*, 2000).

These phase shifters in general consists of 180°, 90°, 45°, 22.5°, 11.25° phase shifters in a cascaded arrangement. The switches are configured in different lines which can be selectively controlled for the propagation of RF signals. The difference between the path length is determined by the phase shift. Pillans *et al.* (1999) demonstrated three-bit and four-bit phase shifters for Ka-band fabricated on silicon. These phase shifters shows a 0° to 315° phase shift with a step of 45° for the three-bit phase shifter and a 0° to 337.5° phase shift with a step of 22.5° for the four-bit phase shifter. The four-bit Ka-band phase shifter, as shown in Figure 6.4(a), demonstrated an average insertion loss of 2.25 dB and the similar three-bit design shown in Figure 6.4(b) has insertion loss of 1.7 dB. Both designs were built on 6-mil high-resistivity silicon substrate. The switching is done by shunt capacitively coupled switches.

6.3.2 Distributed MEMS phase shifters

The capacitive shunt switch consists of a thin metal bridge suspended over the center electrode of a co-planar waveguide (CPW), which moves with a biased dc voltage. A thin dielectric layer such as silicon nitride is deposited on the bottom electrode to reduce

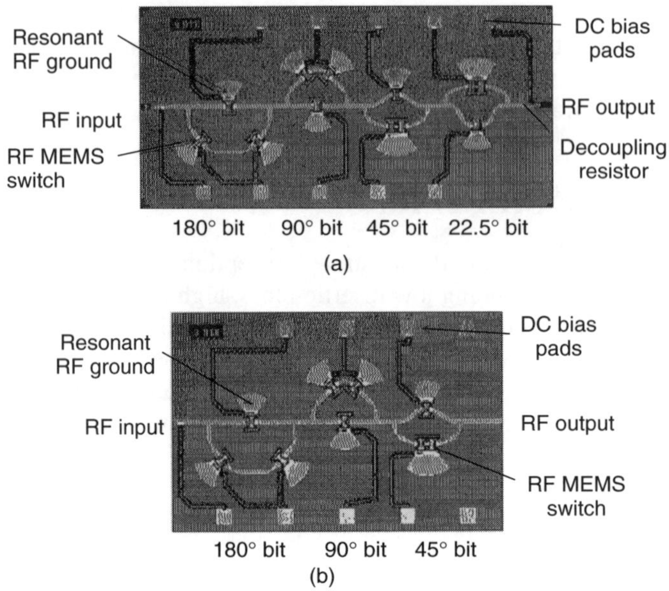

Figure 6.4 Photograph of (a) four-bit MEMS phase shifter; (b) three-bit design. Reproduced from B. Pillans, S. Eshelman, A. Malczewshi, J. Ehmke and C. Goldsmith, 1999, 'Ka-band RF MEMS phase shifters', *IEEE Microwave and Guided Wave Letters* **9**: 520–522, by permission of IEEE, © 1999 IEEE

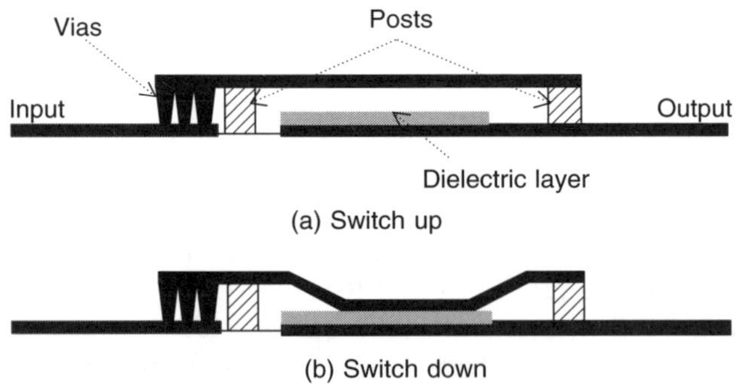

Figure 6.5 Schematic diagram of switch in the (a) up state and (b) down state. Reproduced from C. Goldsmith, J. Randall, S. Eshelman, T.H. Lin, D. Denniston, S. Chen and B. Norvell, 1996, 'Characteristics of micromachined switches at microwave frequencies', in *Proceedings of IEEE MTT-S International Microwave Symposium, Volume 2*, IEEE, Washington, DC: 1141–1144, by permission of IEEE, © 1996 IEEE

stiction and provide isolation between the metal bridge and bottom electrode. Figure 6.5 presents a schematic diagram of a bridge switch. When the bottom electrode is dc biased with respect to the metal bridge, the attractive electrostatic force pulls the metal membrane down toward the bottom electrode, as shown in Figure 6.5. However, when the bias voltage is increased, the system becomes unstable and the bridge collapses suddenly when the deflection reaches one third of the gap height. This voltage, which results in the point of instability, is called the pull-down voltage, given by (Goldsmith *et al.*, 1996)

$$V_p = \left(\frac{8k}{27\varepsilon_0 W w} g_0^3\right)^{-1/2} \tag{6.1}$$

where ε_0 is the free-space permittivity, W is bottom electrode width, w is the width of the MEMS switch, g_0 is the switch height and k is the effective spring constant of the switch, which can be approximated by (Osterberg *et al.*, 1994)

$$k = \frac{32Et^3w}{L^3} + \frac{8\sigma(1-\nu)tw}{L} \tag{6.2}$$

where E is Young's modulus of the bridge material, t is the bridge thickness, L is the switch length, σ is the residual tensile stress in the switch and ν is Poisson's ratio for the switch material.

The applied bias voltage between the MEMS bridges and bottom electrodes changes the height of the MEMS bridges, which in turn varies the distributed MEMS capacitance. This results in a change in the loaded transmission line impedance and phase velocity, which in turn causes phase shift. As shown in Figure 6.6, a structure with several MEMS bridges can act as a phase shifter when the applied bias voltage is less than the pull-down voltage.

The main disadvantages of MEMS shunt switches are slow switching speed (several microseconds) and high actuation voltage (20 to approximately 100 V) compared with PIN diode switches (switching speed ~1 μs, actuation voltage 5 V). However, this switching speed would be sufficient for phased array antenna systems even though it is too slow for transmit and receive switching applications. The actuation voltage can be decreased by changing the gap height between MEMS bridges and bottom electrodes. However, this causes the parasitic capacitance of the bridge in the up state to increase and the impedance of the line to change. Another choice to decrease the actuation voltage of the bridge is to use materials with low Young's modulus (Ji, Vinoy and Varadan, 2001), which is presented in Section 6.3.3.

Barker and Rebeiz (1998) demonstrated a distributed CPW transmission line MEMS phase shifter using MEMS switches, as shown in Figure 6.6. This wide-band device shows a phase shift of 118° for a loss of 2 dB at 60 GHz. Borgioli *et al.*, (2000a, 2000b) fabricated a one-bit K/Ka-band phase shifters (Figure 6.7) with RF MEMS capacitive

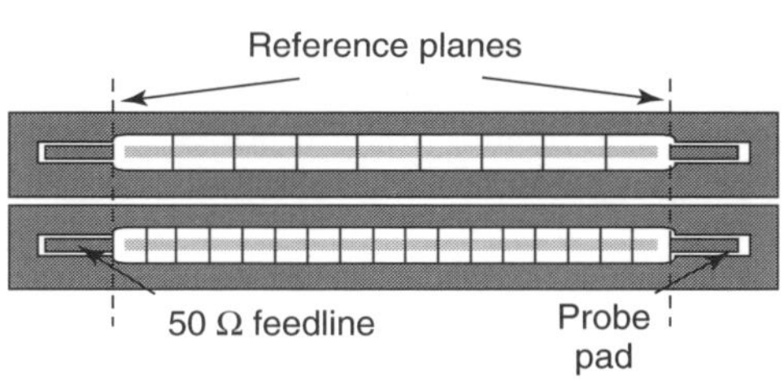

Figure 6.6 Schematic diagram of the fabricated distributed MEMS delay line phase shifter. The width and spacing of the MEMS bridges are 60 μm, 580 μm and 30 μm, respectively, resulting in a total length of 5.2 mm. Reproduced from N.S. Barker and G.M. Rebeiz, 1998, 'Distributed MEMS true-time delay phase shifter and wide-band switches', *IEEE Microwave Guided Wave Letter* (April): 1881–1890, by permission of IEEE, © 1998 IEEE

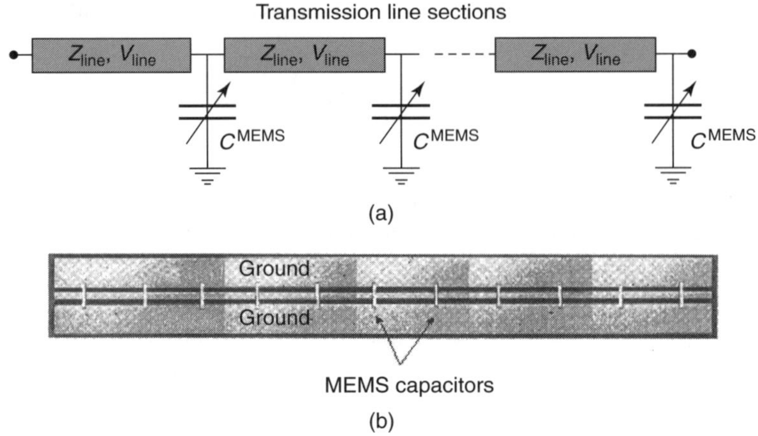

Figure 6.7 (a) Circuit schematic of the phase shifter; (b) photograph of the fabricated phase shifter. The total length of the phase shifter is 8.58 mm. Reproduced from A. Borgioli, Y. Liu, A.S. Nagra and R.A. York, 2000, 'Low loss distributed MEMS phase shifter', *IEEE Microwave and Guided Wave Letters*: 7–9, by permission of IEEE, © 2000 IEEE

switches, which gave 270° phase shift at 35 GHz. In addition to these single-bit phase shifters, multibit MEMS phase shifters have been also reported. Hayden *et al.*, (2000) presented a two-bit distributed CPW phase shifter for X-band. The phase shifter is designed using a series combination of MEMS bridges and MIM capacitors. It has a 90°, 8 MEMS bridge section cascaded with a 180°, 16 MEMS bridge section expecting a phase shift of 0/90/180/270°. It has the advantage to selectively choose either the MEMS bridge capacitance (C_b) or the total lumped capacitance (C_s) by applying voltages on the transmission line. Another design of two-bit and four-bit phase shifters were also presented by Hayden *et al.*, (2001), which consists of cascaded 180° and 90° sections, as shown in Figure 6.8.

Figure 6.9 presents the photograph of a K-band three-bit MEMS phase shifter which consists of three one-bit phase shifters for 180°, 90° and 45° phase shift. Each one-bit phase shifter consists of a co-planar waveguide (CPW) loaded periodically with several shunt MEMS capacitors.

The distributed MEMS phase shifters consist of a high-impedance line (>50 Ω) which is capacitively loaded by the periodic MEMS bridges. This periodic structure has an upper frequency limit called the Bragg frequency, beyond which higher reflection loss occurs. There are several important parameters considered for the design of distributed MEMS transmission line (DMTL) and MEMS bridges. These parameters should be determined keeping the process tolerance for fabrication in mind as well as the circuit performance.

6.3.2.1 Design of distributed MEMS phase shifters

The distributed MEMS phase shifter consists of a high-impedance (Z_0) transmission line periodically loaded with MEMS variable capacitors, as shown in Figure 6.10. This high impedance is indispensable for the unloaded line so that the loaded line including the MEMS bridges can be matched with 50 Ω. The capacitance C_t and inductance L_t of the

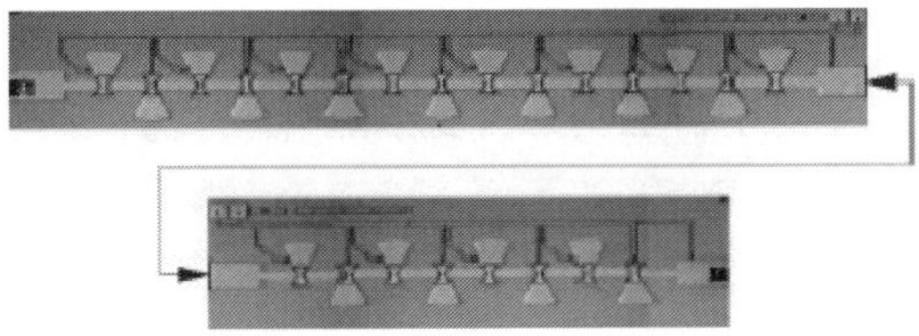

(a)

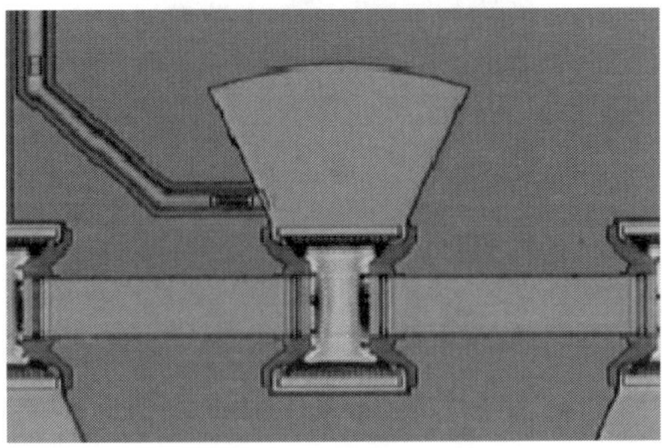

(b)

Figure 6.8 Photograph of (a) the two-bit 180° phase shifter on top and 90° on the bottom; (b) single MEMS section. Reproduced from J.S. Hayden, A. Malczewski, J. Kleber, C.L. Goldsmith and G.M. Rebeiz, 2001, '2 and 4-Bit DC-18 GHz microstrip MEMS distributed phase shifters', in *Proceedings of IEEE MTT-S International Microwave Symposium*, IEEE, Washington, DC: 219–222, by permission of IEEE, © 2001 IEEE

unloaded CPW transmission line can be written as (Barker and Rebeiz, 1998)

$$C_t = \frac{\sqrt{\varepsilon_{r,eff}}}{cZ_0} \tag{6.3}$$

$$L_t = C_t Z_0^2 \tag{6.4}$$

where $\varepsilon_{r,eff}$ is the effective dielectric constant of the unloaded CPW transmission line and c is the free-space velocity. For a CPW line, Z_0 and $\varepsilon_{r,eff}$ are related to its physical parameters as shown in the following equations.

The analytical expression for the quasi-TEM (quasi-transverse-electromagnetic) electrical parameters of the CPW transmission line can be simplified by taking into account the symmetry of transmission lines. The center of symmetry can be replaced by a magnetic wall in even mode propagation. The propagation characteristics of transmission can be analyzed by the conformal mapping method.

The total capacitance of a CPW can be written as the sum of two parallel capacitances due to air C_1 and dielectric substrate C_2 (Ghione and Naldi, 1987):

$$C_1 = 4\varepsilon_0 \frac{12}{23} = 4\varepsilon_0 \frac{K(k_1')}{K(k_1)} \tag{6.5}$$

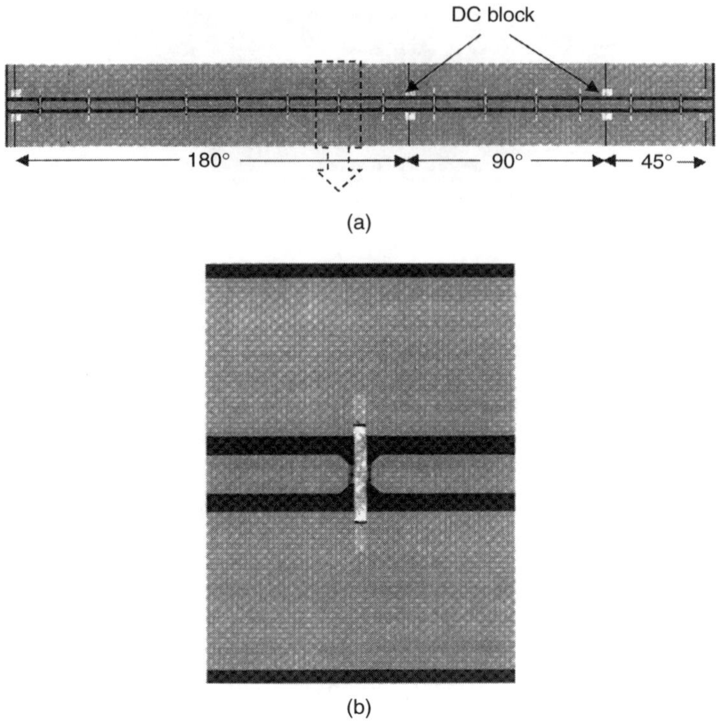

Figure 6.9 (a) Photograph of the distributed MEMS phase shifter; (b) close view of the switch. Reproduced from Y. Liu, A. Borgioli, A.S. Nagra and R.A. York, 2000, 'K-band 3-bit low-loss distributed MEMS phase shifter', *IEEE Microwave and Guided Wave Letters* (October): 415–417, by permission of IEEE, © 2000 IEEE

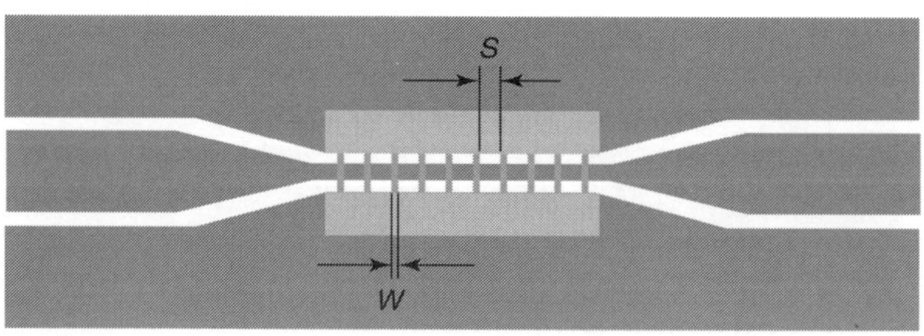

Figure 6.10 Schematic layout of the distributed MEMS transmission line (DMTL) phase shifter

where

$$k_1 = \frac{a}{b}\left[\left(1 - \frac{b^2}{c^2}\right)\left(1 - \frac{a^2}{c^2}\right)^{-1}\right]^{1/2} \tag{6.6}$$

$$k_1' = (1 - k_i^2)^{1/2}, \quad i = 1, 2, \ldots \tag{6.7}$$

and $K(k)$ is the complete elliptic integral of the first kind.

The capacitance C_2 can be written as [28]

$$C_2 = 2\varepsilon_0(\varepsilon_r - 1)\frac{12}{23} = 2\varepsilon_0(\varepsilon_r - 1)\frac{K(k_2')}{K(k_2)} \tag{6.8}$$

where

$$k_2 = \frac{\sinh\left(\frac{\pi a}{2h}\right)}{\sinh\left(\frac{\pi b}{2h}\right)}\sqrt{\frac{1 - \sinh^2\left(\frac{\pi b}{2h}\right)\Big/\sinh^2\left(\frac{\pi c}{2h}\right)}{1 - \sinh^2\left(\frac{\pi a}{2h}\right)\Big/\sinh^2\left(\frac{\pi c}{2h}\right)}} \tag{6.9}$$

Finally, the capacitance per unit length and effective dielectric constant of a CPW can be written as

$$C = C_1 + C_2 = 4\varepsilon_0\frac{K(k_1')}{K(k_1)}\left[1 + \frac{1}{2}(\varepsilon_r - 1)\right]q_1 \tag{6.10}$$

$$\varepsilon_{\text{eff}} = \frac{C(\varepsilon_r)}{C(1)} = 1 + \frac{1}{2}q_1(\varepsilon_r - 1) \tag{6.11}$$

with filling factor q_1:

$$q_1 = \frac{K(k_2')}{K(k_2)}\frac{K(k_1)}{K(k_1')} \tag{6.12}$$

The characteristic impedance of the CPW line is given by

$$Z_0 = \frac{30\pi}{\sqrt{\varepsilon_{\text{eff}}}}\frac{K(k_1')}{K(k_1)} \tag{6.13}$$

For a different configuration of coplanar waveguide, the above conformal mapping method can be expanded to the different boundary conditions including upper shielding and conductor backing.

The CPW line periodically loaded with MEMS capacitors can be modeled as a lumped inductance (L_t) and capacitance (C_t) with a parallel variable capacitor to ground, as shown in Figure 6.11. The characteristic impedance Z_l, and phase velocity v_l of the MEMS bridge loaded CPW transmission line and its Bragg frequency due to the periodic structure can be written as (Barker and Rebeiz, 1998)

$$Z_l = L_t\left(C_t + \frac{C_b}{s}\right)^{-1} \tag{6.14}$$

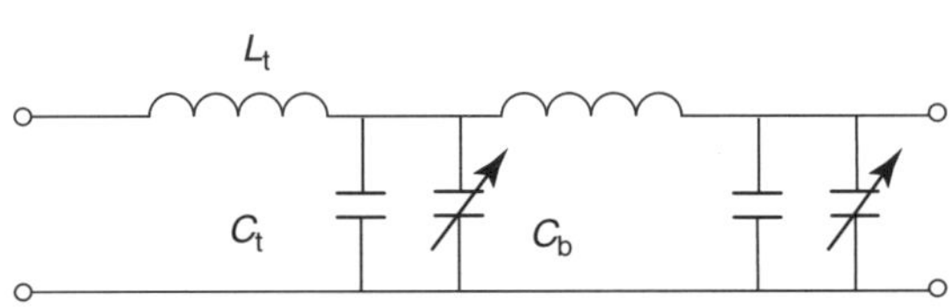

Figure 6.11 Lumped element transmission line model of the loaded distributed MEMS transmission line

$$\nu_l = \left[L_t\left(C_t + \frac{C_b}{s}\right)\right]^{-1/2} \tag{6.15}$$

$$f_{\text{Bragg}} = \frac{\nu_l}{\pi s} \tag{6.16}$$

where C_b/s is the distributed MEMS capacitance on the loaded line.

The phase shift per unit length due to the change of the DMTL characteristic impedance arising from the MEMS bridge capacitance variation by applying a bias voltage can be given as

$$\Delta\phi = \frac{\omega Z_0\sqrt{\varepsilon_{r,\text{eff}}}}{c}\left[\frac{1}{Z_{lu}} - \frac{1}{Z_{ld}}\right] \text{ rad m}^{-1} \tag{6.17}$$

where Z_{lu} and Z_{ld} are the DMTL characteristic impedances for the low and the high bridge capacitance states, respectively. Maximum phase shift can be achieved when the applied voltage increases up to the pull-down voltage given by Equation (6.1).

6.3.3 Polymer-based phase shifters

The MEMS phase shifters have low loss performance because of the bridge structure, which effectively prevents leakage current. However, in spite of these advantages, the reduction of actuation voltage is still a key issue for MEMS structures. Since a MEMS bridge with a height of at least 3 μm is typically required to reduce the parasitic capacitance of the bridges, it results in an actuation voltage of around 100 V with conventional metal bridges. The actuation voltage can be reduced by decreasing the bridge height or adopting bridge materials with a relatively low elastic modulus such as polymer. However, it is not advisable to decrease the height of the MEMS bridges to reduce the actuation voltages because the narrow height decreases fabrication yield as well as increasing the parasitic capacitance of the MEMS bridge. The decrease in yield is due to the difficulties in process with releasing MEMS bridges with a narrow gap from the bottom substrates. There is a need to keep the bridge height as large as possible to increase the fabrication yield. If MEMS bridges can be fabricated with polymers for which the elastic modulus is around 5 GPa, which is much less than metals (50 to approximately 100 GPa), it will be the best choice for decreasing the actuation voltage. From Equation (6.1) it can be seen that for two MEMS bridges with exactly the same dimensions the one with polymer material requires three times less actuation voltage than a metal bridge. The polymer-based MEMS phase shifter fabricated using the microstereolithography (MSL) (Varadan, Jiang and Varadan, 2001) technique has several advantages over conventional lithography such as rapid prototyping for three-dimensional (3D) microfabrication and cost reduction in the early development stage owing to the rapid prototyping. A brief introduction of MSL is given, followed by the description of how to utilize this technique for the fabrication of polymer MEMS bridges.

MSL was introduced to fabricate high aspect ratio and complex 3D microstructures in 1993 (Ikuta and Hirowatari, 1993). In contrast to conventional subtractive micromachining, MSL is an additive process which enables one to fabricate high aspect ratio microstructures with novel smart materials. The MSL process is also, in principle, compatible with silicon processes and capable of batch fabrication (Ikuta and Hirowatari, 1993;

Katagi and Nakajima, 1993). Different MSL systems have been developed aiming at precision and fabrication speed improvement. Basically, scanning MSL (Ikuta *et al.*, 1996) and projection MSL (Bertsch, Lorenz and Renaud, 1998; Monnert *et al.*, 1999; Nakamoto and Yamaguchi, 1996) are the two major approaches being developed. The scanning MSL builds the solid micro-parts in a point-by-point and line-by-line fashion, while projection MSL builds one layer with one exposure, which significantly saves fabrication time.

The MSL fabrication process enables one to make various designs of microstructures without the need for additional photomasks, which is required in conventional silicon surface micromachining. In this approach, the device is built by using computer-aided design (CAD) systems with UV-curable polymers (e.g. SU8). The polymer MEMS bridges have to be coated with metal to use them as bridges which are actuated by an electrostatic force. Therefore, it is very important to establish the coating process for the polymer bridge with metals and to optimize the process tolerance as much as possible, because whole fabrication yield for the polymer-based MEMS phase shifter is significantly dependent on this process step as well as on releasing the MEMS bridge from the bottom substrate.

For a bridge comprising of two different materials such as a metal and a polymer, the Young's modulus of a bridge can be written as (Van Keuls *et al.*, 1997)

$$E_{\mathrm{eff}} = \frac{12(EI)_{\mathrm{eff}}}{w(t_1 + t_2)^3} \tag{6.18}$$

$$(EI)_{\mathrm{eff}} = \left(\frac{wt_1^3}{12}\right)\frac{t_2E_1E_2}{t_2E_2 + t_1E_1}K' \tag{6.19}$$

$$K' = 4 + 6\left(\frac{t_2}{t_1}\right) + \frac{E_1}{E_2}\frac{t_1}{t_2} + 4\left(\frac{t_2}{t_1}\right)^2 + \frac{E_2}{E_1}\left(\frac{t_2}{t_1}\right)^3 \tag{6.20}$$

where w is the bridge width, t_1 and t_2 are the thickness of materials 1 and 2, respectively, and E_1 and E_2 are the Young's modulus of materials 1 and 2, respectively. For example, in the case of the bridge comprising a 10-μm-thick cured SU8 polymer ($E = 5$ GPa) and a 0.1-μm-thick thermal evaporated gold ($E = 61$ GPa) with $L = 1000$ μm and $g_0 = 4.5$ μm assuming $\sigma = 0$, the pull-down voltage is 57 V. This pull-down voltage can be further decreased by reducing the thickness of the polymer bridges without changing the height. A 5-μm-thick polymer bridge would give a pull-down voltage of around 21 V in spite of such a large bridge height of 4.5 μm. Thus, a polymer-based bridge can give a low actuation voltage because of the relatively low Young's modulus of polymer as compared with a metal with the same dimensions.

The fabrication technique of MEMS phase shifter using microstereolithography is shown in Figure 6.12. Fabrication of the distributed MEMS phase shifter begins with preparing transmission line on a substrate such as silicon, sapphire or quartz. The CPW transmission line is defined by evaporating metals and adopting a lift-off or etching process. A dielectric layer is usually deposited on the center electrode which prevents an RF short between the MEMS bridge and the center electrode. A sacrificial layer using photoresists (PRs) or oxides, which determines the height of the MEMS bridge, is deposited on it and patterned. Metal layer for the bridge electrodes is formed using an evaporator or sputter on the sacrificial layer. These fabrication steps are common for silicon surface micromachining and MSL approaches. However, from the next step, these differ significantly.

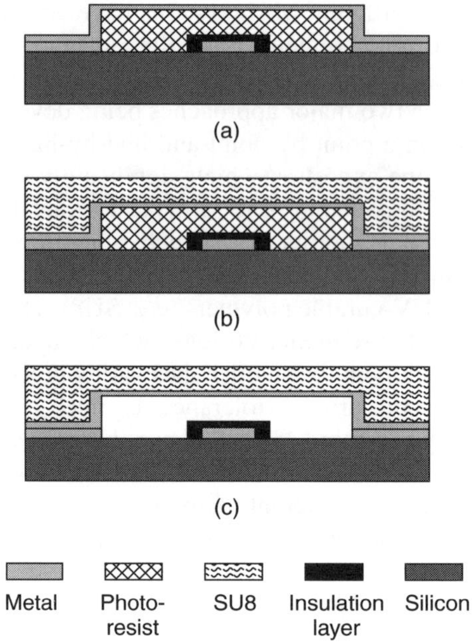

Figure 6.12 Process flow of the microstereolithography (MSL) fabrication technique: (a) metal deposition for the MEMS bridges; (b) polymer MEMS bridge curing by MSL after SU8 deposition; (c) removal of the sacrificial layer and metal together except for the MEMS bridges

UV-curable polymers are used to fabricate the bridge material. UV-curable polymer is spin-coated on the metal layer and cured to form polymer MEMS bridges according to CAD design. SU8 and PR (Shipley 1827) are chosen as the polymer material and sacrificial layer, respectively, because SU8 has a strong adhesion to metal and is resistant to acetone (etchant). The uncured SU8 is developed with PGMEA (propylene glycol methyl ether acetate), followed by removal of the sacrificial layer and metal which is not covered with the SU8 bridges together in acetone. The unexposed metal below the polymer MEMS bridge is used as the electrode to actuate the bridge by applying a bias voltage. The substrate is rinsed with IPA (isopropyl alcohol) and placed in a vacuum chamber for slow drying to release the MEMS bridges.

6.4 FERROELECTRIC PHASE SHIFTERS

Phase shifters using voltage controllable thin films can be fabricated into microstrip or co-planar waveguide configurations. Phase shifters using microstrip transmission lines can be fabricated on bulk ceramic or ferroelectric substrates (Jose *et al.*, 2001; Van Keuls *et al.*, 1997; Varadan *et al.*, 1991, 1995). Monolithic integration of these phase shifters are desirable because of reliability and cost issues. However, microstrip phase shifters using nonsemiconducting substrates has limitations in monolithic integration with another passive circuit and active circuits. Hence, co-planar waveguide phase shifters on semiconductor substrate have been investigated intensively, based on Schottky contact, ferroelectric material or varactor (Acihel and Liu, 2001; Neidert and Krowne, 1985). Also, advancement of microwave CAD has encouraged the widespread use of co-planar waveguides because it can give solutions to complex fringing fields. The design of co-planar waveguides for a phase shifter is presented in Section 6.3.1.

Semiconductor phase shifters have the advantages of fast operating speed and compatibility with monolithic microwave integrated circuits compared with ferroelectric phase shifters. However, the operating frequency of semiconducting phase shifters has been limited because of its low Q value at high frequencies. At higher frequencies, insertion loss of semiconductor devices increases drastically. Owing to these reasons, ferroelectric materials have been regarded as the best choice at high frequencies.

Even though various ferroelectric materials for the microwave phase shifter have been developed, most of them are focused on barium strontium titanate (BST) material because its permittivity can be tuned by applying a dc voltage.

6.4.1 Distributed parallel plate capacitors

The widespread application of ferroelectric capacitors is a result of its improved dielectric loss and voltage tunability. The requirement for high dielectric constant materials for the very-large-scale integrated circuit industry has improved its optimization procedures. BST thin film has shown quite promising results for various applications in memory and microwave devices (Erker and Nagra, 2000; york and Nagra, 2001). Periodically loaded parallel plate and interdigital capacitors based on a co-planar waveguide configuration have shown remarkable progress.

The capacitance of these devices can be varied by applying voltage between two electrodes. Since the electric fields are well confined between the narrow parallel plates, it has the structural advantage of getting higher tenability for a thin sheet. Figure 6.13 presents a schematic diagram of a parallel plate capacitor. BST is inserted between two metal electrodes. Two capacitors are in series with base as common electrode. A narrow gap is preferable between the two top electrodes to reduce the transmission loss.

The applied voltage controls the phase velocity of the transmission line loaded with BST capacitors. This is because of the dependence of the effective dielectric constant of the BST thin film on the applied voltage. The parallel plate, fringing and CPW capacitance can be written as (York and Nagra, 2001):

$$C_{\text{par}} = \frac{1}{2}\frac{\varepsilon_{\text{r}}\varepsilon_0 wl}{t_{\text{BST}}} \tag{6.21}$$

$$C_{\text{fringe}} = \frac{\varepsilon_{\text{SiN}}\varepsilon_0 w(3w + l)}{t_{\text{SiN}}} \tag{6.22}$$

$$C_{\text{CPW}} = 4\varepsilon_0\frac{K(k_1')}{K(k_1)}\left[1 + \frac{1}{2}(\varepsilon_{\text{r}} - 1)\frac{K(k_2')}{K(k_2)}\frac{K(k_1)}{K(k_1')}\right] \tag{6.23}$$

where

$$\varepsilon_{\text{eff}} = \frac{C(\varepsilon_{\text{r}})}{C(1)} = 1 + \frac{1}{2}q_1(\varepsilon_{\text{r}} - 1) \tag{6.24}$$

$$q_1 = \frac{K(k_2')}{K(k_2)}\frac{K(k_1)}{K(k_1')} \tag{6.25}$$

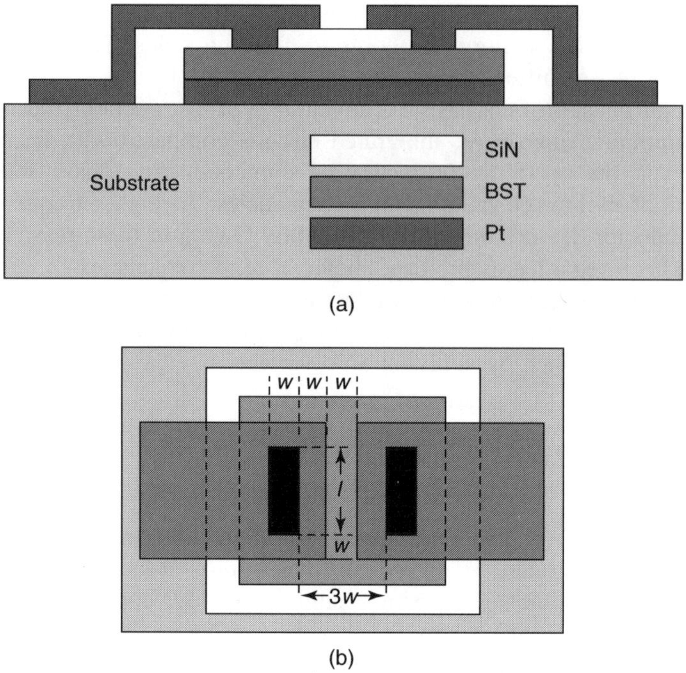

Figure 6.13 (a) Top view and (b) cross-section of the parallel plate varactor structure. Note: BST, barium strontium titanate. Reproduced from R. York and A. Nagra, 2001, 'Microwave integrated circuits using thin-film BST', *Proceedings of IEEE on Applications of Ferroelectrics* **1**: 195–200, by permission of IEEE, © 2001 IEEE

Total capacitance per unit length is the sum of fixed transmission line capacitance and variable capacitance:

$$C_{\text{tot}} = C_{\text{fix}} + \frac{C_{\text{var}}}{l_{\text{sec t}}} \tag{6.26}$$

Also, inductance can be given as

$$L = \frac{l_{\text{sec t}}}{v_i} Z_i \tag{6.27}$$

or

$$L = \frac{\mu}{4} \frac{K(k_1')}{K(k_1)} \tag{6.28}$$

where $K(k)$ is the complete elliptic integrals of the first kind,

$$k_1 = \frac{s - \delta}{(s + g) + \delta} \tag{6.29}$$

and δ is the skin depth.

The phase velocity is changed by change in capacitance and is given by

$$v_{\mathrm{p}} = \lambda f = \frac{\omega}{\beta} \tag{6.30}$$

$$v_{\mathrm{p}} = (L_{\mathrm{t}}C_{\mathrm{tot}})^{-1/2} = \left[L_{\mathrm{t}}\left(C_{\mathrm{fix}} + \frac{C_{\mathrm{var}}}{l_{\mathrm{sec\,t}}}\right)\right]^{-1/2} \tag{6.31}$$

where $\beta = \omega\sqrt{L_{\mathrm{t}}C_{\mathrm{tot}}}$

This periodic structure has an upper frequency limit owing to Bragg reflection. The operating frequency depends on the spacing and size of the capacitors, and the optimum operating frequency is given by

$$f_{\mathrm{Bragg}} = \frac{1}{\pi l_{\mathrm{sec\,t}}}\left[L_{\mathrm{t}}\left(C_{\mathrm{fix}} + \frac{C_{\mathrm{var}}}{l_{\mathrm{sec\,t}}}\right)\right]^{-1/2} \tag{6.32}$$

It is considered that the major part of insertion loss comes from conductor loss of bottom electrode and dielectric loss of BST film. Even though there is a limitation on minimal conductor loss, parallel plate varactors are attractive structures because of the low voltage operation and its compatibility with IC technology. It is quite promising for integration with microwave integrated circuits.

6.4.2 Bilateral interdigital phase shifters

The advantages in cost and reliability of monolithic integration have encouraged research on the interdigitated varactor. For monolithic integration of passive and active microwave devices on a chip, the process has to be compatible with conventional IC technology. The planar interdigital transducer (IDT) varactor is suitable for monolithic integration.

Phase velocity of an RF signal can be controlled by applying voltage across the planar interdigital varactors placed along the transmission line. Figure 6.14 shows a schematic diagram of a phase shifter with uniformly loaded varactor. Owing to the symmetry of transmission line, the line AA′ can be replaced by a magnetic wall so that both sides of IDT capacitance can be calculated separately. Each part of the IDT capacitance on a multilayer substrate can be calculated by the conformal mapping method. The IDT capacitance can be calculated by dividing it into three sections, namely, that for the periodic fingers C_n, external finger C_3, and finger end section, C_{end}. The total capacitance of a bilateral IDT transmission line can be expressed as

$$C_{\mathrm{tot}} = 2(C_n + C_3 + C_{\mathrm{end}}) \tag{6.33}$$

Using the conformal mapping method for a interdigital capacitor on multilayered substrate, the capacitance of the periodic finger section with finite width, as shown in Figure 6.15, can be written as (Gevorgian and Martinsson, 1996)

$$C_n = (n-3)\varepsilon_0\varepsilon_{\mathrm{en}}\frac{K(k_0)}{K(k_0')}l \tag{6.34}$$

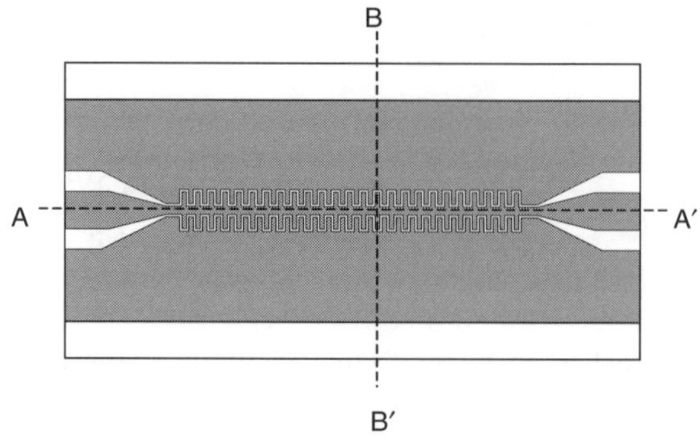

(a)

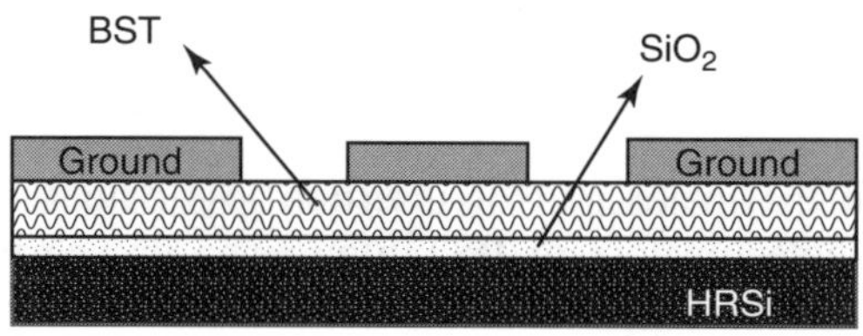

(b)

Figure 6.14 (a) Top view and (b) cross-section of a bilateral interdigital phase shifter. Note: BST, barium strontium titanate; HRSi, high resistivity silicon

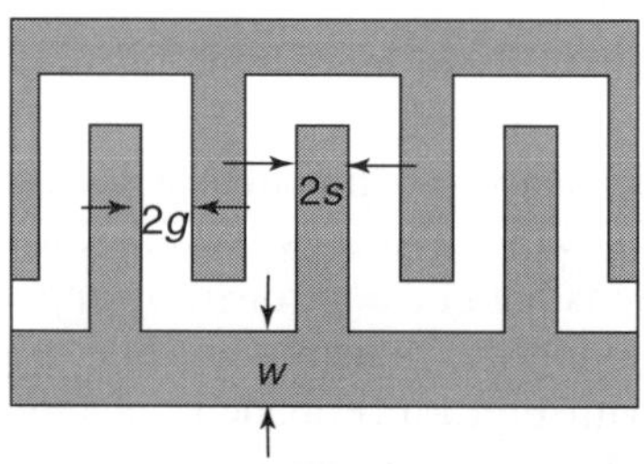

(a)

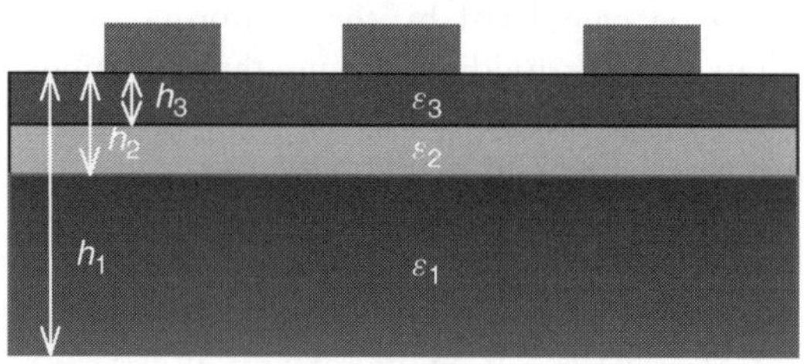

(b)

Figure 6.15 (a) Layout and (b) cross-sectional view of interdigital fingers

where

$$k_0 = \frac{s}{s+g} \tag{6.35}$$

$$k'_0 = (1 - k_0^2)^{1/2} \tag{6.36}$$

$$\varepsilon_{en} = 1 + q_{1n}\frac{\varepsilon_1 - 1}{2} + q_{2n}\frac{\varepsilon_2 - \varepsilon_1}{2} + q_{3n}\frac{\varepsilon_3 - \varepsilon_2}{2} \tag{6.37}$$

$$q_{in} = \frac{K(k_{in})}{K(k'_{in})}\frac{K(k'_0)}{K(k_0)} \tag{6.38}$$

$$k_{in} = \frac{\sinh[\pi s/2h_i]}{\sinh[\pi(s+g)/2h_i]}\left\{\frac{\cosh^2[\pi(s+g)/2h_i] + \sinh^2[\pi(s+g)/2h_i]}{\cosh^2[\pi s/2h_i] + \sinh^2[\pi(s+g)/2h_i]}\right\}^{1/2} \tag{6.39}$$

and $i = 1, 2, 3$.

The capacitance of the external finger section can be calculated from the co-planar waveguide model and is given by

$$C_3 = 4\varepsilon_0\varepsilon_{e3}\frac{K(k'_{03})}{K(k_{03})}l \tag{6.40}$$

where

$$k_{03} = \frac{s}{s+2g}\left\{\left[1 - \left(\frac{s+2g}{s+2s_1+2g}\right)^2\right]\left[1 - \left(\frac{s}{s+2s_1+2g}\right)^2\right]^{-1}\right\}^{1/2} \tag{6.41}$$

$$\varepsilon_{e3} = 1 + q_{13}\frac{\varepsilon_1 - 1}{2} + q_{23}\frac{\varepsilon_2 - \varepsilon_1}{2} + q_{33}\frac{\varepsilon_3 - \varepsilon_2}{2} \tag{6.42}$$

$$q_{i3} = \frac{K(k_{i3})}{K(k'_{i3})}\frac{K(k'_{03})}{K(k_{03})} \tag{6.43}$$

$$k_{i3} = \frac{\sinh[\pi s/2h_i]}{\sinh[\pi(s+2g)/2h_i]}\left\langle\left\{1 - \frac{\sinh^2[\pi(s+2g)/2h_i]}{\sinh^2[\pi(s+2s_1+2g)/2h_i]}\right\}\right.$$
$$\left.\times\left\{1 - \frac{\sinh^2[\pi s/2h_i]}{\sinh^2[\pi(s+2s_1+2g)/2h_i]}\right\}^{-1}\right\rangle^{1/2} \tag{6.44}$$

$$k'_{i3} = (1 - k_{i3}^2)^{1/2} \tag{6.45}$$

and $i = 1, 2, 3$.

The field distribution at the corner of the finger end section is complex because of the fringing field. This capacitance with an approximate value of fringing field can be calculated by conformal mapping:

$$C_{\text{end}} = 2ns(2+\pi)\varepsilon_0\varepsilon_{\text{eend}}\frac{K(k'_{0\text{end}})}{K(k_{0\text{end}})} \tag{6.46}$$

where

$$\varepsilon_{\text{eend}} = 1 + q_{1\text{end}}\frac{\varepsilon_1 - 1}{2} + q_{2\text{end}}\frac{\varepsilon_2 - \varepsilon_1}{2} + q_{3\text{end}}\frac{\varepsilon_3 - \varepsilon_2}{2} \tag{6.47}$$

$$q_{i\text{end}} = \frac{K(k_{i\text{end}})}{K(k'_{i\text{end}})}\frac{K(k'_{0\text{end}})}{K(k_{0\text{end}})} \tag{6.48}$$

$$k_{0\text{end}} = \frac{x}{x + 2g_{\text{end}}}\left\{\left[1 - \left(\frac{x + 2g_{\text{end}}}{x + w + 2g_{\text{end}}}\right)^2\right]\right.$$
$$\left.\times\left[1 - \left(\frac{x}{x + w + 2g_{\text{end}}}\right)^2\right]^{-1}\right\}^{1/2} \tag{6.49}$$

$$k_{i\text{end}} = \frac{\sinh[\pi x/2h_i]}{\sinh[\pi(x + 2g_{\text{end}})/2h_i]}$$
$$\times\left\{\frac{\sinh^2[\pi(x + w + 2g_{\text{end}})/2h_i] - \sinh^2[\pi(x + 2g_{\text{end}})/2h_i]}{\sinh^2[\pi(x + w + 2g_{\text{end}})/2h_i] - \sinh^2[\pi x/2h_i]}\right\} \tag{6.50}$$

The total capacitance is are written as

$$C_{\text{tot}} = 2\left[(n-3)\varepsilon_0\varepsilon_{en}\frac{K(k_0)}{K(k'_0)}l + 4\varepsilon_0\varepsilon_{e3}\frac{K(k'_{03})}{K(k_{03})}l + 2n(2s+g)\varepsilon_0\varepsilon_{\text{eend}}\frac{K(k'_{0\text{end}})}{K(k_{0\text{end}})}\right] \tag{6.51}$$

The effective capacitance is given by

$$\varepsilon_{\text{eff}} = C_{\text{tot}}\left\{2\left[(n-3)\varepsilon_0\frac{K(k_0)}{K(k'_0)}l + 4\varepsilon_0\frac{K(k'_{03})}{K(k_{03})}l + 2n(2s+g)\varepsilon_0\frac{K(k'_{0\text{end}})}{K(k_{0\text{end}})}\right]\right\}^{-1} \tag{6.52}$$

where

$$\varepsilon_{\text{eff}} = \frac{C_{\text{tot}}(\varepsilon_{\text{r}})}{C_{\text{tot}}(\varepsilon_0)} \tag{6.53}$$

The change in permittivity due to the applied voltage gives a change in phase velocity for a uniform loaded bilateral IDT phase shifter. The change in phase velocity is according to the following relation:

$$\nu_{\text{p}} = \frac{c}{\sqrt{\varepsilon_{\text{eff}}}} \tag{6.54}$$

6.4.3 Interdigital capacitor phase shifters

Another configuration of a phase shifters using a planar interdigitated capacitor (IDC) and interdigital finger are shown in Figure 6.16. These IDCs are fabricated with 1 μm gap and are distributed with 340 μm spacing along a transmission line on Al_2O_3 substrate. The planar IDT phase shifter shows a continuous phase shift from 0° to 110° at 20 GHz for 100 V dc bias. A planar coupled microstripline phase shifter (Canedy and Aggarwal, 2000) fabricated on high resistive silicon shows 35° phase shift for a bias voltage of 40 V at

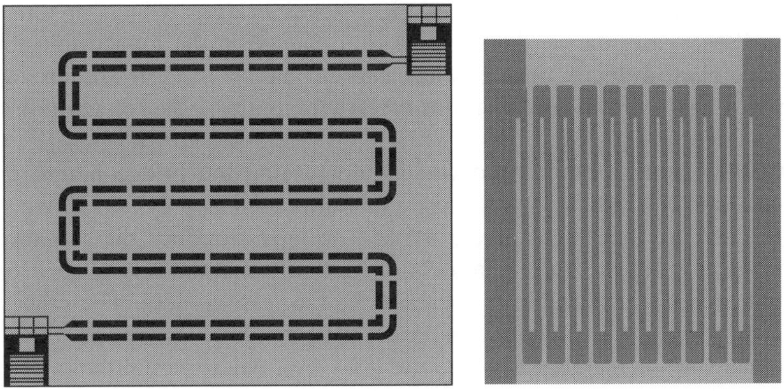

(a) (b)

Figure 6.16 Layout of (a) planar interdigital capacitor phase shifter; (b) interdigital finger phase shifter. Reproduced from Y. Liu, A.S. Nagra, E.G. Erker, P. Periaswamy, T.R. Tayler, J.S. Speck and R.A. York, 2000, '$BaSrTiO_3$ interdigitated capacitors for distributed phase shifter applications', *IEEE Microwave and Guided Wave Letters* (November): 448–450, by permission of IEEE, © 2000 IEEE

15 GHz. A comparison of the performance of the phase shifter on silicon and MgO substrates shows that the phase shifting performance on silicon substrate is superior to that on MgO substrate.

It can be concluded that the IDT configuration is a quite promising technology for high-performance phase shifter designs (Jose *et al.*, 2001). The IDT configuration can easily be integrated with active and passive devices of the microwave integrated circuit.

6.5 APPLICATIONS

Development of adaptable and high gain with polarization diversity antennas is one of the goals of the development of electronically scanned antennas. Constraints on size, cost, polarization and side-lobe levels limit many of the components for further development in phased arrays. MEMS switches and components show exceptional abilities compared with the PIN diode and ferrite phase shifters. RF MEMS switches can be used to provide very-low-loss phase shifters with discrete phase states. Ferroelectric materials such as a thin film of BST can be used for low-loss phase shifters with continuous tuning.

6.6 CONCLUSIONS

The key to the development of low-cost, lightweight phased array antennas and radar systems is the low insertion loss, low drive power and low production cost of phase shifters. The current phase shifter technologies have advanced to a level of being able to provide only one or two of the above key attributes at the same time. The exceptional abilities of MEMS switches includes low insertion loss, low drive power and low intermodulation distortion, leading to the development of MEMS phase shifters utilizing these switches.

The capacitive shunt switch consists of a thin metal bridge suspended over the center of a co-planar waveguide. The applied bias voltage between the MEMS bridges and bottom electrodes changes the height of the MEMS bridges, which in turn varies the distributed MEMS capacitance. This results in a change in the loaded transmission line impedance and phase velocity, which in turn causes phase shift. The structure with several MEMS bridges can act as a phase shifter when the applied bias voltage is less than the pull-down voltage.

The actuation voltage can be reduced by decreasing the bridge height or adopting bridge materials with a relatively low elastic modulus such as polymer. However, it is not advisable to decrease the height of the MEMS bridges to reduce the actuation voltages because the narrow height makes fabrication difficult and there is a need to keep the bridge height as large as possible to increase the fabrication yield. The other alternative is to fabricate bridges with polymers whose elastic modulus is around 5 GPa, which is much less than that of metals (50 to about 100 GPa). Microstereolithography is one of the suitable methods for the fabrication of polymer bridges.

Phase shifters using voltage controllable ferroelectric thin films can provide a continuous phase shift due to change in phase velocity of the RF signal by applying a bias voltage. The capacitance of these devices can be varied by applying voltage between two electrodes. Since the electric fields are well confined in the narrow parallel plates it has the structural advantage of greater tunability for a thin sheet. Monolithic integration of these phase shifters is also possible.

REFERENCES

Acikel, B., Liu, Y., 2001, 'Phase shifters using (Ba,Sr)TiO_3 thin films on sapphire and glass substrates', in *IEEE MTT-S International Microwave Symposium, Volume 2, 2001*, IEEE, Washington, DC: 1191–1194.

Barker, N.S., Rebeiz, G.M., 1998, 'Distributed MEMS true-time delay phase shifter and wide-band switches', *IEEE Microwave Guided Wave Letters* (April): 1881–1890.

Bertsch, A., Lorenz, H., Renaud, P., 1998, 'Combining microstereolithography and thick resist UV lithography for 3D microfabrication micro electro mechanical systems', *Proceedings of IEEE 11th Annual International Workshop on Micro Electro Mechanical Systems (MEMS)*, Jan 1998, Heidelberg, Germany; IEEE, 18–23.

Borgioli, A., Liu, Y., Nagra, A.S., York, R.A., 2000a, 'Low loss distributed MEMS phase shifter', *IEEE Microwave and Guided Wave Letters* **10**(1): 7–9.

Canedy, C.L., Aggarwal, S., 2000, 'Structural and dielectric properties of epitaxial $Ba_{1-x}Sr_xTiO_{3/}$-$Bi_4Ti_3O_{12/}ZrO_2$ heterostructures grown on silicon', *Applied Physics Letters* **77**(10): 1523–1525.

Coget, P., Philippe., P., Pauker., V., Dautriche, P., Jean, P., 1989, 'A multioctave active MMIC quadrature phase shifter', in *Proceedings of IEEE Microwave and Millimeter-wave Monolithic Circuits Symposium*, IEEE, Washington, DC: 1510–1517.

De Flaviis, F., Alexopoulos, N., 1997, 'Planar microwave integrated phase-shifter design with high purity ferroelectric material', *IEEE Transactions on Microwave Theory and Techniques* **45**(6): 963–969.

Erker, E.G., Nagra, A., 2000, 'Monolithic Ka-band phase shifter using voltage tunable $BaSrTiO_3$ parallel plate capacitors', *IEEE Microwave and Guided Wave Letters* **10**: 10–12.

Gevorgian, S.S., Martinsson, T., 1996, 'CAD models for multilayered substrate interdigital capacitors', *IEEE Transactions on Microwave Theory and Techniques* **44**(6): 896–904.

Ghione, G., Naldi, C.U., 1987, 'Coplanar waveguides for MMIC applications: effects of upper shielding, conductor backing, finite-extent ground planes, and line-to-line coupling', *IEEE Transactions on Microwave Theory and Techniques, MTT* **35**(3): 260–267.

Goldsmith, C.L., Lin, T.H., Powers, B., Wu, W.R., Norvell, B., 1995, 'Micromechanical membrane switches for microwave applications', *IEEE Microwave Theory and Techniques Symposium*, IEEE, Washington, DC: 91–94.

Goldsmith, C.L., Randall, J., Eshelman, S., Lin, T.H., Denniston, D., Chen, S., Norvell, B., 1996, Characteristics of micromachined switches at microwave frequencies', *Proceedings of IEEE MTT-S International Microwave Symposium, Volume 2*, IEEE, Washington, DC: 1141–1144.

Goldsmith, C.L., Yao, Z., Eshleman, S., Denniston, D., 1998, 'Performance of low-loss RF MEMS capacitive switches', *IEEE Microwave and Guided Wave Letters* **8**(8): 269–271.

Hayden, J.S., Rebeiz, G.M., 2000, '2-bit MEMS distributed X-band phase shifters', *IEEE Microwave and Guided Wave Letters* **10**(12): 540–541.

Hayden, J.S., Malczewski, A., Kleber, J., Goldsmith, C.L., Rebeiz, G.M., 2001, '2 and 4-bit DC-18 GHz microstrip MEMS distributed phase shifters', in *Proceedings of IEEE MTT-S International Microwave Symposium, Volume 1, 2001*, IEEE, Washington, DC: 219–222.

Ikuta, K., Hirowatari, K., 1993, 'Real three dimensional microfabrication using stereo lithography and metal molding', in *Proceedings IEEE MEMS '93*, IEEE, Washington, DC: 42–47.

Ikuta, K., Ogata, T., Tsubio, M., Kojima, S., 1996, 'Development of mass productive microstereolithography (mass-IH process)', in *Proceedings of IEEE MEMS '96*, IEEE, Washington, DC: 301–305.

Isom, R., Hawkins, M., Richins, J., McEwan, S., Iskabder, M., Grow, R., 2000, 'Comparative evaluation of MEMS and ferroelectric technologies in phase shifter design', in *Proceedings of IEEE Antennas and Propagation Society International Symposium, 2000*, IEEE, Washington, DC: 808–811.

Ji, T.S., Vinoy, K.J., Varadan, V.K., 2001, 'Distributed MEMS phase shifters by microstereolithography on silicon substrates for microwave and millimeter wave applications', *Smart Materials and Structures* **10**: 1224–1229.

Jose, K.A., Yoon, H., Vinoy, K.J., Sharma, P., Varadan, V.K., Varadan, V.V., 2001, 'Low voltage tunable capacitors for RF MEM filters and antenna applications', in *Proceedings of IEEE AP-S International Symposium, Boston Volume 3*, IEEE, Washington, DC: 670–73.

Katagi, T., Nakajima, N., 1993, 'Photoforming applied to fine machining', in *Proceedings IEEE MEMS '93*, IEEE, Washington, DC: 173–178.

Kato, T., Tanaka, Y., Ueda, H., Kano, H., Hashimoto, M., 1992, 'L-band phase shifter with switching FETs for phased array antenna', in *Proceedings of IEEE MTT-S International Microwave Symposium 1992*, IEEE, Washington, DC: 1527–1530.

Koul, S.K., Bhat, B., 1991, *'Microwave and Millimeter Wave Phase Shifters'*, Artech House, London.

Liu, Y., Nagra, A.S., Erker, E.G., Periaswamy, P., Tayler, T.R., Speck, J.S., York, R.A., 2000, '$BaSrTiO_3$ interdigitated capacitors for distributed phase shifter applications', *IEEE Microwave and Guided Wave Letters* (November): 448–450.

Liu, Y., Borgioli, A., Nagra, A.S., York, R.A., 2000b, 'K-band 3-bit low-loss distributed MEMS phase shifter', *IEEE Microwave and Guided Wave Letters* (October): 415–417.

Monnert, S., Loubere, V., Corbel, S., 1999, 'Microstereolithography using a dynamic mask generator and a non-coherent visible light source', *Proceedings of SPIE Design Test and Microfabrication of MEMS and MOEMS*, Paris, SPIE, 553–561.

Muldavin, J.B., Rebeiz, G.M., 2000, 'High-isolation CPW MEMS shunt switches', *IEEE Transactions on Microwave Theory and Techniques* **48**(6): 1045–1056.

Nakamoto, T., Yamaguchi, K., 1996, 'Consideration on the producing of high aspect ratio micro parts using UV sensitive photopolymer', *Proceedings of the Seventh International Symposium on Micro Machine and Human Science, MHS '96*, Nagoya, Japan, 53–58; IEEE, Piscataway, NJ, USA.

Neidert, R.E., Krowne, C.M., 1985, 'Voltage variable microwave phase shifter', *Electronics Letters* **21**(15): 636–638.

Osterberg, P., Yie, H., Cai, X., White, J., Senturia, S., 1994, 'Self-consistent simulation and modeling of electrostatically deformed diaphragms', in *Proceedings of IEEE Microelectromechanical Systems Conference, January 1994*, IEEE, Washington, DC: 28–32.

Pillans, B., Eshelman, S., Malczewski, A., Ehmke, J., Goldsmith, C., 1999, 'Ka-band RF MEMS phase shifters', *IEEE Microwave and Guided Wave Letters* **9**: 520–522.

Pillans, B., Eshelman, S., Malczewski, A., Ehmke, J., Goldsmith, C., 2000, 'Ka-band RF MEMS phase shifters for phased array applications', in *IEEE Radio Frequency Integrated Circuits Symposium, 2000*, IEEE, Washington, DC: 195–198.

Reggia, F., Spencer, E.G., 1957, 'A new technique in ferrite phase shifting for beam scanning of microwave antennas', *Proceedings of IRE* **45**: 1510–1517.

Van Keuls, F.W., Romanofsky, R.R., Bohman, D.Y., Winters, M.D., Miranda, F.A., 1997, '($YBa_2Cu_3O_{7-\delta}$,Au)/$SrTiO_3$/$LaAlO_3$ thin film conductor/ferroelectric coupled microstripline phase shifters for phased array applications', *Applied Physics Letters* **71**: 3075–3077.

Varadan, V.K., Ghodgaonkar, D.K., Varadan, V.V., Kelly, J.F., Gilderdas, P., 1991, 'Ceramic phase shifters for electronically steerable antenna systems', *Microwave Journal* (August): 102–123.

Varadan, V.K., Jose, K.A., Varadan, V.V., Hughes, R., Kelly, J.F., 1995, 'A novel microwave planar phase shifter', *Microwave Journal* (April): 244–254.

Varadan, V.K., Jiang, X.N., Varadan, V.V., 2001, *Micro Stereo Lithography for Fabrication of 3D MEMS*, John Wiley, London.

White, J.F., 1965, 'High power p–i–n diode controlled, microwave transmission phase shifters', *IEEE Transactions on Microwave Theory and Techniques* **13**: 233–242.

Yao, Z.J., Chen, S., Eshelman, S., Denniston, D., Goldsmith, C., 1999, 'Micromachined low-loss microwave switches', *IEEE Journal of Microelectromechanical Systems* **8**(2): 129–134.

York, R., Nagra, A., 2001, 'Microwave integrated circuits using thin-film BST', *Proceedings of IEEE on Applications of Ferroelectrics* **1**: 195–200.

7
Micromachined transmission lines and components

7.1 INTRODUCTION

Radio frequency (RF) planar components and integrated circuits are the central nervous system for many modern portable communication devices. The past decade has witnessed a technological revolution of solid-state devices and their implementation in very-large-scale integrated (VLSI) circuits. This has changed the outlook of most of the communication devices based on classic vacuum tube devices and transistors. Further to it, the planarization of RF interconnections and wires using microstrip lines, strip lines and co-planar waveguides (CPWs) provides compatibility with the solid-state silicon and GaAs components with design flexibility, compactness and reduction of weight with superior performance.

The very high quality factor (tens of thousands) and stability compared with the thermal variations of surface acoustic wave (SAW) resonators and filters are the main reasons for their wide use in frequency selection subsystems in most communication devices. Many high-performance low-power RF transceivers also use devices such as discrete inductors and variable capacitors for tuning it and for coupling the signals from the front-end devices. The use of bulk and SAW filters and tunable tank circuits in cellular applications imposes a significant bottleneck against the miniaturization of transceivers. Even though silicon integrated circuits (ICs) are now operating in the gigahertz frequency range and modern bipolar, CMOS (complementary metal oxide semiconductor) and BiCMOS (bipolar complementary metal oxide semiconductor) processes provide high-frequency silicon RF ICs to compete with GaAs in the low gigahertz frequency regime, the lack of high-quality passive components on silicon has made it a poor choice for high-frequency circuits. Also, the lossy silicon substrate makes the design of high Q reactive components in silicon difficult. There are still many functions that cannot be implemented using conventional IC technology, in particular, the creation of components with high Qs (exceeding 30), required for high-frequency selectivity in communication systems. Also, these planar geometries suffer the drawback of frequency-dependent properties such as parasitic radiation, ohmic losses and dispersion. The parasitic radiation and coupling can be eliminated by sophisticated design of the packages. This may in turn add volume, weight and cost to these devices.

Despite this difficulty, the low cost of silicon IC fabrication techniques over the GaAs IC has the potential for integration of micromachined RF MEMS components in RF circuits, which makes silicon one of the choices. There has been a recent demand for fully integrated planar transmission lines and passive devices for the realization of microelectromechanical systems (MEMS) and monolithic microwave integrated circuits (MMICs). Small size and weight, low power consumption, mass production, reliability and reproducibility are some of the numerous advantages of integration of microwave integrated circuits (MICs) with MEMS.

Many of the recent research efforts have been focused on miniaturizing these RF passive components or on a design procedure to eliminate them from the circuits. This chapter presents the miniaturization efforts of passive devices such as various micromachined transmission line components and devices used for RF applications along with design procedure.

7.2 MICROMACHINED TRANSMISSION LINES

The first transmission line, the stripline, introduced in 1952 in two companion articles (Assadourian and Rimai, 1952; Grieg and Engelmann, 1952), created a new revolutionary hybrid technology. Today, the most commonly used structure to realize MICs are the microstrip. This hybrid technology has evolved to a monolithic one, which drastically increased the operating frequencies by reducing the weight and volume. This planarization of conductors in the form of transmission lines triggered the evolution of other RF components such as directional couplers and eventually at the development of planar antennas of very-large-scale integration for very-high-frequency applications.

Transmission lines in an RF circuit are normally used to carry information from passive elements such as filters, impedance transformers and delay lines and to interconnect them. Multiconductor structures that support the transverse electromagnetic (TEM) wave or non-TEM mode of propagation are commonly referred to as transmission lines. In an electrical circuit, when the transit time of a signal along a connecting strip is similar to a period or close to the duration of a pulse, the circuit must be analyzed in terms of transmission lines and has to be characterized by their geometrical parameters. In such circuits, mismatches and discontinuities may produce multiple reflections. These complex echoes can slow down the signal and add unnecessary delay in a circuit.

Figure 7.1 presents common forms of printed circuit transmission lines generally using low-frequency ranges (microstrip, slotline, co-planar line) and metallic enclosures to prevent the radiation for higher frequency ranges (suspended line, inverted line and fin line) (Gardiol, 1994). A co-planar line is one of the most commonly used transmission lines in which the dominant mode is TEM. The propagation velocity of a TEM wave depends only on the material properties ε and μ. For a microstrip line exposed to air, the wave propagation is through two dielectric mediums. When there are two materials present in a circuit, TEM waves propagate with two velocities, one within the dielectric substrate and the other in the air. The boundary conditions on the interface require a continuity of the tangential components that cannot be satisfied in two-dielectric case. Hence the propagation along a microstrip line is not pure TEM because of the presence of the two dielectric mediums. The field lines between the strip and the ground plane are not contained entirely in the substrate. Therefore, the propagating mode in a microstrip line is not purely TEM but quasi-TEM. The slotline is preferred in circuits requiring high-impedance lines, short circuits, stubs and hybrid combinations of microstrip circuits. Co-planar waveguides can find extensive applications in MICs because adds flexibility to

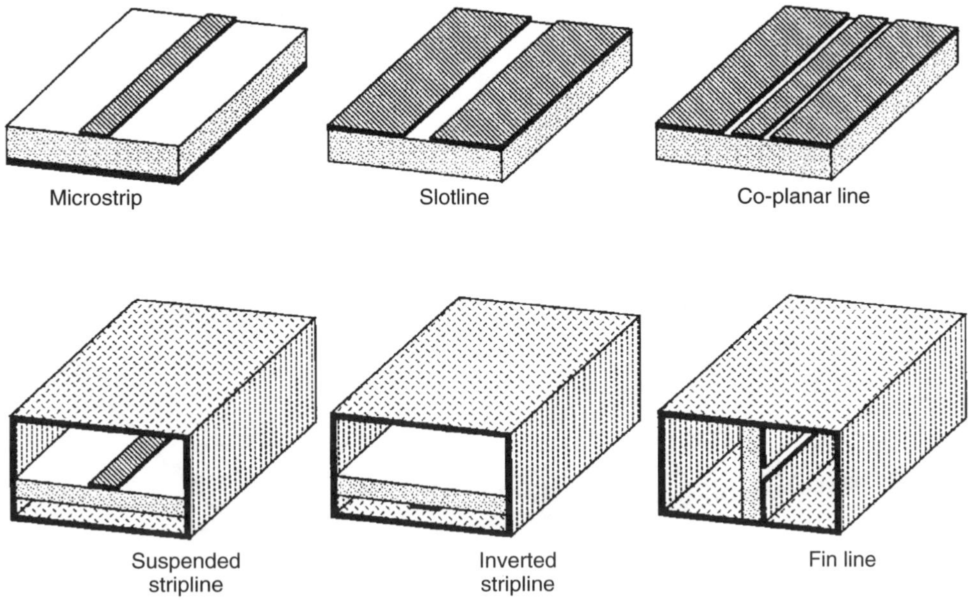

Figure 7.1 Open and shielded microstrip structures. Reproduced from F. Gardiol, 1994, *Microstrip Circuits*, Wiley, New York, by permission of Wiley, © 1994 Wiley

the circuit design. Many other variations of basic microstrip geometry are possible, but the most common one is the covered microstrip, shown in Figure 7.1. The metallic shielding is often used for electrical shielding as well as physical protection of the microstrip circuit. The presence of this metallic ground can, however, perturb the operating characteristics so that its effects may be taken into account during the design. The microstrip and CPW are widely used because their properties are well understood and have advantages for use in MMIC applications.

Figure 7.2 presents the possible means of transmission and radiation in a planar structure due a launching wave. These waves can be cleverly utilized for the application of signal transmission or radiation in a circuit, depending on the design parameters of the conductors, substrate properties and the enclosures. Guided waves may be predominant for a high dielectric thin substrate compared with the wavelength, which is unwanted in case of an antenna. The radiated wave is an unwanted leakage in transmission lines. The leaky waves contribute to radiation under favorable conditions and hence increase apparent antenna size and produce large directive gain. Surface waves also become significant for thick and high-permittivity substrates.

7.2.1 Losses in transmission lines

The inhomogeneities in a microstrip structure, as shown in Figure 7.3, makes analysis difficult. However, the analysis of these structures using integral equations shows results close to the measured values. These inhomogeneities are due to the boundary conditions at the air–dielectric interface, due to partially covered metal surfaces with very thin conducting layers and due to its finite dimensions.

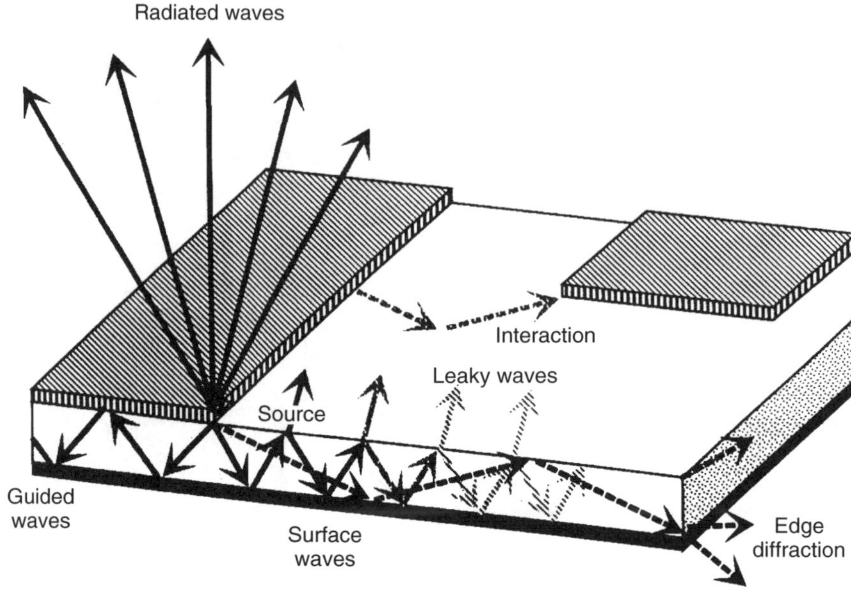

Figure 7.2 Possible radiation in a microstrip structure. Reproduced from F. Gardiol, 1994, *Microstrip Circuits*, Wiley, New York, by permission of Wiley, © 1994 Wiley

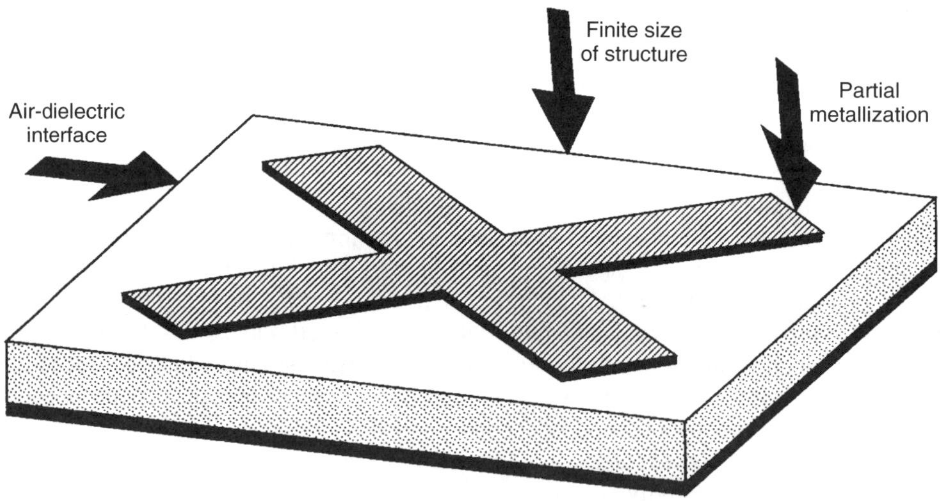

Figure 7.3 Microstrip structure showing general inhomogeneity. Reproduced from F. Gardiol, 1994, *Microstrip Circuits*, Wiley, New York, by permission of Wiley, © 1994 Wiley

The propagation efficiency is an important factor in planar circuit design because of the presence of a dielectric medium between the transmission lines. The substrate tends to induce parasitic currents that impede the line performance. The attenuation of a signal in a transmission line is mainly due to conductor, dielectric and radiation losses. The resistive nature of the conductor forces the signal to penetrate through the conductor. At microwave frequencies, the current density is at a maximum at the surface of the conductor. This current density decreases exponentially with depth into the conductor, known as the skin depth. This generates heat and power loss, called ohmic losses. The skin depth δ, is a function of frequency f and resistivity ρ of the material, and is given by

$$\delta = \frac{1}{2\pi}\left(\frac{\rho}{f}\right)^{1/2} \tag{7.1}$$

which is generally measured in decibels per centimeter. The conductor loss in a microstrip line varies inversely with the width of the line and becomes dominant when the substrate thickness becomes electrically small. Also, the loss can be reduced considerably when the metal increases to many skin depths.

Dielectric loss is introduced in a planar transmission line whenever the excited field is trying to distribute itself inside a substrate. This is mainly due to the tangent loss associated with the dielectric material, which is a result of the inability of the charges instantaneously to follow the changes induced by the electric field. This loss is the dielectric loss measured in decibels per wavelength. The planar transmission also suffers parasitic radiation along the length of the line or at localized discontinuities. Designing the line to operate only in the dominant mode can minimize the radiation loss. However, it is difficult to control the radiation because of discontinuities in the circuit. The dielectric and radiation losses can be reduced by decreasing the dielectric thickness in a microstrip circuit.

The implementation of RF circuits in planar form provided the capability of integration but generated many unwanted effects such as fringing fields, radiation, dispersion and increased ohmic losses, which impose serious limitations at millimeter wave frequencies. The significant advantages of micromachining technology have solved most of these problems in recent years. It is known that the substrate losses become significant as the frequency increases. The micromachining technology started a clever way of reducing these substrate losses by removing portions of the lossy substrate material surrounding a microwave component, similar to high-Q inductors described in Chapter 4. To eliminate substrate effects from transmission line performance, the substrate must be highly resistive or a membrane to separate the lines from the substrate. RF transmission lines enclosed in a micromachined metallized cavity (Drayton and Katehi, 1994) created by anisotropic etching of a low-resistivity silicon is a useful and cost-effective way of minimizing the substrate losses for microwave components. Microwave components developed by micromachining include transmission lines, waveguides (Gedney *et al.*, 1997; McGrath *et al.*, 1993; Pekre *et al.*, 1997; Shenonda and Pearson, 1998), low-pass filter (Din, Harokopus and Katehi, 1991; Weller, 1995; Weller and Katehi, 1996; Weller *et al.*, 1996), micromachined antennas (Lubeche, Mizuno and Rebeiz, 1998), three-dimensional high-frequency distribution networks (Hindreson *et al.*, 2000; Weller *et al.*, 2000), conformal packaging (Dryton and Katehi, 1995; Dryton, Hindreson and Katehi, 1996) and directional couplers (Robertson *et al.*, 1998).

7.2.2 Co-planar transmission lines

One of the ways to reduce the excitation of surface waves is by bringing the ground plane to the proximity of the active device. Co-planar striplines and waveguides fabricated on low-permittivity substrate (Cheng *et al.*, 1994a) and on thin membrane (Cheng *et al.*,

1994b) are capable of supporting ultra-high-frequency pulses. However, this can cause excitation of parallel plate modes and microstrip modes. The finite ground co-planar (FGC) lines provide an alternative to co-planar waveguides for millimeter and submillimeter wave applications (Herrick, Schwarz and Katehi, 1998). The FGC lines are useful for fabricating active circuits owing to the presence of the center conductor and the proximity of the ground lines in the same plane. This avoids the through holes to a plane on the other side of the substrate. It is also possible to narrow the line width to match the lead widths while keeping constant line impedance. The FGC line is printed on high-resistivity silicon with a thin layer of SiO_2. The oxide is removed after metallization. The characteristic impedance of an FGC line depends on the width of the center conductor, the width of the ground plane and the conductor separation. Figure 7.4(a) shows a conventional co-planar waveguide, widely used for many MMIC applications. Figure 7.4(b) shows a CPW with a lower ground plane, and Figure 7.4(c) shows an FGC line. Figure 7.4(d) is the packaged FGC line.

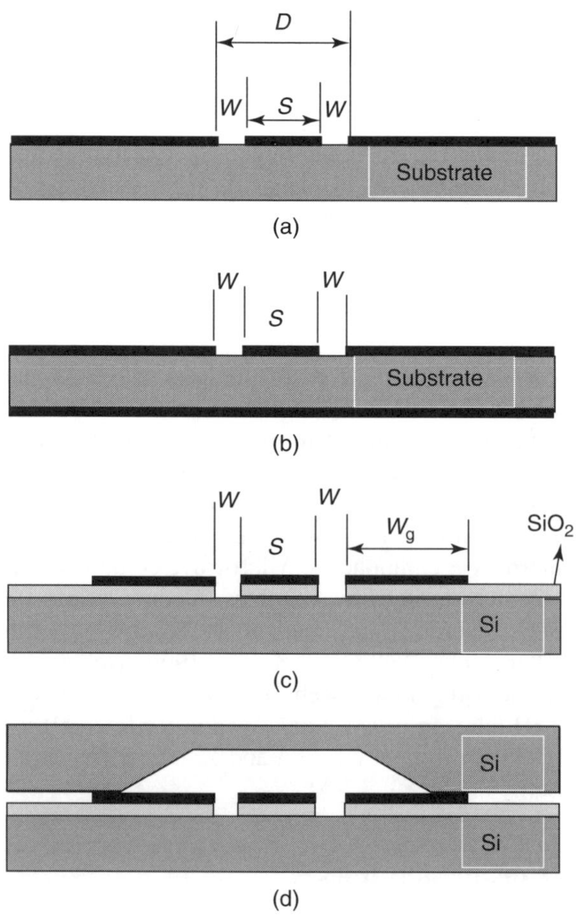

Figure 7.4 Schematic diagram of co-planar waveguides (CPWs): (a) conventional CPW; (b) CPW with lower ground plane; (c) finite ground co-planar (FGC) line on silicon; (d) packaged FGC

7.2.2.1 Design

A CPW consists of a thin metallic strip deposited on the surface of a dielectric film with two conducting ground lines parallel to the strip, as shown in Figure 7.4(a). The ground plane should extend more than $5D$. Also, D should be less than $\lambda/2$ to prevent the propagation of higher-order modes. The characteristic impedance of the line can be written as (Wadell, 1991)

$$Z_c = \frac{Z_0 K(k')}{\sqrt{\varepsilon_e} 4K(k)} \tag{7.2}$$

where Z_0 is the free space impedance, $K(k')$ is the complete elliptical integral of first kind with the modulus $k = S/D$, $D = 2W + S$, and $k' = (1 - k^2)^{1/2}$. W is the space between the conducting strip of width S and the ground plane. Considering the thickness of the strip is negligible, the effective dielectric constant can be written as

$$\varepsilon_e = \frac{\varepsilon_r + 1}{2} \tag{7.3}$$

The finite thickness h of the substrate affects the effective dielectric constant as

$$\varepsilon_e = 1 + \frac{\varepsilon_r - 1}{2} \frac{K(k')}{K(k)} \frac{K(k_t)}{K(k_t')} \tag{7.4}$$

where

$$k_t = \frac{\sinh(\pi W/4h)}{\sinh(\pi D/4h)} \tag{7.5}$$

and

$$k_t' = (1 - k_t^2)^{1/2} \tag{7.6}$$

Microstrip and co-planar waveguides were used to interconnect various circuit elements in an RF circuit. Micromachined FGC lines have geometry similar to conventional FGC lines except that the material under the transmission lines has been removed by etching. The width of the line and the depth of the groves control the cutoff frequency of the line. For a micromachined FGC line of groove size G, the design equation can be written as (Herrick, Schwarz and Katehi, 1998)

$$2(W_g + W) + S < F_g \lambda_{0,h/2} \tag{7.7}$$

where

$$F_g = \frac{1}{\sqrt{\varepsilon_e}} \tag{7.8}$$

which depends on the thickness of the material, as shown in Equation (7.4). The characteristic impedance of the FGC line printed on dielectric substrates depends on their effective dielectric constant, which is approximately 6 for high-resistivity silicon.

The dielectric loss in a CPW (in $\mathrm{Np\,m^{-1}}$) can be written as (Wadell, 1991)

$$\alpha_{\mathrm{d}} = \frac{q\varepsilon_{\mathrm{r}} \tan\delta}{\varepsilon_{\mathrm{e}}\lambda_{\mathrm{g}}} \tag{7.9}$$

where

$$\lambda_{\mathrm{g}} = \frac{c}{f\sqrt{\varepsilon_{\mathrm{r}}}} \tag{7.10}$$

is the wavelength inside the dielectric and q is the ratio of the actual capacitance to its capacitance with air as the dielectric. For a conductor of thickness t, conductor loss (in $\mathrm{Np\,m^{-1}}$) can be written as (Wadell, 1991)

$$\alpha_{\mathrm{c}} = \frac{R_{\mathrm{s}}\sqrt{\varepsilon_{\mathrm{e}}}[\Phi(S) + \Phi(D)]}{480\pi K(k)K(k')} \tag{7.11}$$

where R_{s} is the surface resistivity and

$$\Phi(x) = \frac{\pi}{x^2} \ln\left[\frac{8\pi x(1-k)}{t(1+k)}\right] \tag{7.12}$$

It is observed that the components in transmission lines and CPW lines, tested for higher frequencies, the performance is degraded as a result of parasitic radiation and coupling, along with parasitics from metallized packages. Higher propagation efficiency can be achieved with better confinement of the waves in a transmission line. A micromachined version of the microstrip line (Herrick, Yook and Katehi, 1998) is useful for high-frequency applications.

7.2.3 Microshield and membrane-supported transmission lines

The most common forms of transmission line are the microstrip and CPW, as shown in Figure 7.1. These two geometries are widely used because their properties are well studied. Also, these structures are compatible with MMICs. Despite their advantages in lower-frequency regimes, these transmission lines exhibit potentially significant limitations at millimeter wave frequencies. They exhibit attenuation, dispersion and multimode propagation, all of which are mainly due to the use of high-permittivity substrate. An obvious solution to this problem is to remove the substrate beneath the conducting lines, as shown in Figure 7.5, and to suspend the line on a thin membrane.

There are several advantages of using nearly homogeneous air–dielectric substrate in microshield lines. Owing to the absence of substrate modes, the dispersion can be reduced along with losses due to parasitic radiation. The dispersion is due to the propagation of different frequency components at different phase velocities. The design process of bends, shorts and steps can also be simplified as a result of the absence of high-dielectric substrate.

The microshield microstrip line is also one of the possible membrane-supported geometries in microstrip structure, as shown in Figure 7.6. The microshield lines are characterized by zero dielectric losses and reduced electromagnetic interference and are compatible with microstrip and co-planar waveguides (Katehi *et al.*, 1993). In microshield lines, a

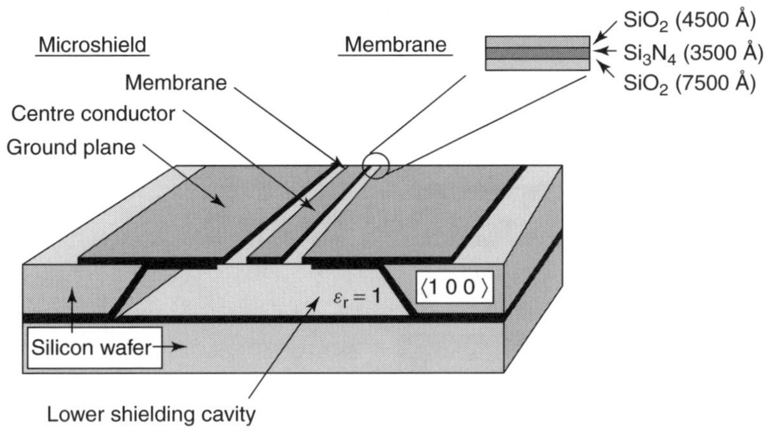

Figure 7.5 Schematic diagram of the microshield line. Reproduced from T.M. Weller, 1995, *Micromachined High Frequency Transmission Lines on Thin Dielectric Membrane*, PhD thesis, University of Michigan, Ann Arbor, MI, by permission of the University of Michigan

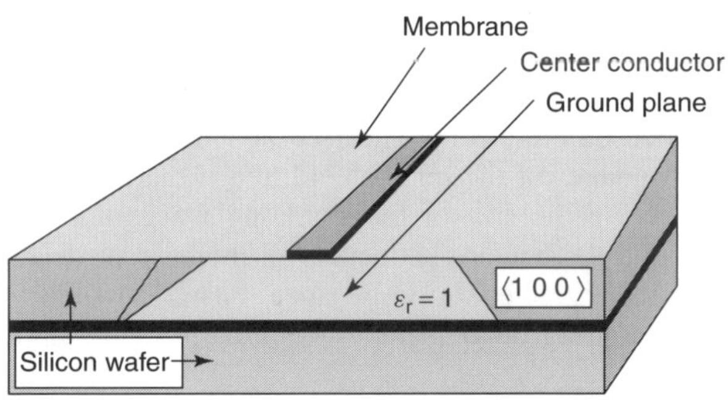

Figure 7.6 Membrane-supported microstrip line. Reproduced from T.M. Weller, 1995, *Micromachined High Frequency Transmission Lines on Thin Dielectric Membrane*, PhD thesis, University of Michigan, Ann Arbor, MI, by permission of the University of Michigan

pure nondispersive TEM wave is propagating though a two-conductor system embedded in a homogeneous medium. A homogeneous medium is achieved by a membrane 1.5-μm thick surrounded by metallized micromachined cavity as a ground. The cavity shielding helps to propagate a pure TEM mode for a frequency band from dc to terahertz, with very low losses and zero dispersion.

The concept of utilizing a thin dielectric membrane to support a high-frequency transmission line was first presented by Katehi and co-workers (Dib and Katehi, 1992; Dib *et al.*, 1991; Weller, Katehi and Rebeiz, 1995), in which they demonstrated that the radiation loss from certain membrane-supported discontinuities is lower than that of a conventional CPW. This allows broadband single-mode operation, without dielectric dispersion and with zero dielectric loss.

The microshield line can be considered an evolution of conventional microstrip or co-planar structures, in which the ground plane has been deformed to totally or partially surround the inner conductor. A microshield line is a partially shielded, quasi-planar

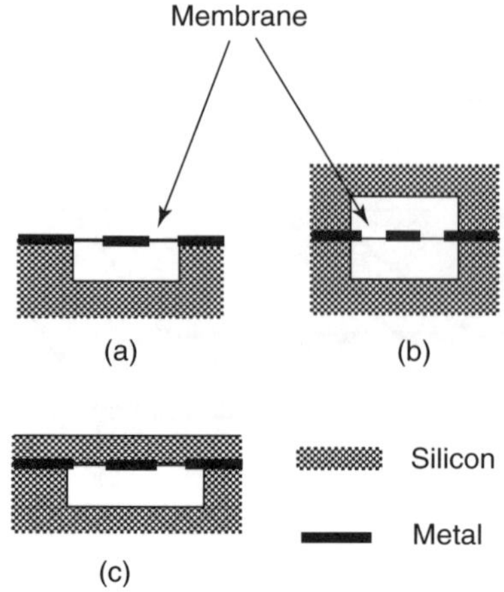

Figure 7.7 Schematic diagram of (a) the microshield transmission lines, (b) the membrane lines and (c) the dielectric microshield line. Reproduced from N.I. Din, W.P. Harokopus and P.B. Katehi, 1991 'Study of a novel planar transmission line', in *Proceedings of IEEE MTT-S Symposium 1991*, IEEE, Washington, DC: 623–626, by permission of IEEE, © 1991 IEEE

transmission line design, which uses a thin dielectric membrane to support the conducting lines or can be suspended on a dielectric sheet, as shown in Figure 7.7.

One of the main advantages of the microshield line is that a wide range of impedance can be achieved by varying the size of the shielding waveguide. Any variation of the conducting ground around the center conductor can increase or decrease in capacitance, resulting in change in its characteristic impedance. Also, the small separation between the center conductor and the ground plane is helpful in single-mode transmission and prevents RF radiation.

The characteristics of the microshield line are found to be well-suited for millimeter-wave and submillimeter-wave applications. Even though the integration of thin dielectric membrane on silicon requires additional steps, the microshield line may eliminate complex steps in fabrication of air bridges.

The microshield line is fabricated on thin dielectric membrane with anisotropic etching on a wafer. The silicon membrane is a $SiO_2-Si_3N_4-SiO_2$ structure. The membrane is slightly under tension to yield flat and rigid self-supporting structure to the line. After depositing and developing the three-layer structure, an opening is defined on the back of the wafer. The silicon is etched until the transparent membrane appears and different microshield circuits are fabricated by attaching them together to form a microshield cavity.

Because the microshield lines use air substrate, the effective dielectric constant is close to 1. However, the use of thin membrane increases the effective dielectric constant slightly. The membrane is a 1.5-mm thick trilayer of $SiO_2-Si_3N_4-SiO_2$. The dielectric constant of oxide is 3.9 and that of nitride is 7.5; the measured effective ε_r changes from 1.09 to 1.15 as the slot width is reduced from 55 to 25 μm, as shown in Figure 7.8.

Figure 7.9 presents the dependence of effective dielectric constant over different geometrical parameters generated using conformal mapping (Weller, 1995). The curves are plotted for different values of K, where $K = S/(S + 2W)$, S is the width of the strip and W is the separation between the ground and the strip line, as shown in Figure 7.4.

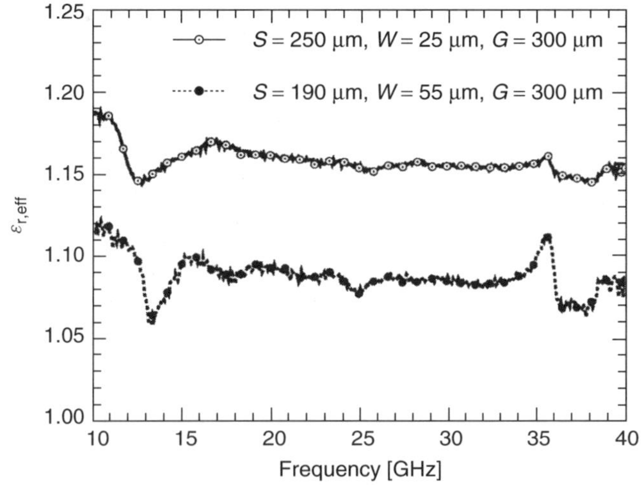

Figure 7.8 Measured effective dielectric constant, $\varepsilon_{\mathrm{reff}}$, of two microshield lines with different aspect ratios. Reproduced from T.M. Weller, L.P.B. Katehi and G.M. Rebeiz, 1995, 'High performance microshield line components', *IEEE Transactions on Microwave Theory and Techniques* **43**(3): 534–543, by permission of IEEE, © 1995 IEEE

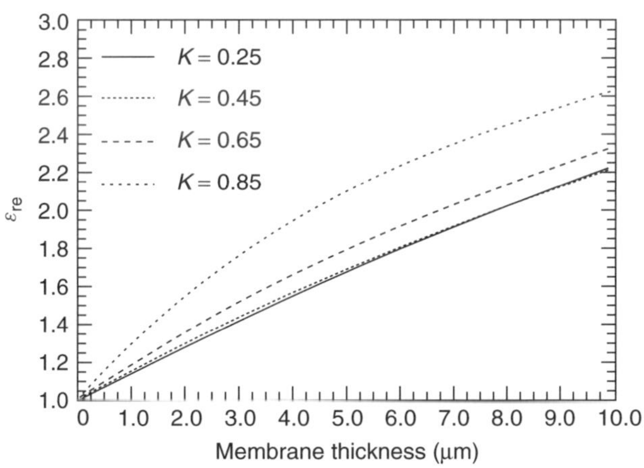

Figure 7.9 Effective dielectric constant, $\varepsilon_{\mathrm{re}}$, versus thickness of the membrane for different aspect ratios, $K = S/(S + 2W)$, where $S + 2W = 100\,\mu\mathrm{m}$. Reproduced from T.M. Weller, 1995, *Micromachined High Frequency Transmission Lines on Thin Dielectric Membrane*, PhD thesis, University of Michigan, Ann Arbor, MI, by permission of the University of Michigan

The development of SiGe heterojunction bipolar transistors (HBTs) operating up to 160 GHz has made unusable the micromachined transmission lines fabricated by CMOS processing on low-resistivity silicon substrate because of the high loss at these frequencies. The solution to this problem is either to develop the circuits on high-resistivity silicon (HRSi) substrate (>2500 Ω cm) or use polyimide layers for wafer planarization on CMOS grade silicon substrate. HRSi can be used to fabricate transmission lines similar to those on GaAs or other good microwave substrates. However, HRSi wafers are more expensive than standard silicon substrate and the standard CMOS fabrication has to be modified to suit the HRSi. RF transmission lines were developed on CMOS grade silicon substrate by depositing a ground plane on top of the silicon substrate, then depositing polyimide layer over it and defining the lines on top of it. Low attenuation is possible for this circuit because the line is defined on a polyimide layer of thickness less than 10 μm and the ground plane is completely shielded from the electromagnetic waves from the lossy silicon wafer. A comparable attenuation to those of the HRSi is obtained for a micromachined CPW on low-resistivity silicon (1 Ω cm) with polyimide interface layer (Ponchak, Margomenos and Katehi, 2001).

Polyimide (PI-1111 from DuPont) of relative dielectric constant 2.8 is deposited on 1 Ω cm silicon wafer. CPW lines with 0.02 μm of titanium and 1.5 μm of gold are fabricated on polyimide using standard lift-off processing. The areas of polymer not protected by the CPW metallization are removed by reactive iron etching (RIE) and the resultant structure is as shown in Figure 7.10.

Elevating the metal lines from the substrate using the micromachining technique also reduces the conductor and dielectric losses. The elevated CPW (ECPW) and overlay CPW (OCPW) designed by Park *et al.* (2000) shows, compared with the conventional CPW lines (2.65-dB cm^{-1} insertion losses at 50 GHz), a measured loss of 1.9 dB cm^{-1} and 1.25 dB cm^{-1} for ECPW and OCPW lines, respectively. Figure 7.11 shows a schematic diagram of the ECPW and OCPW lines. These lines have an overhanging structure separated from the substrate by a sacrificial layer. The lines were fabricated on 560-μm thick glass substrate with electroplated gold structures of thickness 3 μm. The length of the line is 1 cm and the height of the elevated structure is 15 μm from the substrate.

Figure 7.10 Scanning electron micrograph of a micromachined co-planar waveguide line on low-resistiving silicon with an etched polyimide layer of thickness 20.15-μm. Reproduced from G.E. Ponchak, A. Margomenos and L.P.B. Katehi, 2001, 'Low-loss CPW on low-resistivity Si substrates with a micromachined polyimide interface layer for RFIC interconnections', *IEEE Transactions on Microwave Theory and Techniques* **49**(5): 866–870, by permission of IEEE, © 2001 IEEE

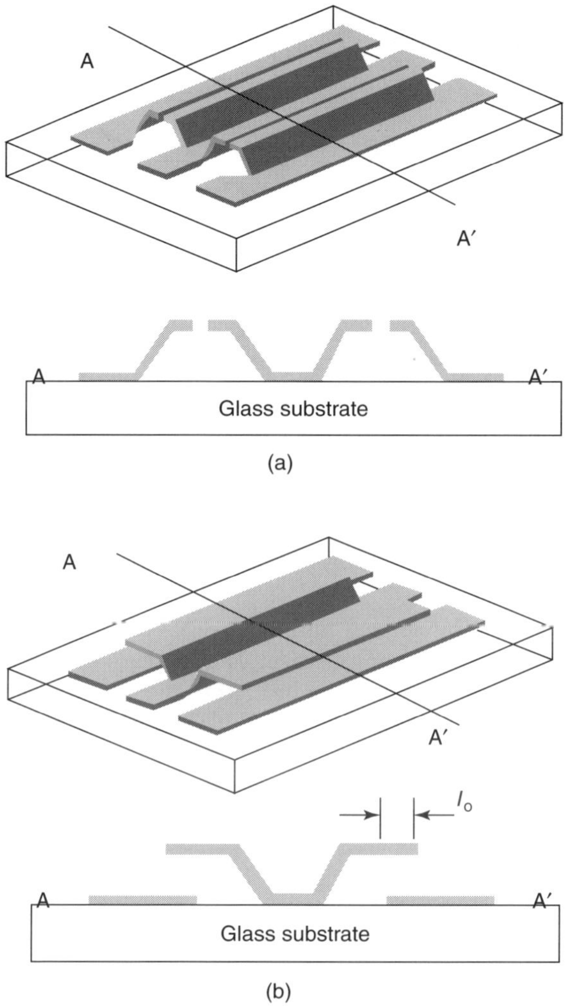

Figure 7.11 Schematic diagram of (a) elevated and (b) overlay co-planar waveguide transmission lines. Reproduced from J.Y. Park, C.W. Baek, S. Jung, H.T. Kim, Y. Kwon and Y.K. Kim, 2000, 'Novel micromachined coplanar waveguide transmission lines for applications in millimeter-wave circuits', *Journal of Applied Physics* **39**: 7120–7124, by permission of the Japanese Journal of Applied Physics

Figure 7.12 shows a comparison of conventional CPW line, which has 2.65 dB cm^{-1} insertion loss at 50 GHz and the losses of ECPW and OCPW lines, in which the losses are found to be reduced at 50 GHz. The insertion loss of the OCPW line is half that of the conventional CPW line as a result of the reduction of conductor and the substrate losses.

7.2.4 Microshield circuit components

The simplicity of fabrication and the planar nature allowing integration of series and shunt elements made FGC transmission lines widely used in MMICs. Using micromachining

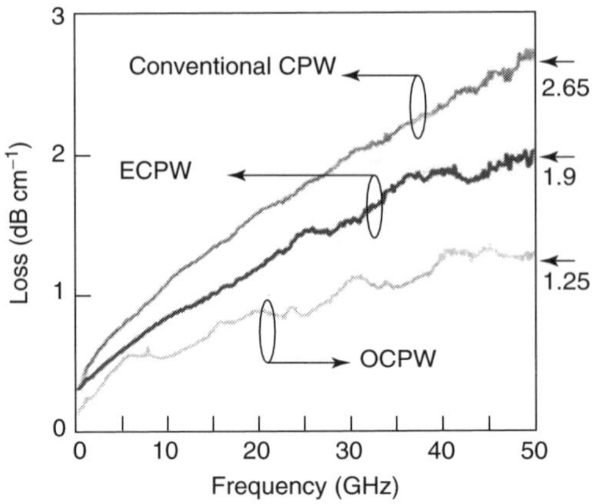

Figure 7.12 Measured losses of three types of 40-Ω transmission lines. Note: CPW, co-planar waveguide; ECPW, elevated co-planar waveguide; OCPW, overlay co-planar waveguide. Reproduced from J.Y. Park, C.W. Baek, S. Jung, H.T. Kim, Y. Kwon and Y.K. Kim, 2000, 'Novel micromachined coplanar waveguide transmission lines for applications in millimeter-wave circuits', *Journal of Applied Physics* **39**: 7120–7124, by permission of the Japanese Journal of Applied Physics

technology, it is possible to implement millimeter-wave one-chip integrated circuits with high performance, low cost and compact size. The increased interest in micromachining and integration of MEMS into RF circuits has led to the advancement of interconnecting components to connect various micromachined waveguides to planar transmission lines. Transitions between microshield and grounded co-planar waveguide (GCPW)(Weller, 1995), FGC line to microstrip transition (Gildas, Katehi and Reseiz, 1998), co-planar to waveguide transition (Becker and Katehi, 1999; Becker *et al.*, 2001), FGC line circuit elements (Goverdhanam, Simons and Katehi, 1999; Herrick and Katehi, 1997; Margomenos *et al.*, 2000; Ponchale, Downey and Katehi, 1997) wafer-to-wafer transitions (Herrick and Katehi, 2000, 2001) are some of the components that need to be integrated in printed or solid-state circuit applications. Figure 7.13 presents the schematic diagram of a 75 Ω microshield line to CPW transition. This transition is the most direct approach for coupling the power into or out of a microshield circuit.

Figure 7.14 presents the measure scattering parameters of two transitions of microshield to grounded CPW line. The widths of the center conductor and slot are 250 and 25 μm, respectively. In curve A, the GCPW dimensions are $S = 50\,\mu\text{m}$, $W = 125\,\mu\text{m}$, with the length of the line $L = 1460\,\mu\text{m}$ (for illustration of dimensions, see Figure 7.13). In curve B, the GCPW dimensions are $S = 30\,\mu\text{m}$, $W = 80\,\mu\text{m}$, and $S = 3500\,\mu\text{m}$.

The common circuit elements such as right-angle bend, filters, open-end and short-end series stubs and a few other realizations of microshield components gave very good performance in the frequency band from 10 GHz to 70 GHz (Weller, Katehi and Rescic, 1995). The difference between the microshield line and a substrate-supported line is that the former has larger circuit dimensions because of the low dielectric constant and the use of a thin dielectric membrane. Also, very broadband operation with a wide range of characteristic impedance is possible for microshield lines.

Figure 7.15 presents top and side views of a 50–73–106 Ω transition in which the 106 Ω is a membrane-supported microstrip line. The width of the lines are $A = 322\,\mu\text{m}$, $B = 122\,\mu\text{m}$ and $C = 513\,\mu\text{m}$.

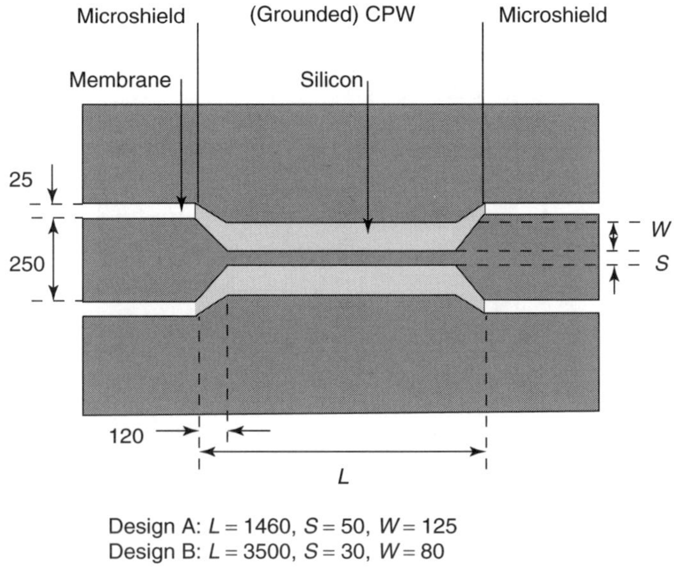

Figure 7.13 Schematic diagram of the microshield to grounded co-planar waveguide (CPW) transition. Dimensions are in microns. Reproduced from T.M. Weller, 1995, *Micromachined High Frequency Transmission Lines on Thin Dielectric Membrane*, PhD thesis, University of Michigan, Ann Arbor, MI, by permission of the University of Michigan

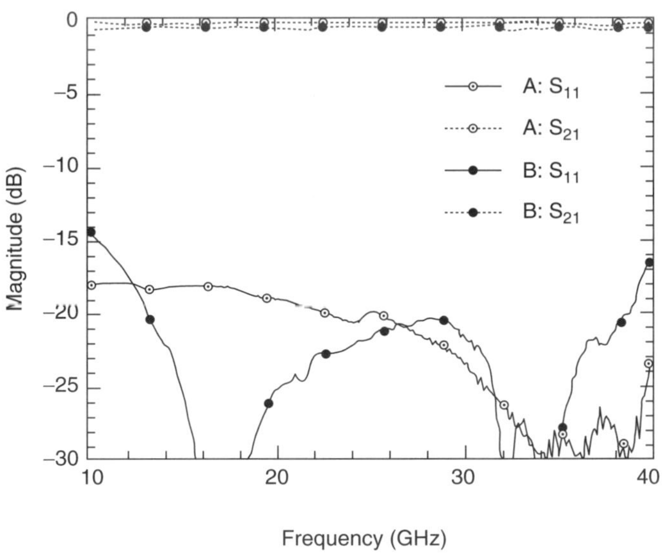

Figure 7.14 Measured scattering parameters of two designs of 75-Ω transition of microshield line to grounded co-planar waveguide transitions. Reproduced from T.M. Weller, 1995, *Micromachined High Frequency Transmission Lines on Thin Dielectric Membrane*, PhD thesis, University of Michigan, Ann Arbor, MI, by permission of the University of Michigan

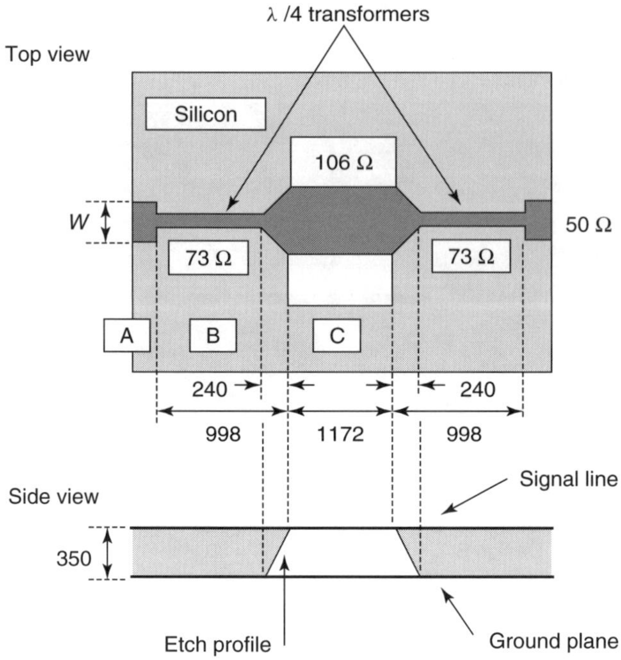

Figure 7.15 Micromachined 50–73–106 Ω transition. Reproduced from T.M. Weller, 1995, *Micromachined High Frequency Transmission Lines on Thin Dielectric Membrane*, PhD thesis, University of Michigan, Ann Arbor, MI, by permission of the University of Michigan

The design of a microshield band pass filter (Weller, 1995) is shown in Figure 7.16 using open-ended series stubs. The metallization shown in Figure 7.16(a) is for $L = 250\,\mu\text{m}$, $S = 50\,\mu\text{m}$, $W = 20\,\mu\text{m}$. The membrane and filter metallization, cavity walls and complete lower ground plane comprises three sections of silicon wafer. Figure 7.16(b) shows the cross-sectional view, with $H = 200\,\mu\text{m}$, $W_1 = 320\,\mu\text{m}$, $W_2 = 40\,\mu\text{m}$.

A micromachined Wilkinson power divider/combiner with a center frequency of 20 GHz, realized on an 8×8 mm membrane (Weller, 1995) is shown in Figure 7.17. Micromachining was done on patterned wafer, simultaneously etching the scribe lines such that the circuits were connected only by 100-μm wide silicon struts. Figure 7.18 shows the Ka-band MMIC power amplifier that shows a gain of 5.2 dB and output power of 0.85 W.

7.2.5 Micromachined waveguide components

Many modern communication and test instruments prefer waveguide components at millimeter wave frequencies because of their low-loss performance and easy fabrication. However, when frequencies are above a few hundred gigahertz, the conventional fabrication of these waveguides in submillimeter size becomes difficult. Also, the recent advancements of microelectronics demand electronic and optoelectronic components with a hundredth of gigahertz bandwidth. The enormous high-frequency signal distortion due to permittivity mismatch between the substrate and air and the losses due to excitation of surface wave modes at these frequencies make planar transmission lines useless at

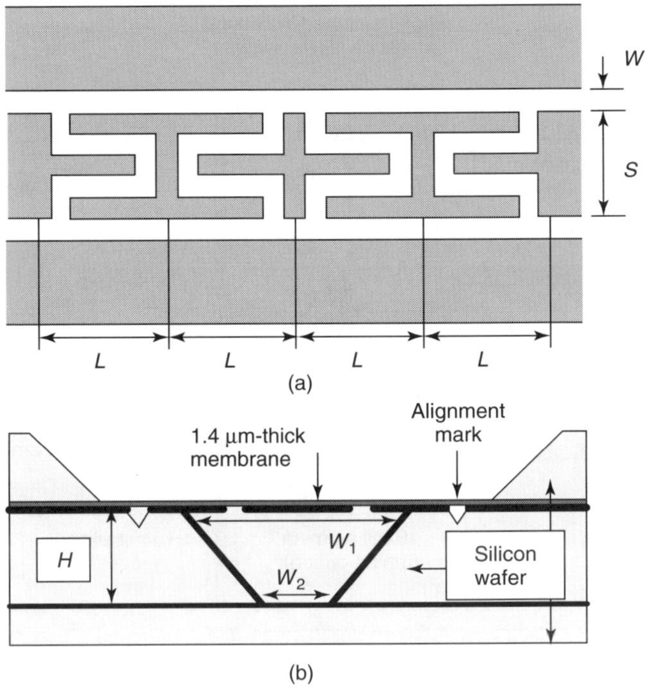

Figure 7.16 Schematic diagram of the four-section microshield bandpass filter: (a) metallization section; (b) cross-sectional view. Reproduced from T.M. Weller, 1995, *Micromachined High Frequency Transmission Lines on Thin Dielectric Membrane*, PhD thesis, University of Michigan, Ann Arbor, MI, by permission of the University of Michigan

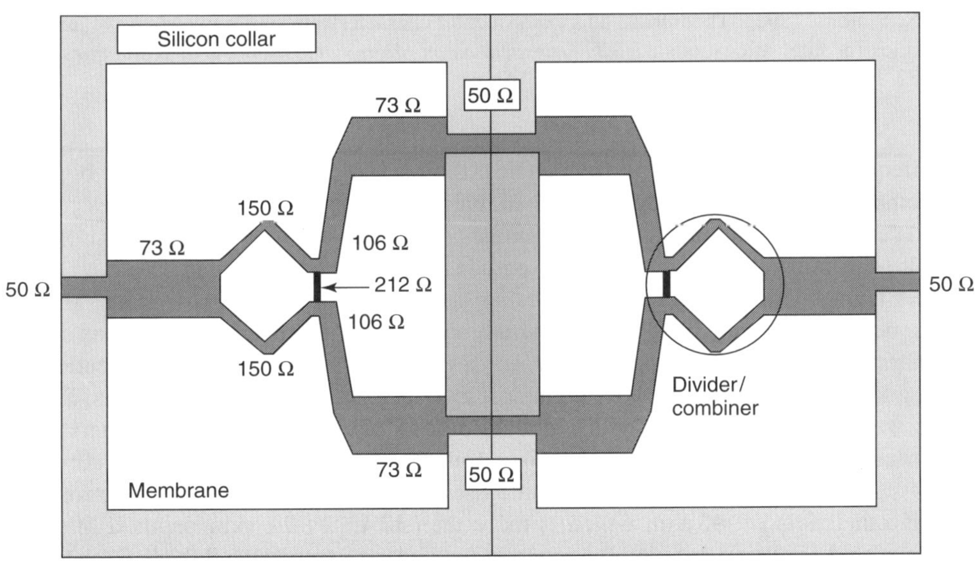

Figure 7.17 Layout of the membrane-supported Wilkinson power divider/combiner. Reproduced from N.I. Dib, L.P.B. Katehi, G.E. Ponchak and R.N. Simons, 1991, 'Theoretical and experimental characterization of coplanar waveguide discontinuities for filter allocations', *IEEE Transactions on Microwave Theory and Techniques* **39**(5): 874–882, by permission of IEEE, © 1991 IEEE

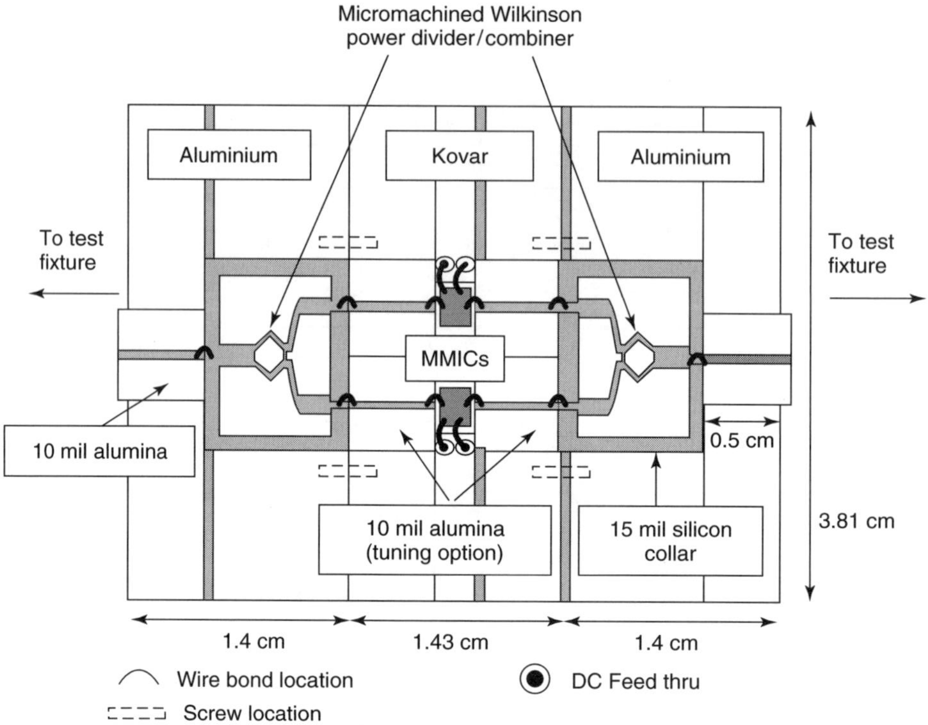

Figure 7.18 Ka-band monolithic microwave integrated circuit (MMIC) power amplifier using alpha MESFET (metal–semiconductor field effect transistor) based chips (AA035P2-00) utilizing a micromachined power divider/combiner. Reproduced from N.I. Dib, L.P.B. Katehi, G.E. Ponchak and R.N. Simons, 1991, 'Theoretical and experimental characterization of coplanar waveguide discontinuities for filter allocations', *IEEE Transactions on Microwave Theory and Techniques* **39**(5): 874–882, by permission of IEEE, © 1991 IEEE

these frequencies. Also, at these frequencies, the passive components become bigger in size compared with the waveguide and mounting components such as diodes and filters. Recent silicon micromachining techniques have been applied to solve most of these difficulties in transmission lines and waveguides.

The silicon micromachining techniques are used to create silicon waveguides that can operate between 100 and 1000 GHz (McGrath *et al.*, 1993; Shenouda and Pearson, 1998). The fabrication process is simplified by making the waveguide in two half sections by splitting them along the broadwall and later integrating them with planar circuits. Smooth groves were etched on silicon with identical dimensions. Metallization is done then by first depositing a thin (200 Å) chrome layer followed by thicker gold layer (5000 Å) on the waveguide walls. The RF conduction losses were reduced further by electroplating gold to a thickness of $\sim 3\,\mu m$, which is more than 12 times the skin depth at 100 GHz. The measured insertion loss of the micromachined waveguide is 0.04 dB/λ, which is comparable to commercially available waveguides.

The difficulties associated with the silicon micromachined waveguides integrating with active devices led to the development of photolithographic techniques for micromachining

(Digby *et al.*, 2000) and CMOS fabrication of transmission lines (Milanouic *et al.*, 1996, 1997). Rectangular waveguide components have been successfully fabricated in W and G band. Figure 7.19 summarizes the lithographic micromachining fabrication process. The waveguides and horns are developed from an initial layer of titanium of 30 nm followed by evaporating gold onto the substrate as the bottom wall of the waveguide. Waveguide and horn structures are defined by exposing and developing a 100-μm thick photoresist [step (iii), Figure 7.19]. A second layer of gold is evaporated and is then electroplated for extra strength. The slotted horn antennas are created by another thin layer of photoresist, exposed and developed using a second mask. Removal of all photoresist leaves an air-filled rectangular waveguide integrated with slotted horn antenna [step (vii)]. A comparison of the inside walls of micromachined and commercially available W-band waveguides shows that a micromachined waveguide has no obvious features greater than 1 μm in size. Such a surface finish implies very low loss performance for micromachined waveguides. The measurements at 75–110 GHz bandwidth shows that the loss of TE_{10} mode propagation is 0.2 dB per wavelength.

Traditionally used silicon technologies such as CMOS and BiCMOS can also be applied for the fabrication of RF MEMS components, including inductors, thermally actuated capacitors, co-planar transmission lines and couplers. The simplest CMOS micromachining involves partial removal of silicon substrate by wet etching. CPWs fabricated in CMOS technology consist of glass-encapsulated metal conductor strips fully suspended by selective etching of silicon substrate. The transmission lines are formed with conductors encapsulated in silicon dioxide thin films. The minimum etching for the proper operation of a micromachined transmission line can be determined using an isolation criterion (Milanouic *et al.*, 2000). The total membrane thickness after the removal of silicon substrate is usually less than 5 μm. Figure 7.20 shows the cross-sectional view of a fully suspended CPW. This process has proved to be feasible for passive microwave components such as inductors, capacitors, transmission lines and couplers. Maximum quality factor achieved for 11-nH inductor is 14.2 (Ozgur and Zaghloul, 2001).

7.2.6 Micromachined directional couplers

A 20-dB directional coupler was designed developed on a thin dielectric membrane using micromachining techniques, as shown in Figure 7.21. Wide-band-extremely-low-noise performance is obtained by suspending the signal conductors in air on a very thin dielectric membrane, which has a thickness of approximately 1.5 μm. A separate shielded wafer is used to self-package the shielded membrane microstrip. This allows the selection of ground plane height independent of the wafer thickness. The circuit wafer and the shield wafer are bonded together and the assembly is attached to a metallized support. Figure 7.22 presents the photograph of the micromachined coupler. The membrane-supported region appears darker than the surrounding silicon. This configuration was used to measure the power from coupled port 3 output. Port 2 is direct and port 4 is the isolated port.

7.2.7 Micromachined mixer

Traditional millimeter and submillimeter wave components such as multipliers and mixers are fabricated by direct machining the waveguides and components. Even though the

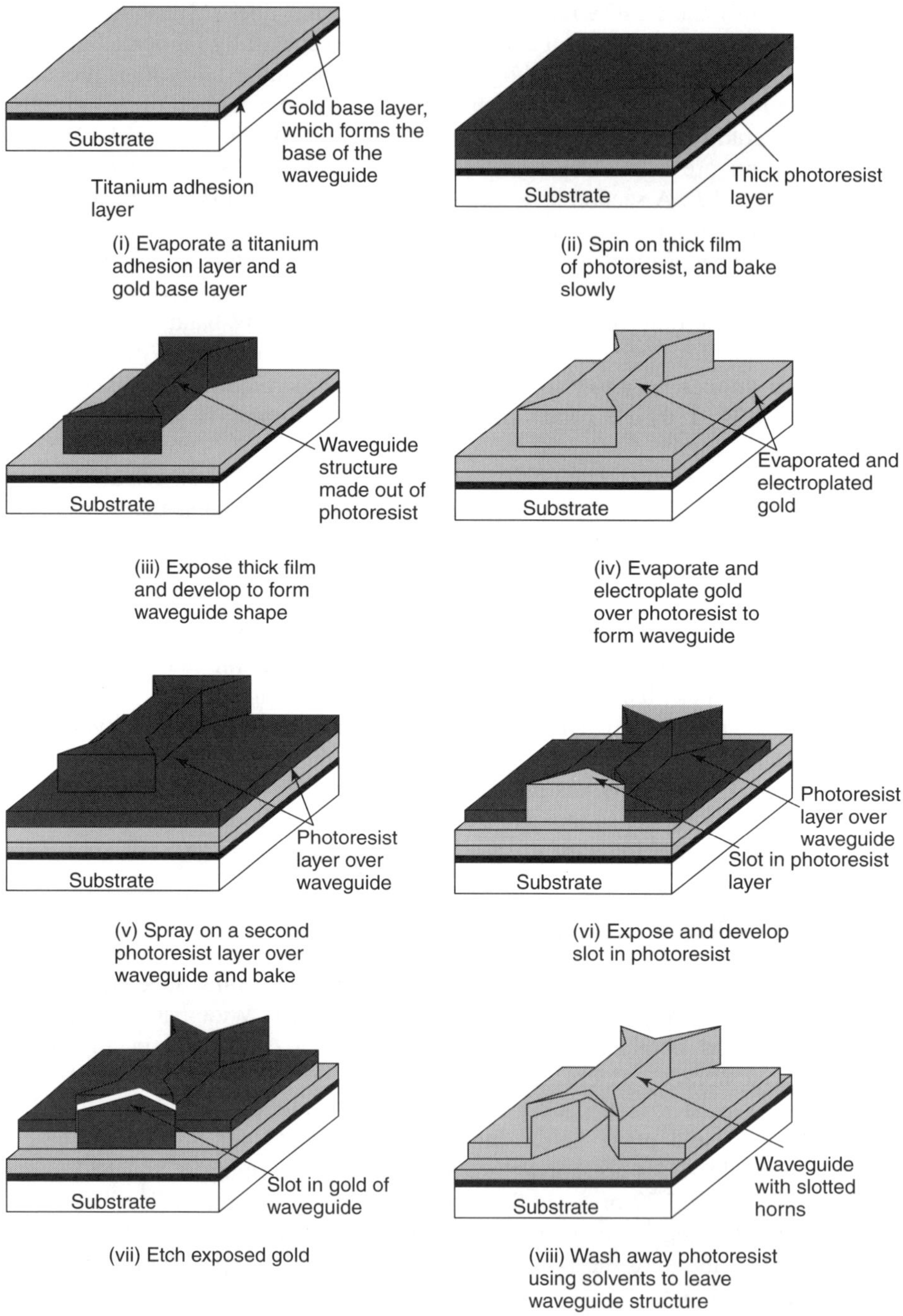

Figure 7.19 Summary of the fabrication process of lithographic micromachining. Reproduced from J.W. Digby, C.E. McIntosh, G.M. Parkhurst, B.M. Towlson, S. Hadjiloucas, J.W. Bowen, J.M. Chamberlain, R.D. Pollard, R.E. Miles, D.P. Steenson, L.S. Karatzas, N.J. Cronin and S.R. Davis, 2000, 'Fabrication and characterization of micromachined rectangular waveguide components for use at millimeter-wave and terahertz frequencies', *IEEE Transactions on Microwave Theory and Techniques* **48**(8): 1293–1302, by permission of IEEE, © 2000 IEEE

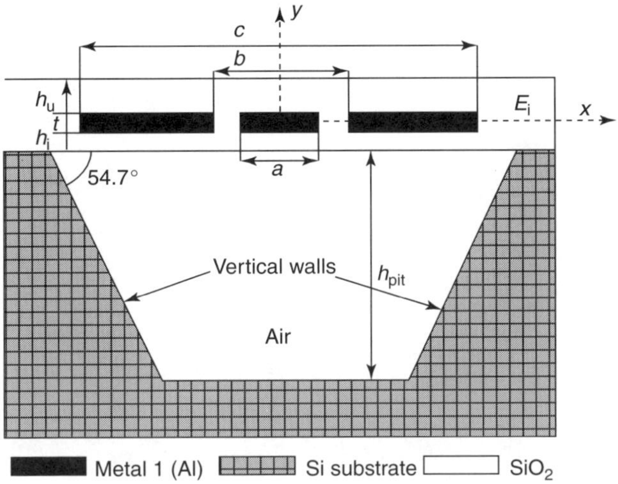

Figure 7.20 Cross-Sectional view of the CMOS (complementary metal oxide semiconductor) micromachined co-planar waveguide. Reproduced from M. Ozgur and M.E. Zaghloul, 2001, 'RF components for wireless communication using CM-CMOS Technology', *International Journal of RF and Microwave CAE* **11**: 330–340, by permission of IEEE, © 2001 IEEE

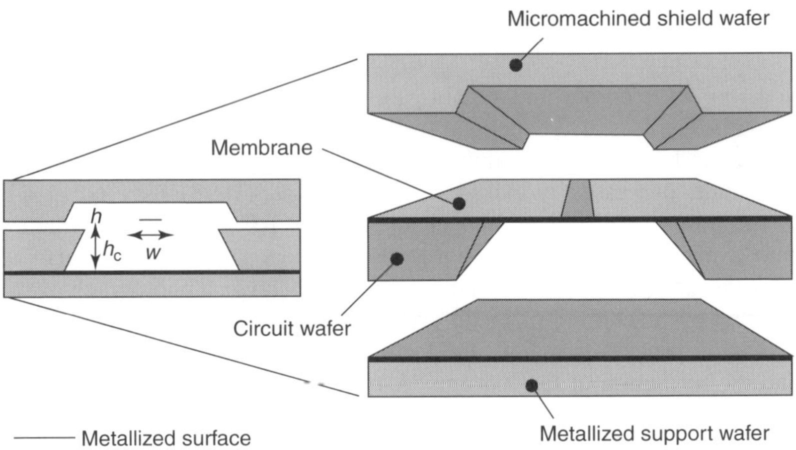

Figure 7.21 Schematic diagram of the micromachined directional coupler. Reproduced from S.V. Robertson, A.R. Brown, L.P.B. Katehi and G.M. Rebeiz, 1998, 'A 10–60 GHz micromachined directional coupler', *IEEE Transactions on Microwave Theory and Techniques* **46**(11): 1845–1149, by permission of IEEE, © 2001 IEEE

machining on metal from these mixer and multiplier elements benefits from reliable, straightforward design and easy integration, it also suffers disadvantages. The machining of such small waveguides becomes very expensive in terms of equipment and personnel, especially for very high frequencies and systems with arrays of components.

Recent micromachining technology has alleviated most of these cost-constraint problems in small mechanical structures because micromachining on silicon has become a

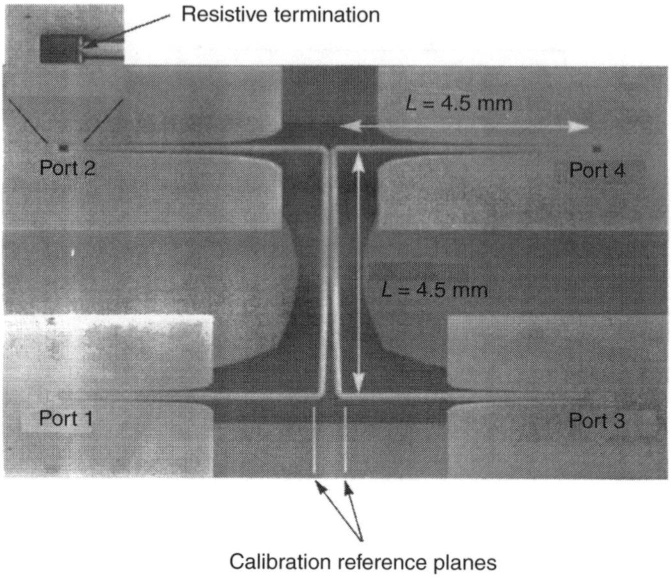

Figure 7.22 Photograph of the micromachined coupler. Reproduced from S.V. Robertson, A.R. Brown, L.P.B. Katehi and G.M. Rebeiz, 1998, 'A 10–60 GHz micromachined directional coupler', *IEEE Transactions on Microwave Theory and Techniques* **46**(11): 1845–1149, by permission of IEEE, © 2001 IEEE

widely used tool in accurate and reliable fabrication of small parts for electrical and mechanical purposes. It is also beneficial for large-scale applications because many duplicate structures can be fabricated simultaneously on a single structure.

The fabrication processes proposed by Mann *et al.* (1998) uses novel silicon etch to process the horns, waveguides and channels. The RF and local oscillator power is coupled into a mixer by a diagonal horn which tapers to a $200 \times 400\,\mu m$ waveguide which propagates 450 to 700 GHz signals. Figure 7.23 shows a photograph of the Hesler's mixer fabricated by machining of metal and its micromachined form (Phipps, 1999).

The fabrication process is mainly the formation of the horn using preferential silicon etching, and the formation of the waveguide, channel structure and alignment pins using epoxy-based photoresist SU-8, and gold plating and dicing of the wafer. The orthogonal aperture of the horn is formed on silicon by preferential etching techniques (Phipps, 1999) using EDP (ethylene–diamine–pyrocatechol) and water. The mixer structure is created using epoxy-based negative tone photoresist SU-8. Chrome and gold are sputtered on to the wafer to form a seed layer; on to it 2-μm gold is plated. The wafer is diced and assembled into a mixer, as shown in Figure 7.24. Figure 7.25 shows the radiation pattern of the micromachined horn antenna.

7.2.8 Passive components: resonators and filters

Filters and resonators fabricated using MEMS exhibits the desired parameters such as narrow bandwidth, low loss and good stability. However, when the dimensions become smaller and smaller, the coupling coefficients linking the electrical and mechanical become

Figure 7.23 Micromachined and Hesler's original mixer. Reproduced from E.S. Phipps, 1999, 'Micromachined waveguide components for submillimeter-wave applications', masters thesis, University of Virginia, Charlottesville, VA, by permission of the University of Virginia

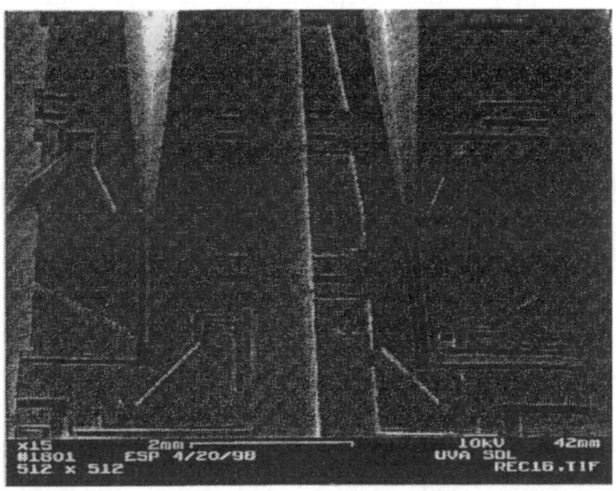

Figure 7.24 Micromachined mixer. Reproduced from E.S. Phipps, 1999, 'Micromachined waveguide components for submillimeter-wave applications', masters thesis, University of Virginia, Charlottesville, VA, by permission of the University of Virginia

critical because of the smaller gap. The signal power handling capability of a filter, for example, is another important design parameter. The packaging of mechanical filters is very important in this respect because hermaticity is the key parameter to achieve a high Q value and long-term stability. The performance of mechanical resonators and filters rely on their mechanical characteristics to produce the desired electrical functions. In short, the mechanical Q for a filter is also same as the electrical Q. Unlike other RF MEMS components, the design of resonators and filters should consider first the mechanical aspect in order to obtain a better electrical performance.

Since the phase noise is inversely proportional to the square of Q, the improvement in Q can minimize the phase noise. A micromachined cavity resonator (Kim *et al.*,

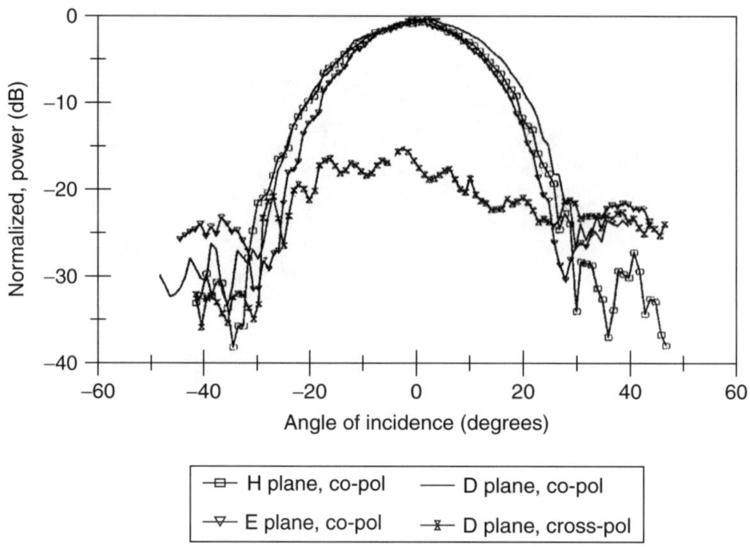

Figure 7.25 Measured radiation pattern of micromachined horn antenna. Note: co-pol, co-polarization; cross-pol, cross-polarization. Reproduced from E.S. Phipps, 1999, 'Micromachined waveguide components for submillimeter-wave applications', masters thesis, University of Virginia, Charlottesville, VA, by permission of the University of Virginia

1999) with a Q of 120 has demonstrated an improvement in phase noise of more than $10\,\mathrm{dBc\,Hz^{-1}}$ (noise power is in decibel scale, the reference being the carrier power, denoted as dBc).

Detailed design and analysis of micromachined and MEMS filters are presented in Chapter 5.

7.2.9 Micromachined antennas

Micromachined antennas have tremendous potential in today's communication systems to meet the demands of high-speed, reduced-size and lightweight electronic component requirements. The feasibility of coupling silicon micromachined rectangular waveguides to a horn antenna (Shenouda *et al.*, 1996; Veidt *et al.*, 1995), thin-film waveguides with integrated horn antennas (Lubeche, Mizuno and Rebeiz, 1998). MEMS reconfigurable V-antennas (Chiao *et al.*, 1999), microstrip antennas (Drayton, 1995; Wong *et al.*, 2000), Terahertz antennas (Lubeche, Mizuno and Rebeiz, 1998) are a few of the recent developments in MEMS antennas.

Microstrip antennas are widely used in communications systems because of their light weight and planar shape. However, the patch antenna efficiency and bandwidth deteriorate as a result of excitation of substrate modes when such patches are fabricated on modern IC-compatible materials such as silicon and GaAs. Using the techniques employed to develop high-frequency micromachined circuits, the suppression or elimination of these modes are possible by selectively altering the substrate material under the antenna (Drayton, 1995), as shown in Figure 7.26. This reduces the effective dielectric constant and the amount of material available to support the surface waves under the antenna.

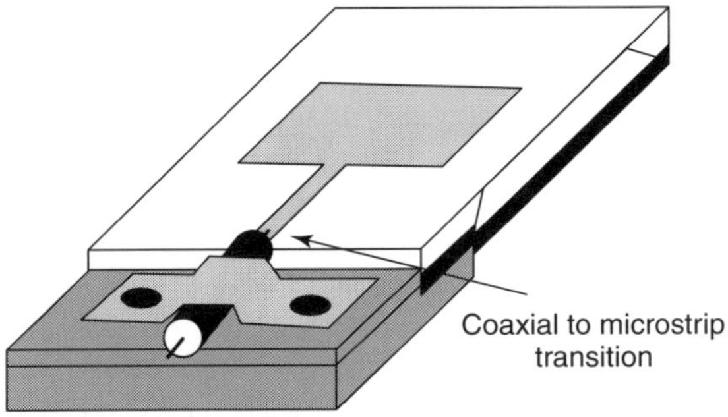

Figure 7.26 Schematic diagram of the micromachined antenna with feed. Reproduced from R.F. Drayton, 1995, *The Development and Characterization of Self-packages using Micromachining Techniques for High Frequency Circuit Applications*, PhD thesis, University of Michigan, Ann Arbor, MI, by permission of the University of Michigan

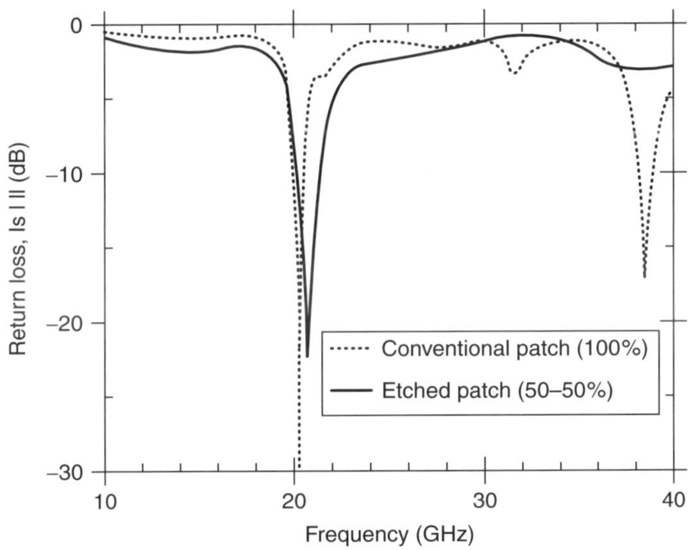

Figure 7.27 Measured return loss of the antenna on silicon with no etching and 50% etching. Reproduced from R.F. Drayton, 1995, *The Development and Characterization of Self-packages using Micromachining Techniques for High Frequency Circuit Applications*, PhD thesis, University of Michigan, Ann Arbor, MI, by permission of the University of Michigan

A micromachined antenna can be fabricated on silicon by etching the silicon underneath the patch. Figure 7.27 shows the measured return loss of the antenna on silicon along with that of a micromachined antenna with 50% of the material removed under the patch. The patch on unaltered material shows the effect of higher-order modes in addition to the excitation of surface waves observed as nonsymmetric response at 20.4 GHz. These higher-order modes are eliminated by the surface micromachining, as evident from the figure. Measured E and H plane patterns are presented in Figure 7.28.

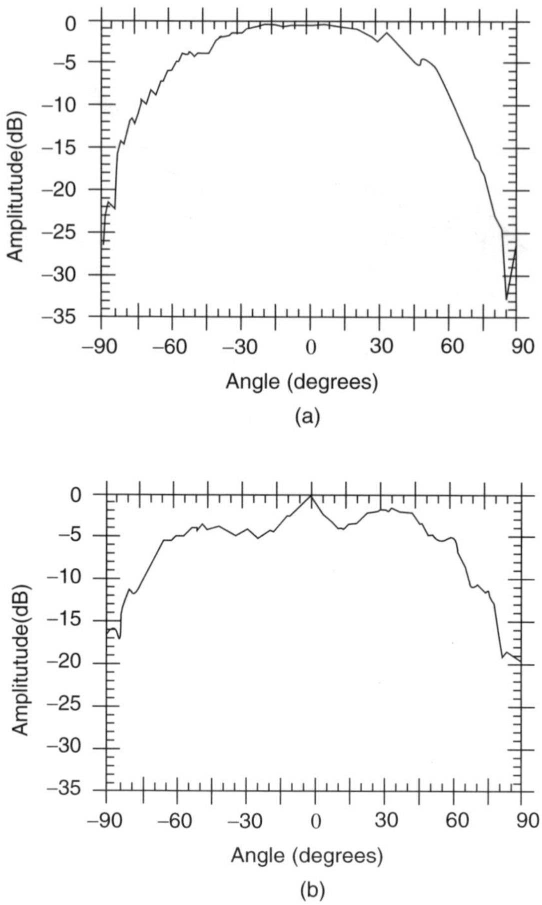

Figure 7.28 Measured radiation (a) E-plane pattern; (b) H-plane pattern of the micromachined antenna. Reproduced from R.F. Drayton, 1995, *The Development and Characterization of Self-packages using Micromachining Techniques for High Frequency Circuit Applications*, PhD thesis, University of Michigan, Ann Arbor, MI, by permission of the University of Michigan

7.3 DESIGN, FABRICATION AND MEASUREMENT

The recent developments in low-cost radio frequency integrated circuits (RFICs) integrated with digital circuits have created a strong interest in silicon as a substrate for RF and microwave applications. Powerful theoretical characterization combined with the current sophisticated solid-state processing techniques greatly enhanced both the understanding and capability of micromachined circuits. Despite the desirable characteristics, however, there exist potentially significant limitations to the use of full-wave analysis. The primary solution is to use approximate models and later optimize with a time-constraint full-wave approach. The main issues in design, fabrication and evaluation of micromachined components for RF applications are discussed in the following sections.

7.3.1 Design

High-frequency micromachined circuit development in general can be divided into two: a design and analysis phase and a fabrication and measurement phase. A full-wave analysis would offer most accurate modeling of the high-frequency circuits. However, this is not an appropriate design tool because of its complexity and time-intensive computer requirements. Therefore, the initial circuit design has to be achieved using quasi-static models based on transverse electromagnetic (TEM) approximations. The quasi-TEM approach can provide an entry-level circuit, which can later be optimized using the full-wave approach, because it requires the exact geometry. Performance improvements can be obtained by an iterative approach to fine tune the circuit dimensions to meet the specifications. The circuit can then be fabricated and evaluated using Network Analyzer and probe stations.

The electrical response of the circuit can be simulated using typical computer-aided design (CAD) tools based on quasi-static or semi-empirical formulations as explained for a co-planar line in Section 7.1. Once the approximate dimensions are obtained, the full-wave analysis, such as the spectral domain integral equation method (Dib and Katehi, 1991; El-Shandwily and Dib, 1990) or the finite-difference time-domain (FDTD) method (Drayton, Dib and Katehi, 1995; Kunz and Luebbers, 1993; Sheer *et al.*, 1990; Zhang and Mei, 1988), can be used for its optimization. The quasi-static model (Dib and Katehi, 1992) based on conformal mapping can be used to predict the impedance characteristics of the circuit, while FDTD offers most versatility in modeling near-exact geometries and dimensions.

The discrepancies between theoretical and experimental results are mainly due to the fabrication tolerances and the approximations made in the model.

7.3.2 Fabrication

The micromachining of CPWs ensures quasi-free-space propagation behavior of the RF signals. The mechanical properties of the membranes are also important because they must exhibit both low-level stress and an approximate thickness typically greater that 1 μm to withstand stress while cutting, transport, vacuum packaging and assembling using flip-chips with integrated circuits. The stress can be minimized by depositing a thin membrane with composite structures (SiO_2/Si_3N_4). The typical sequence of fabrication is explained in Figure 7.29.

The first layer deposited on silicon is a 800-nm thick thermal oxide. To compensate the mismatch provided by this layer, a 600-nm thick nitride layer is grown by low-pressure chemical vapor deposition. Gold is deposited up to 8 μm to reduce the ohmic loss. Backside silicon is removed later by reactive ion etching (RIE) followed by anisotropic chemical etch using potassium hydroxide (KOH) solution. Figure 7.30 presents the typical dimensions of a micromachined transmission line.

7.3.3 Evaluation

Characterization and evaluation of the micromachined lines can be done via test fixtures and coaxial feeds with Network Analyzer using the probing techniques (Godshalk,

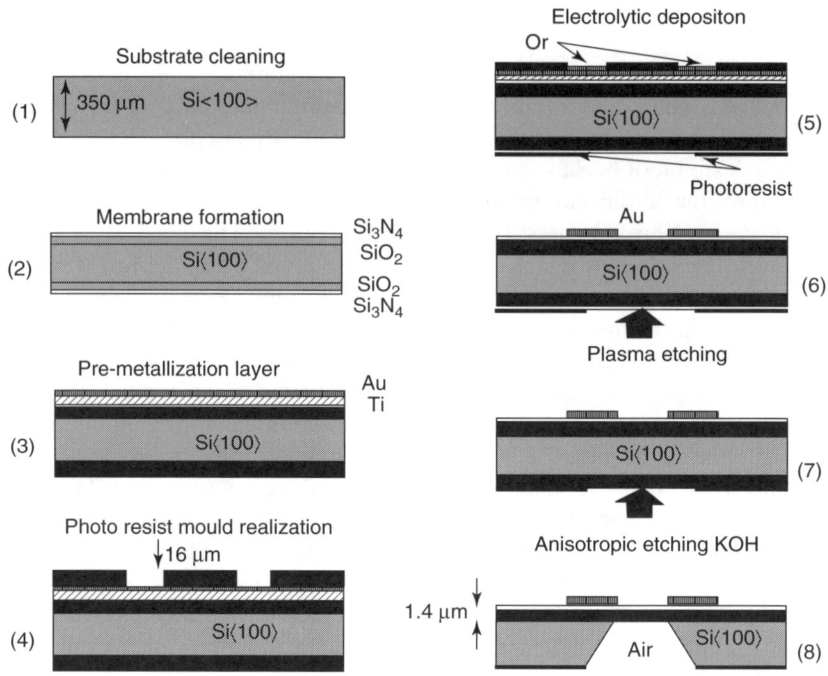

Figure 7.29 Micromachined transmission line fabrication process. Reproduced from B. Gullion, K. Grenier, P. Pons, J.L. Cazaux, J.C. Lalaurie, D. Cros and R. Plana, 2000, 'Silicon micromachining for millimeter-wave applications', *Journal of Vacuum Science Technology A* **18**(2): 743–745, by permission of AVS Publications, © 2000 AVS Publications

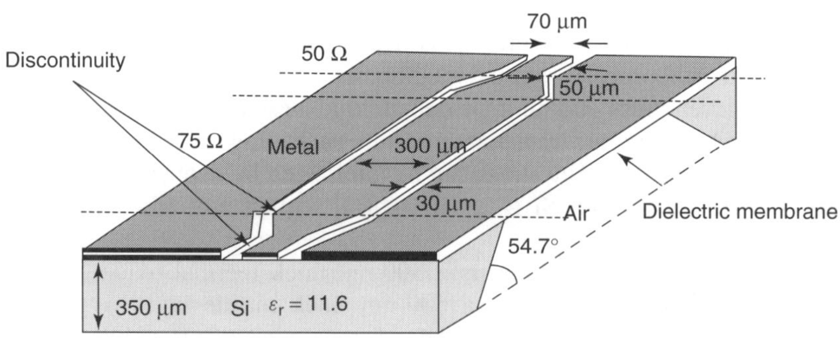

Figure 7.30 Micromachined transmission line. Reproduced from B. Gullion, K. Grenier, P. Pons, J.L. Cazaux, J.C. Lalaurie, D. Cros and R. Plana, 2000, 'Silicon micromachining for millimeter-wave applications', *Journal of Vacuum Science Technology A* **18**(2): 743–45, by permission of AVS Publications, © 2000 AVS Publications

1991, 1993). Earlier, high-frequency probes were used primarily based on co-planar waveguide geometries. However, owing to the recent demand on MMIC processing techniques on wafer probe stations such as Alessi REL-4300 from Cascade Microtech Inc. (www.cmicro.com) probes or Suss Microtec probes (www.suss.com) can be used for high-frequency evaluations of the planar circuits.

For an accurate characterization of micromachined devices, the losses caused by the cables connecting with the device under test (DUT) and the measurement system has to be minimized. The probe calibration is more complex because of the size and the nature of the contact points; the approach is typically characterized as a one-tier or two-tier de-embedding technique. Basically, the measurement plane is transferred from a known location at the test set interface to a predetermined location; this provides the principle of the calibration. Different methods available for on-wafer calibration includes short–open–load–thru (SOLT), line–reflect–match (LRM), line–reflect–line (LRL) and thru–reflect–line (TRL) (Ebgen and Hoer, 1979; Hewlett Packard, 1996; Marks, 1991; Maury, March and Simpson, 1987).

7.4 CONCLUSIONS

This chapter has presented developments in micromachining technology for the design and fabrication of planar components. The most common form of transmission line, the microstrip and CPW, is widely used because its properties are well studied and these structures are compatible with MMICs. Despite their advantages in a lower-frequency regime, they exhibit potentially significant limitations at millimeter-wave frequencies. These transmission lines suffer attenuation, dispersion and multimode propagation, all of which are mainly a result of the use of high-permittivity substrate. An obvious solution to this problem is to remove the substrate beneath the conducting lines and suspend this line on a thin membrane.

Micromachined membrane technology can provide a localized area that has an effective dielectric constant close to one. This localized low-dielectric-constant environment has advantages such as the reduction of parasitic capacitance and improvements in resonance frequencies, reduction in dielectric and dispersion losses and support of a pure TEM wave, which is very useful for filter and passive element applications.

A wide range of impedances can be achieved using microshield lines by varying the size of the shielding waveguide. Any variation of the conducting ground around the center conductor can increase or decrease its capacitance, resulting in a change in its characteristic impedance. Elevating the conductor lines above the lossy dielectric substrate by the micromachining technique can also reduce the conductor and dielectric losses. The design of elevated CPW and overlay CPW lines shows a reduction of losses at 50 GHz compared with a conventional CPW line.

Many modern communication and test instruments prefer waveguide components at millimeter-wave frequencies because of its low-loss performance and ease of fabrication. The enormous high-frequency signal distortion due to permittivity mismatch between the substrate and air and the losses due to excitation of surface wave modes at these frequencies make planar transmission lines useless at these frequencies. Recent silicon micromachining techniques have been applied to solve most of these difficulties in transmission lines and waveguides. The micromachining techniques have been used to create silicon waveguides that can operate between 100 and 1000 GHz.

Micromachined antennas have tremendous potential in today's communication systems to meet the demands of high-speed, reduced-size and lightweight electronic component requirements. The feasibility of coupling silicon micromachined rectangular waveguides to horn antennas, thin-film waveguides with integrated horn antennas, MEMS reconfigurable V-antennas and terahertz antennas are a few of the recent developments in MEMS antennas.

This chapter concluded with a discussion on the fabrication and evaluation of micromachined passive components.

REFERENCES

Assadourian, F., Rimai, E., 1952, 'Simplified theory of microstrip transmission systems', *Proceedings of IRE* **40**: 1651–1657.

Becker, J.P., Katehi, L.P.B., 1999, 'Toward a novel planar circuit compatible silicon micromachined waveguide', in *Proceedings of IEEE Conference on Electrical Performance of Electronic Packaging*, IEEE, Washington, DC: 221–224.

Becker, J.P., Lee, Y., East, J.R., Katehi, L.P.B., 2001, 'A finite ground coplanar line-to-silicon micromachined waveguide transition', *IEEE Transactions on Microwave Theory and Techniques* **49**(10): 1671–1676.

Cheng, H.J., Whitaker, J.F., Weller, T.M., Katehi, L.P.B., 1994a, 'Terahertz bandwidth characteristics of coplanar transmission line on low permittivity substrates', *IEEE Transactions on Microwave Theory and Techniques* **42**(12): 2399–2406.

Cheng, H., Whitaker, J.F., Weller, T.M., Katehi, L.P.B., 1994b, 'Terahertz bandwidth pulse propagation on a coplanar stripline fabricated on a thin membrane', *IEEE Microwave and Guided Wave Letters* **4**(3): 89–91.

Chiao, J.C., Fu, Y., Chio, I.M., DeLosio, M., Lin, L.Y., 1999, 'MEMS reconfigurable V-antenna', in *Proceedings of IEEE MTT-S Symposium Volume 4*, IEEE, Washington, DC: 1515–1518.

Dib, N.I., Katehi, L.P.B., Ponchak, G.E., Simons, R.N., 1991, 'Theoretical and experimental characterization of coplanar waveguide discontinuities for filter allocations', *IEEE Transactions on Microwave Theory and Techniques* **39**(5): 874–882.

Dib, N., Katehi, L.P.B., 1991, 'Modeling of shielded CPW discontinuities using the space domain integral equation method', *Journal of Electromagnetic Waves and Applications* **5**(4/5): 502–523.

Dib, N.I., Katehi, L.P.B., 1992, 'Impedance calculation for the microshield line', *IEEE Microwave and Guided Wave Letters* **2**(10): 406–408.

Digby, J.W., McIntosh, C.E., Parkhurst, G.M., Towlson, B.M., Hadjiloucas, S., Bowen, J.W., Chamberlain, J.M., Pollard, R.D., Miles, R.E., Steenson, D.P., Karatzas, L.S., Cronin, N.J., Davis, S.R., 2000, 'Fabrication and characterization of micromachined rectangular waveguide components for use at millimeter-wave and terahertz frequencies', *IEEE Transactions on Microwave Theory and Techniques* **48**(8): 1293–1302.

Din, N.I., Harokopus, W.P., Katehi, P.B., 1991, 'Study of a novel planar transmission line', in *Proceedings of IEEE MTT-S Symposium, 1991*, IEEE, Washington, DC: 623–626.

Drayton, R.F., Katehi, L.P.B., 1994, 'Development of miniature microwave circuit components using micromachining techniques', in *Proceedings of IEEE MTT-S Symposium*, IEEE, Washington, DC: 225–228.

Drayton, R.F., Dib., N.I., Katehi, L.P.B., 1995, 'Design of micromachined high frequency components', *International Journal of Microcircuits and Electronic Packaging* **18**(1): 19–26.

Drayton, R.F., 1995, *The Development and Characterization of Self-packages using Micromachining Techniques for High Frequency Circuit Applications*, PhD Thesis, University of Michigan, Ann Arbor, MI.

Dryton, R.F., Katehi, L.P.B., 1995, 'Micromachined conformal packages for microwave and millimeter wave applications', in *Proceedings of IEEE MTT-S Symposium*, IEEE, Washington, DC: 1387–1390.

Dryton, R.F., Hindreson, R.M., Katehi, L.P.B., 1996, 'Advanced monolithic packaging concepts for high performance circuits and antennas', in *Proceedings of IEEE MTT-S Symposium*, IEEE, Washington, DC: 1615–1618.

Ebgen, G., Hoer, C., 1979, 'Thru–reflect–line: an improved technique for calibrating the six-port automatic network analyzer', *IEEE Transactions on Microwave Theory and Techniques* **27**(12): 987–993.

El-Shandwily, M., Dib, N., 1990, 'Spectral domain analysis of finlines with composite ferrite-dielectric substrate', *International Journal of Electronics* **68**(4): 571–583.

Gardiol, F., 1994, *Microstrip Circuits*, Wiley, New York.

Gedney, S.D., Petre, P., Matloubian, M., Kihm, R.T., 1997, 'Simulation and performance of passive millimeter wave coplanar waveguide circuit devices', in *Proceedings of IEEE Wireless Communications conference*, IEEE, Washington, DC: 27–31.

Gildas, P., Katehi, L.P.B., Rebeiz, G.M., 1998, 'W-band finite ground coplanar waveguide (FGCPW) to microstrip line transition', in *Proceedings of IEEE MTT-S Symposium*, IEEE, Washington, DC: 107–109.

Godshalk, E.M., 1991, 'A V-band wafer probe using ridge–trough waveguide', *IEEE Transactions on Microwave Theory and Techniques* **39**(12): 2218–2228.

Godshalk, E.M., 1993, 'A W-band wafer probe', in *Proceedings of IEEE MTT-S Symposium*, IEEE, Washington, DC: 171–174.

Goverdhanam, K., Simons, R.N., Katehi, L.P.B., 1999, 'Novel three-dimensional vertical interconnect technology for microwave and RF applications', in *Proceedings of IEEE MTT-S Symposium*, IEEE, Washington, DC: 641–644.

Grieg, D.D., Engelmann, H.F., 1952, 'Microstrip – a new transmission technique for the kilomegacycle range', *Proceedings of IRE* **40**: 1644–1650.

Gullion, B., Grenier, K., Pons, P., Cazaux, J.L., Lalaurie, J.C., Cros, D., Plana, R., 2000, 'Silicon micromachining for millimeter-wave applications', *Journal of Vacuum Science and Technology A* **18**(2): 743–745.

Herrick, K.J., Katehi, L.P.B., 1997, 'W-band micromachined finite ground coplanar (FGC) line circuit elements', in *Proceedings of IEEE MTT-S Symposium*, IEEE, Washington, DC: 269–272.

Herrick, K.J., Schwarz, T.A., Katehi, L.P.B., 1998, 'Si-micromachined coplanar waveguides for use in high frequency circuits', *IEEE Transactions on Microwave Theory and Techniques* **46**(6): 762–768.

Herrick, K.J., Yook, J.G., Katehi, L.P.B., 1998, 'Microtechnology in the development of three-dimensional circuits', *IEEE Transactions on Microwave Theory and Techniques* **46**(11): 1832–1844.

Herrick, K.J., Katehi, L.P.B., 2000, 'Micromachined circuit combining networks for W-band applications', in *Proceedings of IEEE MTT-S Symposium*, IEEE, Washington, DC: 295–298.

Herrick, K.J., Katehi, L.P.B., 2001, 'RF W-band wafer-to-wafer transition', *IEEE Transactions on Microwave Theory and Techniques* **49**(4): 600–608.

Hewlett Packard, 1996, 'HP Product Note 8510-6': on-wafer measurements using HP 8510 Network Analyzer and Cascade Microtech wafer probes, Agilent Technologies, CA, May 1996.

Hindreson, R.M., Herrick, K.J., Weller, T.M., Robertson, S.V., Kihm, R.T., Katehi, L.P.B., 2000, 'Three-dimensional high frequency distribution network, part II: packaging and integration', *IEEE Transactions on Microwave Theory and Techniques* **48**(10): 1643–1651.

Katehi, L.P.B., Rebeiz, G.M., Weller, T.M., Drayton, R.F., Cheng, H.J., Whitaker, J.F., 1993, 'Micromachined circuits for millimeter and sub-millimeter wave applications', *IEEE Antennas and Propagation Magazine* **35**(5): 9–17.

Kim, C., Song, I., Song, C., Cheon, C., Kwon, Y., Lee, S., 1999, 'A micromachined cavity resonator for mm-wave oscillator applications', in *Technical Digest, 10th International Conference on Solid State Sensor and Actuators*: Elsevier Sequoia SA, Lausanne, Switzerland, 1268–1271.

Kunz, K., Luebbers, R., 1993, in *The Finite Difference Time Domain Method for Electromagnetics*, CRC Press, Boca Raton, FL.

Lubecke, V.M., Mizuno, K., Rebei, G.M., 1998, 'Micromachining for terahertz applications', *IEEE Transactions on Microwave Theory and Techniques* **46**(11): 1821–1831.

Margomenos, A., Valas, S., Herman, M.I., Katehi, L.P.B., 2000, 'Isolation in three dimensional integrated circuits', in *Proceedings of IEEE MTT-S Symposium*, IEEE, Washington, DC: 1875–1878.

Marks, R.B., 1991, 'A multilane method of network analyzer calibration', *IEEE Transactions on Microwave Theory and Techniques* **39**(7): 1205–1215.

Maury, M., March, S., Simpson, G., 1987, 'LEL calibration of vector automatic network analyzers', *Microwave Journal* (May): 387–391.

McGrath, W.R., Walker, C., Yap, M., Tai, Y.C., 1993, 'Silicon micromachined waveguides for millimeter-wave and submillimeter-wave frequencies', *IEEE Microwave and Guided Wave Letters* **3**(3): 61–63.

Milanovic, V., Gaitan, M., Bowen, E.D., Zaghloul, M.E., 1996, 'Micromachined coplanar waveguides in CMOS technology', *IEEE Microwave and Guided Wave Letters* **6**(10): 380–382.

Milanovic, V., Gaitan, M., Bowen, E.D., Zaghloul, M.E., 1997, 'Micromachined microwave transmission lines in CMOS technology', *IEEE Transactions on Microwave Theory and Techniques* **45**(5): 630–635.

Milanovic, V., Gaitan, M., Bowen, E.D., Zaghloul, M.E., 2000, 'Quasi-TEM characteristic impedance of micromachined CMOS coplanar waveguides', *IEEE Transactions on Microwave Theory and Techniques* **48**(5): 852–854.

Moon, S.W., Mann, C.M., Maddison, B.J., Turcu, I.C.E., Allot, U., Hug, S.E., Lisi, N., 1996, 'Terahertz waveguide components fabricated using a 3D X-ray microfabrication technique', *Electronics Letters* **32**(19, 12): 1794–1795.

Ozgur, M., Zaghloul, M.E., 2001, 'RF components for wireless communication using CM-CMOS technology', *International Journal of RF and Microwave CAE* **11**: 330–340.

Park, J.Y., Baek, C.W., Jung, S., Kim, H.T., Kwon, Y., Kim, Y.K., 2000, 'Novel micromachined coplanar waveguide transmission lines for applications in millimeter-wave circuits', *Journal of Applied Physics* **39**: 7120–7124.

Petre, P., Matloubian, M., Kihm, R.T., Gedney, S.D., 1997, 'Simulation and performance of passive microwave and millimeter wave coplanar circuit devices with flip chip packaging', in *Proceedings of IEEE 6th Topical Meeting on Electrical Performance of Electronic Packaging*, IEEE, Washington, DC: 203–206.

Phipps, E.S., 1999, 'Micromachined waveguide components for submillimeter-wave applications', masters thesis, Electrical Engineering Department, University of Virginia, Charlottesville, VA.

Ponchak, G.E., Downey, A.N., Katehi, L.P.B., 1997, 'High frequency interconnects on silicon substrates', in *Proceedings of IEEE Radio Frequency Integrated Circuits Symposium*, IEEE, Washington, DC: 101–104.

Ponchak, G.E., Margomenos, A., Katehi, L.P.B., 2001, 'Low-loss CPW on low-resistivity Si substrates with a micromachined polyimide interface layer for RFIC interconnections', *IEEE Transactions on Microwave Theory and Techniques* **49**(5): 866–870.

Robertson, S.V., Brown, A.R., Katehi, L.P.B., Rebeiz, G.M., 1998, 'A 10–60 GHz micromachined directional coupler', *IEEE Transactions on Microwave Theory and Techniques* **46**(11): 1845–1849.

Sheen, D., Sli, S., Abouzahra, M., Kong, J., 1990, 'Finite difference time domain method to the analysis of planar microstrip circuits', *IEEE Transactions on Microwave Theory and Techniques* **38**(7): 849–857.

Shenouda, B., Pearson, L.W., Harriss, J.E., Wang, W., Guo, Y., 1996, 'Etched-silicon micromachined waveguides and horn antennas at 94 GHz', in *Proceedings of IEEE AP-S International Symposium Volume 2*, IEEE, Washington, DC: 988–991.

Shenouda, B.A., Pearson, L.W., 1998, 'Micromachined waveguide for millimeter wave applications', in *Proceedings of International Conference on Microwave and Millimeter Wave Technology*: IEEE, 615–618.

Veidt, B., Kornelsen, K., Vaneldik, J.F., Routledge, D., Brett, M.J., 1995, 'Diagonal horn integrated with micromachined waveguides for submillimeter applications', *Electronics Letters* **31**(16): 1307–1309.

Wadell, B.C., 1991, in *Transmission Line Design Handbook*, Artech House, Boston, MA.

Weller, T.M., Katehi, L.P.B., Rebeiz, G.M., 1995, 'High performance microshield line components', *IEEE Transactions on Microwave Theory and Techniques* **43**(3): 534–543.

Weller, T.M., 1995, *Micromachined High Frequency Transmission Lines on Thin Dielectric Membrane*, PhD thesis, Electrical Engineering Department, University of Michigan, Ann Arbor, MI.

Weller, T.M., Katehi, L.P.B., Herman, M.I., Wamhof, P.D., Lee, K., Kolawa, E.A., Tai, B.H., 1996, 'New results using membrane supported circuits: a Ka-band power amplifier and survivability testing', *IEEE Transactions on Microwave Theory and Techniques* **44**(9): 1603–1606.

Weller, T.M., Katehi, L.P.B., 1996, 'A millimeter wave micromachined low pass filter using lumped elements', in *Proceedings of IEEE MTT-S Symposium*, IEEE, Washington, DC: 631–634.

Weller, T.M., Hinderson, R.M., Herrick, K.J., Robertson, S.V., Kihm, R.T., Katehi, L.P.B., 2000, 'Three-dimensional high frequency distribution network, part I: optimization of CPW discontinuities', *IEEE Transactions on Microwave Theory and Techniques* **48**(10): 1635–1642.

Wong, S., Ooi, B.L., Kooi, P.S., Ng, T.H., Liu, A.Q., 2000, 'Optimization of surface micromachined patch antenna', in *Proceedings of SPIE-4176: Micromachined Devices and Components*: 204–213.

Zhang, X., Mei, K., 1988, 'Time-domain finite difference approach to the calculation of the frequency-dependent characteristics of microstrip discontinuities', *IEEE Transactions on Microwave Theory and Techniques* **36**(12): 1775–1781.

www.aa.com

8
Micromachined antennas

8.1 INTRODUCTION

In the preceding chapters we have discussed various microelectromechanical and micromachined systems and components that could be used at different functional blocks in telecommunications equipment. The antenna is an important component in any wireless communication system as it is an efficient interface between the electronics inside the system and the outside world. In keeping with the trend towards using higher frequencies in communications [e.g. local multipoint distribution systems (LMDSs) networks at 28 GHz, industrial, scientific and medical (ISM) networks and wireless local area networks (WLANs) at around 60 GHz] the antenna technology needs to meet new requirements. It is obvious that the size of the antenna gets smaller as the frequency increases (e.g. at millimeter wave frequencies). This poses several manufacturing issues, some of which could be overcome with the use of micromachining techniques described in this chapter.

One antenna configuration that finds increasing application in several telecommunications systems is based on microstrip technology. In the next section we present a brief overview of the working principles and the design of microstrip antennas. These antennas, currently used in several microwave bands, cannot easily be scaled to millimeter-wave frequencies, and beyond. Hence many of the micromachining approaches discussed in the chapter aim at overcoming these difficulties. A second, but closely related, motivation is better explained in the context of the higher level of system integration. As more and more components are integrated into circuits, and chip levels, semiconductor substrates are increasingly being used in antennas. Some modifications need to be made as these substrates have high dielectric constants and are generally thin compared with common antenna materials. Section 8.3 presents these micromachining approaches to improve the performance of the microstrip antenna.

As the antenna size shrinks with increasing frequency, the fabrication tolerance also gets smaller. This leads to high fabrication cost, as conventional fabrication techniques are not amenable to batch production. Micromachining technique offers tremendous potential for fabrication of these small antennas. Section 8.4 reviews some investigations presented in recent literature on effectively utilizing micromachining techniques as a fabrication tool for very small antennas at extremely high frequencies. Attempts have also been made to incorporate micromachined actuators within the antenna itself to facilitate special features to the antenna such as beam shaping and reconfigurability. Some investigations on the integration of actuators with antenna structures are described in Section 8.5. The ideas evolved in this chapter are summarized in Section 8.6.

8.2 OVERVIEW OF MICROSTRIP ANTENNAS

In contrast to the popular perception that microstrip antennas are fairly recent, their origins date back to the 1950s. However, significant advancements were not made until the early 1970s. By this time improved photolithographic technologies and lower-cost solid-state microwave sources at higher frequencies enabled the widespread use of these antennas in several microwave applications. A schematic of one such antenna shown in Figure 8.1 consists of a metallic patch separated from a ground plane by a dielectric medium. Several geometric shapes have been used as the patch in these antennas, some of which are presented in Figure 8.2 (Balanis, 1997). The choice of the geometry largely depends on the type of radiation characteristics expected of the antenna. For example, rectangular, square or circular patch antennas are often used because of their good radiation characteristics. However, where larger bandwidth is required, dipoles are preferred. All these geometries lead to conformal antenna configurations, increasing their aesthetic value. These can be placed on virtually any metallic surface, planar or nonplanar, and hence have found several newer applications in wireless communications in the recent past.

Despite their wide acceptability as a result of their conformal nature, these antennas, however, are limited by their inherent limitations. These disadvantages include their small impedance bandwidth (high Q) and low gain and efficiency. Accordingly, much of the research efforts in these areas has been towards improving these characteristics. Before discussing these it is essential to have a brief idea of the principle of operation of these antennas.

8.2.1 Basic characteristics of microstrip antennas

As mentioned earlier, the microstrip antenna consists of a printed metallic patch above a conducting ground plane, separated by a low-loss dielectric substrate. Some of the common feed configurations of this type of antenna are shown in Figure 8.3. The location

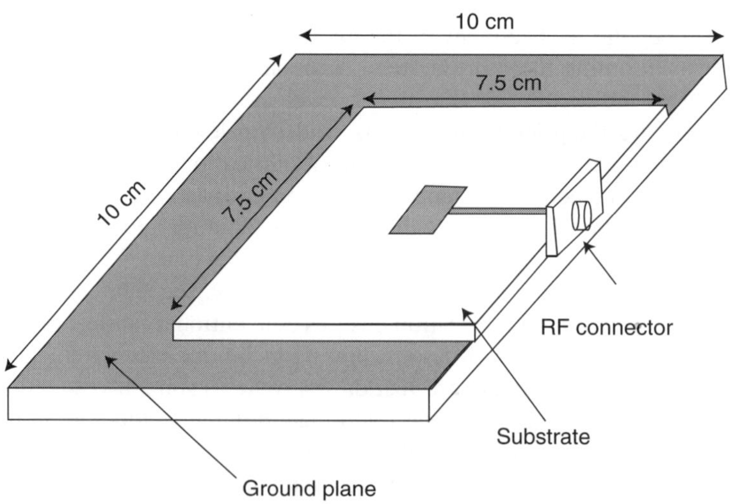

Figure 8.1 Schematic of a microstrip patch antenna. Reproduced from C.A. Balanis, 1997, *Antenna Theory: Analysis and Design*, Wiley, Chichester, by permission of Wiley, © 1997 Wiley

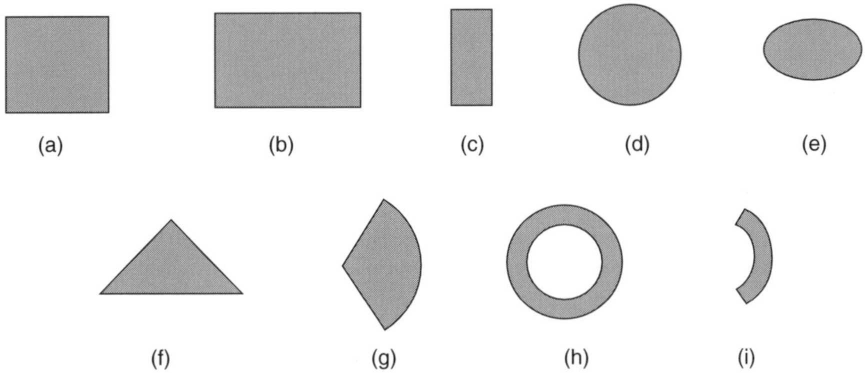

Figure 8.2 Typical patch geometries used for microstrip antennas: (a) square, (b) rectangular, (c) dipole, (d) circular, (e) elliptical, (f) triangular, (g) disc sector, (h) circular ring and (i) ring sector. Reproduced from C.A. Balanis, 1997, *Antenna Theory: Analysis and Design*, Wiley, Chichester, by permission of Wiley, © 1997 Wiley

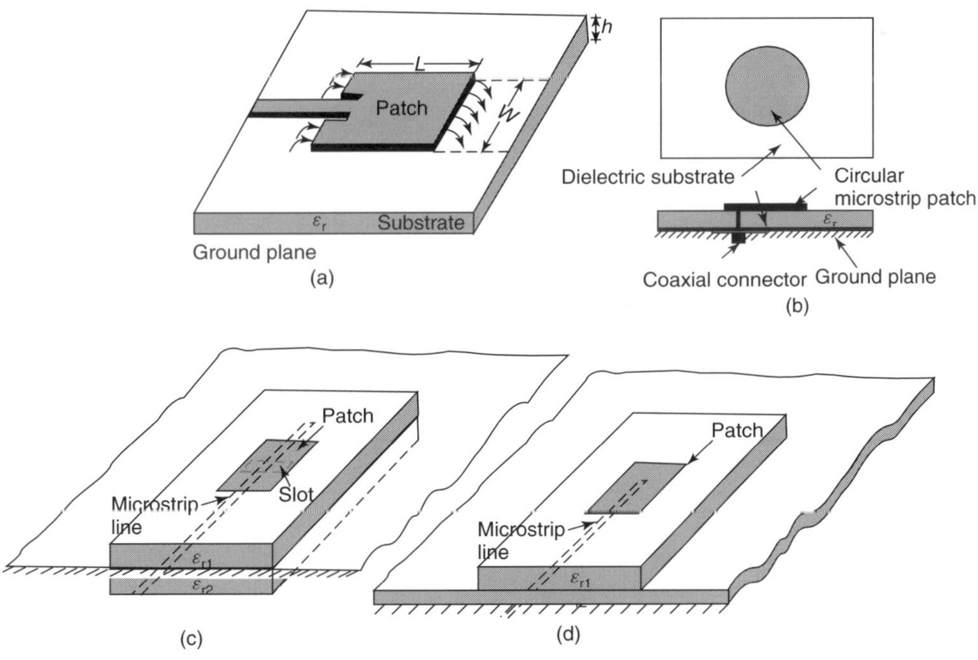

Figure 8.3 Some feed configurations used in microstrip antennas: (a) microstrip line feed, (b) probe feed, (c) aperture-coupled feed, (d) proximity-coupled feed. Reproduced from C.A. Balanis, 1997, *Antenna Theory: Analysis and Design*, Wiley, Chichester, by permission of Wiley, © 1997 Wiley

of the feed with respect to the patch geometry is determined by the radiation characteristics of the antenna which in essence is the mode of the field distribution at the antenna. Several models have been suggested for this type of antenna. The resonance phenomenon of this type of antenna is easily understood with the cavity model (Bahl and Bhatia, 1980).

In this model, the upper and lower electric walls of the cavity are defined by the patch and the ground plane. The sidewalls of the cavity are defined in terms of magnetic walls which are equivalent to, but more convenient than, open-circuited electric boundaries. These boundary conditions lead to the definition of the cavity model of the antenna. In the analysis, one would generally start with an ideal, lossless system and then introduce loss mechanisms such as losses within the antenna and radiation, often represented in the circuit models as radiation resistance.

The thickness h of the substrate conventionally used in microstrips is generally very small compared with wavelength, typically $0.003\lambda < h < 0.05\lambda$. Because of this, only a small fraction of the waves generated within the cavity volume defined by the patch reaches its edges to be radiated. This causes poor efficiency for the microstrip antenna. The fields within can be assumed normal to the patch, with no variations in the vertical direction, because of this small thickness. All these assumptions lead to the restriction that only TM (transverse magnetic) modes are considered for the field distribution.

Using the wave equations for the cavity volume with the boundary conditions mentioned above, one can derive the resonant frequency of the antenna as:

$$f_{0,n,p}^{\mathrm{res}} = \frac{1}{2\pi(\mu\varepsilon)^{1/2}}\left[\left(\frac{n\pi}{L}\right)^2 + \left(\frac{p\pi}{w}\right)^2\right]^{1/2} \tag{8.1}$$

where the dimensions of the antenna and the rectangular coordinate system used are marked in Figure 8.4. The fields within the cavity are

$$E_x = \frac{k}{j\omega\mu\varepsilon} A_{0,n,p}\cos(k_y y)\cos(k_z z) \tag{8.2}$$

$$E_y = E_z = 0 \tag{8.3}$$

$$H_x = 0 \tag{8.4}$$

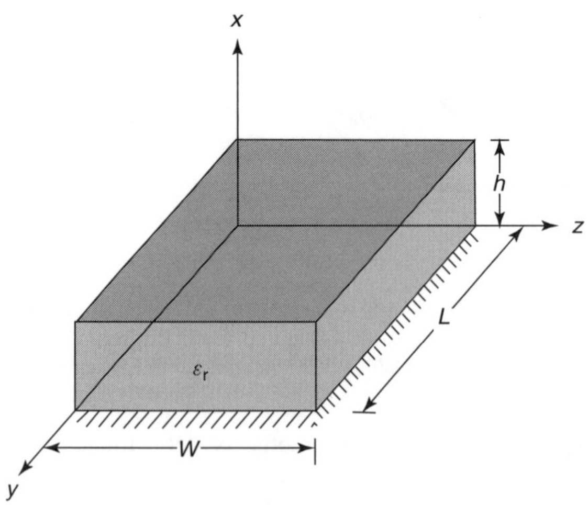

Figure 8.4 Coordinate system used in the cavity model. Reproduced from C.A. Balanis, 1997, *Antenna Theory: Analysis and Design*, Wiley, Chichester, by permission of Wiley, © 1997 Wiley

$$H_y = \frac{k_z}{\mu} A_{0,n,p} \cos(k_y y) \sin(k_z z) \tag{8.5}$$

$$H_z = \frac{k_y}{\mu} A_{0,n,p} \sin(k_y y) \cos(k_z z) \tag{8.6}$$

The cavity model further suggests that the antenna can be represented by four slots defined by the sidewalls of the cavity. However, only two of these, separated by a distance L, account for most of the radiation. These are called radiating slots, and the radiation from the other two nonradiating slots cancel each other along the principal planes.

The radiation from these two relevant slots can be computed by considering them as a two-element array separated by a distance L. The radiated field in far zone for such case can be derived as

$$E_\phi = j\frac{2hE_0}{\pi}\frac{e^{-jkr}}{r}\tan\theta\sin\left(\frac{kw}{2}\cos\theta\right)\cos\left(\frac{kL_{\text{eff}}}{2}\sin\theta\sin\phi\right) \tag{8.7}$$

The other field components are negligible at the far zone region.

8.2.2 Design parameters of microstrip antennas

Compared with the cavity model, the transmission line model of microstrip antennas offers better insights into their design. The patch is treated here as a line resonator of approximately half wavelength, and the radiation occurs as a result of the fringing fields at its open-circuited ends (Bahl and Bhatia, 1980). Hence increasing the width of the antenna increases its radiation efficiency. But beyond an optimum width, higher-order modes get generated, hampering the radiation characteristics. The optimum width of a rectangular patch microstrip antenna is given by

$$W = \frac{c}{2f_0}\left(\frac{\varepsilon_r + 1}{2}\right)^{-1/2} \tag{8.8}$$

The medium surrounding the patch is not homogeneous. On one side of the patch is the dielectric material, while it is air on the other side. Hence the effective dielectric constant ε_e for the patch is approximately estimated to be

$$\varepsilon_e = \frac{\varepsilon_r + 1}{2} + \frac{\varepsilon_r - 1}{2}\left(\frac{1 + 12h}{W}\right)^{-1/2} \tag{8.9}$$

The length of the patch is

$$L = \frac{c}{2f_0\sqrt{\varepsilon_e}} - 2\Delta l \tag{8.10}$$

where Δl is the change in length due to fringing fields at the ends, and ε_e is the effective dielectric constant. Δl is approximately given by

$$\Delta l = 0.412h\frac{(\varepsilon_e + 0.3)\left(\dfrac{W}{h} + 0.264\right)}{(\varepsilon_e - 0.258)\left(\dfrac{W}{h} + 0.8\right)} \tag{8.11}$$

As is clear from the antenna model, the patch length plays a significant role in deciding the resonant frequency of the antenna. However, other factors such as the dielectric constant and height of the substrate play key roles in its bandwidth (or Q factor) and efficiency. The losses in the antenna typically include radiation, conduction, dielectric and surface wave components. Thus the total Q factor (Q_t) for the antenna can be written as:

$$\frac{1}{Q_t} = \frac{1}{Q_r} + \frac{1}{Q_c} + \frac{1}{Q_d} + \frac{1}{Q_s} \tag{8.12}$$

where the subscripts on the right-hand side correspond to radiation, conduction, dielectric and surface wave contribution, respectively. Approximate expressions for an antenna in the dominant mode are (Carver and Mink, 1981):

$$Q_r = \frac{2\omega\varepsilon_r}{hG_{t/L}}\frac{L}{4} \tag{8.13}$$

$$Q_c = h(\pi f \mu\sigma)^{1/2} \tag{8.14}$$

$$Q_d = \frac{1}{\tan\delta} \tag{8.15}$$

where

$$G_{t/L} = \frac{G_{rad}}{W} \tag{8.16}$$

and G_{rad} is the radiation conductance for the antenna. The surface wave component can be neglected for most practical situations if the frequency of operation is less than the cut-off frequency of higher-order modes:

$$f_n = \frac{nc}{4h(\varepsilon_r - 1)^{1/2}} \tag{8.17}$$

Odd and even indices (n) denote transverse electric (TE) and transverse magnetic (TM) wave modes for the antenna. Therefore $n = 1, 3, 5 \ldots$ for TE_n modes and $n = 2, 4, 6, \ldots$ for TM_n modes. The power coupled by TM_0 substrate mode between adjacent elements of a dipole array can be given approximately as (Griffin, 1995; Sayyah and Griffin, 1997):

$$P_{TM_0} = 3.33\frac{d}{\lambda_m}P_{total} \tag{8.18}$$

where d is the thickness of the substrate, and λ_m is the wavelength inside the medium of the substrate material.

The bandwidth of the antenna can then be expressed in terms of the total quality factor, the resonant frequency f_0, and for a maximum voltage standing wave ratio (VSWR) of S_{max} is:

$$\Delta f = \frac{f_0}{Q_t}\frac{S_{max} - 1}{\sqrt{S_{max}}} \tag{8.19}$$

The bandwidth is inversely proportional to the square root of the dielectric constant of the substrate (Balanis, 1997). The bandwidth increases with the substrate height.

In the cavity model the radiation mechanism of the antenna consists of fields emanating from two radiating slots at the ends of the antenna length. The directivity of each of these slots is given as

$$D = \frac{4W^2\pi^2}{I_1\lambda_0^2} \tag{8.20}$$

where

$$I_1 = \int_0^{\pi} \sin^2\left(\frac{kW\cos\theta}{2}\right)\tan^2\theta\sin\theta\,\mathrm{d}\theta \tag{8.21}$$

The total directivity of the patch antenna is then given by

$$D_\mathrm{W} = \frac{2D}{1+g_{12}} \tag{8.22}$$

where the mutual conductance g_{12} is

$$g_{12} = \frac{1}{120\pi^2}\int_0^{\pi} \frac{\sin^2\left(\dfrac{\pi W\cos\theta}{\lambda}\right)\tan^2\theta\sin\theta J_0\left(\dfrac{2\pi L}{\lambda}\sin\theta\right)}{G}\,\mathrm{d}\theta \tag{8.23}$$

In terms of the quality factors described earlier, the efficiency of the antenna is

$$\eta = \frac{Q_\mathrm{t}}{Q_\mathrm{r}} \tag{8.24}$$

The dependence of the efficiency and bandwidth of the antenna on the substrate height in wavelengths is shown in Figure 8.5.

Parameters discussed so far show the dependence of substrate properties on the performance of a single element microstrip antenna. However, one of the primary advantages of microstrip antennas is the ease of formation of linear and planar arrays. One additional performance parameter that needs to be studied in this context is the coupling between the array elements. This affects the array properties such as the side lobe level and hence is an important parameter to be controlled in array design.

The coupling between adjacent elements depends on their relative arrangement. The mutual coupling between two elements as a function of their separation in two orthogonal planes is shown in Figure 8.6. The mutual coupling is due to fields that exist along the dielectric air interface and are contributed by space waves, surface waves and leaky waves traveling between the elements. For electrically thick substrates, the surface wave contribution can be significant in the E-plane, especially for large separation.

The discussion in this section was constrained to rectangular patch microstrip antennas only. However, the basic principles evolved can easily be translated to other geometries as well. However, their behavior to dielectric substrates remains generally the same, and is the primary concern in the discussions that follow.

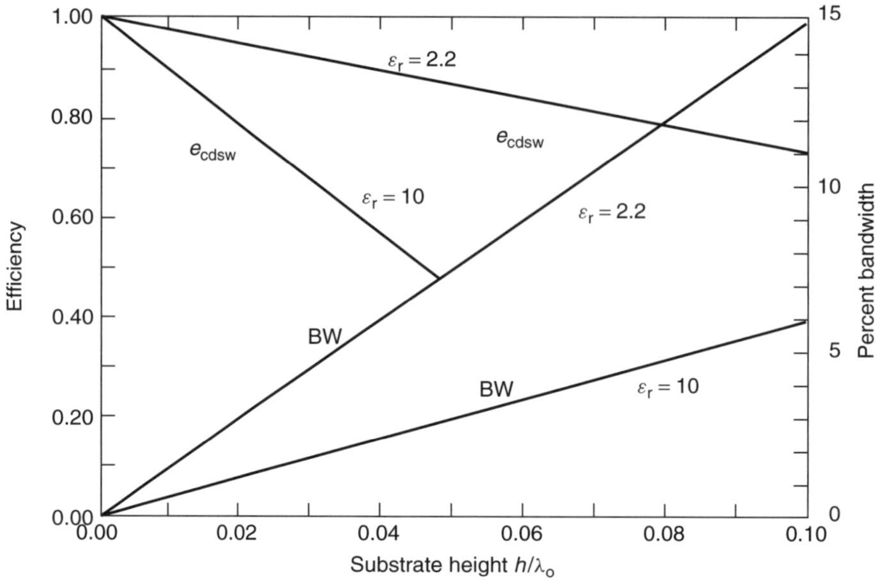

Figure 8.5 Efficiency and bandwidth of microstrip antenna for different substrate heights. Note: ε_r, relative permittivity, BW, bandwidth, e_{cdsw}, efficiency. Reproduced from C.A. Balanis, 1997, *Antenna Theory: Analysis and Design*, Wiley, Chichester, by permission of Wiley, © 1997 Wiley

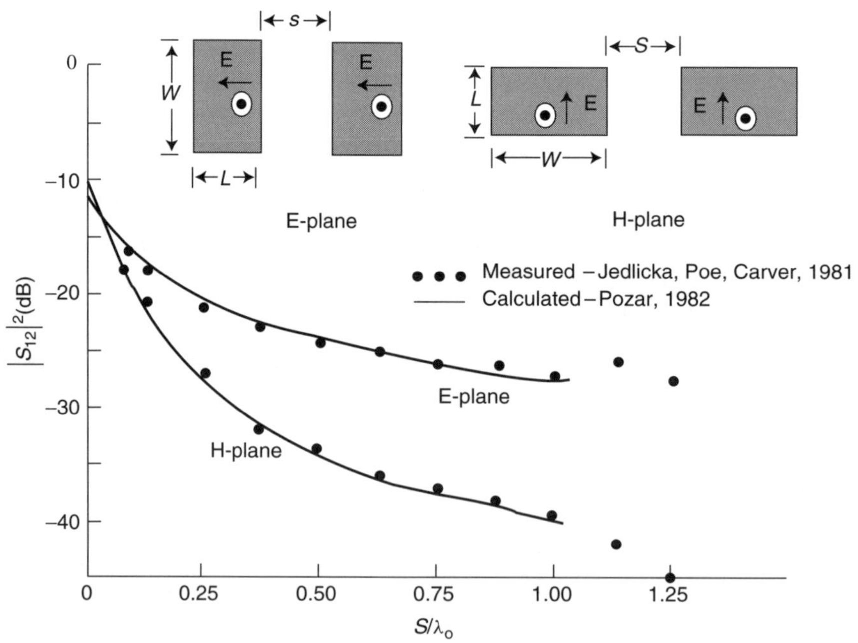

Figure 8.6 Mutual coupling of two coaxial-fed microstrip antennas for E-plane and H-plane arrangements. Reproduced from D.M. Pozar, 1982, 'Input impedance and mutual coupling of rectangular microstrip antenna', *IEEE Transactions on Antennas and Propagation* **30**: 1191–1196 by permission of IEEE, © 1982 IEEE

8.3 MICROMACHINING TECHNIQUES TO IMPROVE ANTENNA PERFORMANCE

Having described the performance parameters of the microstrip antenna, and some of the design parameters affecting these, we now proceed to discuss how these goals of performance improvement can be achieved by micromachining. This assumes great significance in the context of fully integrated systems, as described below.

The complete integration of a system onto a single board entails that all components of the system be made compatible. At microwave frequencies this approach poses an additional difficulty, in selecting the optimum value for the dielectric constant of the substrate. From the circuits point of view it is always preferred to use materials with a high value of dielectric constant. This reduces the overall size, as well as helping to reduce losses such as those from radiation. On the contrary, if an antenna were to be integrated onto the subject system, questions arise on using such a substrate. As discussed in the previous section, the radiation characteristics of antennas and antenna arrays are degraded by using such high values of dielectric constant for substrate materials. Hence a compromise may have to be reached on the choice of the substrate material.

One approach is to use two different substrates, one for the circuits and another for antennas. But, in general, monolithic substrates are generally pursued to minimize the overall cost. Micromachining offers an alternative scheme that satisfies the demands of the antenna and circuits perspectives, using the same substrate (Yook and Katehi, 1998).

With the increased use of higher millimeter-wave frequencies in communication systems, an increased possibility for system integration is now in sight. Antennas and other components at these frequencies are relatively small. Hence it may not be impossible to integrate an entire system, including the antennas, into a single chip. This would require use of semiconductor substrates for antennas as well as for active and passive devices that form part of the circuits. Such a system is envisaged to have limited, or no, off-chip components.

The substrate materials suitable for active devices are of high dielectric constant, typically around 12. This leads to electrically thick substrates, especially at high microwave and millimeter-wave frequencies. For example, the commercially available 200 μm GaAs or silicon substrate is $\lambda_m/20$ at 20 GHz, and reaches $\lambda_m/4$ at 100 GHz, where λ_m is the wavelength inside the substrate medium (Papapolymerou, Drayton, and Katehi, 1998). Higher thickness leads to increased surface waves, and hence losses. Obviously, the thin substrate assumption used previously in the derivation of the antenna characteristics is no longer valid. Although the thickness can be reduced further by thinning procedures, this leads to fragile substrates, which becomes a fabrication issue for large-area substrates as demanded by millimeter-wave circuits. Hence artificial methods to reduce its effective dielectric constant, especially below the antenna, assume significance. With these perspectives in mind, we now discuss micromachining techniques for semiconductor substrates and their advantages in antenna design.

The idea of micromachining substrates to improve antenna properties derives from similar approaches attempted on regular antenna substrates such as duroid by closely drilled holes. The presence of these air-filled holes primarily reduces the effective dielectric constant of the medium surrounding the antenna. A simplified approach to study this effect uses the capacitance model. Based on the parallel plate capacitance between the patch and the ground plane, the dielectric constant of the perforated dielectric medium

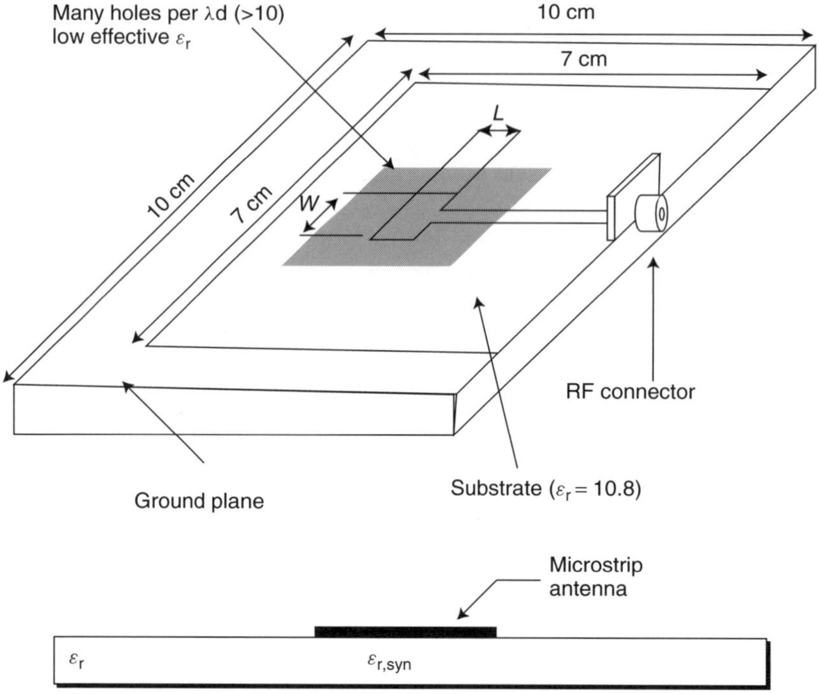

Figure 8.7 Microstrip antenna on a synthesized substrate with holes drilled. The synthesized dielectric constant of the substrate depends on the hole size and hole density. Reproduced from G.P. Gauthier, A. Courtay and G.M. Rabeiz, 1997, 'Microstrip antennas on synthesized low dielectric constant substrates', *IEEE Transactions on Antennas and Propagation* **45**: 1310–1314, by permission of IEEE, © 1997 IEEE

$\varepsilon_{r,\text{syn}}$ can be approximated as (Gauthier, Courtay and Rabeiz, 1997):

$$\varepsilon_{r,\text{syn}} = \varepsilon_r \left(1 - \frac{\pi d_h^2}{2d_c^2}\right) + \frac{\pi d_h^2}{2d_c^2} \tag{8.25}$$

where d_h is the diameter of the hole, and d_c is the center-to-center spacing of the triangular lattice grid. The triangular lattice for these holes (Figure 8.7) enables maximum effectiveness. By this process, the radiation efficiency of the antenna is improved by as much as 25%. The same approach can be extended to semiconductor substrates by using micromachining techniques instead of drilling holes. The shape of these micromachined holes depends on the etching method, etching duration and the crystal orientation of the substrate (Figure 8.8; Bean, 1978). However, owing to the improved mechanical characteristics of the semiconductor substrates, it is even possible to remove the substrate material almost entirely below the antenna (Figure 8.9; see also Zheng *et al.*, 1998). Assuming vertical walls for the cavity, the effective dielectric constant is estimated to be (Papapolymerou, Drayton and Katehi, 1998):

$$\varepsilon_{r,\text{eff}} = \frac{\varepsilon_{\text{cavity}}}{L + 2\Delta L}\left(L + 2\Delta L \frac{\varepsilon_{\text{fringe}}}{\varepsilon_{\text{cavity}}}\right) \tag{8.26}$$

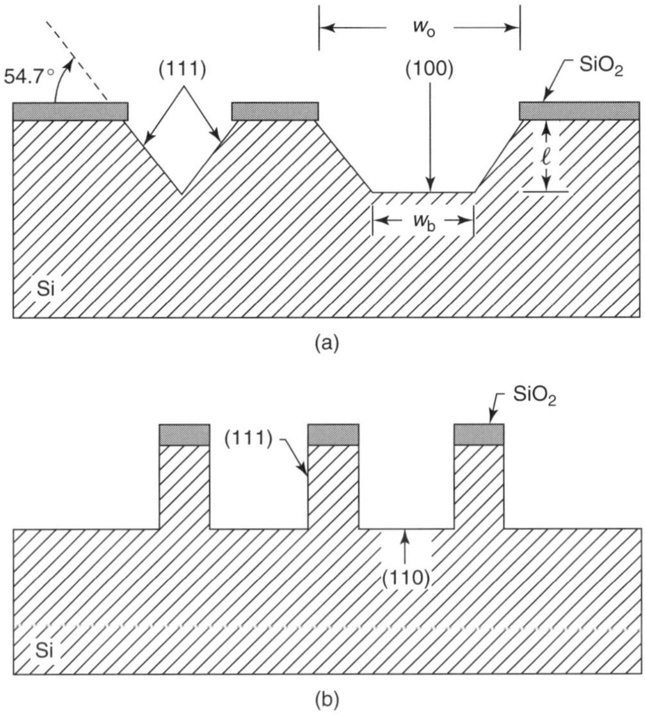

Figure 8.8 Orientation dependence of anisotropic wet etching of silicon: (a) ⟨1 0 0⟩ silicon and (b) ⟨1 1 0⟩ silicon. Reproduced from K.E. Bean, 1978, 'Anisotropic etching of silicon', *IEEE Transactions in Electronic Devices* **25**: 1185–1193, by permission of IEEE, © 1978 IEEE

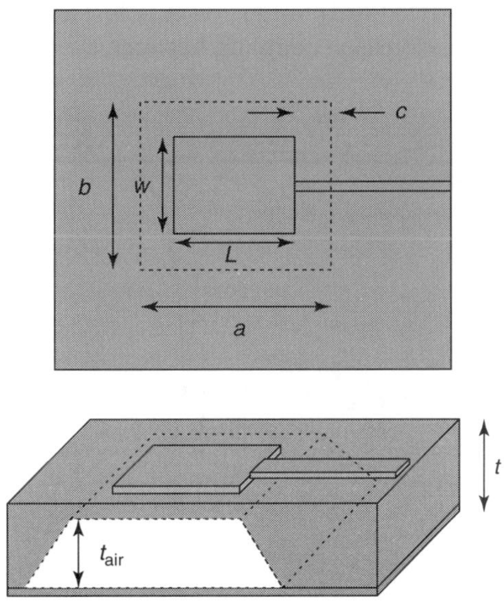

Figure 8.9 Micromachined microstrip antenna with a portion of substrate material below the patch removed by backside etching. Reproduced from I. Papapolymerou, R.F. Drayton and L.P.B. Katehi, 1998, 'Micromachined patch antennas', *IEEE Transactions on Antennas and Propagation* **46**: 275–283, by permission of IEEE, © 1998 IEEE

where the dielectric constant for the fringing field region and the mixed substrate cavity region are given by:

$$\frac{\varepsilon_{\text{fringe}}}{\varepsilon_{\text{cavity}}} = \frac{\varepsilon_{\text{air}} + (\varepsilon_{\text{sub}} - \varepsilon_{\text{air}})x_{\text{air}}}{\varepsilon_{\text{air}} + (\varepsilon_{\text{sub}} - \varepsilon_{\text{air}})x_{\text{fringe}}} \tag{8.27}$$

$$\varepsilon_{\text{cavity}} = \frac{\varepsilon_{\text{air}}\varepsilon_{\text{sub}}}{\varepsilon_{\text{air}} + (\varepsilon_{\text{sub}} - \varepsilon_{\text{air}})x_{\text{air}}} \tag{8.28}$$

where x_{air} and x_{fringe} are the ratios of air to substrate thickness in the mixed and fringing field regions. This approach has resulted in significant improvements in the antenna bandwidth as well as radiation efficiency. The antenna bandwidth is improved by over 60%, compared with conventional substrate (Figure 8.10; Papapolymerou, Drayton, and Katehi, 1998). The cavity below the patch are designed to have resonant frequency close that of the patch in an effort to improve the bandwidth.

In another research effort, instead of forming a cavity below the patch, trenches were formed below the radiating edges so that the conductor of the patch was overhanging, to improve the antenna characteristics (Chen *et al.*, 1998). Backside etching of the ⟨1 0 0⟩ high-resistivity silicon wafer results in trenches at an inclination angle of 55° (Figure 8.11). A 40% improvement in bandwidth and a marked improvement in radiation efficiency is achieved for an antenna designed at 13.8 GHz by this approach. Extension of the approach to microstrip antenna with a co-planar waveguide (CPW) feed for 94 GHz is presented in (Gauthier *et al.*, 1999).

A similar micromachining approach is also followed for array elements to reduce mutual coupling (Figure 8.12). As described previously, the surface waves traveling between the patch elements are the primary contributors for their mutual coupling. Surface waves are excited on the substrate above a cut-off frequency determined by its thickness. It is estimated that more than two thirds of the power is lost as surface waves on a 200-μm thick substrate. This leads to increased coupling between array elements. Cavities formed

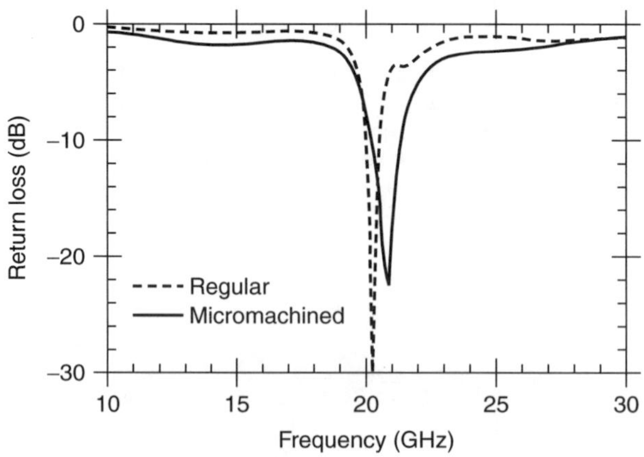

Figure 8.10 Performance improvement of micromachined microstrip antenna compared with one on a regular substrate. Reproduced from I. Papapolymerou, R.F. Drayton and L.P.B. Katehi, 1998, 'Micromachined patch antennas', *IEEE Transactions on Antennas and Propagation* **46**: 275–283, by permission of IEEE, © 1998 IEEE

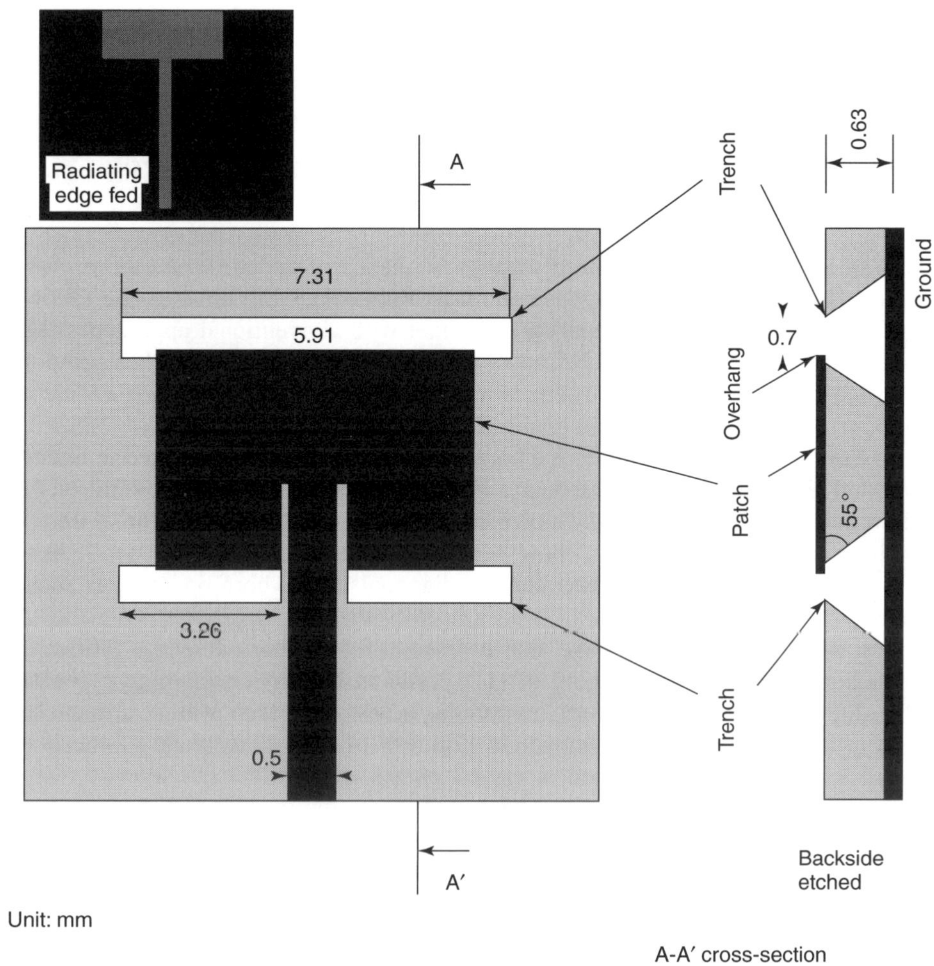

Figure 8.11 Microstrip antenna with micromachined trenches below its radiating edges. Reproduced from Q. Chen, V.F. Fusco, M. Zheng and P.S. Hall, 1998, 'Micromachined silicon antennas', *Proceedings of the International Conference on Microwave and Millimeter Wave Technology*, IEEE, Washington, DC, by permission of IEEE, © 1998 IEEE

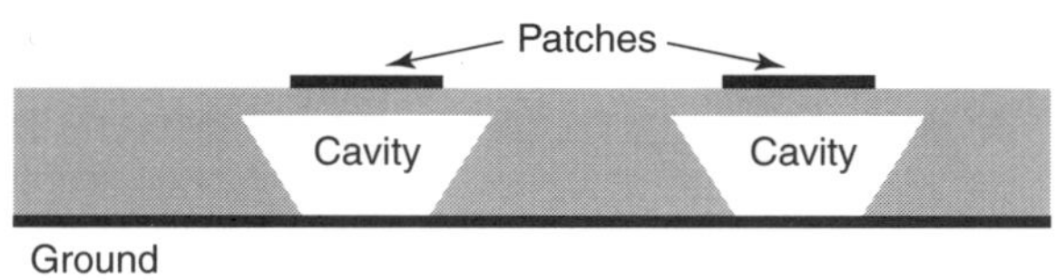

Figure 8.12 A two-element microstrip array with cavities formed below the patches to reduce mutual coupling. Reproduced from J.W. Yook, L.P.B. Katehi, 2001, 'Micromachined microstrip patch antenna with controlled mutual coupling and surface waves', *IEEE Transactions on Antennas and Propagation* **49**: 1282–1289, by permission of IEEE, © 2001 IEEE

below these patches reduce the surface wave contribution. Numerical studies for various geometries for the cavity have been presented in Yook and Katehi (2001). The depth of micromachining is optimized to improve the mutual coupling characteristics.

8.4 MICROMACHINING AS A FABRICATION PROCESS FOR SMALL ANTENNAS

The size of an antenna depends largely on its frequency of operation. As the operational frequencies increase such that the wavelength is shorter than a millimeter, the fabrication tolerance levels become too demanding to be met with conventional approaches. Micromachining offers an alternative fabrication approach for antennas and other components at such high frequencies. These structures are not necessarily planar, hence a wide range of components can be built by this approach.

As an example, micromachining of a horn antenna for W-band is reported in Shenouda, Pearson and Harriss (2001). Two separate V-shaped grooved sections are built on (1 0 0) silicon wafer. It may, however, be noted that the crystal structure of the wafer determines the wedge angle of each of these sections. For example, the (1 0 0) silicon wafer results in an aspect ratio of 1.42 between the diagonals at the horn opening, as shown in Figure 8.13.

Several other configurations have also been reported for horn antennas fabricated by micromachining. Anisotropic etching of (1 0 0) silicon by ethylenediamine pyrocatechol (EDP) solution can be used to form pyramidal holes. Two such wafers are stacked to form an array of horn antennas shown in Figure 8.14. The flare angle of the horn is, however, decided by the orientation of crystal planes of the wafer (angle $= 70.6^\circ$ in this case). These fabrication approaches make use of the matured semiconductor wet etching techniques for precise and well-defined cavities.

In yet another approach the flare angle of the horn is defined lithographically, to optimize for antenna parameters such as gain and beam shape (Hesler *et al.*, 2001). This octagonal horn antenna is designed for 585 GHz for possible applications in chemical

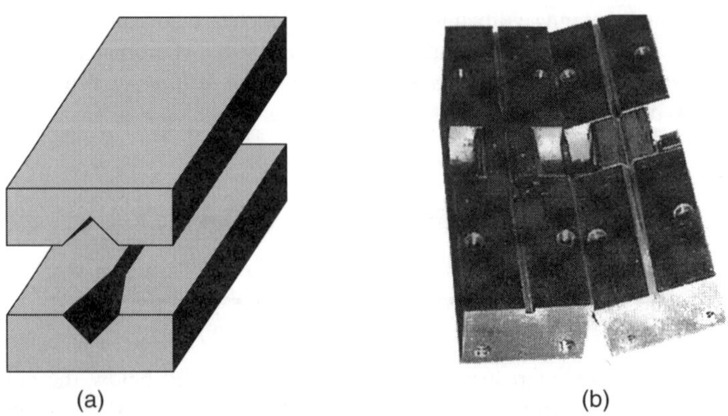

Figure 8.13 A micromachined horn antenna for W-band application: (a) schematic; (b) photograph of fabricated antenna in test jig. Reproduced with permission from Shenouda *et al.*, 2001, by permission of IEEE, © 2001 IEEE

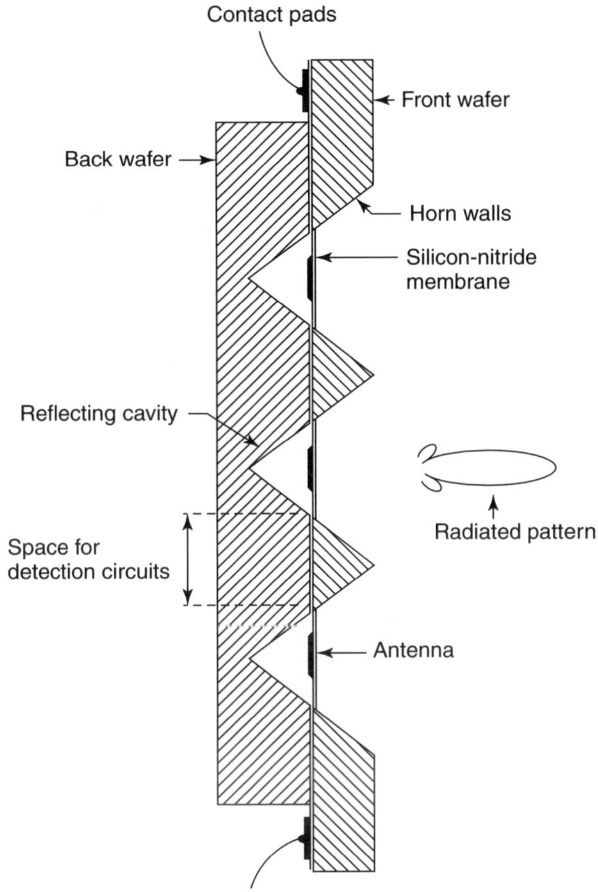

Figure 8.14 Cross-section of a two-dimensional array of micromachined horn antenna fabricated by stacking two wafers. Reproduced from G.M. Rebeiz, D.P. Kasilingam, Y. Guo, P.A. Stimson and D.B. Tutledge, 1990, 'Monolithic millimeter-wave two-dimensional horn imaging array', *IEEE Transactions on Antennas and Propagation* **38**: 1473–1482, by permission of IEEE, © 1990 IEEE

spectroscopy and radio astronomy. The micromachining process used in its fabrication is shown in Figure 8.15. The flare angle of the horn is defined by patterning an oxide layer using photolithography. Silicon substrate is then etched through this triangular window using EDP solution. After removal of the oxide, a triangular aperture horn is formed. An ultrathick layer of photoresist is applied on top of the substrate with a uniform thickness. The surface is metalized by sputtering and electroplating. Two halves are then joined together to form the horn (Hesler *et al.*, 2001).

Micromachining techniques have also enabled the integrated fabrication of antennas and waveguides at millimeter wave frequencies. An exponentially flared slotted horn is fabricated using a thick photoresist as the sacrificial layer for the hollow part of the waveguide and antenna (Digby *et al.*, 1997). The antenna shown in Figure 8.16 has 23° and 31° beam widths in horizontal and vertical planes at 200 GHz. The antenna is designed for G-band (140 GHz to 220 GHz).

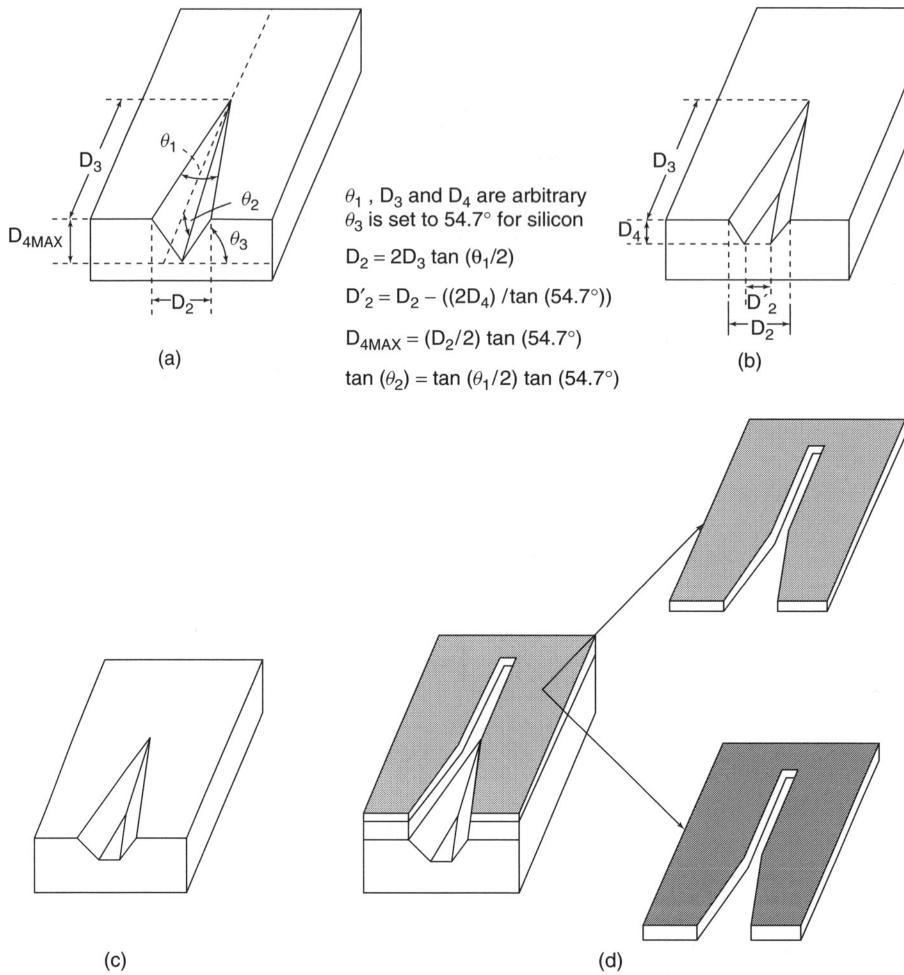

Figure 8.15 A micromachined octagonal horn antenna for submillimeter wave band applications. (a) Shows the triangular profile after completion of etching process; (b) a flat bottom surface is obtained by controlled etching; (c) top section and (d) exploded views of bottom section of octagonal horn antenna. Reproduced from J.L. Hesler, K. Hui, R.K. Dahlstrom, R.M. Weikle, T.W. Crowe, C.M. Mann and H.B. Wallace, 2001, 'Analysis of an octagonal micromachined horn antenna for submillimeter-wave applications', *IEEE Transactions on Antennas and Propagation* **49**: 997–1001, by permission of IEEE, © 2001 IEEE

Compared with the traditional silicon micromachining processes used in the fabrication of the previous antennas, the capability of LIGA (a German acronym for *Lithographie, Galvanoformung, Abformung*) to etch very thick metal structures has also been utilized in the fabrication of millimeter-wave antennas. This consists of tapered slot antennas (TSAs) on GaAs substrate for 94 and 184 GHz (Gearhart and Willke, 1998). If these antennas are made directly on a substrate, the thickness of the substrate should meet the following criterion:

$$0.005 \leq \frac{t(\varepsilon_r - 1)^{1/2}}{\lambda} \leq 0.03 \tag{8.29}$$

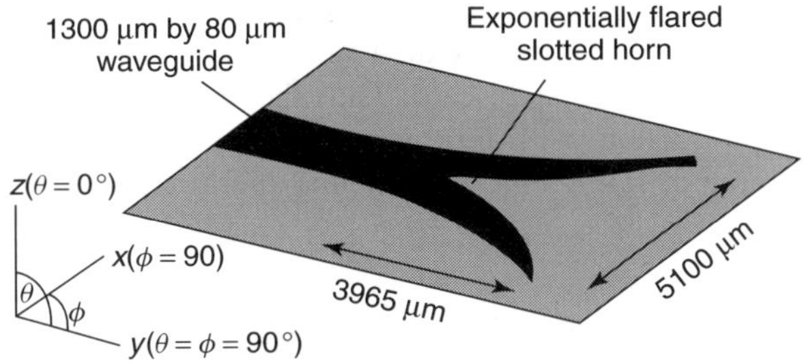

Figure 8.16 Exponentially flared slotted horn antenna integrated with G-band waveguide. Reproduced from J.W. Digby, C.E. Collins, B.M. Towlson, L.S. Karatzas, G.M. Parkhurst, J.M. Chamberlain, J.W. Bowen, R.D. Pollard, R.E. Miles, D.P. Steenson, D.A. Brown and N.J. Cronin, 1997, 'Integrated micro-matched antenna for 200 GHz operation', in *Proceedings of the 1997 IEEE MTT-S International Microwave Symposium, Volume 2, June 8–13 1997*, IEEE, Piscataway, NJ: 561–564, by permission of IEEE, © 1997 IEEE

Commonly used materials such as GaAs require a thickness less than 20 μm to meet this criterion at 184 GHz. Synthesized substrates mentioned previously with microstrip antennas have enabled the use of regular antenna substrates for use in TSAs in Ka-band frequencies (Muldavin, Ellis and Rebeiz, 1997). Holes are micromachined on the substrate to achieve a low synthesized dielectric constant. The diameter and packing of these holes determines the reduction in dielectric constant.

However, alternate schemes are required for higher frequencies. In another approach, first a thick metal film of approximately 150 μm is deposited on the GaAs substrate and is patterned using the LIGA process (Figure 8.17). It may be recalled from Section 1.3.4

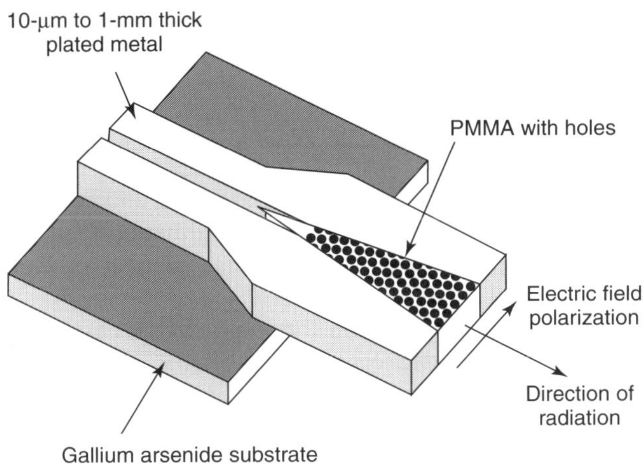

Figure 8.17 A tapered slot antenna fabricated by LIGA micromachining. Note: LIGA, Lithographie, Gavanoformung, Abformung; PMMA, polymethylmethacrylate. Reproduced with permission from S.G. Gearhart and T. Willke, 1998, 'Integrated antennas and filters fabricated using micromachining techniques', in *IEEE Aerospace Applications Conference Proceedings, Volume 3, 1998, March 21–28 1998*, IEEE Computer Society, Los Alamitos, CA: 249–254, by permission of IEEE, © 1998 IEEE

that LIGA process is extremely useful in X-ray etching of thick vertical structures. The substrate below the metallic structures is then removed by backside etching, essentially making the substrate thickness close to zero. Thus the antenna radiates with a wide main beam in the broadside direction, with little substrate losses (Gearhart and Willke, 1998). A portion of the substrate is left behind the antenna aperture to accommodate for the rest of the circuits.

8.5 MICROMACHINED RECONFIGURABLE ANTENNAS

Antennas capable of adaptively changing their characteristics are called reconfigurable antennas. Reconfigurable antennas are primarily used as satellite broadcast antennas. The improvement in the manufacturing technology of satellite systems has resulted in increased lifetime for satellites. Communication antennas are major payloads and hence are designed in the early stages of the satellite launch programme. The beam coverage expected at the initial design stages and at satellite launch, or during its lifetime, may not necessarily match. This could be because of a reorganization of the coverage requirements of the satellites in a scheme or because of the repositioning of the satellite in space. The only means of adapting to these variations is to redesign the antenna before the launch, or relocate the orbital position of the satellite. Both these are very costly options and would be attempted only as a last resort. Furthermore, it may also be required that these satellite antennas be capable of dynamically switching their beam configurations to target different geographic zones on the earth's surface to match the communication traffic as a result of changes in time zones. A viable solution to this varying directional requirement of the antenna radiation is to use a reconfigurable antenna in the satellite (Balcewicz, 1983). When designed properly, the shape and direction of their beams could be modified by ground control.

Satellite-based reconfigurable antennas are generally based on a reflector antenna with a single feed, or a cluster of offset feeds. The beam-forming network in a typical configuration involving a large number of feed elements is fairly complex and heavy. The high-power components used, such as power dividers and switches, are based on semiconductor or electromechanical systems. A much simpler approach is to control the relative phase of the inputs to the array elements (Bucci, Mazzarella and Panariello, 1991). Such a scheme is expected to be more flexible and efficient. Furthermore, the gain of the antenna can be considerably increased without adding a large number of elements to the feed array, thus making the overall system less complicated.

In an effort to replace the complex beam-forming networks associated with the array-fed reflector antennas a team of researchers led by Clarricoats (Clarricoats and Zhou, 1991) proposed the idea of pattern reconfiguration by changing the surface of the reflector. Both rigid-plate-type or grid-type reflectors could be used for the purpose. For easy manipulation, they used a reflector consisting of plates with knitted mesh of gold-plated molybdenum wires. The mesh profile is controlled by motorized actuators, interconnected via cords, making the antenna reconfigurable (Hai, Clarricoats and Monk, 1991). Deforming reflector antenna reconfiguration has also been proposed using a piezoelectric actuator instead of electric motors. One such study has been made to explore the possibility of the use of thin strips of PVDF bonded to the antenna surface to achieve the movement required (Washington, 1996).

These modifications address antennas on geo-synchronous communication satellites that require high gain. MEMS structures are limited by size and power-handling capability and hence are incompatible in such requirements. However, reflectors with novel feed arrangements and various other antenna configurations can incorporate arrays of systems based on MEMS technology. Another aspect worth mentioning is that since existing systems would not be capable of handling the exponential increase in traffic, reliance on higher-frequency systems is bound to increase. This necessitates incorporation of reconfigurable components such as switches and phase shifters in the feeding networks of these antennas operating at millimeter-wave and higher frequencies. RF-MEMS-based systems and many MEMS actuators assume tremendous significance in this context.

MEMS and micromachining has also been used in the design of reconfigurable antennas (Chauvel *et al.*, 1997; Chiao *et al.*, 1999). Detailed fabrication and characterization steps for a micromachined microstrip antenna whose patch radiator is rotated about a fulcrum to get beam-steering capabilities is discussed in Chauvel *et al.*, (1997). The antenna is patterned on 100 μm fused quartz atop 500 μm silicon (1 1 0) substrate, suspended by a pair of torsion springs (Figure 8.18; Chauvel *et al.*, 1997). The rotation is achieved by electrostatic forces by activating fixed electrodes on the substrate. Detailed static and dynamic analyzes have been performed on the proposed structure; however, experimental proof of the concept is still unavailable.

MEMS actuators have also been used to adjust the angles of the arms of a Vee antenna operative at 17.5 GHz (Figure 8.19; Chiao *et al.*, 1999). When the angle between the arms of the antenna is changed the beam shape is altered. In contrast, when both the arms are rotated together in one direction, the beam is scanned. A prototype antenna has been built and the experimental results proved the concept. The advantages of using MEMS technology in this case include elimination of costlier, less lossy and bulkier

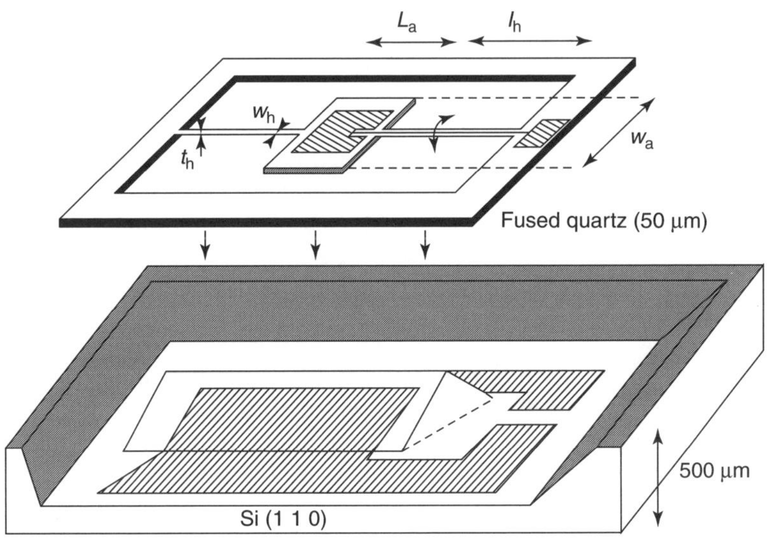

Figure 8.18 Micromachined microstrip antenna for beam-steering applications. Reproduced from D. Chauvel, N. Haese, P.-A. Rolland, D. Collard and H. Fujita, 1997, 'A micromachined microwave antenna integrated with its electrostatic spatial scanning', in *IEEE Microelectromechanical Systems Conference Proceedings*, IEEE, Washington, DC: 84–87, by permission of IEEE, © 1997 IEEE

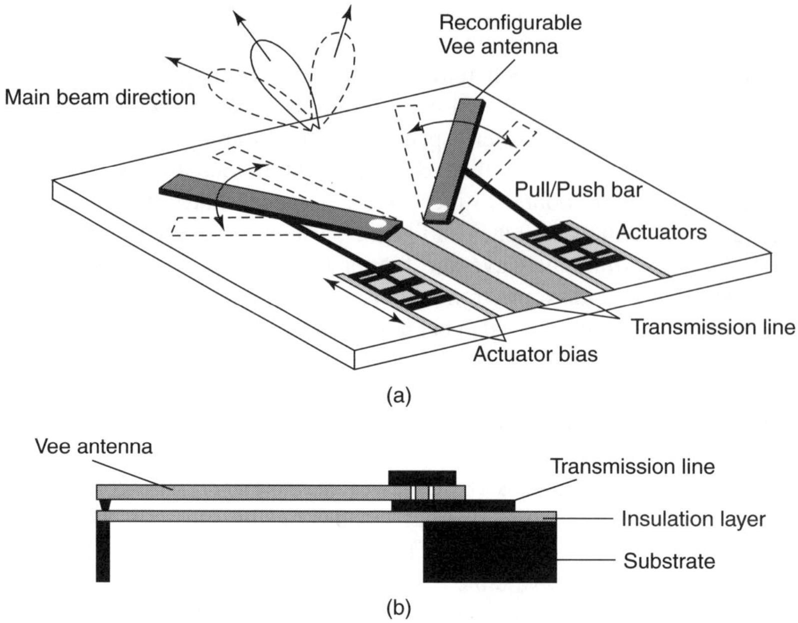

Figure 8.19 Micromachined Vee antenna for beam shaping and beam steering. Reproduced from J.C. Chiao, Y. Fu, I.M. Chio, M. DeLisio and L.-Y. Lin, 1999, 'MEMS reconfigurable Vee antenna', in *IEEE Microwave Theory and Techniques Symposium, Digest*, IEEE, Washington, DC: 1515–1518, by permission of IEEE, © 1999 IEEE

phase shifters for scanning antennas and the possibility of fabrication of a high-frequency array on a single chip. Such arrays are expected to be useful in future reconfigurable wireless and satellite communication networks, seekers for smart weapons and in collision-avoidance radars.

These MEMS-based approaches are in their initial stages, and it is expected that further improvements can be accomplished with the incorporation of MEMS actuators into antenna structures. However, it may also be mentioned that because of the small size of the components involved this technology is largely restricted to millimeter-wave frequencies.

8.6 SUMMARY

Several approaches to incorporate the fast-evolving micromachining concepts into the realm of antenna engineering have been presented in this chapter. With the increased reliance of modern communications equipment on higher frequencies, at millimeter-wave bands and above, the antenna size gets smaller, leading to several difficulties in their fabrication by conventional means. Use of micromachining as a means of fabrication for such small antennas enables increased accuracy and mass production, thus reducing the cost of such systems.

A planar microstrip antenna is an attractive choice for millimeter-wave applications because of its low profile. To improve the bandwidth and radiation efficiency of the antenna, it is preferred that the substrate be thick and have a low dielectric constant. On the contrary, to reduce the size, and losses, microwave circuits generally use thin high-dielectric-constant substrates. Efforts to integrate antennas with the rest of circuits could be prevented by their two contradicting requirements. Micromachining techniques such as back-etching below the antenna, presented in this chapter, can reduce the effective dielectric constant (at the antenna) while at the same time offering original high dielectric constant where circuits are located on the substrate. This approach is also useful as one proceeds towards designing fully integrated systems on semiconductor substrates such as 'system on a wafer'.

REFERENCES

Bahl, I.J., Bhatia, P., 1981, *Microstrip Antennas*, Artech House, Dedham, MA.

Balanis, C.A., 1997, *A Antenna Theory: Analysis and Design*, Wiley, New York.

Balcewicz, J.F., 1983, 'In-orbit reconfigurable communications-satellite antennas', *RCA Engineer* **28**(2): 36–41.

Bean, K.E., 1978, 'Anisotropic etching of silicon', *IEEE Transactions on Electronic Devices* **25**: 1185–1193.

Bucci, O.M., Mazzarella, G., Panariello, G., 1991, 'Reconfigurable antennas by phase-only control', *IEEE Transactions on Antennas and Propagation* **39**: 919–926.

Carver, K.R., Mink, J.W., 1981, 'Microstrip antenna technology', *IEEE Transaction on Antennas and Propagation* **29**: 2–24.

Chauvel, D., Haese, N., Rolland, P.-A., Collard, D., Fujita, H., 1997, 'A micromachined microwave antenna integrated with its electrostatic spatial scanning', in *IEEE Microelectromechanical Systems Conference, Proceedings*, IEEE, Washington, DC: 84–87.

Chen, Q., Fusco, V.F., Zheng, M., Hall, P.S., 1998, 'Micromachined silicon antennas', *Proceedings of the International Conference on Microwave and Millimeter Wave Technology*, IEEE, Washington, DC: 289–292.

Chiao, J.C., Fu, Y., Chio, I.M., DeLisio, M., Lin, L.-Y., 1999, 'MEMS reconfigurable Vee antenna', in *IEEE Microwave Theory and Techniques Symposium, Digest*, IEEE, Washington, DC: 1515–1518.

Clarricoats, P.J.B., Zhou, H., 1991, 'Design and performance of a reconfigurable mesh reflector antenna, part 1: antenna design', *IEEE Proceedings Part. H: Microwave, Antennas and Propagation* **138**: 485–492.

Digby, J.W., Collins, C.E., Towlson, B.M., Karatzas, L.S., Parkhurst, G.M., Chamberlain, J.M., Bowen, J.W., Pollard, R.D., Miles, R.E., Steenson, D.P., Brown, D.A., Cronin, N.J., 1997, 'Integrated micro-matched antenna for 200 GHz operation', in *Proceedings of the 1997 IEEE MTT-S International Microwave Symposium, Volume 2, June 8–13 1997*, IEEE, Piscataway, NJ: 561–564.

Gauthier, G.P., Courtay, A., Rebeiz, G.M., 1997, 'Microstrip antennas on synthesized low dielectric constant substrates', *IEEE Transactions on Antennas and Propagation* **45**: 1310–1314.

Gauthier, G.P., Raskin, J.P., Katehi, L.P.B., Rebeiz, G.M., 1999, 'A 94-GHz aperture-coupled micromachined microstrip antenna', *IEEE Transactions on Antennas and Propagation* **47**: 1761–1766.

Gearhart, S.G., Willke, T., 1998, 'Integrated antennas and filters fabricated using micromachining techniques', *IEEE Aerospace Applications Conference Proceedings, Volume 3, 1998, Mar 21–28 1998*, IEEE Computer Society, Los Alamitos, CA: 249–254.

Griffin, D.W., 1995, 'Monolithic active array limitations due to substrate modes', in *IEEE Antennas and Propagation Society International Symposium, 1995, AP-S. Digest, Volume 2*, IEEE, Washington, DC: 1300–1303.

Hai, Z., Clarricoats, P.J.B., Monk, A., 1991, 'Experimental verification of an electronically controlled reconfigurable reflector antenna', *Electronics Letters* **27**: 64–65.

Hesler, J.L., Hui, K., Dahlstrom, R.K., Weikle, R.M., Crowe, T.W., Mann, C.M., Wallace, H.B., 2001, 'Analysis of an octagonal micromachined horn antenna for submillimeter-wave applications', *IEEE Transactions on Antennas and Propagation* **49**: 997–1001.

Jedlicka, R.P., Poe, M.T., Carver, K.R., 1981, 'Measured mutual coupling between microstrip antennas', in *IEEE Transactions on Antennae and Propagation*, **AP-29**: 147–149.

Muldavin, J.B., Ellis, T.J., Rebeiz, G.M., 1997, 'Tapered slot antennas on thick dielectric substrates using micromachining techniques', in *Proceedings of the 1997 IEEE Antennas and Propagation Society International Symposium, Volume 2, July 13–18 1997*, IEEE, Piscataway, NJ: 1110–1113.

Papapolymerou, I., Drayton, R.F., Katehi, L.P.B., 1998, 'Micromachined patch antennas', *IEEE Transactions on Antennas and Propagation* **46**: 275–283.

Pozar, D.M., 1982 'Input impedance and mutual coupling of rectangular microstrip antenna', *IEEE Transactions on Antennas and Propagation* **30**: 1191–1196.

Rebeiz, G.M., Kasilingam, D.P., Guo, Y., Stimson, P.A., Tutledge, D.B., 1990, 'Monolithic millimeter-wave two-dimensional horn imaging array', *IEEE Transactions on Antennas and Propagation* **38**: 1473–1482.

Sayyah, A.A., Griffin, D.W., 1997, 'Thick substrate mode transmitting antennas for monolithic millimeter wave development', in *Proceedings of the 1997 IEEE Antennas and Propagation Society International Symposium, Part 4, July 13–18 1997*, IEEE, Piscataway, NJ: 2472–2474.

Shenouda, B.A., Pearson, L.W., Harriss, J.E., 2001, 'Etched-silicon micromachined W-band waveguides and horn antennas', *IEEE Transactions on Microwave Theory and Techniques* **49**: 724–727.

Washington, G., 1996, 'Smart aperture antennas', *Smart Material Structures* **5**: 801–805.

Yook, J.-G.; Katehi, L.P.B., 1998 'Suppression of surface waves using the micromachining technique', in *IEEE Antennas and Propagation Society International Symposium, 1998, Volume 2*, IEEE, Washington, DC: 652–655.

Yook, J.W., Katehi, L.P.B., 2001, 'Micromachined microstrip patch antenna with controlled mutual coupling and surface waves', *IEEE Transactions on Antennas and Propagation* **49**: 1282–1289.

Zheng, M., Chen, Q., Hall, P.S., Fusco, V.F., 1998, 'Broadband microstrip patch antenna on micromachined silicon substrates', *Electronics Letters* **34**: 3–4.

9

Integration and packaging for RF MEMS devices

9.1 INTRODUCTION

MEMS is a relatively new field which developed so closely with silicon processing that most of the early packaging technologies were borrowed from the microelectronics field. Packaging of a micromachine is the science of establishing interconnections between the systems and providing an appropriate operating environment for the electromechanical circuits to process the gathered information. Most MEMS devices need physical access to the outside world to react mechanically with an external parameter or to sense a physical variable. MEMS not only condition the signals but also move, which requires care in handling. The state-of-the-art of current sensing technologies is that the device normally accesses the outside world via electrical connections alone and the rest of the systems are totally sealed and isolated. Inertial and optical devices are sometimes special cases, but, in general, the packaging approach of MEMS is fundamentally different from microelectronic packaging. Unlike electronic packaging, where most of the standard packages can be used for a wide variety of applications, the MEMS packaging therefore tends to be customized to the specific applications, which can be summed up in three words: cost, performance and reliability.

Packaging can span from consumer to midrange systems to high-performance weapon-grade applications. No sharp boundaries exist between these classes. However, the gradual shift of optimization parameters, controls the performance, reliability and cost. The size of the package, the choice of its shape and material, the alignment of the device, the mounting for the isolation of shock and vibration, and the seal are some of the many concerns in MEMS packaging. Many important lessons that have seen learned throughout years of experience in the microelectronics industry can be adapted to the packaging of MEMS devices. A MEMS package contains many electrical and mechanical components, which need to be interconnected. Electrical inputs need to be interfaced with the circuits. MEMS can be extremely fragile. They must be protected from mechanical damage and hostile environments. MEMS packaging involves the components of mechanical and electrical structures and the combination of there to form a system.

Webster's Collegiate Dictionary (10th edition) defines *package* as a 'commodity or a unit of a product uniformly wrapped or sealed'. This chapter presents the fundamentals of microelectronic packaging adapted by MEMS technology for its packaging along with the state-of-the-art in customized MEMS packages. The theme continues to be *smaller–faster–cheaper–optimal* systems.

The key issue facing the packaging of the MEMS device is die separation. The current standard die separation method adopted for silicon is to cut the wafer using a diamond-impregnated blade. The blade and the wafer are flooded with high-purity water while the blade spins at 45 000 rpm. This creates no problem for standard integrated circuits (ICs) because the surface is essentially sealed against the effects of water and silicon dust. However, if the MEMS device is exposed to water and debris, the system may break or become clogged and the moisture may have adverse effects, for example in case of radio frequency (RF) switches. Efforts to protect these surfaces with photoresist and other coatings have had only limited success.

9.2 ROLE OF MEMS PACKAGES

The aim of a package is to facilitate the integration of all components such that it minimizes the cost, mass and complexity. The main functions of a MEMS package can be summarized as providing: mechanical support, an electrical interface to the other system components and protection from the environment. The packaging provides an interface between the chip and the physical world. The package should protect the device, at the same time letting it perform its intended functions with less attenuation of signal in a given environment at low cost (Blackwell, 2000; Elwenspoek and Wiegerink, 2001). The packaging becomes more expensive when protection is required for relatively fragile structures integrated into the device. For a standard integrated circuit, the packaging process can take up to 95% of the total manufacturing cost. Issues in MEMS packaging are much more difficult to solve because of stringent requirements in processing, handling and the nature of fragile microstructures; the diversity also complicates the packaging problem.

Many MEMS sensors often require a sensing media interface with a sensing area. For example, a pressure sensor packaging requires incorporation of a pressure port to transmit fluid pressure to the sensor. This makes a major difference between standard semiconductor device packages and MEMS packages.

9.2.1 Mechanical support

Owing to the fundamental nature of MEMS as a mechanical device, the protection and isolation of the device from thermal and mechanical shock, vibration, acceleration and other physical damage during operation become critical. The mechanical stress affecting a system depends on the application. For example, for the same space-borne application, the device package for a military aircraft is different from those used in communication satellites because the operating environments are different. The coefficient of thermal expansion of the package should be equal to that of silicon for reliability because the thermal cycle may cause cracking or delamination if the materials are unmatched. If the packaging solution is creating excessive stress in the sensing structure, it can cause a change of device performance. Once the MEMS devices are wire bonded and other electrical connections are made, the assembly must be protected by covering the base or by encapsulating the assembly in plastic or ceramic materials since the electrical connections are usually made through the walls. Managing package-induced stress in the device becomes important for MEMS package design.

9.2.2 Electrical interface

Wire bonds and other electrical connections to the device should be made by protecting the device from scratches and other physical damages. Direct current (dc) and RF signals to the MEMS systems are given through these connections to interface the MEMS device with the systems. Also, these packages should be able to distribute RF signals to other components inside the package. High-frequency RF signals can be introduced into the package by metal transmission lines or coaxial lines or the function can be electromagnetically coupled into the device. The final connection between the MEMS and the RF lines is usually made with wire bonds or flip-chip die attachments and multilayer interconnections.

9.2.3 Protection from the environment

Many of the MEMS devices and sensors are designed to measure outside variables from the surrounding environment. The hermetic packaging generally applicable to microelectronic devices is not suitable in many cases of MEMS devices. These devices might be integrated with the circuits or mounted to a circuit board and protected from mechanical damage. Only special attention to packaging will protect a micromachined device from aggressive surroundings. Protection starts at the dice level (Sparks, 2001). Elements that cause corrosion or physical damage to the metal lines as well as other components such as moisture remains a concern for many MEMS devices. The moisture that may be introduced into the package during fabrication and before sealing can damage the materials. For example, aluminum lines can corrode quickly in presence of moisture, and gold lines degrade slowly in moisture. Junctions of dissimilar metals can also corrode in the presence of moisture. MEMS packages need to be hermetic, with good barriers against liquids and gases.

In most space-borne applications, the parts are hermetically sealed to give a perceived increase in reliability and to minimize outgassing. When epoxies or cyanate esters are used for die attach, they outgas when they cure. Outgassing is a concern for many devices since the particles could deposit onto components and reduce device performance. For example, outgassing leads to stiction and corrosion of the device. Die attach materials with a low Young's modulus allow the chip to move during the ultrasonic wire bonding, resulting in low bond strength.

9.2.4 Thermal considerations

The MEMS devices used for current applications do not have a high power dissipation requirement. The thermal dissipation from MEMS devices is not a serious problem since the temperature of the MEMS devices usually does not increase substantially during the operation. However, as the integration of MEMS with other high-power devices such as amplifiers in a single package increases, the need for heat dissipation will have to be addressed to protect the MEMS device from high temperatures. This thermal management can place a high design consideration on package design.

9.3 TYPES OF MEMS PACKAGES

Methods of packaging of very small mechanical devices are not a new topic. The aerospace industry has performed well in this respect over half a century, and the watch industry for more than that. Each MEMS application usually requires new package design, depending on the application and optimization procedures. In general, the possible group of packages can be categorized into four types: (1) all-metal, (2) ceramic, (3) plastic and (4) multilayer.

9.3.1 Metal packages

IC packaging using metal packages is well advanced because of the wide applications of ICs, excellent thermal dissipation and electromagnetic shielding. Metal packages are also often used in monolithic microwave integrated circuits (MMICs) and hybrid circuits. Materials such as CuW (10/90), Silver (Ni-Fe), CuMo (15/85) and CuW (15/85) are good thermal conductors and have a higher coefficient of thermal expansion (CTE) than silicon. All these metals, with copper, gold or silver plating are good choices for MEMS packages.

9.3.2 Ceramic packages

One of the most common packages used in the microelectronics industry is the ceramic package because of features such as low mass, low cost and ease of mass production. The ceramic packages can be made hermetic, adapted to multilayer designs and can be easily integrated for the signal feedthrough lines. Multilayer packages reduce the size and cost of integration of multiple MEMS into a single package. The electrical performances of the packages can be tailored by incorporating multilayer ceramics and interconnect lines. These types of packages are generally referred as *co-fired multilayer ceramic packages*.

Co-fired ceramic packages are constructed from individual pieces of thin films in the 'green' or unfired state. Metal lines are deposited in each film by thick-film processing, such as screen printing, and via holes for interconnections are drilled. After these lines and interconnecting holes are done, the unfired layers are stacked and aligned and laminated together and fired at high temperature. MEMS and the necessary component are then attached using epoxy, or solder, and wire bonds are made the same as the metal packages.

There are several problems associated with ceramic packaging. The green state shrinks during the firing process and the amount of shrinkage depends on the number of via holes and wells cut into each layer. The ceramic-to-metal adhesion is not strong as ceramic-to-ceramic adhesion. The processing temperature of ceramics limits the choice of metal lines, and the metal should not react with the ceramic during the firing process. In *low-temperature co-fired ceramic* (LTCC), the most frequently used metal lines are tungsten and molybdenum, and the conductors are silver, gold and AuPt.

9.3.3 Plastic packages

Plastic packages are common in the electronic industry because of their low manufacturing cost. However, hermetic seals are not possible with plastic packages, which is generally required for highly reliable applications. Plastic packages are also susceptible to cracking during temperature cycling.

9.3.4 Multilayer packages

Figure 9.1 shows a cross-sectional view of a three-dimensional multilayered packaging for MEMS structures on silicon substrate. Passive elements such as filters and matching circuits are formed in each layer and active devices are assembled on the top layer using flip-chip technology. The structure is a three-dimensional hybrid IC using silicon, which is more cost-effective than GaAs. Figure 9.2 shows a 25.0-GHz receiver front-end incorporating a built-in micromachined filter along with the measured responses (Takahashi *et al.*, 2000). The whole down converter and filter were built into a size of 11×11 mm with overall conversion gain of 22 dB and a noise figure less that 4 dB.

9.3.5 Embedded overlay

An embedded overlay (Butler and Bright, 2000) concept for packaging of micro-opto-electromechanical systems (MOEMS) and RF MEMS devices is derived from chip-on-flex (COF) process currently used for microelectronics packaging. COF is a high-performance multichip packaging technology in which dies are encased in a moulded plastic substrate and interconnections are made via a thin-film structure formed over the components. The electrical interconnections are made through a patterned overlay while the die is embedded in a plastic substrate, as shown in Figure 9.3. Chips are attached face down on the COF overlay using polyimide or thermoplastic adhesives. The substrate is formed after bonding the chips around the components using a plastic mould-forming process such as transfer, compression or injection moulding at 210 °C. The electrical connections are made by drilling via holes using a continuous argon ion laser at 35 m nm. Ti/Cu metallization is sputtered and patterned to form the electrical interconnections. The use

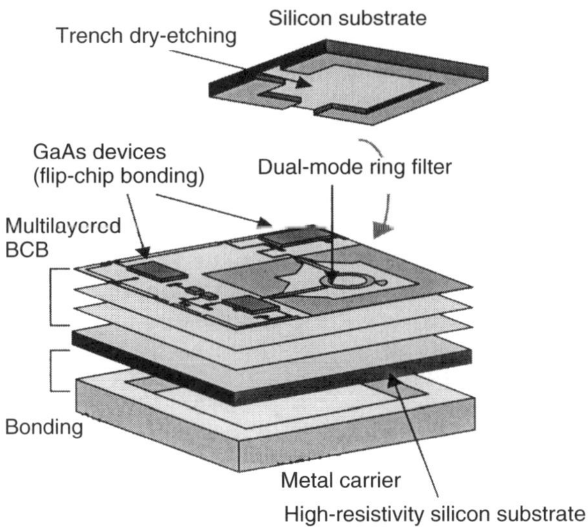

Figure 9.1 Three-dimensional millimeter-wave MEMS integrated circuit. Reproduced from K. Takahashi, U. Sangawa, S. Fujita, M. Matsuo, T. Urabe, H. Ogura and H. Yubuki, 2001, 'Packaging using microelectromechanical technologies and planar components', *IEEE Transactions or Microwave Theory and Techniques* **49**(11): 2099–2104, by permission of IEEE, © 2001 IEEE

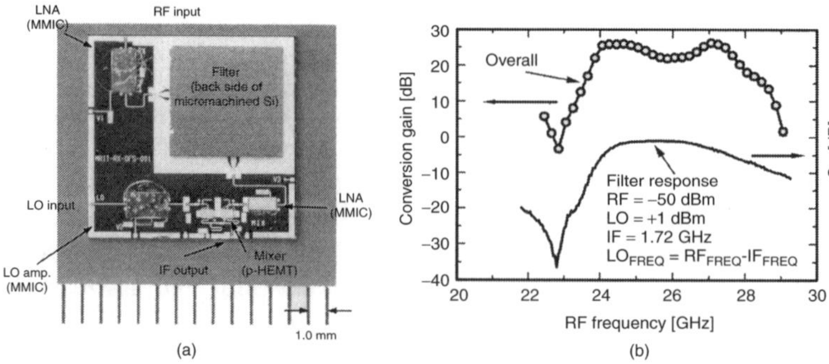

Figure 9.2 (a) Fabricated 25-GHz receiver front-end integrated circuit with micromachined filter and (b) measured response. Note: LNA, low noise amplifier; MMIC, monolithic microwave integrated circuit; RF, radio frequency; LO, local oscillator; IF, intermediate frequency; HEMT, high electron mobility transistor. Reproduced from K. Takahashi, U. Sangawa, S. Fujita, K. Goho, T. Uare, H. Ogura and H. Yabuki, 2000, 'Packaging using MEMS technologies and planar components', in *2000 Asia Pacific Microwave Conference*, IEEE, Washington, DC: 904–907, by permission of IEEE, © 2000 IEEE

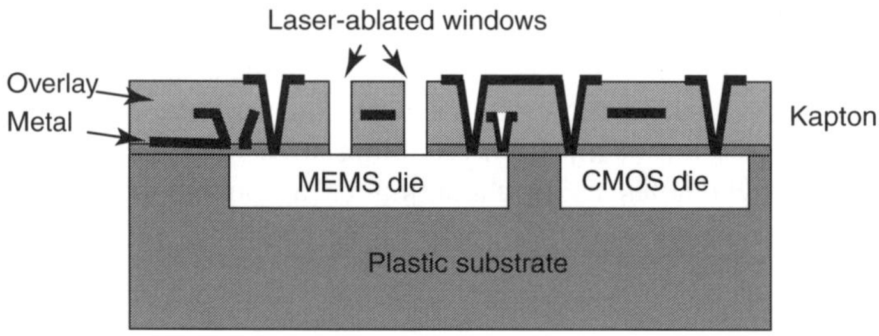

Figure 9.3 Chip-on-flex MEMS packaging concept. Reproduced from J.T. Butler and V.M. Bright, 2000, 'An embedded overlay concept for microsystems packaging', *IEEE Transactions on Advanced Packaging* **23**(4): 617–622, by permission of IEEE, © 2000 IEEE

of varying laser ablation power levels with plasma cleaning and high-pressure water scrubs provide an effective means of removing the COF overlay without damaging the embedded MEMS devices. Figure 9.4 shows the 5 × 5 array of micromirrors packaged in COF/MEMS modules with integrated micromirror control circuitry.

9.3.6 Wafer-level packaging

MEMS packaging should be considered from the beginning of device development. Cost-efficient MEMS packaging focuses on wafer-level packaging (Gilleo, 2001a, Reichal and Grosser, 2001). Designing the packaging schemes and incorporating then into the device manufacturing process itself can reduce the cost. Versatile packaging may be needed for many devices in which MEMS and microelectronics are on a single chip. Each MEMS device may have its own packaging methods, which may be absolutely suitable for its functioning. Since MEMS devices have movable structures on the surface of the wafer, addition of a cap wafer on the silicon substrate makes them suitable for many applications.

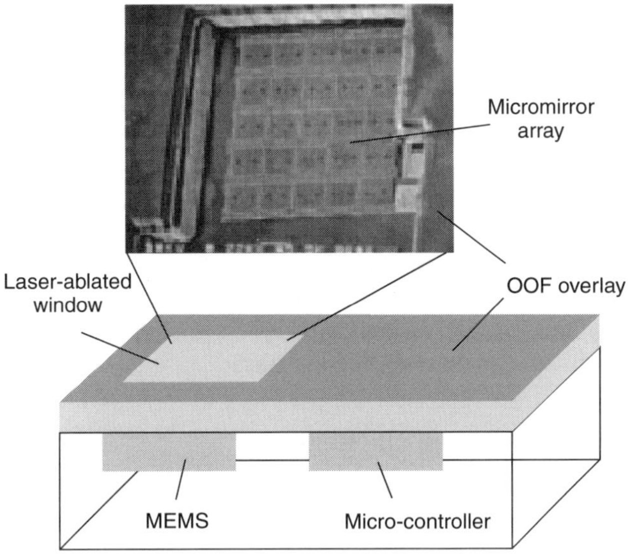

Figure 9.4 COF/MEMS package of 5 × 5 array of micromirrors. Note: COF, chip-on-flex. Reproduced from J.T. Butler and V.M. Bright, 2000, 'An embedded overlay concept for microsystems packaging', *IEEE Transactions on Advanced Packaging* **23**(4): 617–622, by permission of IEEE, © 2000 IEEE

The cap provides protection against handling damage as well as avoiding atmospheric damping. This is done by bonding the substrate with an active device to a second wafer, either of the same material or of different material. The bonding is done by using glass frit or by anodic bond created by electrical potential. Precision-aligned wafer bonding is the key technology for high-volume, low-cost packaging of MEMS devices (Helsel *et al.*, 2001; Mirza, 2000). State-of-the-art silicon wafer bonding can provide assembly level packaging solutions for many MEMS devices.

The wafer-level package, which protects the device at the wafer stage itself, is a clear choice to make at the product design stage itself. This involves an extra fabrication process, where a micromachined wafer has to be bonded to a second wafer with appropriate cavities etched on it. Figure 9.5 shows a schematic diagram of wafer level packaging. This enables the MEMS device to move freely in vacuum or inert atmosphere with hermetic bonding, which prevents any contamination of the structure. Etching the cavities in blank silicon wafer and placing it over the MEMS device and bonding them together can make a hermetic seal.

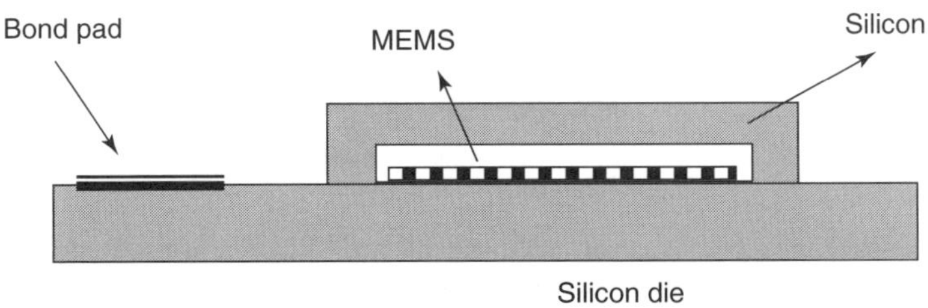

Figure 9.5 Silicon wafer-level packaging of RF MEMS

Anisotrophic wet etching of bulk silicon along certain crystal planes using strong alkaline solutions such as KOH can create thin diaphragms, through-wafer via holes and V-grooves. The fastest etch rates for the silicon are the ⟨1 0 0⟩ and ⟨1 1 0⟩ crystal planes and the slowest is for the ⟨1 1 1⟩ plane with typical masking layers such as silicon dioxide or low-pressure chemical vapor deposition (LPCVD) silicon nitride. Examples of successful development and packaging using silicon micromachining are the ink-jet heads and silicon piezoresistive pressure sensors for automotive and industrial control applications. Many of these devices require silicon wafer bonding to another substrate as a first-level packaging solution. Anodic (electrostatic) bonding of silicon to glass, low-temperature glass-frit bonding of silicon to silicon, silicon direct wafer bonding, eutectic bonding and epoxy bonding are examples of a few methods available to bond silicon wafer to other silicon, as explained in Section 9.6.1.

9.3.7 Microshielding and self-packaging

The micromachining technology has proved a flexible approach for the development of low-loss transmission lines (Dryton, 1995) as well as micropackages that provide

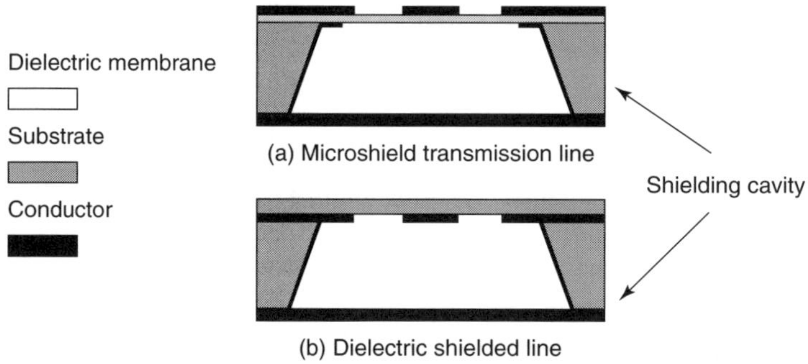

Figure 9.6 Topology of self-packaging transmission lines: (a) dielectric membrane supported line; (b) dielectric shielded line. Reproduced from R.F. Dryton, 1995, *The Development and Characterization of Self-packages using Micromachining Techniques for High Frequency Circuit Applications*, PhD dissertation, University of Michigan, Ann Arbor, MI, by permission of the University of Michigan

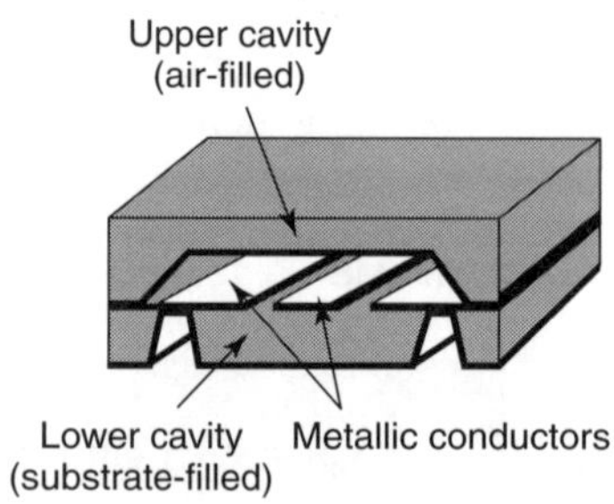

Figure 9.7 Self-packaged circuit constructed out of two silicon wafers. Reproduced from R.F. Dryton, 1995, *The Development and Characterization of Self-packages using Micromachining Techniques for High Frequency Circuit Applications*, PhD dissertation, University of Michigan, Ann Arbor, MI, by permission of the University of Michigan

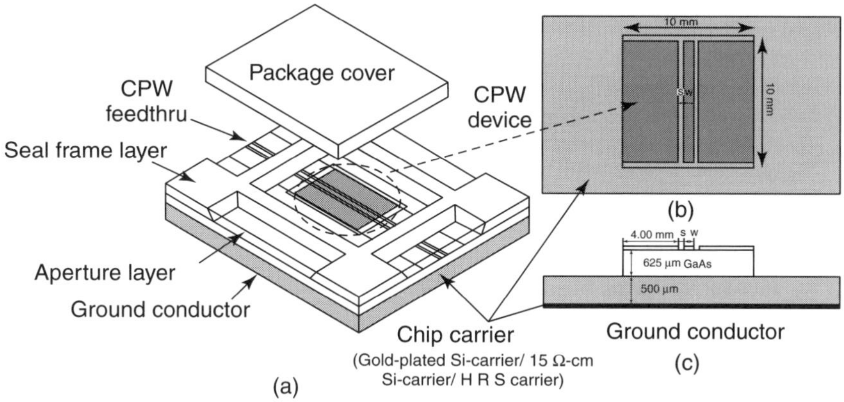

Figure 9.8 (a) Typical MEMS packaging with co-planar waveguide (CPW) line; (b) top view and (c) side view Note: HRS, high-resistivity silicon. Reproduced from S.J. Kim, Y.S. Kwon and H.Y. Lee, 2000, 'Silicon MEMS Packages for coplanar MMICs', in *Proceedings of 2000 Asia-Pacific Microwave Conference, Australia, December 2000*, IEEE, Washington, DC, by permission of IEEE, © 2000 IEEE

self-packaging (Hindreson *et al.*, 2000) to individual planar circuit components. As shown in Figure 9.6, the metal conductors are supported by membrane and a lower cavity is below the conducting line. In Figures 9.7 and 9.8, the upper wafer has an air-filled cavity that is mounted over the metallic conductors. Integration of both upper and lower shielded circuits results in a self-packaged RF circuit.

9.4 FLIP-CHIP ASSEMBLY

Flip-chip is the most favored assembly technology for high-frequency applications because the short bump interconnect can reduce parasitic impedances. In flip chips, an IC die is placed on a circuit board with bond pads facing down and directly joining the bare die with the substrate. The bumps form electrical contact as well as a mechanical joint to the die. This reduces the electrical path length and the associated capacitance and inductance, which is particularly suited for high-density RF applications. The minimization of parasitic capacitance and inductance can reduce the signal delay in high-speed circuits.

Flip-chip bonding involves the bonding of die, top-face down on a package substrate. Electrical connections are made by means of plated solder bumps between bond pads on the die and metal pads on the substrate (Oppermann *et al.*, 2000). The attachment is intimate with relatively small spacing ($\sim 100\,\mu m$) between the die and the substrate. In flip-chip assemblies the bumps form the electrical contacts to the substrate as well as serving as a mechanical joint.

Figure 9.9 shows the flip-chip design of a MEMS package. Since the active surface of the MEMS is placed towards the substrate, the cavity will protect the movable MEMS. The stand-off distance can be accurately controlled by the bump height. Flip-chip technology is a very flexible assembly method for different applications. Another concept in wafer-level packaging is to apply a microcap to the device and then package with standard procedures. Figure 9.10 shows the concept of cap-on-chip packaging

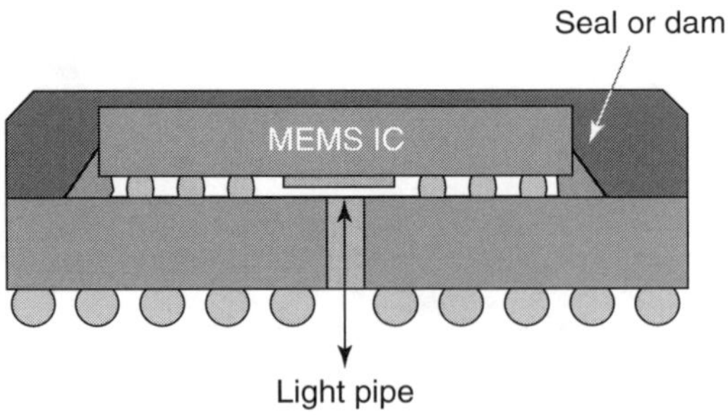

Figure 9.9 Flip-chip MEMS package. Reproduced from S.J. Kim, Y.S. Kwon and H.Y. Lee, 2000, 'Silicon MEMS Packages for coplanar MMICs', in *Proceedings of 2000 Asia-Pacific Microwave Conference, Australia, December 2000*, IEEE, Washington, DC, by permission of IEEE, © 2000 IEEE

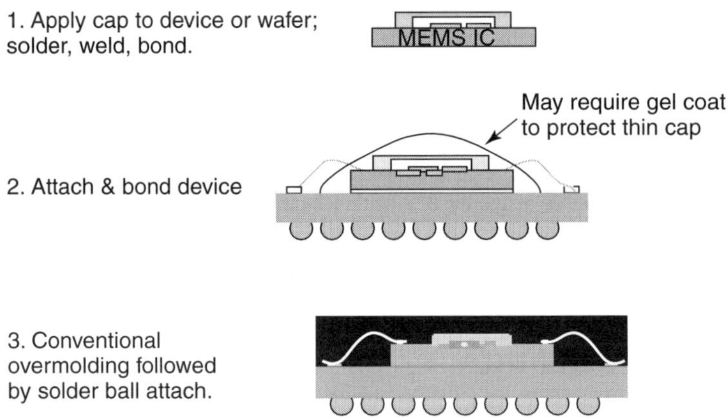

Figure 9.10 Cap-on-chip packaging. Reproduced from K. Gilleo, 2001b, 'MEMS packaging issues and materials', in *Proceedings of IEEE International Symposium on Advanced Packaging: Process, Properties and Interfaces*, IEEE, Washington, DC: 1–5, by permission of IEEE, © 2001 IEEE

for MEMS. Figure 9.11(a) presents the flip-chip bonding process on a ceramic-based (alumna) substrate and (Figure 9.11(b)) shows the gold bumps formed on pads of the substrate. The bump with an acute tail makes it easy to deform and to make the bonding area more stable under thermal conditions.

Flip-chip bonding is attractive to the MEMS industry because of its ability to package closely a number of dice on a single package substrate with multiple levels of electrical traces. Similar systems can built with wire bonding, but the area usage will be greater and the number of gold wires within the package may present a reliability issue. However, flip-chip may not be compatible with the packaging of MEMS that include microstructures exposed to the open environment.

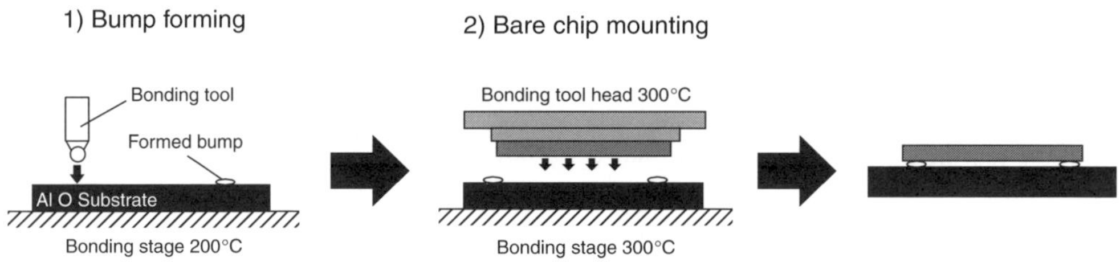

(a)

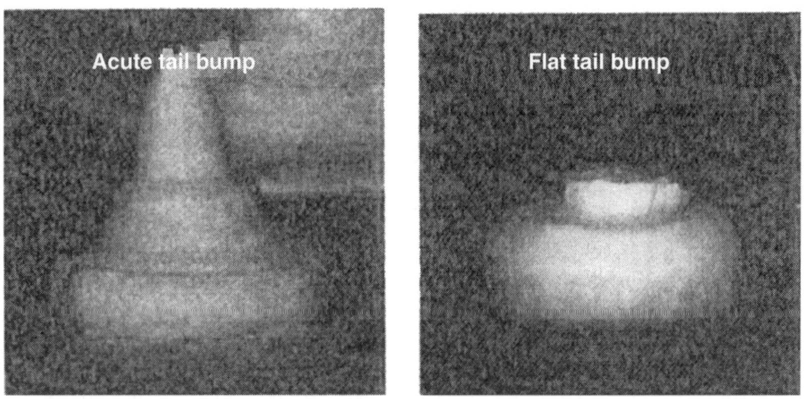

(b)

Figure 9.11 (a) Flip-chip bonding procedure; (b) photograph of acute and flat-tail bump used for flip-chip bonding. Reproduced from H. Kusamitsu, Y. Morishita, K. Murushashi, M. Ito and K. Ohata, 1999, 'The flip-chip bump interconnection for millimeter wave GaAs MMIC', *IEEE Transactions on Electronics Packaging and Manufacturing* **22**(1): 23–28, by permission of IEEE, © 1999 IEEE

9.5 MULTICHIP MODULE PACKAGING

The incompatibilities in fabrication of MEMS and ICs make them difficult for monolithic integration. Multichip module (MCM) packaging provides an efficient solution to integrate MEMS with microelectronic circuits because it supports a variety of die types in a common substrate without the need for many changes in either the MEMS or microelectronics fabrication processes. It adopts the high-density interconnect (HDI) process, consisting of embedding bare die into premilled substrate.

The micro module system (MMS) multichip module-deposited (MCM-D) process is the more traditional approach. The interconnect layers are first deposited on the substrate, and the die are mounted above the interconnect layers. The interconnect is mainly done by wire bonding (Butler, Bright and Comtios, 1997; Butler *et al.*, 1998; Cohn *et al.*, 1998; Coogan, 1990; Sardborn, Swaminathan and Subramanian, 2000).

Modifying the HDI process allows physical access to MEMS devices. Figure 9.12(a) shows the HDI process flow and Figure 9.12(b) shows an augmented HDI process for MEMS packaging by an additional laser ablation step to allow physical access to the MEMS die. The windows in the dielectric overlay above the MEMS device were selectively etched using laser ablation. Figure 9.13 shows a photograph of an MCM-D/MEMS a package.

Among various types of MCMs, the MCM-C (ceramic-based multichip module) is a multiplayer substrate based on aluminum oxide, and MCM-V (Gotz *et al.*, 2001) is the vertical multichip module. The lines and vias are printed on different layers. All the layers are then co-fired (high-temperature co-fired ceramic) at the same time at high temperature. The metal parts, such as lead frames and heat sinks if necessary, can be soldered with eutectic.

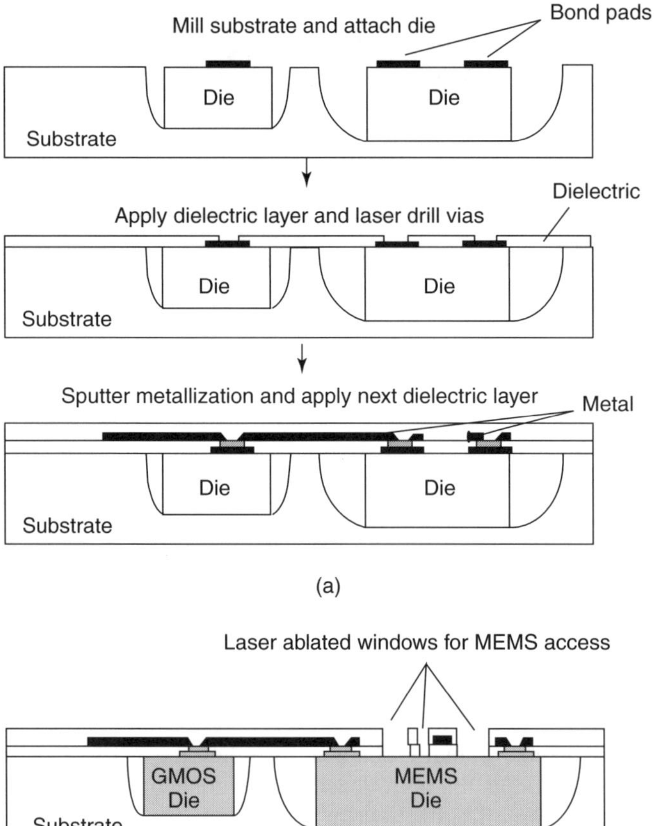

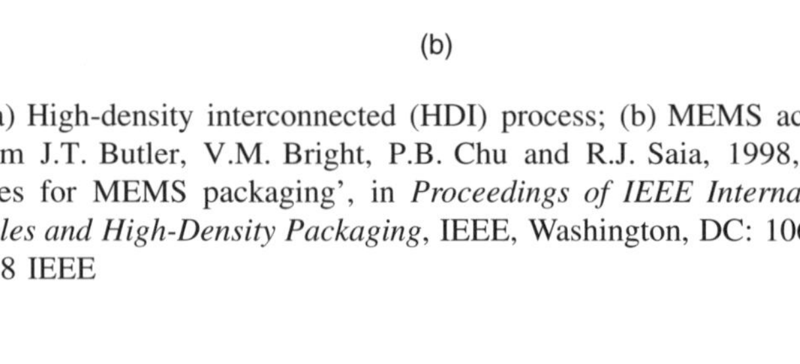

Figure 9.12 (a) High-density interconnected (HDI) process; (b) MEMS access in HDI process. Reproduced from J.T. Butler, V.M. Bright, P.B. Chu and R.J. Saia, 1998, 'Adapting multichip module foundries for MEMS packaging', in *Proceedings of IEEE International Conference on Multichip Modules and High-Density Packaging*, IEEE, Washington, DC: 106–111, by permission of IEEE, © 1998 IEEE

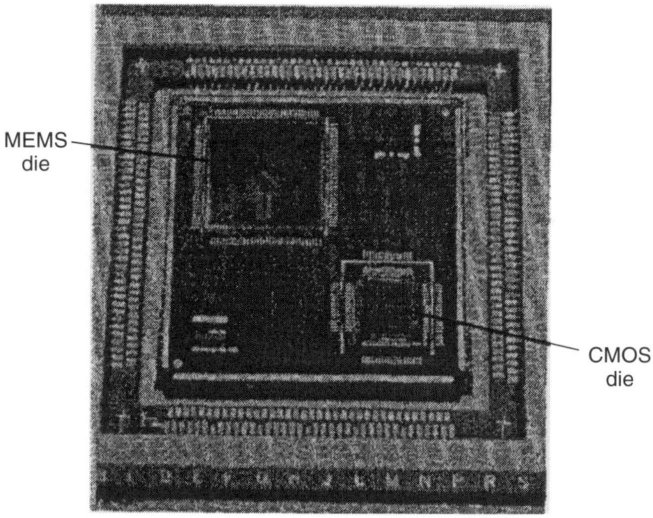

Figure 9.13 MCM-D/MEMS package. Note: CMOS, complementary metal oxide semiconductor; MEMS, microelectromechanical system; MCM-D, multichip module-deposited. Reproduced from J.T. Butler, V.M. Bright and J.H. Comtios, 1997, 'Advanced multichip module packaging of microelectromechanical systems', in *Transducers '97*, IEEE, Washington DC: 261–264, by permission of IEEE, © 1997 IEEE

9.5.1 Wafer bonding

MEMS packaging can also be done by bonding a recessed cap onto a micromachined wafer. However, conventional wafer bondings such as line fusion or anodic bonding cannot be employed because the micromechanical circuits can be damaged by to high temperatures or high electric fields. The low-temperature bonding techniques may increase the cost of packaging. If MEMS devices can be packaged at the device level first then the remaining packaging can be done with the IC packaging using common procedures. Microscale riveting (Lin, 1993; Shivkumar and Kim, 1997) or eutectic bonding (Cheng, Lin and Najafi, 2000) can be performed by directional etching of silicon for the rivet moulds and directional electroplating in an electric field for rivet formation. The wafer joining can be done at room temperature and low voltages. The protected devices after microriveting can be treated the same as the IC wafer during the dicing process. Once the joining is complete, the resulting chips can be handled in the same way as IC chips during the remaining packaging steps, such as wire bonding and moulding for plastic packages.

Figure 9.14 shows the concept of a protected chip with MEMS device. Rivets are formed all around the cap wafer to hold the cap–base pair together. Figure 9.15 shows the prepared cap and base wafers and the electroplating setup. Nickel can easily be electroplated as a rivet material.

A seed layer of 125 Å of Chromium and 750 Å of Nickel is deposited on the surface of the base wafer by thermal evaporation. The cap and the base wafers are held together during the plating process so that the plating can start at the exposed area of the seed layer, grow through the rivet hole in the cap wafer and form the rivet. Simple mechanical clamping of the wafer together in the electrolyte is sufficient to rivet them together since electroplating does not occur in the microscopic wafer gap.

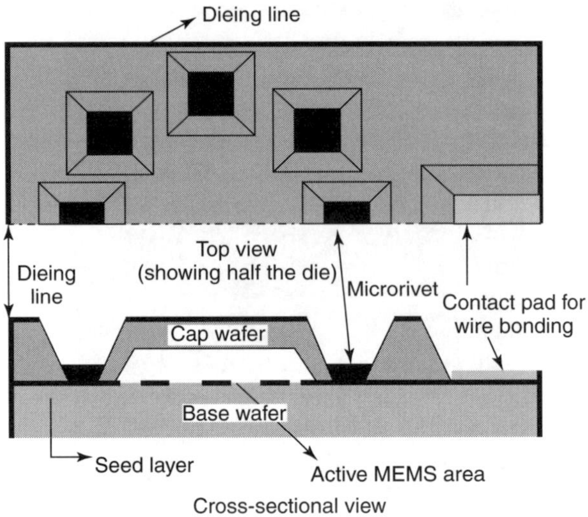

Figure 9.14 View of a packaged chip using microrivets. Reproduced from B. Shivkumar and C.J. Kim, 1997, 'Microrivets for MEMS packaging: concept, fabrication and strength testing', *Journal of Microelectromechanical Systems* **6**(3): 217–225, by permission of IEEE, © 1997 IEEE

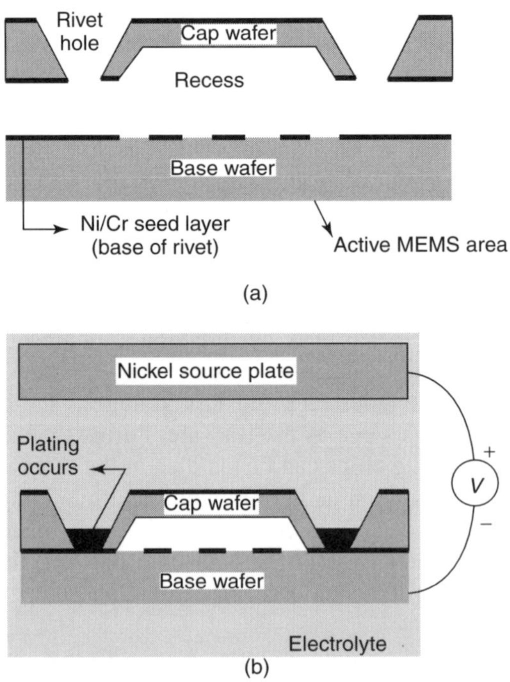

Figure 9.15 Schematic diagram of (a) the prepared cap and (b) the electroplating setup. Reproduced from B. Shivkumar and C.J. Kim, 1997, 'Microrivets for MEMS packaging: concept, fabrication and strength testing', *Journal of Microelectromechanical Systems* **6**(3): 217–225, by permission of IEEE, © 1997 IEEE

In fusion bonding, polysilicon is deposited and patterned as the heating and bonding material. Fusion bonding is used mostly in silicon-on-insulator (SOI) technology such as Si-SiO_2 (Laskey, 1986; Li *et al.*, 2002; Mirza, 2000) and silicon bonding (shimbo *et al.*, 1986). Aluminum-to-glass (Cheng, Lin and Najafi, 2001) bonding using localized heating can be applied for hermetic packaging. In eutectic bonding, gold resistive heaters are sputtered and used as heating and bonding material. The temperature of the microheater rises upon the flow of current, which activates the bonding process. The principle of localized bonding is shown in Figure 9.16. The effectiveness of the microheater depends on the selection of materials and the design of the geometrical shape of the structure. For example, a high temperature of 1000 °C can be created using microheaters, while the temperature at less than 2 μm away can drop to 100 °C (Mirza, 2000). Figure 9.17 shows the experimental setup for localized eutectic bonding. The conventional bonding takes one hour to reach the temperature, while localized eutectic heating will take only less than 5 minutes.

Phosphorous-doped polysilicon and gold resistive heaters are used in the silicon-to-glass fusion and the silicon-to-gold eutectic bonding process, respectively. Both processes can be accomplished in less than 5 minutes.

The aligned wafer-bonding process typically consists of two separate steps (Table 9.1). The wafers are aligned initially to each other in a bond aligner. This system can align a mask to a wafer for conventional photolithography as well as being able to align two wafers to each other. The aligned wafers are clamped with an appropriate separation gap

Table 9.1 Typical wafer bonding process conditions for anodic, glass frit and silicon direct wafer (DW) bonding

Condition	Bonding process		
	anodic	glass frit	DW
Temperature (°C)	300–500	400–500	1000
Pressure (bar)	N/A	1	N/A
Voltage (kV)	0.1–1	N/A	N/A
Surface roughness (nm)	20	N/A	0.5
Precise gap?	Yes	No	Yes
Hermetic seal?	Yes	Yes	Yes
Vacuum level during bond (Torr)	10^{-5}	10	10^{-3}

N/A Not applicable.
Source: Mirza and Ayon, 1998.

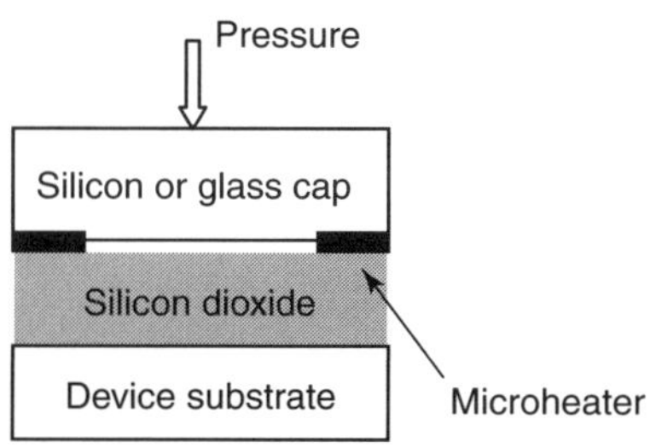

Figure 9.16 Schematic diagram of the localized microheater setup

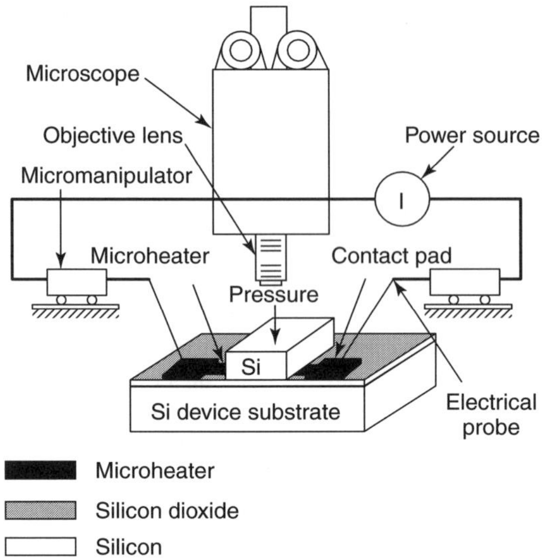

Figure 9.17 Experimental setup for the localized heating and bonding test. Reproduced from L. Lin, 2000, 'MEMS post-packaging by localized heating and bonding', *IEEE Transactions on advanced Packaging* **23**(4): 608–616, by permission of IEEE, © 2000 IEEE

between them in a bond fixture. The next step is to load the bond fixture into a vacuum bond chamber where the wafers are contacted together.

9.6 RF MEMS PACKAGING: RELIABILITY ISSUES

9.6.1 Packaging materials

Since MEMS devices have also to be fabricated other than silicon substrate, the compatibility with materials other than silicon and manufacturing in a silicon IC foundry is a major issue. One of the major capital investments needed is the equipment for automated packaging. For example, for automotive sensors, the environment in which the devices are going to operate must be considered at the beginning of package design. Table 9.2 shows the conditions in which most automotive components operate.

9.6.2 Integration of MEMS devices with microelectronics

The integration of a MEMS sensor with electronics has advantages, in particular when dealing with small signals. However, in such cases it is important that the process used for MEMS fabrication does not adversely affect the added electronics, required for the device to function correctly. MEMS devices can be fabricated as pre- or post-processing modules, which are integrated within the standard processing. The choice of whether or not to integrate depends on the application of the sensors and different aspects of the implementation technology. The state-of-the-art in MEMS is combining MEMS with ICs and utilizing advanced packaging techniques to create system-on-a-package (SOP) or system-on-a-chip (SIP) (Malshe *et al.*, 2001).

Table 9.2 Operating parameters of automobile sensors

Environment	Parameter value
Temperature (°C)	
driver interior	40–85
under the bonnet	125
on the engine	150
in the exhaust and combustion area	200–600
Mechanical shock (g)	
assembly (drop test)	3000
on vehicle	50–500
Mechanical vibration at 15g (Hz)	100–2000
Electromagnetic impulses ($V\,m^{-1}$)	100–200

Note: depending on the application, there may also be exposure to humidity, salt spray, fuel, oil, break fluid, transmission fluid, ethylene glycol, freon and exhaust gas.
Source: Sparks, Chang and Eddy, 1998.

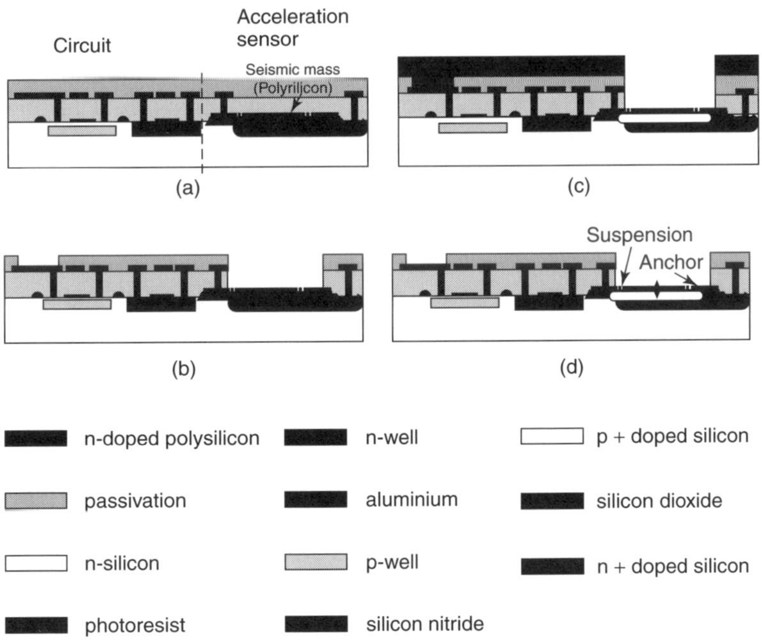

Figure 9.18 Integration of surface micromachining with CMOS. Reproduced from P.J. French, 1999, 'Integration of MEMS devices', in *Proceedings of SPIE Device and Process Technologies for MEMS and Microelectroncis, Queensland Australia*, SPIE volume 3892: 39–48, by permission of SPIE

The simplest form of integrated MEMS device is where existing layers are used for mechanical and sacrificial layers (French, 1999; Hsu, 2000; Ramesham and Ghaffarian, 2000). Standard processes have a number of layers on top of the wafer such as oxide, polysilicon, metal and nitride. This requires only the additional steps of masking and etching, as explained in Figure 9.18. Surface micromachining using post-processing additional layers is but maintaining standard processing by adding depositions at the end of processing. This may cause limitations on the thermal budget if aluminum is used as the metallization. Plasma-enhanced chemical vapor deposition (PECVD) can lower the temperature compatible with aluminum metallization.

In general, there are three main methods that have been used for monolithic integration of CMOS and MEMS; (a) electronics first (University of California, Berkeley, CA), (b) MEMS in the middle (Analog Devices, Cambridge, MA), and (c) MEMS first (Sandia National Laboratories, Livermore, CA) (O'Neal *et al.*, 1999). Each of these methods has its own advantages as well as disadvantages. Sandia fabricated MEMS first and etched a trench and covered it with sacrificial oxide, which protects the MEMS devices from the CMOS processing steps. After the trench is completely filled with SiO_2, the surface is planarized, which serves as the starting material for CMOS foundry. The sacrificial oxide covering the MEMS device is removed after the fabrication of the CMOS device.

The alternative approach for monolithic integration with MEMS is the multi-chip-module (MCM) in which IC and MEMS dice can be placed in the same package. Several sensors, actuators or a combination can be combined in a single chip using the MCM technique (Butler *et al.*, 1998). The main disadvantage is the probable signal loss due to parasitic effects between the components and the apparent added packaging expenses.

Co-planar MMICs packaged using a silicon (1 to $\sim 300\,\Omega\,\text{cm}$) substrate is found to reduce the parasitic effects, coupling and resonance compared with the unpackaged devices (Kim, Kwon and Lee, 2000). Common resistive silicon without gold plating can be an ideal packaging solution for low-cost and high-performance co-planar lines.

9.6.3 Wiring and interconnections

MEMS packages must protect the micromachined parts from environments and at the same time provide interconnections to electrical signals as well as access to and interaction with external environments. In hermetic packages, the electrical interconnections through a package must confirm hermetic sealing. Wire bonding is the popular technique to connect the die to the package electrically. Bonding of gold wires is easier than bonding of aluminum wires. The use of wire bonding has serious limitations in MEMS packaging because of the application of ultrasonic energy at a frequency between 50 and 100 kHz. Unfortunately, these frequencies may simulate oscillation of microstructures. Since most microstructures have resonant frequencies in the same range, the chance of structural failure during the wire bonding is high (Maluf, 2000).

9.6.4 Reliability and key failure mechanisms

Reliability requirements for various MEMS will be significantly different for different applications, especially with systems with unique MEMS devices. Hence standard reliability testing is not possible until a common set of reliability requirements is developed. The understanding of reliability of the systems comes from the knowledge of failure behavior and the failure mechanisms. The main failure mechanisms of MEMS devices are summarized as follows.

- Stiction: stiction and wear are the real concern and cause for most of the failure of MEMS. Stiction occurs as a result of microscopic adhesion when two surfaces come into contact. Wear due to corrosive environment is another aspect of failure.

- Delamination: MEMS may fail because of the delamination of bonded thin-film materials. Bond failure of dissimilar and similar materials such as wafer-to-wafer bonding can also cause delamination (Sandborn, Swaminathan and Subramanian, 2000).
- Dampening: dampening is critical for MEMS because of the mechanical nature of the parts and the resonant frequency. Dampening can be caused by many variables, including atmospheric gases. Good sealing is critical for MEMS devices. Since MEMS devices have mechanical moving parts, they are more susceptible to environmental failure than are packaging systems.
- Mechanical failure: the changes in elastic properties affect the resonant and damping characteristics of the beam and that will change the sensor performance.

9.7 THERMAL ISSUES

Heat-transfer analysis and thermal management become more complex by packing different functional components into a tight space. The miniaturization also raises issues such as coupling between system configurations and the overall heat dissipation to environment. The configuration of the system shell becomes important for the heat dissipation from system to the environment (Lin, 2000; Nakayama, 2000). Heat spreading in a thin space is one of the most important modes of heat transfer in compact electronic equipment and microsystems. As the system shrinks, the space available for installation of a fan or pump inside the system shell disappears and the generated heat has to be dissipated through the shell to the surrounding environment. In general, strategies of heat transfer in a microsystem can be presented as: first, to diffuse heat as rapidly as possible from the heat source; second, to maximize the heat dissipation from system shell to the environment.

9.8 CONCLUSIONS

The three levels of packaging strategy may be adaptable for MEMS packaging. There are: (1) die level, (2) device level and (3) system level. Die-level packaging involves the passivation and isolation of the delicate and fragile devices. These devices have to be diced and wire bonded. The device-level packaging involves connection of the power supply, signal and interconnection lines. System-level packaging integrates MEMS devices with signal conditioning circuitry or ASICs (application-specific integrated circuits) for custom applications.

The major barriers in the MEMS packaging technology can be attributed to lack of information and standards of materials and a shortage of cross-disciplinary knowledge and experience in the electrical, mechanical, RF, optics, materials, processing, analysis and software fields. Microsystem packaging is more a combination of engineering and science, which must share and exchange experiences and information in a dedicated fashion. Table 9.3 presents different challenges and solutions faced during microsystem packaging.

Packaging design standards should be unified. Apart from certain types of pressure and inertial sensors used by the automotive industry, most MEMS devices are custom built. A standardized design and packaging methodology is virtually impossible at this time because of the lack of data available in these areas. However, the joint efforts of industry and academic and research institutions can develop sets of standards for the design of

Table 9.3 Current packaging parameters, challenges and suggested solutions

Packaging parameters	Challenges	Possible solutions
Release etch and dry	Stiction of devices	Use freeze drying; use supercritical CO_2 drying; roughen contact surfaces such as dimples and nonstick coatings
Dicing and Cleaving	Contamination risks, elimination of particles generated	Release dice after dicing; cleave wafers; use laser swing; use waferlevel encapsulation
Die handling	Device failure, top die face is very sensitive to contact	Use fixtures that hold the MEMS dice by the sides rather than by the top face
Stress	Performance degradation and resonant frequency shifts	Use low-modulus die attach; use annealing; use compatible CTE match-ups
Outgassing	Stiction, corrosion	Use low-outgassing epoxies, cyanate esters, low-modulus solders, new die-attach materials, remove outgassing vapors
Testing	Applying nonelectric stimuli to devices	Test all that is possible using wafer-scale probing, and finish with cost-effective specially modified test systems

Note: CTE, coefficient of thermal expansion.
Source: Malshe *et al.*, 2001.

microsystems. Also, the thin-film mechanics that includes constitutive relations of thin-film materials used in the FEM (finite element method) and other numerical analysis systems need to be thoroughly investigated.

REFERENCES

Blackwell, G.R. (Ed.), 2000, *The Electronic Packaging Handbook*, CRC Press, Boca Raton, FL.

Butler, J.T., Bright, V.M., Comtios, J.H., 1997, 'Advanced multichip module packaging of microelectromechanical systems', in *Transducers '97*, IEEE, Washington, DC: 261–264.

Butler, J.T., Bright, V.M., Chu, P.B., Saia, R.J., 1998, 'Adapting multichip module foundries for MEMS packaging', in *Proceedings of IEEE International Conference on Multichip Modules and High Density Packaging*, IEEE, Washington, DC: 106–111.

Butler, J.T., Bright, V.M., 2000, 'An embedded overlay concept for microsystems packaging', *IEEE Transactions on Advanced Packaging* **23**(4): 617–622.

Cheng, Y.T., Lin, L., Najafi, K., 2000, 'Localized silicon fusion and eutectic bonding for MEMS fabrication and packaging', *Journal of Microelectromechanical Systems* **9**(1): 3–8.

Cheng, Y.-Y., Lin, L., Najafi, K., 2001, 'A hermetic glass-silicon package formed using localized aluminum/silicon-glass bonding', *Journal of Microelectromechanical Systems* **10**(3): 392–399.

Cohn, M.B., Bohringer, K.F., Noworolski, J.M., Singh, A., Keller, C.G., Goldberg, K.Y., Howe, R.T., 1998, 'Microassembly technologies for MEMS', *Proceedings of SPIE Conference on Microfluidic Devices and Systems* **3515**(September): 2–16.

Coogan, S.A., 1990, 'System engineering: a summary of electronics packaging techniques available for present and future systems', in *Proceedings of Third Annual IEEE ASIC Seminar and Exhibit*, IEEE, Washington, DC: P4-3.1–3.4.

Dryton, R.F., 1995, The Development and Characterization of Self-packages using Micromachining Techniques for High Frequency Circuit Applications, Ph.D. dissertation, University of Michigan, Ann Arbor, MI.

Elwenspoek, M., Wiegerink, R., 2001, *Mechanical Microsensors*, Springer, Berlin.

French, P.J., 1999, 'Integration of MEMS devices', *Proceedings of SPIE Device and Process Technologies for MEMS and Microelectronics; Queensland Australia, SPIE* **3892**: 39–48.

Gilleo, K., 2001a, 'Overview of new packages, materials and Processes', *IEEE International Symposium on Advanced Packaging Materials*, IEEE, Washington, DC: 1–5.

Gilleo, K., 2001b, 'MEMS packaging issues and materials', in *Proceedings of IEEE International Symposium on Advanced Packaging: Process, Properties and Interfaces*, IEEE, Washington, DC: 1–5.

Gotz, A., Garcia, I., Cane, C., Morrissey, A., Aldreman, J., 2001, 'Manufacturing and packaging of sensors for their integration in a vertical MCM microsystem for biomedical applications', *Journal of Microelectromechanical Systems* **10**(4): 569–579.

Helsel, M.P., Berger, J.D., Wine, D.W., Osborn, T.D., 2001, 'Wafer scale packaging for a MEMS video scanner', in *Proceedings of SPIE Symposium on MEMS Design, Fabrication, Characterization and Packaging* **4407**: 214–220.

Hindreson, R.M., Herrick, K.J., Weller, T.M., Robertson, S.V., Kihm, R.T., Katehi, L.P.B., 2000, Three-dimensional high frequency distribution network, part II: packaging and integration, *IEEE Transactions on Microwave Theory and Techniques* **48**(10): 1643–1651.

Hsu, T.-R., 2000, 'Packaging design of microsystems and meso-scale devices', *IEEE Transactions on Advanced Packaging* **23**(4): 596–601.

Kim, S.J., Kwon, Y.S., Lee, H.Y., 2000, 'Silicon MEMS packages for coplanar MMICs', in *Proceedings of 2000 Asia-Pacific Microwave Conference, Australia, December 2000*, IEEE, Washington, DC: 17–20.

Kusamitsu, H., Morishita, Y., Marushashi, K., Ito, M., Ohata, K., 1999, 'The flip-chip bump interconnection for millimeter wave GaAs MMIC', *IEEE Transactions on Electronics Packaging and Manufacturing* **22**(1): 23–28.

Laskey, J., 1986, 'Wafer bonding for silicon-on-insulator technologies', *Applied Physics Letters* **48**(1): 78–80.

Li, Z., Hao, Y., Zhang, D., Li, T., Wu, G., 2002, 'An SOI–MEMS technology using substrate layer and bonded glass as wafer level package', *Sensors and Actuators A* **96**: 34–42.

Lin, L., 1993, *Selective Encapsulations of MEMS: Micro Channels, Needles, Resonators and Electromechanical Filters, Ph.D. dissertation, University of California at Berkeley, Berkeley, CA.*

Lin, L., 2000, 'MEMS post-packaging by localized heating and bonding', *IEEE Transactions on Advanced Packaging* **23**(4): 608–616.

Malshe, A.P., O'Neal, C., Singh, S., Brown, W.D., 2001, 'Packaging and integration of MEMS and related microsystems for system-on-a-package (SOP)', *Proceedings of SPIE Symposium on Smart Structures and Devices* **4235**: 198–208.

Maluf, N., 2000, *An Introduction to Micromechanical System Engineering*, Artech House, Boston, MA.

Mirza, A.R., Ayon, A.A., 1998, 'Silicon wafer bonding', *Sensors* December: 24–33.

Mirza, A.R., 2000, 'One micron precision wafer-level aligned bonding for interconnect, MEMS and packaging applications', in *Proceedings of IEEE 2000 Electronic Components and Technology Conference*: 676–680.

Mirza, A.R., 2000, 'Wafer level packaging technology for MEMS', in *Proceedings of IEEE 2000 Inter Society Conference on Thermal Phenomena*, IEEE, Washington, DC: 113–119.

Nakayama, W., 2000, 'Thermal issues in microsystems packaging', *IEEE Transactions on Advanced Packaging* **23**(4): 602–607.

O'Neal, C.B., Malshe, A.P., Singh, S.B., Brown, W.D., 1999, 'Challenges in packaging of MEMS', *IEEE International Symposium on Advanced Packaging Materials*, IEEE, Washington, DC: 41–47.

Oppermann, H.H., Kallmayer, C., Klein, C., Aschenbrenner, R., Reichl, H., 2000, 'Advanced flip chip technologies in RF, microwave and MEMS applications', *Proceedings of SPIE Design, Test, Integration and Packaging of MEMS/MOEMS* **4019**: 308–314.

Ramesham, R., Ghaffarian, R., 2000, 'Challenges in interconnection and packaging of microelectromechanical systems (MEMS)', in *Proceedings of 2000 Electronic components and Technology Conference*, IEEE, Washington, DC: 666–675.

Reichal, H., Grosser, V., 2001, 'Overview and development trends in the field of MEMS packaging', in *Proceedings of the 14th IEEE International Conference on MEMS*, 2001, IEEE, Washington, DC: 1–5.

Sandborn, P., Swaminathan, R., Subramanian, G., 2000, 'Test and evaluation of chip-to-chip attachment of MEMS devices', in *Proceedings of IEEE 2000 Inter Society Conference on Thermal Phenomena*, IEEE, Washington, DC: 133–140.

Shimbo, M., Furukawa, K., Fukuda, F., Tanzawa, K., 1986, 'Silicon-to-silicon direct bonding method', *Journal of Applied Physics Letters* **60**: 2987–2989.

Shivkumar, B., Kim, C.J., 1997, 'Microrivets for MEMS packaging: concept, fabrication and strength testing', *Journal of Microelectromechanical Systems* **6**(3): 217–225.

Sparks, D.R., Chang, S.C., Eddy, D.S., 1998, 'Application of MEMS technology in automotive sensors and actuators', in *Proceedings of IEEE International Symposium on Micromechatronics and Human Science*, IEEE, Washington, DC: 9–15.

Sparks, D.R., 2001, 'Packaging of microsystems for harsh environments', *IEEE Instrumentation and Measurement Magazine*: 30–33.

Takahashi, K., Sangawa, U., Fujita, S., Goho, K., Urabe, T., Ogura, H., Yabuki, H., 2000, 'Packaging using MEMS technologies and planar components', in *Proceedings of 2000 Asia Pacific Microwave Conference, Australia, December 2000*, IEEE, Washington, DC: 904–907.

Takahashi, K., Sangawa, U., Fujita, S., Matsuo, M., Urabe, T., Ogura, H., Yabuki, H., 2001, 'Packaging using Microelectromechanical technologies and planar components', *IEEE Transactions on Microwave Theory and Techniques* **49**(11): 2099–2104.

Index